T0310096

Ecology and Evolution of Dung Beetles

Ecology and Evolution of Dung Beetles

Edited by
Leigh W. Simmons & T. James Ridsdill-Smith

A John Wiley & Sons, Ltd., Publication

Blackwell Publishing was acquired by John Wiley & Sons in February 2007. Blackwell's publishing program has been merged with Wiley's global Scientific, Technical and Medical business to form Wiley-Blackwell.

Registered office: John Wiley & Sons Ltd, The Atrium, Southern Gate, Chichester, West Sussex, PO19 8SQ, UK

Editorial offices: 9600 Garsington Road, Oxford, OX4 2DQ, UK
The Atrium, Southern Gate, Chichester, West Sussex, PO19 8SQ, UK
111 River Street, Hoboken, NJ 07030-5774, USA

For details of our global editorial offices, for customer services and for information about how to apply for permission to reuse the copyright material in this book please see our website at www.wiley.com/wiley-blackwell

Library of Congress Cataloguing-in-Publication Data

Ecology and evolution of dung beetles / edited by Leigh W. Simmons & T. James Ridsdill-Smith.
p. cm.
Includes index.
ISBN 978-1-4443-3315-2 (hardback)
1. Dung beetles–Ecology. 2. Dung beetles–Evolution. I. Simmons, Leigh W., 1960- II. Ridsdill-Smith, J., 1942-
QL596.S3E26 2011
595.76'49–dc22

2010046392

A catalogue record for this book is available from the British Library.

This book is published in the following electronic formats: eBook 9781444341973; Wiley Online Library 9781444342000; ePub 9781444341980; MobiPocket 9781444341997

Set in 10.5/12pt, Classical Garamond by Thomson Digital, Noida, India
Printed and bound in Malaysia by Vivar Printing Sdn Bhd

1 2011

Contents

Preface

Scarabaeine dung beetles feed on the dung of herbivores as adults, and bury dung masses as provisions for their offspring. The subfamily contains about 6,000 species and is found in all continents except Antarctica. Beetles of different species are attracted to the same pad of fresh dung, but they occupy many different niches, thus reducing competition. Activity of the beetles is clearly visible to the casual observer and it fascinated the early Egyptians and Greeks, who considered the rolling of dung balls as representing the sun being rolled across the sky.

In the 19th century, J.H. Fabre described cooperation between male and female beetles in the formation of brood balls, the female role in oviposition and, in some cases, brood care, while Charles Darwin used the horns of adult male beetles to illustrate his theory of sexual selection. The biology and taxonomy of many species continued to be described through the 20th century, and books have been published summarising dung beetle natural history by Halffter & Matthews (1966), reproductive biology by Halffter & Edmonds (1982), ecology by Hanski & Cambefort (1991) and, most recently, a general overview of their evolutionary biology and conservation by Scholtz, Davis & Kryger (2009).

Our thesis in this book is that the wealth of information now available on dung beetles elevates them to the status of 'model system'. Dung beetles have proved remarkably useful for broad-scale ecological studies that address fundamental issues in community and population ecology and its extension to conservation biology. At the same time, they are providing valuable laboratory tools to explore fundamental questions in evolutionary biology; Darwin's theories of sexual selection have been validated through work on dung beetles and they are contributing to our understanding of the evolution of parental care. Moreover, their utility for studies of phenotypic plasticity is contributing to emerging research fields of evolutionary developmental biology ('evo-devo') and ecological developmental biology ('eco-devo').

The development of genomic tools for dung beetles will no doubt invigorate future research on this important taxon. Thus, our aim with this book is to provide detailed and focused reviews of the important contributions dung beetles continue to provide in evolutionary and ecological research.

Leigh W. Simmons and T. James Ridsdill-Smith
December 2010, Perth, Western Australia

Acknowledgements

We would like to thank our co-authors and the following individuals for reviewing chapters of the manuscript:

John Alcock, Andy Austin, Bruno Buzatto, Paul Cooper, Saul Cunningham, Vincent Debat, Raphael Didham, Mark Elgar, Doug Emlen, Federico Escobar, John Evans, Francisco García-Gonzáles, Mark Harvey, Richard Hobbs, Peter Holter, Geoff Parker, Alexander Shingleton, Per Smiseth, Steve Trumbo, Melissa Thomas, Craig White, Phil Whithers, and Jochem Zeal. We are indebted to Ward Cooper of Wiley-Blackwell for his enthusiasm for the project.

Contributing Authors

Barend (Ben) V Burger
Laboratory for Ecological Chemistry, Department of Chemistry, Stellenbosch University, Stellenbosch 7600, South Africa

Marcus Byrne
School of Animal, Plant and Environmental Sciences, University of the Witwatersrand, Johannesburg 2050, South Africa.

Steven L. Chown
Centre for Invasion Biology, Department of Botany and Zoology, Stellenbosch University, Private Bag X1, Matieland 7602, South Africa.

Marie Dacke
Vision Group, Zoology, Sölvegatan 35, 223 62 Lund, Sweden.

Penny B. Edwards
PO Box 865, Maleny, Queensland 4552, Australia.

Toby A. Gardner
Department of Zoology, University of Cambridge, Downing Street, Cambridge, CB2 3EJ, UK.

Wade Hazel
Department of Biology, DePauw University, Greencastle, IN 46135, USA.

Clarissa House
Centre for Ecology and Conservation, School of Biosciences, The University of Exeter, Tremough Campus, Penryn, TR10 9EZ, Cornwall, UK.

John Hunt
Centre for Ecology and Conservation, School of Biosciences, The University of Exeter, Tremough Campus, Penryn, TR10 9EZ, Cornwall, UK.

C. Jaco Klok
School of Life Sciences, Arizona State University, PO Box 874601, Tempe, AZ 85287-4601, USA.

Robert Knell
School of Biological and Chemical Sciences, Queen Mary University of London, Mile End Road, London E1 4NS, UK.

Armin Moczek
Department of Biology, Indiana University, 915 E. Third Street, Myers Hall 150, Bloomington, IN 47405-7107, USA.

Elizabeth S. Nichols
Center for Biodiversity and Conservation, Invertebrate Conservation Program, American Museum of Natural History, Central Park West at 79th St., New York, NY 10024-5193, USA.

T. Keith Philips
Systematics and Evolution Laboratory, Department of Biology, Western Kentucky University, 1906 College Heights Blvd., Bowling Green, KY 42101-3576, USA.

T. James Ridsdill-Smith
School of Animal Biology, University of Western Australia, Crawley 6009, Crawley, Western Australia.

Tomas Roslin
Department of Agricultural Sciences, PO Box 27, FI-00014 University of Helsinki, Finland.

Leigh W. Simmons
Centre for Evolutionary Biology, School of Animal Biology, University of Western Australia, Crawley, 6009, Crawley, Western Australia.

Joseph L. Tomkins
Centre for Evolutionary Biology, School of Animal Biology, University of Western Australia, Crawley, 6009, Crawley, Western Australia.

Geoffery D. Tribe
ARC-Plant Protection Research Institute, Private Bag X5017, Stellenbosch 7599, South Africa.

Heidi Viljanen
Metapopulation Research Group, Department of Biological and Environmental Sciences, PO Box 65, FI-00014 University of Helsinki, Finland.

1

Reproductive Competition and Its Impact on the Evolution and Ecology of Dung Beetles

Leigh W. Simmons[1,2] and T. James Ridsdill-Smith[2]

[1]*Centre for Evolutionary Biology, The University of Western Australia, Crawley, Western Australia*

[2]*School of Animal Biology, The University of Western Australia, Crawley, Western Australia*

1.1 Introduction

Beetles make up one quarter of all described animal species, with over 300,000 named species of Coleoptera, making them the most speciose taxon on planet earth (Hunt *et al.*, 2007). One of the larger groups is the Scarabaeoidea, with approximately 35,000 known species including the stag beetles, the scarabs and the dung beetles (Scarabaeinae) (Hunt *et al.*, 2007). Currently there are 6,000 known species and 257+ genera of dung beetles distributed across every continent on earth with the sole exception of Antarctica (Chapter 2). What better taxon could there be for the study of biodiversity, and the evolutionary and ecological processes that generate that biodiversity? Given their abundance and species richness, it is little wonder that dung beetles have attracted significant attention both from early naturalists and contemporary scientists. As we shall see throughout this volume, the unique biology of dung beetles makes them outstanding empirical models with which to explore general concepts in ecology and evolution.

The extreme diversity of beetles generally appears due to the early origin, during the Jurassic period (approx. 206–144 million years ago) of numerous lineages that have survived and diversified into a wide range of niches (Hunt *et al.*, 2007). In Chapter 2 Keith Phillips reviews our current understanding of the phylogenetic history of the dung beetles, which seem to have appeared during the Mesozoic era (around 145 million years ago), in the region of Gondwana that would later become Southern Africa.

Ecology and Evolution of Dung Beetles, First Edition. Edited by Leigh W. Simmons and T. James Ridsdill-Smith. Published 2011 by Blackwell Publishing Ltd.

The majority of extant species of dung beetles feed predominantly on the dung of herbivorous or omnivorous mammals. There was probably a single origin of specialist dung-feeding (coprophagy) from detritus- (saprophagy) or fungus- (fungivory) feeding ancestors, and the dung beetles are likely to have then co-radiated with the diversifying mammalian fauna (Cambefort, 1991b; Davis *et al.*, 2002b). However, throughout the dung beetle phylogeny there are numerous evolutionary transitions to alternative feeding modes, ranging from fungivory to predation (see Chapter 2), reflecting the divergence into new niches that characterizes the evolutionary radiation of beetles generally (Hunt *et al.*, 2007).

In this volume, we highlight the extraordinary evolutionary lability of dung beetles, arguing that much of their radiation is driven by reproductive competition. In their work on dung beetle ecology, Hanski & Cambefort (1991) argued that competition for resources was a major driver of the population and community dynamics of dung beetles. However, they noted the paucity of empirical studies available at that time which had actually examined reproductive competition.

Much progress has since been made. The chapters in this volume examine how reproductive competition affects organism fitness at the individual, species, population and community levels, and thereby illustrates the consequences of reproductive competition for evolutionary divergence and speciation. In this first chapter, we provide an overview of the evolution and ecology of dung beetles and introduce the detailed treatments of our co-authors that constitute the majority of the volume. While the often unique behaviour and morphology of dung beetles make them interesting taxa in their own right, the chapters highlight how dung beetles have proved to be model organisms for testing general theory, and how they have, and will, continue to contribute to our general understanding of evolutionary and ecological processes.

1.2 Competition for mates and the evolution of morphological diversity

A striking morphological feature of the Scarabaeoidea is the presence in males of exaggerated secondary sexual traits. Among the 6,000 known species of dung beetles, the males of many species possess horns (Emlen *et al.*, 2007). Darwin (1871) was the first to note the extraordinary evolutionary radiation in dung beetle horns and the general patterns of sexual dimorphism. If horns are present in females at all, they are generally – though not always – rudimentary structures compared with those possessed by the males of the species (Figure 1.1). Darwin (1871) argued that contest competition between males and female choice of males bearing attractive secondary sexual traits are general mechanisms by which sexual selection drives the evolutionary divergence of male secondary sexual traits. There is now considerable theoretical and empirical evidence to support his view that sexual selection can drive rapid evolutionary divergence among populations of animals (Lande, 1981; West-Eberhard, 1983; Andersson, 1994).

Emlen *et al.*'s studies (2005a; 2005b; 2007) of the genus *Onthophagus* have taught us much about the evolutionary diversification of horns in what is one of the most species-rich genera of life on Earth (there are already more than 2,000 species

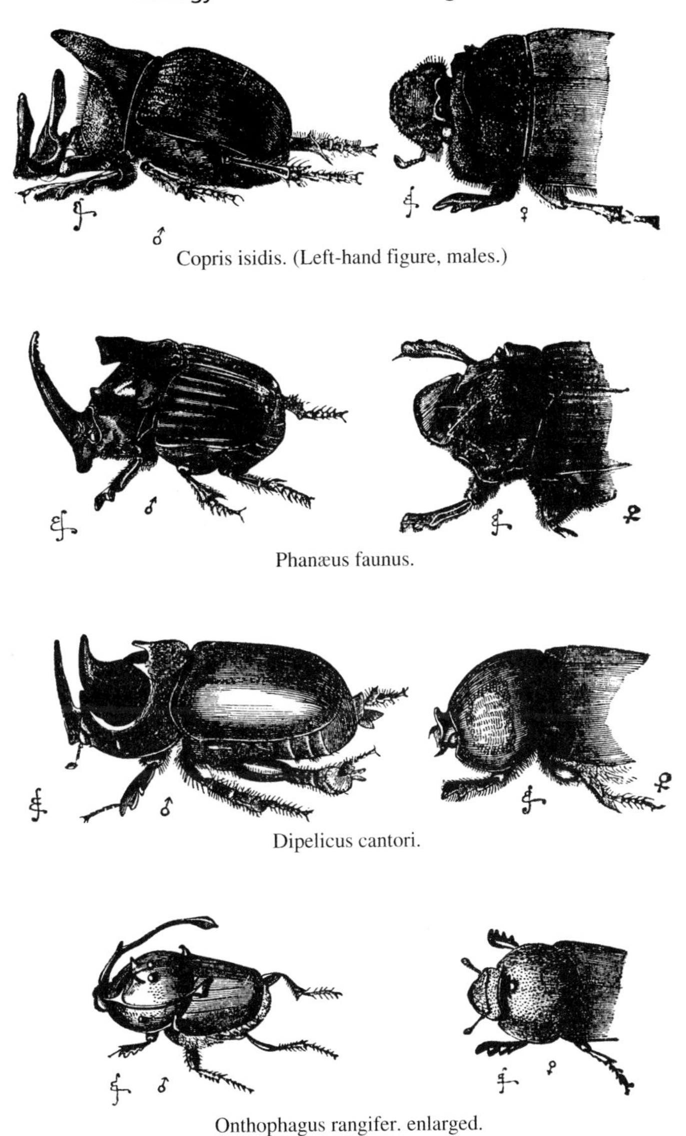

Fig. 1.1 Darwin (1871) argued that sexual selection was responsible for the evolutionary diversification of secondary sexual traits such as dung beetle horns, and he used these species of beetles to illustrate the sexual dimorphism that might be expected from selection by female choice. We now know that sexual selection via contest competition can favour the evolution of horns in males and females of tunnelling species, while female choice has not yet been shown to be important for horn evolution.

of described onthophagines). Based on a phylogeny of just 48 species – a mere 2 per cent of this genus – Emlen *et al.* (2005b) identified over 25 evolutionary changes in the physical location of horns on adult male beetles (Figure 1.2a). Moreover, from the reconstructed ancestral head horn shape (a single triangular horn arising from

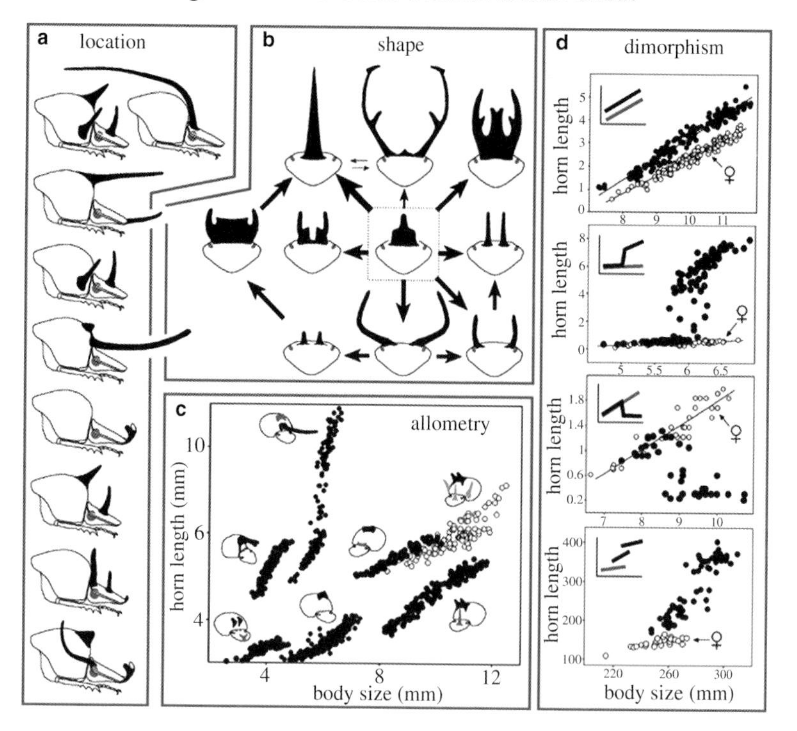

Fig. 1.2 Four trajectories of beetle horn evolution. **a**: Species differ in the location of horns; side-views of nine species of *Onthophagus* (Scarabaeinae) shown. **b**: Species differ in horn shape. Head horns shown for ten *Onthophagus* species; arrows indicate relative frequencies of changes as reconstructed from a phylogeny (from Emlen *et al.*, 2005b). **c**: Species differ in horn allometry, the slopes, intercepts, and even the shapes of the scaling relationships between horn length and body size. Data for thoracic horns of seven *Onthophagus* species shown. **d**: Species differ in the presence and nature of dimorphism in horn expression (males=closed circles; females=open circles). Top to bottom: sexual dimorphism (*O. pentacanthus*); male dimorphism and sexual dimorphism (*O. nigriventris*); reversed male & sexual dimorphism (*O. sloanei*); male dimorphism and sexual dimorphism (*Enema pan* (Dynastinae); unpublished data, JM Rowland). From Emlen *et al.* (2007); reprinted by permission of Macmillan Publishers Ltd, copyright 2007.

the centre of the vortex), there have been at least seven variant forms, several of which have themselves radiated into additional forms (Figure 1.2b).

Darwin (1871) noted that while dung beetle horns often exhibited sexual dimorphism, there was considerable within-species variation in this pattern. Indeed, in their study of 31 species of *Onthophagus*, Emlen *et al.* (2005a) identified at least 7 gains and 13 losses of sexual dimorphism. In one species, *O. sagittarius*, the horns of males are qualitatively different from the horns of females; males possess a pair of short horns at the sides of the frons and an enlarged thoracic ridge, while females possess a single long horn in the centre of the frons and a second single long horn in the centre of the thorax (Emlen *et al.*, 2005a). Thus, horn morphology in dung beetles appears to exhibit extraordinary evolutionary lability in the size, shape and number of horns, and in the degree and nature of sexual dimorphism (see Figure 1.2 and Chapter 3, Figure 3.1).

Early researchers rejected Darwin's (1871) argument that sexual selection was responsible for the evolutionary radiation of beetle horns, and the idea of sexual selection generally, arguing that beetle horns were more likely to function as protective structures against predators (Wallace, 1891) or to arise as a correlated response to evolutionary increases in body size (Arrow, 1951). However, there is now considerable evidence that dung beetle horns *are* subject to sexual selection through their use in contest competition.

In Chapter 3, Robert Knell provides an overview of the functional significance of dung beetle horns. Among the dung beetles, there appears to be a close evolutionary association between tunnelling behaviour and the possession of horns. As we shall see, dung beetles can be broadly classified into tunnellers that nest in the soil below the dung, and rollers that construct balls of dung which they roll away from the dung pad for burial elsewhere (Section 1.3 and Chapter 2). The available phylogeny suggests that tunnelling was the ancestral behaviour pattern, and that there have been numerous evolutionary transitions to rolling behaviour (Chapter 2). Horns function primarily in blocking access to the confined spaces within tunnels, allowing males to monopolize access to breeding females (Chapter 3). In contrast, for rollers operating in an open above-ground environment, horns would be unlikely to contribute to a male's ability to monopolize access to females and/or breeding resources (Emlen & Philips, 2006).

Based on a phylogeny of 46 species from 45 genera, Emlen and Phillips (2006) showed how all of eight evolutionary origins of horns were on lineages of tunnellers, while not a single lineage of rollers included an evolutionary gain of horns (see Figure 3.4).

The monopolizability of mates and/or breeding resources is thought to be a major factor moderating the strength of sexual selection (Emlen & Oring, 1977). In Chapter 3, Knell shows how the density of breeding beetles impacts the evolution of horns even within tunnelling species. Tunnelling dung beetles that live in highly crowded environments, where their ability to control access to breeding resources is limited, are significantly less likely to have evolved horns than species from less crowded environments, where the monopolizability of mates and resources is easier (Pomfret & Knell, 2008).

Importantly, there are now several within-species studies from a number of genera which confirm that horn size is a strong predictor of the outcome of disputes between competing males (see Chapter 3). Moreover, the form of sexual selection, estimated from the slope of male reproductive success on horn length, has been shown to be directional for increasing horn length within experimental populations of *O. taurus* (Hunt & Simmons, 2001) (see Figure 6.1b). Interestingly, directional positive linear selection has also recently been documented for horn length in female *O. sagittarius*. In this species, females compete for dung with which to build brood masses, and differences in horn length predict the amount of dung females can monopolize and, therefore, the number of offspring they are able to produce (Watson & Simmons, 2010b). This study represents the first demonstration of selection acting on female secondary sexual traits for any species, and it suggests that sexual selection is likely to be

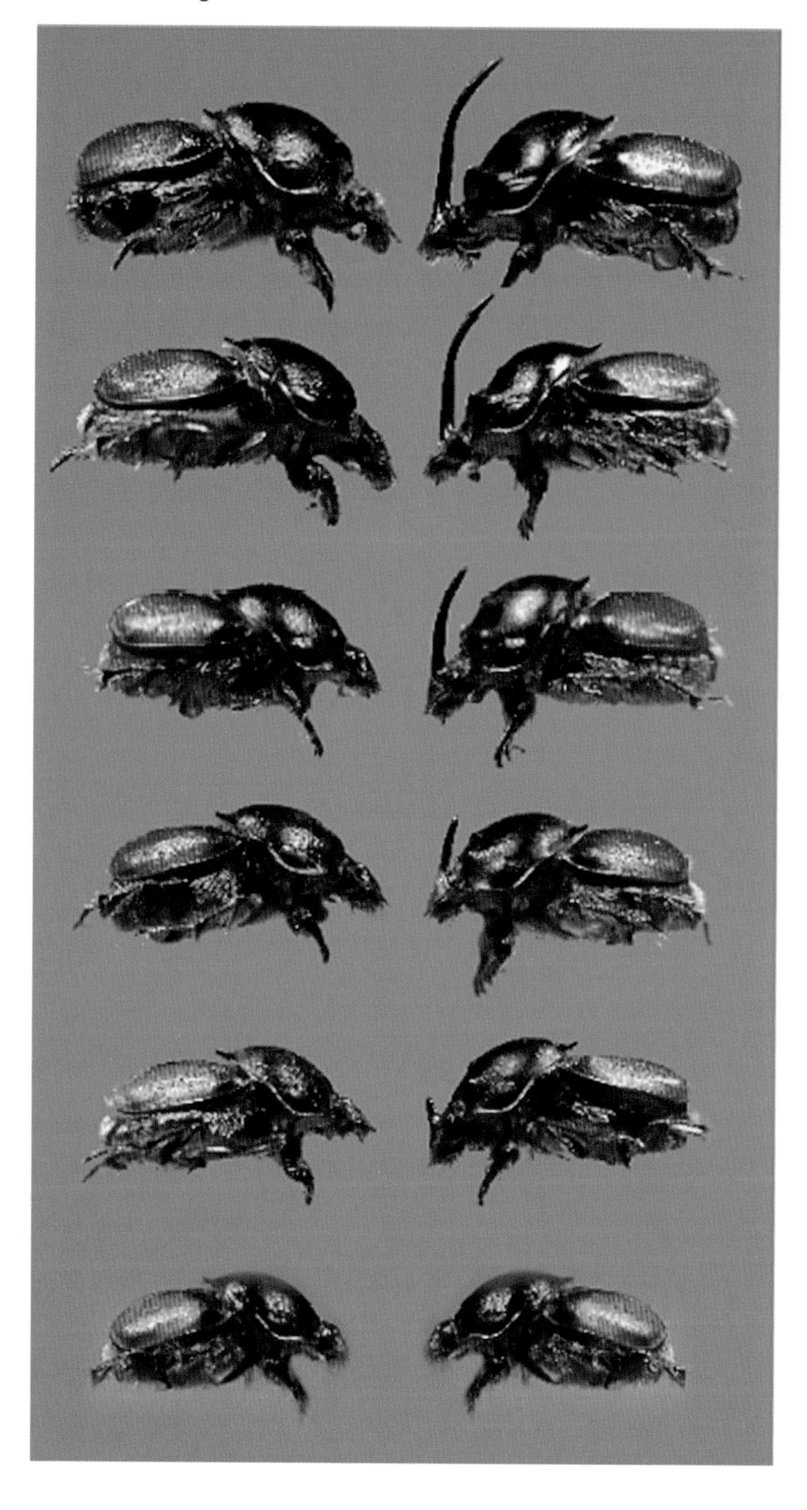

important in the many evolutionary origins of female horns in dung beetles (Emlen *et al.*, 2005a).

Darwin (1871) noted that horn morphology could be just as variable within species as it was among species. Thus, in discussing onthophagines, he noted that, 'in almost all cases, the horns are remarkable from their excessive variability; so that a graduated series can be formed, from the most highly developed males to others so degenerate that they can barely be distinguished from the females.' (Figure 1.3). This extreme morphological variability is now known to be associated with alternative mate-securing tactics, in which minor males remain hornless and sneak matings with females guarded by horned males. The tactic adopted depends critically on the amount of dung provided by a male's parents when they provisioned his brood mass. Thus, brood size influences adult body size, and males exceeding a threshold body size develop horns and adopt the fighting and mate-guarding tactic (see Figs. 1.2d and 7.3).

In Chapter 6, Joseph Tomkins and Wade Hazel provide an overview of the general theoretical issues surrounding the evolution of such phenotypic plasticity and show how dung beetles have contributed significantly to our understanding of this area of developmental biology. They demonstrate how an interaction between environmental cues and genetic variation can influence the expression of alternative male phenotypes in onthophagine dung beetles, and specifically the position of the body size threshold at which males switch between alternative phenotypes, thereby generating variation within and among populations in the proportion of males that adopt the horned fighting tactic.

In Chapter 7, Armin Moczek penetrates this subject to the genetic level, using the latest genomic techniques to identify the genes responsible for horn development and to reveal the signalling pathways responsible for switching the developmental trajectories that lead to the horned and hornless phenotypes. These studies of *Onthophagus* are providing us with detailed insights into the developmental mechanisms that underpin morphological diversity in dung beetles, while at the same time contributing to the emergence of the cross-disciplinary research fields of evolutionary developmental biology and ecological developmental biology (Chapter 7).

Moczek shows us that beneath the apparently extreme evolutionary lability in phenotypic diversity among onthophagine dung beetles lies a rather small and conserved set of regulatory pathways. These pathways can readily account for the multiple evolutionary gains and losses of horns within and between the sexes, and for the phenotypic plasticity and nutrient sensitive growth that collectively generate

Fig. 1.3 Darwin (1871) noted the extreme variability in horn development within species of dung beetles, as illustrated by these images of *Proagoderus (Onthophagus) lanistra*, which show both sexual dimorphism and male dimorphism. Females (left) do not develop horns. Large males (majors) develop exaggerated horns, while small males (minors) remain hornless, resembling females. These alternative phenotypes are associated with different mating tactics whereby major males fight for females and assist with brood production, while minor males sneak copulations when major males are collecting dung or fighting with other major males for the possession of females. From Emlen *et al.* (2007). Copyright (2007) National Academy of Sciences, USA.

the extraordinary phenotypic diversity which characterizes the genus *Onthophagus* (Figure 1.2).

The adoption of sneak mating behaviour by a subset of the male population generates a sexual selection pressure that was not appreciated by Darwin – that of sperm competition (Parker, 1970; Simmons, 2001). Whenever a female mates with two or more males, the sperm from those males will compete to fertilize the few eggs that she produces during her lifetime.

Sexual selection is predicted to favour any morphology, physiology or behaviour that enhances a male's success in competitive fertilization. In Chapter 4, Leigh Simmons reviews sperm competition theory and shows how dung beetles in the genus *Onthophagus* have been important in its empirical evaluation. Within the onthophagines, the considerable among-species variation in the proportion of males adopting the sneaking tactic generates variation in the strength of sexual selection arising from sperm competition and provides an opportunity to test the theoretical expectation that sperm competition should influence the evolution of male investment in sperm production. Thus, across a phylogeny of 18 species of *Onthophagus*, evolutionary increases in the proportion of males adopting the sneaking tactic were found to be positively associated with evolutionary increases in male investment into their testes (Chapter 4). Moreover, within species, by virtue of their mating tactic, sneaks are always subject to sperm competition and tend to invest more in testes growth than do horned fighters (Simmons *et al*., 2007).

Interestingly, these studies have revealed important nutrient allocation trade-offs between traits involved in competition for mating opportunities (horns) and competition for fertilizations (testes). Both within and among species, males that invest more in their testes tend to invest less in horn expression (Chapter 4).

Nutrient allocation trade-offs are likely to contribute greatly to the evolutionary diversification of dung beetle horns. Morphological traits that develop in close proximity will compete for the same pool of resources, thereby constraining each other's patterns of growth (Emlen, 2001). The strength of selection acting on one trait is then expected to shape the allocation of resources to the other.

For example, thoracic horns develop in closer proximity to testes than do head horns, and Simmons & Emlen (2006) found that novel gains of thoracic horns were far less likely in lineages in which there were alternative sneak tactics (and thus intense sperm competition) than in lineages without sneak tactics. Thus, pre- and post-copulatory processes of sexual selection can interact in determining the evolutionary diversification of male morphology.

In a similar manner, during development, horns at the rear of the head compete for resources with eyes, while those at the front of the head compete for resources with antennae, and thoracic horns compete for resources with wings (Emlen, 2001). In Chapter 9, Marcus Byrne and Marie Dacke provide an extensive survey of the visual ecology of dung beetles, illustrating the considerable evolutionary diversification in dung beetle eye morphology and visual acuity. They point out how nutrient allocation trade-offs between horns and eyes may dictate the evolutionary response to sexual selection. Indeed, across a phylogeny of 48 species of *Onthophagus*, Emlen *et al*. (2005b) found losses of horns located at the rear of the head, where horn development results in reduced eye size, were concentrated on lineages that have switched from diurnal to nocturnal flight behaviour, where greater visual acuity would be required.

As noted in Chapter 5, the detection of olfactory cues is also critical for locating ephemeral resources. Gains in horns at the front of the head tend to be associated with forest-dwelling lineages, where odour plumes from dung are perhaps more likely to persist and trade-offs with antennae are therefore less costly compared to open pastures (Emlen *et al.*, 2005b). Much more work is required in this area, but the data clearly suggest that ecology plays an important role in modulating the evolutionary responses in male weaponry to sexual selection.

Ironically, in the absence of firm evidence for competition among males, Darwin (1871) thought that sexual selection through female choice was likely to be the more powerful selective force in the evolution of beetle horns. It is becoming clear, however, that while female dung beetles do exercise mate choice, they do not appear to use male horns as cues to mate quality. Thus, studies of several species of *Onthophagus* suggest that females choose among males based on their overall genetic and phenotypic condition, not on the length of their horns (Kotiaho *et al.*, 2001; Kotiaho, 2002; Watson & Simmons, 2010a; Simmons & Kotiaho, 2007a). As Simmons shows in Chapter 4, females rely on pre-copulatory (courtship) and post-copulatory (sperm competitiveness) performance as predictors of male genetic quality, and in so doing they are able to produce offspring that are more likely to reach reproductive maturity.

However, female choice in dung beetles remains poorly explored. In Chapter 5, Geoff Tribe and Ben Burger review what is available on the olfactory ecology of dung beetles, and in so doing they reveal a rich area for future research. They show how pheromone signalling is a key component of the breeding biology of ball-rolling species. While much is known of the chemical composition of the sex attraction pheromone in the genus *Kheper*, little is known of other species. We know nothing of within-species variability in pheromone composition or signalling effort.

Pheromone signalling has been shown to be subject to intense sexual selection in other insect groups (Wyatt, 2003; Johansson & Jones, 2007), so it is highly likely to be an important aspect of reproductive competition in dung beetles as well, at least among ball-rollers, where males often attract a female to a location somewhat removed from the dung source (Chapter 5). Almost nothing is known of semiochemicals in tunnelling species, but the occurrence of sexually dimorphic chemical-producing glands on the cuticle suggest that here, too, chemical signals are likely to play an important role in species mate recognition and mate choice.

1.3 Competition for resources and the evolution of breeding strategies

The breeding behaviour of dung beetles is perhaps the most conspicuous aspect of their biology. The early Egyptians observed dung beetles emerging from the soil in spring, which they believed represented reincarnation, and when beetles made and rolled perfect spheres of dung it represented to them their god Kheper, rolling the sun across the sky (Ridsdill-Smith & Simmons, 2009). They revered the beetles as symbolizing rebirth; scarab amulets are found on paintings and in tombs to simulate reincarnation and they were used by the living to bring good luck. Also, identifiable beetles are often found preserved in tombs.

The breeding biology of several dung beetle species was described in exquisite detail in the works of the early French naturalist, J. H. Fabre. Fabre (1918) studied representatives from most of the major genera, including *Scarabaeus*, *Gymnopleurus*, *Copris*, *Onthophagus*, *Oniticellus*, *Onitis*, *Geotrupes* and *Sisyphus*. Not only did he describe the major nest-building behaviours and the patterns of parental care, but he also made the first detailed observations on the developmental biology of many of the species he studied.

For example, in his studies of the ontology of *O. taurus*, Fabre discussed extensively the pupal horns and their loss prior to adulthood. He was at a loss to explain the functional significance of these structures, asking, 'What is the meaning of those horny preparations, which are always blighted before they come to anything? With no great shame I confess that I have not the slightest idea.' As Moczek describes in Chapter 7, we now know that pupal horns probably function in releasing the head capsule during the pupal moult; they are not always lost, being the precursors of thoracic horns in the adults of some species.

Fabre's important observations were followed by the formal classification system of Halffter and his colleagues (Halffter & Mathews, 1966; Halffter & Edmonds, 1982). The nesting behaviour of dung beetles can be broadly classified into telecoprid (the rollers), paracoprid (the tunnellers), and endocoprid (the dwellers). These can be further classified on the complexities of brood mass and/or nest construction and the extent of parental care (Chapter 2 and Figure 1.4):

- Paracoprids dig tunnels in the soil beneath the dropping and carry fragments of dung to the blind ends of those tunnels, where they are packed into brood masses. A single egg is laid in an egg chamber and the brood mass sealed with dung (Halffter & Edmonds, 1982).
- The males of telecoprids fashion a ball of dung before emitting a pheromone signal to attract a female, either at the dropping or after rolling the ball away from the dropping and burying it in a chamber below ground (Chapter 5 and Figure 1.4). The female enters the chamber to fashion a brood ball with the supplied dung, and in some species she will remain with the brood until the adult offspring emerge (Halffter and Edmonds, 1982).
- Endocoprids fashion brood balls within the dropping (Figure 1.4).

As noted above, current evidence suggests that tunnelling is the ancestral nesting behaviour of dung beetles and that there have been several evolutionary gains of telecoprid behaviour (Chapter 2). There have also been several evolutionary gains of brood parasitism or kleptoparasitism, in which females deposit their eggs into the broods provisioned by telecoprid or paracoprid species (Hanski & Cambefort, 1991; González-Megías & Sánchez-Piñero, 2003; 2004).

Reproductive competition for dung has undoubtedly played an important role in the evolutionary diversification of breeding behaviour. Hanski and Cambefort (1991) suggested a competitive hierarchy among dung beetle species in which rollers and fast tunnellers are competitively superior to slow tunnellers, who are

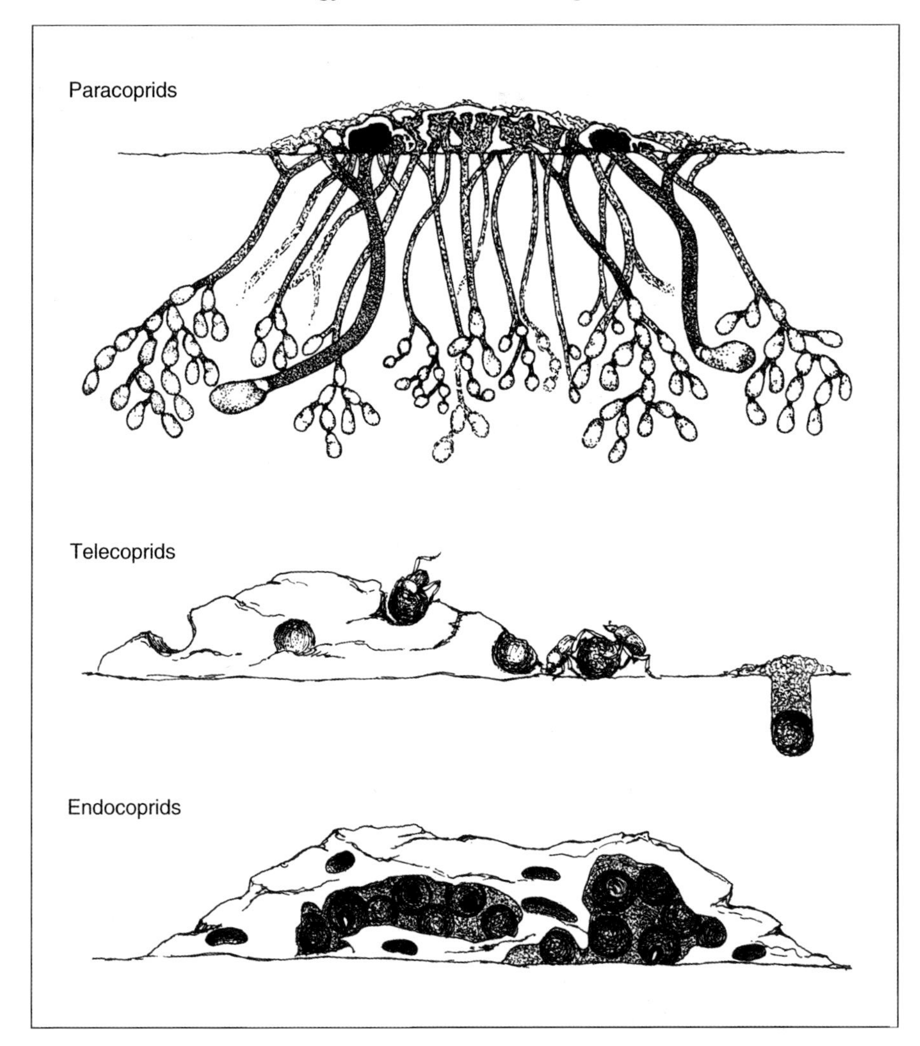

Fig. 1.4 Nesting behaviours of scarabaeine dung beetles can be broadly classified into three major types. In tunnelling or paracoprid species (**a**), beetles dig tunnels beneath the dung pad and pack fragments of dung that they bring from the surface into the blind ends of tunnels before laying a single egg into the brood chamber. A brood mass provides all the resources for the development to adulthood of a single offspring. In rolling or telecoprid species (**b**), beetles build a dung ball and roll it away from the pad before burying it in the soil. The dung ball can be used as food for the adults or fashioned into one or more brood balls. In dwelling or endocoprid species (**c**), beetles build broods within the dung pad itself (reproduced from Bornemissza, 1976).

competitively superior to dwellers (see Section 1.4), and it is certainly easy to imagine how telecoprid behaviour might arise in response to competition among paracoprid species that are rapidly burying dung in the soil beneath the dropping.

The African scarab *Scarabaeus catenatus* appears to adopt both tunnelling and rolling tactics (Sato, 1997; 1998b). When tunnelling, a pair of beetles will dig a nest within 1 m of the dropping, and will move back and forth from the dropping with small fragments of dung to provision the nest. Alternatively, the male may roll a ball of dung up to 15 metres from the dropping to establish a nest, a behaviour more typical for a telecoprid.

Sato (1998b) observed that male competition was far greater for those adopting the tunnelling tactic because of interference from other tunnellers for dung and space around the dropping. Males adopting the rolling tactic did not suffer from competition but, because they did not return to the dropping, they obtained a smaller share of dung for brood production. The average reproductive success obtained from the two tactics was equal for males, but not for females, who fared better when adopting the tunnelling tactic (Sato, 1998b). Such differences in reproductive pay-offs are predicted to generate sexual conflict between males and females over which breeding tactic to adopt (Arnqvist & Rowe, 2005).

Perhaps the most interesting aspect of dung beetle breeding biology is the often extensive level of parental care that limits their lifetime fecundity to as few as three offspring in the rolling *Kheper* (Edwards, 1988), and over 100 in the tunnelling *Onthophagus* (Hunt *et al.*, 2002; Simmons & Emlen, 2008) (see Table 3.2 in Hanski & Camberfort, 1991). It is often the case that males and females cooperate in brood production. In both *Kheper* and the tunnelling *Copris*, males and females will cooperate in excavating a nest and supplying it with dung (Edwards & Aschenborn, 1988; Halffter *et al.*, 1996; Sato, 1988; 1998a; Sato & Hiramatsu, 1993). Cooperation may have arisen in response to the need to sequester dung quickly in the face of intense intraspecific and interspecific competition for the limited resource.

Paternal care appears to cease after the nest is provisioned with dung. The female will use the dung provisions to build brood masses and will remain with her broods and tend them until the adult offspring emerge. Female *Copris lunaris* keep the brood balls upright and will repair them should they break open during the development of the larvae (Klemperer, 1982).

Olfactory communication may be important in interactions between females and their developing young. For example, in *C. lunaris*, females will not right or repair broods that do not contain larvae unless dichloromethane extracts from *C. lunaris* broods have been added (Klemperer, 1982). Moreover, female *C. diversus* have been shown to reallocate dung from broods within which an offspring has died to viable broods, so that the size of surviving adult offspring is increased (Tyndale-Biscoe, 1984).

Numerous experimental removal studies have shown that brood survival is dependent on maternal care. Thus, in *K. nigroaeneus*, maternal care increases egg-to-larva survival by 20 per cent, larva-to-pupa survival by 39 per cent and post-feeding survival by 20 per cent (Edwards and Aschenborn, 1989). Likewise, egg-to-adult survival is increased by maternal care in several species of *Copris* (Klemperer, 1982; Tyndale-Biscoe, 1984; Halffter *et al.*, 1996). Female *Copris* spend a considerable proportion of their time tending to brood balls, compacting and smoothing their surfaces (Halffter *et al.*, 1996). Broods that do not receive maternal care appear vulnerable to invasion by fungi *Metarrhizium anisoplae* and *Cephalosporium* sp. (Halffter *et al.*, 1996) and also to predation by other soil invertebrates (Sato, 1997).

Maternal care is also likely to be an important guard against reproductive competition from brood parasites. Thus, the brood parasite *Aphodius* reduces host brood survival by as much as 68 per cent, with 12 per cent of *S. puncticollis* nests being parasitized (González-Megías & Sánchez-Piñero, 2003). Klemperer (1982) observed that female *C. lunaris* would attack and kill *Aphodius* larvae when they were encountered in the nest.

Dung beetles have proved to be ideal model organisms with which to test empirically the extensive theoretical models that have been developed around the evolution of parental care. In Chapter 8, John Hunt and Clarissa House review the extensive and detailed work on biparental care in *Onthophagus* and show how the study of this genus has contributed to our general understanding of parental care. Biparental care is common in this genus, where horned males assist females by delivering fragments of dung to the brood chamber where the female constructs the brood mass. Although females can construct broods alone, male assistance increases the number and weight of broods produced, thereby improving female and offspring reproductive fitness (Palestrini & Rolando, 2001; Hunt & Simmons, 2000; Sowig, 1996a; Lee & Peng, 1981).

Unlike *Kheper* and *Copris,* neither sex of *Onthophagus* provide care after oviposition is completed. Nonetheless, biparental provisioning of the brood mass has dramatic effects on offspring fitness. In Chapter 8, Hunt and House show how parental provisioning is optimized, depending on the costs and benefits of provisions to offspring and parental fitness. Behavioural interactions between male and female *O. taurus* during provisioning influences the relative amounts of dung that each parent provides, as well as how males adjust their investment facultatively to the risk of sperm competition from sneak males, and thus their confidence in paternity of offspring they help to provision.

Hunt and House also show how brood provisioning, rather than egg production, represents the major cost of reproduction for *Onthophagus*, and how male assistance can ameliorate the female's costs of reproduction. This finding is consistent with the fact that ovariole development is inhibited, and the terminal oocyte resorbed, during the period when females are provisioning and caring for their offspring (Klemperer, 1983; Sato & Imamori, 1987, Anduaga *et al.*, 1987). In other words, females spend much more of their resources on caring for young than they do in manufacturing eggs.

The amount of maternal and paternal provisions are an important source of environmental effects that contribute to offspring fitness. Where provisioning has an underlying genetic basis, these parental effects can generate evolutionary responses to selection in traits that they affect, such as offspring body size, even when there is little or no additive genetic variance for those traits (Wolf *et al.*, 1998). As Hunt and House point out, parental care can thereby have important, yet unappreciated, implications for the evolutionary diversification of dung beetles.

The very different environments in which dung beetles must operate will also generate different selection pressures on their morphology. Rollers are often characterized by adaptations to the hind tibia for ball construction and rolling (seen in its extreme in the hind legs of *Neosisyphus*), while the tunnellers have relatively short robust forelegs and specialized structures on the head for moving soil (see Figure 17.2 in Hanski & Cambefort, 1991). Moreover, as we have discussed above, both sexes of tunnellers can have horns with which to defend their tunnels, an adaptation that comes at a cost to visual acuity.

In Chapter 9, Marcus Byrne and Marie Dacke show us how the morphology of the eyes vary between tunnellers and rollers, and between diurnal and nocturnal species. Indeed, they show us how well the eyes of rollers are adapted to the need to roll balls of dung away from the source of resource competition. The dorsal rim of the eye is adapted to function as a polarizing compass that allows the beetles to follow an accurate bearing when rolling a ball away from the dropping – and, more importantly perhaps, for those flightless species, to return to their nests by the quickest straight-line path once they have secured additional pieces of dung (Chapter 9).

1.4 Ecological consequences of intraspecific and interspecific competition

Intraspecific interference competition is common in the scarabaeine dung beetles (Hanski & Cambefort, 1991). The annual peak adult activity of scarabaeine dung beetles tends to occur for short periods. For species active in summer, these periods follow rainfall events in months when temperatures are highest. As a result, large numbers of dung beetles of many species can arrive at the same fresh dung pads (Figure 1.5). Over 1,000 beetles can be caught in one dung-baited trap over 24 hours (Hanski & Cambefort, 1991; and see Tables 12.1 and 12.2 in this volume). There is not sufficient dung for all females in the pad to breed, and oviposition is affected by competition.

However, intraspecific interference competition between beetles can occur in pads long before any shortage of dung generates exploitation competition (Ridsdill-Smith, 1991). For example, a negative exponential curve described the fall in number of eggs per female per week with increasing beetle density from 2 to 100 *Onthophagus binodis* on one litre of cattle dung (Ridsdill-Smith *et al.*, 1982). Dung burial, calculated from the volume of each brood mass, reached a maximum of

Photo P. Edwards

Fig. 1.5 Dung beetles competing for dung in Mkuzi Park in Southern Africa. Main beetles are *Pachylomera femoralis* (large) and *Allogymnopleurus thalassinus* (smaller).

45 per cent with 20–30 beetles. Egg production of both *Onthophagus ferox* and *O. binodis* was greatly reduced by intraspecific competition (71 per cent and 85 per cent reduction respectively, between low and high density populations) (Ridsdill-Smith, 1993b).

In Chapter 12, James Ridsdill-Smith and Penny Edwards describe the serial introduction of exotic dung beetle species to pasture sites where there was a surplus of cattle dung. They show how in single-species populations, the large native species, *O. ferox*, was unable to increase its population size to utilize more than 30 per cent of the available dung, while the smaller exotic species, *O. binodis*, used only 50 per cent of the available dung. Over 14 years, the total number of beetles trapped increased with the number of exotic species present (Figure 12.6), and they presumably used more of the available dung. Intraspecific competition thus appears to be a more important factor limiting the growth of dung beetle populations than the supply of fresh dung.

Most of the examples of interspecific interference competition given by Hanski & Cambefort (1991) are for rollers, where it is relatively easy to observe contests over dung balls. In general, larger species capture dung balls from small species (Hanski & Cambefort, 1991). In laboratory studies, large tunnelling species bury more dung and show asymmetric competition with smaller species (see Chapters 12 and 13). Egg production of the large tunnelling species, *Copris elphenor* and *Catharsius tricornutus* were not affected by the smaller species *Onitis alexis*, but egg production of *O. alexis* was reduced in the presence of the larger beetles (Giller & Doube, 1989). Similarly, at high beetle densities, egg production of the larger species *O. ferox* was unaffected by the smaller species *O. binodis,* but egg production of *O. binodis* was reduced in the presence of the larger *O. ferox* (Ridsdill-Smith, 1993b).

In both of these studies, the larger species showed pre-emptive dung burial behaviour, burying relatively more dung on the first day, which was then used to produce brood masses on subsequent days. In contrast, the smaller species buried only enough dung for one brood mass on the first day, and then again on subsequent days. In the field, the pad can be disturbed very quickly when beetles are very abundant, resulting in interspecific as well as intraspecific exploitation competition, so that any beetles returning to the pad from their tunnel in the soil after a day or two are unable to obtain dung to produce any further brood masses (Ridsdill-Smith, 1991).

1.4.1 Niche expansion

Competition for resources will reduce individual fitness and generate selection on traits that reduce the intensity of competition. For example, when two species exploit the same resource, those individuals within each species that compete least with members of the other species are expected to have a higher fitness, generating disruptive selection that can drive niche divergence (Slatkin, 1980; Day & Young, 2004; Rundle & Nosil, 2005; Abrams *et al.*, 2008). Likewise, when individuals within a species compete for resources, divergent selection is expected to favour individuals that compete least, i.e. individuals who differ from the average competitor phenotype. Thus, both interspecific and intraspecific competition can drive phenotypic divergence and promote niche expansion and subsequent speciation (Schluter, 1994; Pfennig *et al.*, 2007; Agashe & Bolnick, 2010).

A very striking feature of the scarabaeine dung beetles is the niche separation of co-existing species (Chapter 2). Different species have evolved to fill the same niches in different geographical regions, and different species within the same regions have evolved differences in diet, nesting behaviour, thermal tolerances or visual acuities to fill different niches.

Most scarabaeine beetles fly upwind to locate fresh dung pads, attracted by the volatile odours given off by the dung, in particular, 2-butanone (Chapter 5). Beetles can also distinguish between dung from different mammals and, although they do not specialize on any one dung type, they can show clear preferences when presented with alternatives (Chapter 5; Dormont *et al.*, 2007).

Other species feed on alternative food resources such as carrion, fungi, millipedes or fruit, and they use different volatile cues to find each resource. Seeds of plants may be present in dung, and dung beetles can also be attracted to volatiles from the seeds. For example, *Pachylomera femoralis* is attracted to seeds of spineless monkey orange trees, and Tribe and Burger have identified volatiles from the seeds that will attract the beetles (Chapter 5). In the European flightless species *Thorectes lusitanicus* (Geotrupidae), adult beetles are attracted by oak acorns, and feeding on acorns can increase female fitness through enhanced ovarian development (Verdú *et al.*, 2010). Thus, dung beetles can exhibit considerable variation in the types of resources they exploit, both within and among species.

We saw in Section 1.3 that dung beetles have evolved three major patterns of nesting behaviour: tunnelling, rolling and dwelling (Figure 1.4). Within these groups there is much variation, based on how deep the brood masses are placed in the soil under the pad, the speed with which the dung is buried and the amount buried (Doube, 1990). Some species bury all the dung in the first two days, in pre-emptive dung burial, while other species bury dung over longer periods (Doube *et al.*, 1988a; Ridsdill-Smith, 1993b). In *S. catenatus*, beetles can adopt either the rolling or tunnelling nesting tactic (Sato, 1997) and the amount of brood provisions per offspring can vary, depending on the type of dung exploited or whether males cooperate in brood provisioning (Hunt & Simmons, 2004; see Chapter 8).

It is reasonable to expect that dietary preferences and/or nesting behaviours harbour underlying genetic variation which would, when coupled with intraspecific or interspecific competition, facilitate niche evolution (Agashe & Bolnick, 2010). Indeed, brood provisioning has a genetic basis in *O. taurus*, and Hunt and House propose that plasticity in brood provisioning, particularly in response to environmental factors such as soil moisture, has the potential to play an important role in promoting niche expansion in onthophagines (Chapter 8).

Another striking feature of the scarabaeine dung beetles is the intensity of sexual selection, a form of intraspecific competition that favours traits, such as increased body size and condition, that contribute to success in reproductive competition (Chapters 3 and 4). Theory suggests that sexual selection can play an important role in niche expansion. Under good gene models of sexual selection, traits that females find attractive in males, or which give males a competitive advantage over other males, are reflective of the underlying genetic quality or condition of an individual, so that these individuals also have higher non-sexual fitness (Chapter 4). As such, sexual selection can accelerate the fixation of advantageous alleles (Proulx, 1999; 2002) and the purging of disadvantageous alleles (Whitlock & Agrawal, 2009), processes that can, in theory, accelerate the rate of adaptation to new niches (Lorch *et al.*, 2003).

Empirical tests of this idea are few, and the evidence for a role of sexual selection in promoting adaptation to new niches is contradictory (Candolin & Heuschele, 2008). Studies of *Drosophila* suggest that sexual selection may not influence adaptation to new thermal or resource environments (Holland, 2002; Rundle *et al.*, 2006), but studies of bruchid beetles, *Callosobruchus maculatus*, demonstrate clearly that adaptation to a novel resource is accelerated by sexual selection (Fricke & Arnqvist, 2007). The later study is informative because the authors found evidence that the costs of sexual selection, bought about by sexual conflict, may depress population fitness once a species has adapted to its new niche.

An important fitness trait in the context of sexual selection is adult body size (Chapters 3, 4, 6 and 8). In Chapter 10, Steven Chown and Jaco Klok explore the importance of body size from an ecophysiological perspective. They demonstrate a significant physiological advantage to beetle size that has important ramifications for species richness and the structuring of beetle communities. For example, they describe the ability of species over 2 g in weight to be endothermic regulators in flight, while species below this mass have thoracic temperatures similar to ambient. Elevated body temperatures will give the larger beetles considerable advantage in exploiting a range of foraging options that might otherwise not be open to them, and in making and rolling balls of dung faster than species at ambient temperature (Heinrich & Bartholomew, 1979). Ball-rolling species occupying similar trophic habits showed very different thermal niches (Verdú *et al.*, 2007a).

Chown and Klok also point out how thermal tolerance influences the range of new habitats that species are able to exploit, and how thermal tolerance can account for the altitudinal and latitudinal gradients of species richness and abundance, thus playing a role in niche separation and the reduction of interspecific competition. Climatic and abiotic factors influence the structure of local communities at a regional scale (Hanski & Cambefort, 1991). Species richness of dung beetles tends to be greater near the equator and to decrease with latitude (Fig 11.2 in Chapter 11), as predicted from their ecophysiology (Chapter 10).

1.4.2 Regional distribution and seasonal activity

In Chapter 11, Tomas Roslin and Heidi Viljanen provide a broad overview of the factors thought to underlie the distribution and abundances of dung beetle species, contrasting the dung beetle fauna of Finland and Madagascar to illustrate broad geographical patterns. Like Chown and Klok, they identify an important role of body size associated with the regional distribution of dung beetle species. Using mark recapture data, they show how the large *Aphodius fossor* can move over a much larger spatial scale than the small *Aphodius pusillus*, illustrating how body size can influence a species ability to expand its range and, potentially, exploit new niches.

The dung beetle fauna of open and forested habitats differ markedly. The low species richness of scarabaeine dung beetles in pastures, compared with the high species richness in forests, is particularly evident in warmer regions nearer the equator. Examples of this are given in this volume from Madagascar (Chapter 11), Australia (Chapter 12) and South America (Chapter 13). Indeed, the low species richness in pastures was the basis for the biological control programme which introduced exotic dung beetles to utilize cattle dung in Australia (Chapter 12). Pasture species can disperse rapidly (Chapter 12). Most forest species, however, remain strictly confined to the forest habitat and, as Roslin and Viljanen show, habitat discontinuities provide a strong barrier to dispersal (Chapter 11).

The dung beetle faunas of adjacent forest and savannah are completely different, but finer-scale subdivision of habitats can also influence the composition of the dung beetle community (Chapter 13). These broad-scale ecological patterns are reflected in the population genetic structuring of species. For example, the Madagascan forest dwellers have large and stable populations, with restricted gene flow, but for species in Finnish pastures there has been a recent and rapid expansion of populations, with pasture species having a larger range size and a strong ongoing gene flow (Chapter 11). These patterns suggest that the opportunities for niche expansion and speciation are far greater in forest than pasture habitats.

Seasonal activity of dung beetles is influenced by interactions between seasonal rainfall and temperature. Most adult dung beetles are active in summer and, while immatures can spend dry seasons in the soil, they do not survive in cold, wet seasons. For species in winter rainfall areas, the winter is spent in the adult stage.

Beetles breed in spring or after the commencement of rain, when dung quality is high. Seasonal changes in dung quality as a result of changing patterns of plant growth can have a substantial impact on the rate of egg production of many dung beetle species (Chapter 12). It is possible for winter active and summer active beetle species to co-exist at the same sites (Chapter 12), but this does not always occur, and the mechanisms involving interspecific competition that allow coexistence between these species are not well understood.

1.4.3 Community dynamics

Despite competition, it is possible for many species to coexist in a dung beetle community and, as we have seen, in any one community, the different species occupy many different niches. In addition, dung beetle distribution between pads tends to be aggregated (Hanski & Cambefort, 1991; Lobo & Montes de Oca, 1997; Slade *et al.*, 2007), so that high numbers of beetles of different species may not occur in the same pads. Small species tend to be more aggregated than large species, possibly as a result of having smaller niche differences.

The high species richness of many tropical communities makes an analysis of the factors influencing community structure hard to assess. At a broad level, Hanski & Cambefort (1991) note that large rollers and fast-burying tunnellers are usually the top competitors, while dwellers are the weakest competitors.

The dung beetle community in pastures grazed by cattle supports fewer species and is dominated by small species (Chapter 13). In Australian pastures, even though both large and small exotic species have become established, the dominant species are *Euoniticellus intermedius*, *Digitonthophagus gazella* and *O. taurus*, all considered small species (Chapter 12). Indeed, these species are far more abundant in Australian pastures than they are in their country of origin. However, large species have established and, as noted, they remove more dung proportionally than do small species. If their abundance could be enhanced, they would increase an ecosystem function by burying more dung, with pasture productivity and fly control benefits. While this could occur naturally over a much longer time period, it has been proposed that there is a need to manage the structure of the pasture dung beetle community, perhaps by increasing the habitat complexity of vegetation (Chapter 12).

1.5 Conservation

Given their extraordinary evolutionary radiation, the dung beetles offer a good model taxon with which to address the many problems associated with the conservation of global biodiversity. This is a difficult area of research, because it involves considerable elements of human judgement, so conservation efforts need to be compatible with other aspects of human use of the environment. For example, in Spain it has been noted that the abundance of the 11 species of rollers (mostly larger dung beetles, including *Scarabaeus sacer*) collected in the Iberian peninsula has decreased from 24 per cent to 6 per cent during the 20th century, particularly as a result of the loss of coastal sandy country to urban development (Lobo, 2001). An example of the use of biodiversity to measure conservation need comes from data on temperate and tropical systems, which indicates that a regional scale decline and loss of medium to large mammals has severely disrupted the diversity and abundance of dung beetle communities (Nichols *et al.*, 2009; see also Figure 13.6).

The conservation of insects has two important components. One is the conservation of key species which are a focus for particular concern, and the second is the use of insect biodiversity to indicate the general health of an environment. In Chapter 13, Elizabeth Nichols and Toby Gardner describe the use of scarabaeine dung beetles as an ecological indicator taxon for the conservation of biodiversity.

The characteristics needed of a group to be an ecological disturbance indicator are that they must have viability, reliability and interpretability (Chapter 13). A taxonomic database is available for the scarabaeine beetles, managed by members of The Scarabaeine Research Network (www.scarabnet.org). Sampling methods using dung-baited pitfall traps have been widely tested and represent a cost-effective way to detect the effects of management on dung beetles (Figures 13.4 and 13.5). Also, dung beetles have proved very responsive to habitat disturbance of tropical forests (Figures 13.2 and 13.3, and Chapter 11).

Nichols and Gardner separate species traits into response traits, which relate resource or environment needs to species performance, and effect-based traits which, as the name suggests, are those affecting the impact of the species on the environment. The effect-based traits are listed as hard traits against the more easily measured soft traits which are commonly used (Table 13.1). Distinguishing how soft and hard traits interact with human activities is one of the cutting-edge issues in understanding the meaning of biodiversity (Chapter 13).

A key soft trait is beetle body size (a key factor in inter- and intraspecific competition), which has a disproportionate effect on the amount of dung buried. However, large beetles also have the greatest risk of local extinction, so consideration should be given as to how to conserve them. In their meta-analysis, Nichols and Gardner show the importance of structurally complex habitats in maintaining dung beetle biodiversity (Chapter 13). They suggest that increasing habitat complexity in pastures could assist building communities with more large beetles and give a list of examples where dung beetles are being used as indicators of ecological disturbance, helping to develop a better understanding of the key factors.

Individual Coleoptera are underrepresented on the IUCN red list, but they are included in the Sampled Red List Index Program thanks to the Scarabaeine Research Network, who estimate that over 12 per cent of all dung beetles are threatened with extinction, with a further 9 per cent vulnerable to extinction. Dung beetle survival requires both the maintenance of intact mammal communities and tight habitat control. The reality of the threat of extinction is illustrated strikingly in Chapter 11, where the loss of 90 per cent of habitat in both Madagascar and in Finland has resulted in about 40 per cent of species disappearing in each case (Chapter 11, Fig 11.5). The dung beetles are thus proving an excellent taxon with which to investigate conservation needs of the world's tropical forests.

1.6 Concluding remarks

This volume summarizes a rich history of research on scarabaeine dung beetles. Mostly, researchers choose their subjects because of a passion or admiration for the animals in their own right, but they are also driven by research agendas, using their chosen taxon to test or advance some general scientific theory. With the accumulation of knowledge, some taxa become recognized as 'model systems' (Dugatkin, 2001). The humble fruit fly *Drosophila melanogaster* is perhaps one of the best known model systems, and much of our knowledge of genetics, developmental and cell biology, life history and evolution comes from research on this one species – research that has been invigorated by the publication of the full genome in 2000 (Powell, 1997; Markow & O'Grady, 2006).

The cumulative research available on scarabaeine dung beetles now covers a broad array of disciplines. It has given us insights into the private lives of these fascinating and endearing creatures. More importantly, it has been instrumental in developing our understanding of broad ecological processes and how they shape the evolution of biological diversity. We are seeing further than ever before, with our research efforts yielding new information at all levels of analysis from functional genomics to developmental biology; comparative morphology; physiology; behaviour; and population and community ecology.

Research on dung beetles is shedding light on the ultimate goal of how best to document and conserve the world's biodiversity. With this volume, dung beetles emerge as a model system that will continue to deliver important progress in evolutionary and ecological research. To quote Fabre (1918): 'Notwithstanding their filthy trade, the dung-beetles occupy a very respectable rank.'

2

The Evolutionary History and Diversification of Dung Beetles

T. Keith Philips

Systematics and Evolution Laboratory, Department of Biology, Western Kentucky University, KY, USA

2.1 Introduction

The scarabaeine dung beetles are a group of insects well known to many due to their daytime activity around dung pats and their often bright body colours. They are also one of the more morphologically diverse groups of animals and include nearly 6,000 species and 257+ genera (ScarabNet taxon database). This is a large group, but not overwhelmingly so, for us to understand the evolution of its global biodiversity. Importantly, too, they are of great use in conservation; they possess a number of characteristics that can enable one to sample their diversity relatively easily and use these data to assess the health of ecosystems, including those considered biodiversity hotspots (Spector, 2006; Chapter 13).

Biodiversity can be defined as the variation among living organisms from all sources and the ecological complexes of which they are a part, including the diversity within species, among species, and of ecosystems (Heywood, 1995). Although it is not stated explicitly, one key goal of this research is to understand how the amazing diversity of life came about via the course of evolution. The process of understanding dung beetle diversification is part of a larger desire by systematists to reconstruct the 'tree of life'. We are now truly at a point where we have a reasonable idea of how dung beetles evolved, even though numerous details remain to be discovered. However, one could argue that this research has been ongoing since the first naturalists started documenting the diversity on Earth, perhaps as far back as Aristotle, circa 300 BC!

The modern field of systematics goes much further than simply documenting the existence of species. It is now a science of comparison of taxa and the discovery and explanation of the pattern of evolution. Only when we understand how species are

Ecology and Evolution of Dung Beetles, First Edition. Edited by Leigh W. Simmons and T. James Ridsdill-Smith. Published 2011 by Blackwell Publishing Ltd.

related to each other via a well-supported hypothesis of relationships can we begin to have a true understanding of organisms.

The process of creating phylogenetic trees gives us knowledge of the patterns of diversification of life. These results are incredibly powerful for three reasons:

- first, because they are based on scientific evidence that can later be supported or refuted with additional evidence;
- second, because these trees answer questions on the pattern of evolution of various structures, genes or behaviours; and
- third, they enable us to predict likely or plausible features or characteristics in taxa for which we know little.

Halffter & Matthews (1966) suggested that, 'a phylogeny [evolutionary history] of the Scarabaeinae cannot be constructed at present because of the absence of a sound morphological basis for doing so.' Nevertheless, they proposed a very plausible scenario of an ancestral saprophage dung beetle species in the Mesozoic era that split or diverged into two lineages – one that lead to all 'rolling' species and the other that lead to all 'tunnelling' species. This intuitive approach was perhaps a gestalt impression laden with assumptions about the evolution of dung beetles, and it was no doubt based in part on morphological similarity of these lineages, as well the idea of 'parsimony' or the simplest path of evolution being the most likely.

Recently there has been a great increase in the number of phylogenetic studies using either morphological or molecular evidence, analyzed with more advanced methods to hypothesize relationships. These have shown that our first impression of the diversification of dung beetles was clearly incorrect, and dung beetles provide a classic example of how evolution is not always parsimonious; we now realize that various lineages of ancestral tunnellers lead to the evolution of many clades of rollers.

This chapter will discuss the various phylogenies available that show relationships among the currently recognized tribes. It will emphasize two studies that include representatives of all tribes and that are fairly congruent, even though they use two different sources of data (Philips *et al.*, 2004a; Monaghan *et al.*, 2007). It will also present a general consensus of dung beetle evolution from all existing studies (see Figures 2.2–2.4). Finally, this chapter will discuss the evolutionary origins of various clades and behaviours and will illustrate the amazing degree of evolutionary plasticity in this group. Conclusions will be summarized and future directions for research suggested.

2.2 Scarabaeinae diversity and tribal classification issues

The Scarabaeinae is a cosmopolitan group, with species found on all continents except Antarctica. The majority of taxa are found in tropical regions. The greatest diversity is in the land masses that once comprised Gondwana, and all of the basal groups occur in sub-Saharan Africa, tropical South America, the Australian region and Madagascar (Table 2.1).

Table 2.1 Biogeographical diversity of the Scarabaeinae.

Taxon richness	*Palaearctic*	*Nearctic*	*Oriental*	*Afrotropical*	*Neotropical*	*Australasian*	*Madagascar*	*World*
Total tribes	8	8	9	9	8	3	3	12
Total tribes (revised)	8	8	12	15	9	3	3	23
Total genera	17	11	42	104	68	30	19	234
Total species	335	88	690	2001	1163	441	215	4989

Data from Davis and Scholtz (2001). The total number of revised tribes is based on individual lineages in Figure 2.4.

Table 2.2 Current tribes showing known dung manipulation behaviour and possible clade origin. Ages at this stage are highly speculative due to lack of fossil or molecular clock evidence.

TRIBE	*Traditional grouping*	*Actual biology*	*Traditional Age*	*Clade origin?*
Canthonini	Roller	Rolling, dragging, tunnelling	Old	Old to Modern
Coprini	Tunneller	Tunnelling	Modern	Old
Dichotomiini	Tunneller	Tunnelling, dwelling, kleptoparasitism	Old	Old to Modern
Eucraniini	Roller	Tunnelling, carrying	Intermediate	Modern
Eurysternini	Roller	Tunnelling, dwelling	Intermediate	Intermediate
Gymnopleurini	Roller	Rolling	Intermediate	Intermediate
Helictopleurini	Tunneller	Tunnelling	Modern	Modern
Oniticellini	Tunneller	Tunnelling, dwelling	Modern	Modern
Onitini	Tunneller	Tunnelling	Modern	Intermediate
Onthophagini	Tunneller	Tunnelling, kleptoparasitism	Intermediate	Modern
Phanaeini	Tunneller	Tunnelling	Intermediate	Modern
Scarabaeini	Roller	Rolling, dragging, tunnelling	Intermediate	Intermediate
Sisyphini	Roller	Rolling	Modern	Intermediate

The most commonly used tribal level classification is that of Balthasar (1963a and b). Of 12 tribes, six are broadly classified as rollers while the others are tunnellers. This simplified view, however, neither reflects the evolution of the group nor true food manipulation behaviour; 'rolling' in dung beetles has evolved to encompass many different behaviours that clearly have separate origins (Table 2.2, Figure 2.2).

It is possible that a minimum of 23 tribes based on both monophyly and, to a lesser extent, food habits, should be recognized within the Scarabaeinae (see Figure 2.4). Alternatively, and perhaps less desirable, one could drastically modify the current definitions by increasing the size of most tribes. From evidence in existing phylogenies, a consensus tree on the evolution of dung beetles is hypothesized in Figure 2.4. Most of the following discussion will be based in large part upon this single phylogeny, based on evidence from multiple sources. While the upper clade with generic names is shown as sister to the remaining dung beetles, this group may ultimately be shown to be composed of some or all of the oldest lineages of dung beetles. This part of the tree will therefore be used as a reference point for some discussion on possibly the most ancient groups of dung beetles.

2.2.1 Dichotomiini and Coprini

There is little evidence for monophyly in these tribes. A clade of dichotomiines and/or canthonines is likely the oldest lineage of dung beetles and sister to all others. The separation of the coprines from the dichotomiines also appears quite difficult to justify as evidenced, for example, by the former appearing in two clades as derived dichotomiines in Vaz-de-Mello, (2007a). Further, this study, with over 40 taxa of dichotomiines *sensu lato*, (i.e. including the coprines), shows at least eight clades that could be given tribal status. Even though the two main phylogenies have arguably sparse taxonomic representation, members of these tribes are found in nine (Philips *et al*., 2004a) and 13 (Monaghan *et al*., 2007) separate lineages. In Sole & Scholtz (2010), the African members can be viewed as forming five separate clades.

2.2.2 Canthonini

Most canthonines are relatively derived, and many members with a 'roller' body shape do not even roll dung! All canthonines evolved directly from an ancestral dichotomiine *sensu lato*. Evidence for this pattern is seen in Monaghan *et al*. (2007) where, for example, the canthonine *Amphistomus* is sister to the dichotomiines *Coptodactyla* and *Dermarziella* and all form a clade with the canthonine *Diorygopyx*. In Philips *et al*. (2004a), three clades of basal canthonines are also seen to have separate origins from ancestral tunnellers. Lastly, the canthonine *Odontoloma* may be derived from within a dichotomiine clade (Figure 2.4). Some lineages of canthonines are monophyletic, and many are located within a single region. Several of these lineages should probably be considered as separate tribes.

This polyphyletic evolutionary pattern has contributed to many possibly misclassified taxa. For example, the Neotropical genera *Anomiopus*, *Hypocanthidium* and *Scatonomus*, long considered dichotomiines, were placed in the canthonines by Vaz-de-Mello (2008). The highly derived morphological features of many species associated with ants and termites have also caused difficulties in tribal placement. For example, the genus *Haroldius* (see Krikken and Huijbregts, 2006) or the junior synonym *Afroharoldius* have been considered members of the Canthonini or the once-used Allocelini (e.g. Paulian, 1993). Morphological evidence suggests that these ant-associated taxa are derived onthophagines (Philips, 2005) or even a suggested new tribe Ateuchini (Montreuil, 2010). Other African genera of canthonines (e.g. *Hammondantus* and *Odontoloma*) may also be onthophagines (Scholtz and Howden, 1987).

2.2.3 Eucraniini

Endemic to Argentina, the eucraniines are composed of only four genera and are overwhelmingly supported as a monophyletic clade (Philips *et al*., 2002; Ocampo & Hawks 2006). The only controversial genus in the tribe is the basal monotypic *Ennearabdus*, a taxon that was once placed in its own tribe (Martínez, 1959). However, there is little doubt that this is sister to the three more derived genera based on morphological, molecular and behavioural evidence (see Ocampo, 2004; 2005).

2.2.4 Phanaeini

The New World phanaeines are strongly supported as a monophyletic group, again with both morphological (Philips *et al*., 2004b) and molecular data (Ocampo & Hawks, 2006). The only controversial generic inclusions are *Oruscatus* and

Gromphas, both of which were excluded by Edmonds (1972) in his tribal definition. The study of Philips *et al.* (2004b), however, including a wide selection of outgroups, found that these two genera share at least 10 synapomorphies with the phanaeines and should be considered as part of this tribe.

2.2.5 Phanaeini + Eucraniini

This relationship is strongly supported by both morphological (Philips *et al.*, 2004b) and molecular evidence (Ocampo & Hawks, 2006). This clade in turn is sister to a lineage of New World dichotomiines, likely *Dichotomius*, perhaps, together with *Holocephalus* and additional genera. The close relationship of the eucraniines with the phanaeines was also noted by Zunino (1983; 1985), based mainly on study of the genitalia.

2.2.6 Scarabaeini

This tribe is composed of a group of morphologically similar taxa distributed mainly in Africa but also ranging through the western to eastern Palaearctic, India and Madagascar. There is good evidence that this Old World clade is indeed a monophyletic assemblage (e.g. Villalba *et al.*, 2002; Philips *et al.*, 2004a; Forgie *et al.*, 2005; Sole *et al.*, 2005; Forgie *et al.*, 2006; Sole *et al.*, 2007; Monaghan *et al.* 2007). The canthonine genus *Circellium* should be placed within this tribe based on both Philips *et al.* (2004a) and Monaghan *et al.* (2007).

2.2.7 Gymnopleurini

This unquestionably monophyletic tribe (Monaghan *et al.*, 2007) is composed of four very similar genera that probably originated in Africa and then spread throughout the western and central Palaearctic and Oriental regions.

2.2.8 Eurysternini

The Eurysternini is composed of a single genus, *Eurysternus*, found in the Neotropics. The relationships of this tribe to other taxa are quite uncertain. Molecular evidence (Monaghan *et al.*, 2007) places the group as sister to a species of *Canthidium* and also as sister to *Ontherus*, both dichotomiines and a plausible relationship as all are New World groups. In contrast, Philips *et al.* (2004a) indicates a sister relationship with the sisyphines, onitines, oniticellines and onthophagines.

2.2.9 Sisyphini

Species in this tribe are relatively widespread, with two genera found in Africa and Asia while one genus is endemic to Mauritius. Species are also found in Central America. Definitely a monophyletic group, the relationship of this tribe to the others is similar regardless of the evidence examined. Morphology (Philips *et al.*, 2004a) places this tribe as sister to the eurysternines, onitines, oniticellines and onthophagines, while DNA (Monaghan *et al.*, 2007) has this clade as sister to the same group excluding the eurysternines. Based on distribution, the latter relationship is more likely.

2.2.10 Onitini

Although this Old World tribe appears to be a closely related group of genera, there is evidence that it is paraphyletic without the oniticellines and onthophagines (Philips, unpublished). Assuming monophyly, the latter clade of two tribes is the sister group based on evidence from the two main phylogenies, as well as Ocampo & Hawks (2006).

2.2.11 Oniticellini

The tribe evolved in Africa but spread into the Palaearctic and the Orient. There are also several isolated New World taxa, while the sub-tribe Helictopleurini is endemic to Madagascar. Some evidence supports the monophyly of this tribe (Philips *et al.*, 2004a; Monaghan *et al.*, 2007), although two studies suggest it is paraphyletic without the inclusion of the Onthophagini (Villalba *et al.*, 2002; Philips, 2005). A third study suggests paraphyly without some members of both the Onthophagini and even the Onitini (Wirta *et al.*, 2008), although the latter tribal inclusion seems less likely.

Traditionally the Oniticellini has been divided into three sub-tribes: the Oniticellina, the Drepanocerina and the Helictopleurina (Halffter & Matthews, 1966). The only sub-tribe as currently defined that can be justified as monophyletic is the Helictopleurina, a group of two genera that currently includes all the native oniticelline species known from Madagascar. The genus *Helictopleurus* (with about 60 species), together with the monotypic genus *Heterosyphus* (which appears to be nothing more than a derived *Helictopleurus* (Wirta *et al.*, 2008)) is strongly supported as monophyletic. This clade likely evolved from within the remaining oniticellines.

2.2.12 Onthophagini

The Onthophagini is the largest clade of scarabaeines; with over 2,500 species, it includes close to half of all known taxa. With a probable origin in Africa (basal lineages are generally Afrotropical in Philips, 2005 and Monaghan *et al.*, 2007), species then spread to all tropical and temperate regions in the world. While this tribe is fairly well-defined and probably monophyletic (Philips *et al.*, 2004a; Philips 2005), the former study in one topology places the onthophagines as an ancestral lineage to the onitines (plus eurysternines) and the oniticellines. There is also evidence in Monaghan *et al.* (2007) that some basal lineages may belong in the Onitini or Oniticellini. Although the onthophagines appear to be monophyletic, it would not be surprising if this tribe was ultimately found to be polyphyletic.

2.3 Scarabaeine dung beetle phylogenies

Probably the first attempt to create a tree reflecting evolutionary relationships within dung beetles was that of Paulian (1933) for the Scarabaeini (i.e., all 'rollers' at that time) and included 17 taxa mainly of what now are considered canthonines (Figure 2.1).

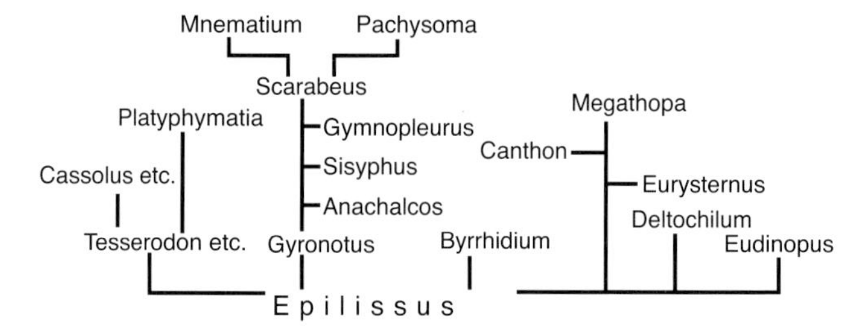

Fig. 2.1 The first tree of dung beetle evolution, proposed by Renaud Paulian (1933) for the rolling species.

Fifty years later Zunino (1983) published the first phylogenetic study, including all 12 tribes but with only 18 taxa, using genitalic characters. Although some relationships have been supported, such as the relatively basal position of *Coptorhina* and the close relationship of the oniticellines and onthophagines, others, such as the onitines as derived phanaeines, are not. A similar, but smaller, study using the same types of characters, not surprisingly produced a similar tree (Luzzatto, 1994). Although rooted with a group that should properly polarize characters (the Aphodiinae), it showed the clearly derived eucraniines at the tree base and sister to all other scarabaeines, and the onitines as polyphyletic. Both of these studies reflect the potential pitfall of relying on a limited set of characters, regardless of whether the data are morphological or molecular.

A more thorough study of the tunnelling dichotomiines and coprines, with species representing 27 genera and using 42 morphological characters, resulted in the proposal of a newly defined Coprini, the Ateuchini and the elimination of the Dichotomiini (Montreuil, 1998). A minor conundrum was the rooting of the tree using *Onitis* and *Onthophagus*, two taxa that are now known to be almost certainly derived. Also, as most (if not all) of the canthonines and all other dung beetles are derived lineages of the dichotomiines and coprines, it does not clarify where these more derived clades appear and, therefore, where further subdivisions within these ancient tunnelling clades are needed to reflect monophyly. Regardless, rooting this tree using *Coptorhina* + *Delopleurus* (which may represent one of the oldest extant lineages of dung beetles) results in at least three major clades – that of the root and two others.

More recently, Villalba *et al.*, (2002) studied relationships among seven tribes found in the Iberian Peninsula. This phylogeny was the first to use molecular sequence (CO1 and CO2) data. Interestingly, their results support the long-accepted dogma of a single origin for both tunnelling and rolling behaviour, although, as they note, with a loss of rolling and reversion to tunnelling in *Copris*. Alternatively, one should consider their result as support for two origins of rolling, both in the sisyphines and in the scarabaeines + gymnopleurines. No doubt a much different picture would have been presented if the authors had included some basal dichotomiine and 'rolling' canthonine taxa.

A second molecular study (Ocampo & Hawks, 2006) with 18S and 28S rDNA data, including ten tribes (gymnopleurines and sisyphines were absent), shows the basal origin of many of the dichotomiine and coprine clades, the more recent evolution of some of the canthonine lineages and the derived nature of the eucraniines, phanaeines, onitines, oniticellines, and onthophagines. This study also supports rolling behaviour evolving at least twice.

Vaz-de-Mello (2007a), using a fairly broad array of 87 taxa and a large amount of evidence from 297 morphological characters, examined a wide range of dichotomiine, coprine and canthonine taxa, as well as small numbers of all other tribes. This study shows evidence for some basal dung beetle clades supported in other studies, proposes three new tribes, redefines the Ateuchini of Montreuil (1998) and defines two new sub-tribes. Some unexpected relationships seen in this topology are due to the problems of convergence, and include a species of *Eurysternus* (Eurysternini) and two Oniticellini as the basal clade and sister to all other dung beetles, as well as the sister relationship of a species each of *Onthophagus* (Onthophagini) + *Onitis* (Onitini) as sister to the Phanaeini + Eucraniini.

A fairly comprehensive study of the evolution of the dung beetles was that of Philips *et al.* (2004a). Relationships among species representing all 12 tribes and 48 genera were hypothesized using morphological evidence from 200 characters, although canthonines were poorly represented (Figure 2.2). Regardless, this topology was the first to show that the idea of a tunnelling clade and a rolling clade was very weakly supported; rolling behaviour evolved at least five separate times from ancestral tunnelling clades. The basal position of the dichotomiines (plus coprines) is again seen, as in the molecular studies of Villalba *et al.* (2002) and Ocampo & Hawks (2006).

A later more taxonomically comprehensive study, which included all tribes and nearly twice as many genera as Philips *et al.* (2004a), is Monaghan *et al.* (2007), that used molecular evidence from three genes (CO1, 28S, rrnL). This tree topology again supported the earlier conclusions of several origins of rolling and the non-monophyly of the mainly basal dichotomiine, coprine and canthonine tribes (Figure 2.3). In addition, it showed how biogeographical distributions at least in part reflect common ancestry. Importantly, their inclusion of more canthonine taxa indicated for the first time the possibility that one or more of these clades might be basal.

Both Philips *et al.* (2004a) and Monaghan *et al.* (2007) show implausible relationships due to the effects of convergence. Regardless of these problems, both studies are relatively robust (in terms of taxa and data) and there is a fair amount of congruence between the two data sets, giving us reasonable confidence in the overall general pattern of evolution shown.

The most recent study is that of Sole & Scholtz (2010) on the African fauna. This phylogeny includes eight of nine dichotomiine and 17 of the 23 canthonine genera found in this region. Evidence was based on five genes and about 3,030 nucleotides, compared to three regions and about 1670 bases in Monaghan *et al.*'s (2007) study. Unfortunately, selection of taxa from only one region is problematic for an overall picture of dung beetle evolution; but regardless of this, their tree shows a small clade of four dichotomiine genera plus one canthonine as sister to all other taxa. Moreover, there are some similarities with the phylogenies presented in Philips *et al.* (2004a), Vaz-de-Mello (2007a), and Monaghan *et al.* (2007), including the basal nature of a clade containing *Coptorhina*, as well as some canthonines such as *Dicranocara*.

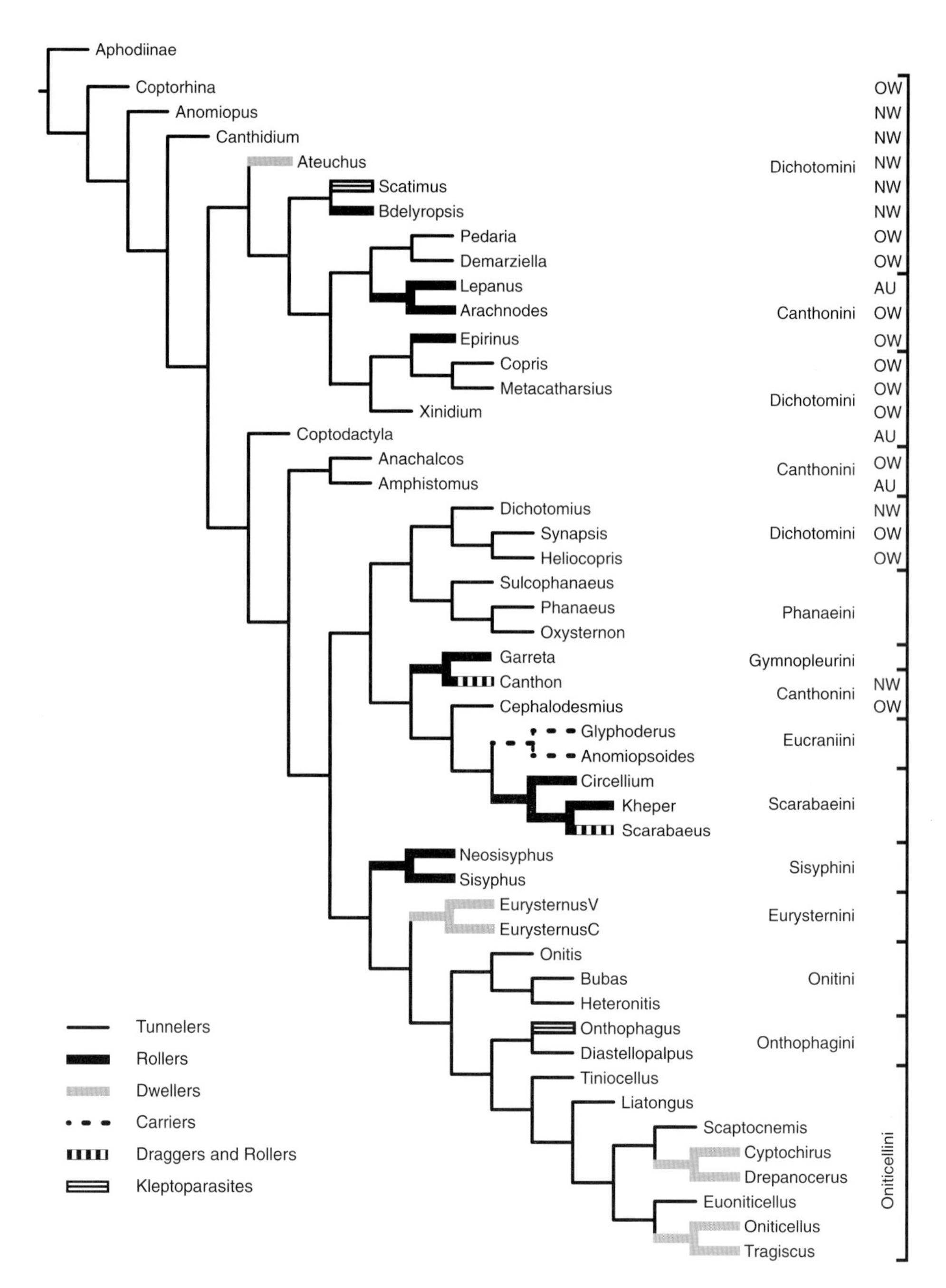

Fig. 2.2 The phylogeny of Philips et al. (2004a), with tribal affiliation indicated and dung manipulation behaviour plotted on the tree. Dwelling in the dichotomiines is represented by *Ateuchus* and kleptoparasitism in the onthophagines by *Onthophagus*. OW = Old World excluding Australasia, NW = New World, AUST = Australasia.

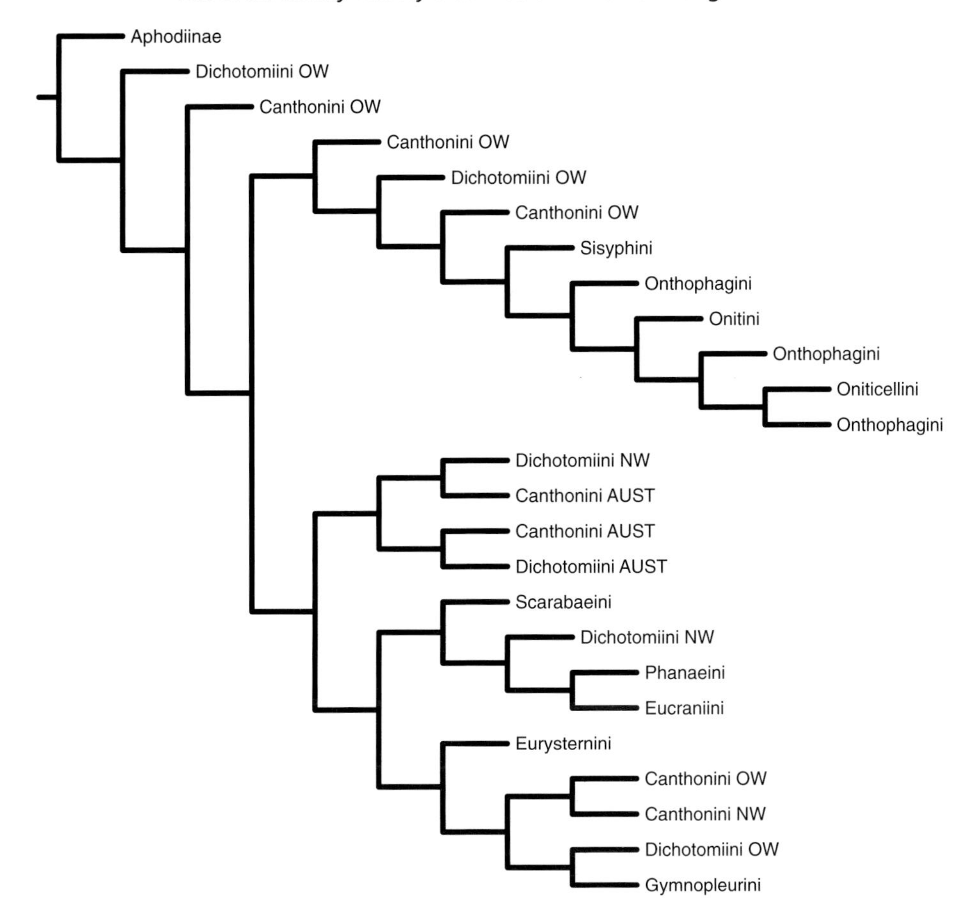

Fig. 2.3 The phylogeny of Monaghan et al. (2004) based on their Bayesian analysis result. The tree has been simplified by converting species to tribal affiliations and eliminating potential outlier taxa that appear to be misplaced. OW = Old World excluding Australasia, NW = New World, AUST = Australasia.

2.4 The sister clade to the Scarabaeinae

In studying the evolution of dung beetles, it is critical to first examine the likely sister group, as this will help elucidate ancestral behaviours such as both food sources and food manipulation behaviours. If we know the group that shares a common ancestor with dung beetles, and the successive sister clades above this lineage (as well as those below), we can find evidence for the biology of the first true dung beetle.

Interestingly, there is evidence that the first dung beetles were not actually utilizing dung for food but were possibly saprophagous, as suggested by Halffter & Matthews (1966), or perhaps even macro-fungivorous, a possibility first noted by Philips *et al.* (2004a).

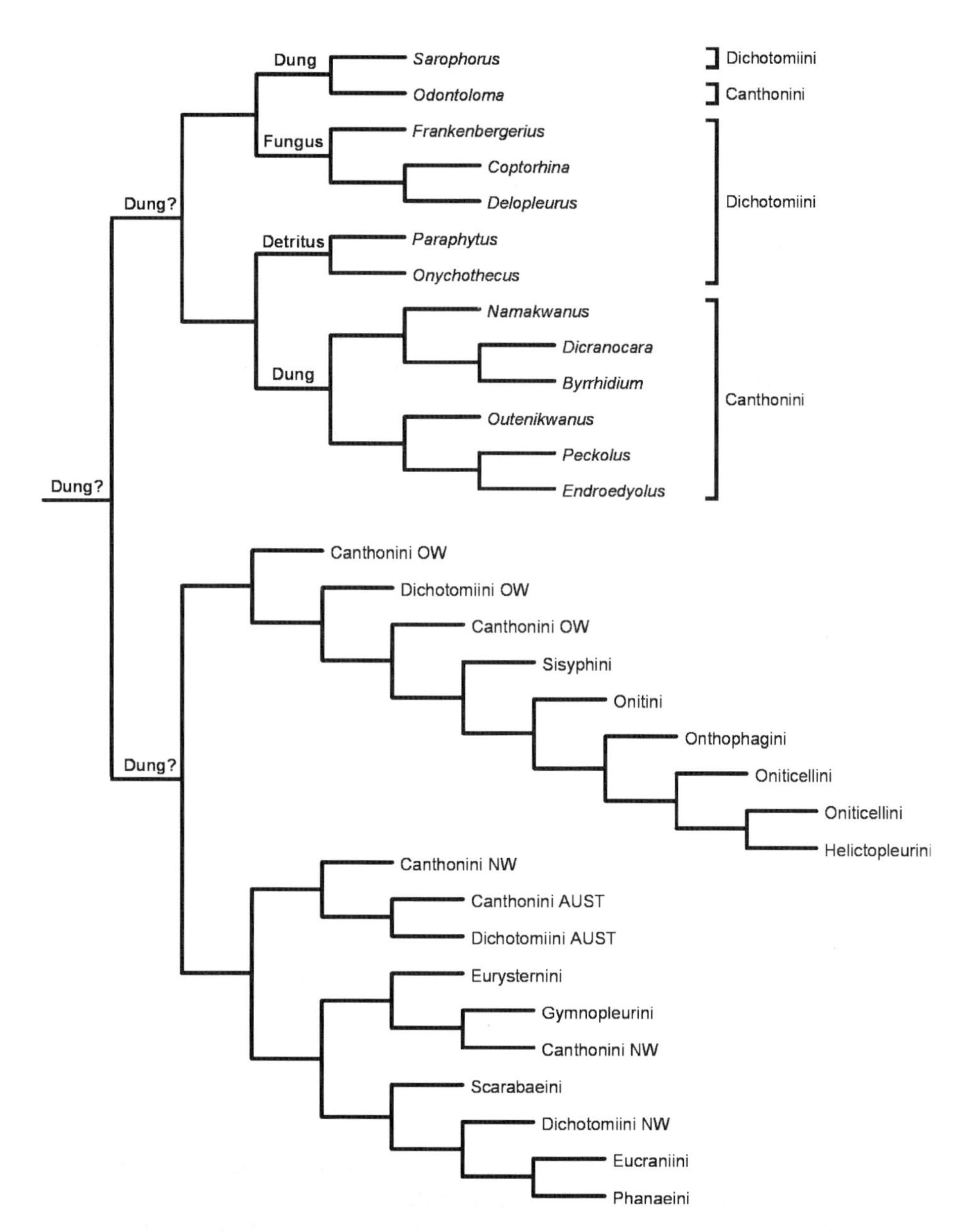

Fig. 2.4 The hypothesized phylogeny of the scarabaeine dung beetles with tribal affiliation indicated. All genera included are African, with the exception of the Asian *Onychothecus*, and may represent one or more basal clades. The Dichotomiini includes the Coprini and both the dichotomiine and canthonine lineages have been divided into Old World (NW), New World (NW) and the Australian region (AUST). This tree is based on evidence from the studies of Philips *et al.* (2004a), Ocampo & Hawks (2006), Monaghan *et al.* (2007), Vaz-de-Mello (2007a), Wirta *et al.* (2008) and Sole & Scholtz (2010).

Nearly all of the data suggests that the sister clade of the Scarabaeinae is the Aphodiinae, a group of small, elongate-bodied scarabs. Support for this relationship is seen via morphological evidence from Browne & Scholtz (1998) and 18S and 28S DNA sequence evidence from Smith *et al.* (2006). The only contrasting view is that of Hunt *et al.* (2007) in a study of the Coleoptera using 16S, 18S and CO1 sequences, showing the scarabaeines as sister to the Glaresidae, and this clade in turn sister to the Hybosoridae, and all three clades as sister to the Aphodiinae. This is likely an anomalous result, possibly due to difficulties in alignment. Regardless, it is quite likely that the Aphodiinae is paraphyletic due in particular to the problematic status of Aegialiini and Aulonocnemini (see Cambefort, 1987; Dellacasa, 1987; Stebnicka, 1985; Matthews & Stebnicka, 1986 and molecular evidence of Ocampo & Hawks, 2006 and Smith *et al.*, 2006).

So, as the sister group is probably the Aphodiinae, at least in part, what are their habits? While the majority of species are known to feed on dung, others have a range of lifestyles similar to that seen in the scarabaeine dung beetles. For example, species are known to feed on detritus, including that found in *Atta* leaf-cutter ant or termite nests (e.g. Howden & Storey, 1992; Stebnicka & Skelley, 2009) or may even be predators (Skelley & Gordon, 2002). Others are reported to feed on decaying plants or on roots (Hanski, 1991). Interestingly, most of the Aphodiinae tribes are saprophagous (Cambefort, 1991a; 1991b), suggesting that strict coprophagy may be a derived trait within this group of beetles (e.g. Verdú & Galante, 1999). Unfortunately, phylogenies of the aphodiines needed to address this question are very limited in taxonomic scope (see Cabrero-Sañudo & Zardoya, 2004; Cabrero-Sañudo, 2007) and hinder our attempts to understand the ancestral feeding habits of the Scarabaeinae.

2.5 The origin of the dung beetles

Based on the fossil record, Krell (2006) dates the Aphodiinae-Scarabaeinae divergence around the boundary between the Jurassic and Cretaceous periods, about 140 million years ago. A preliminary study using a molecular clock approach dates the origin of dung beetles at 110 m.y.a. (F.C. Ocampo, personal communication). One fossil identified as a scarabaeine dung beetle, *Prionocephale deplanate*, is known from the Upper Cretaceous 92–83.5 m.y.a. (Krell, 2006). Dinosaur coprolites with feeding traces from coprophages in the Cretaceous exist, but the precise identity of the organism making tunnels or other marks is not clear and it may be due to geotrupid scarabs (e.g. Chin & Gill, 1996).

Possible ichnofossils of dung beetle pupal chambers indicate their existence from as early as the late Cretaceous (Genise *et al.*, 2000). In contrast, others have proposed that dung beetles did not arise until the Cenozoic era, around 50 m.y.a. (Scholtz & Chown, 1995). A recent study estimates the divergence time of the aphodiine-scarabaeine split at about 56 m.y.a. (Sole & Scholtz, 2010), correlating well with the latter hypothesis. Lastly, Scholtz *et al.* (2009), using molecular ageing results for *Helictopleurus* (from Wirta *et al.*, 2008) to calibrate a relative age scale for the Scarabaeinae provided by Monaghan *et al.* (2007), yields ages of 71–44 m.y.a. for the origin of dung beetles.

During the Mesozoic era, dinosaurs were the dominant and largest terrestrial vertebrates and there must have been abundant faeces, although arguments have been made that dinosaur dung would not have been very attractive due to the mixing of dung and nitrogenous waste in the cloaca (Arillo & Ortuño, 2008). Similar extant cloacal dung is poorly studied regarding attractiveness, although records of dung beetles using both reptile and bird dung exist and include dichotomiines, canthonines, and sisyphines (e.g. Vinson, 1951; Martinez, 1952; Matthews, 1965; Young, 1981).

Regardless of how attractive dinosaur dung or that from other cloacal animals actually is, it may not be critical in regards to knowing how early dung feeding scarabaeines evolved. Therian mammals (those lacking a cloaca) were present in the Upper Triassic and, although diversity was low, the group as a whole was widespread (e.g. Kielan-Jaworowska *et al*, 2004).

To conclude, clock data may not be as reliable as one might hope, and unfortunately fossils needed to clarify the age of dung beetles are rare. However, there is no doubt that the diversification of mammals in the Cenozoic, and especially the grassland-adapted artiodactyls, led to an incredible radiation of dung beetles.

2.6 The oldest lineages and their geographical origin

The earliest dung beetles were likely small, with elongate bodies, and similar to their sister group, the Aphodiinae. An examination of generic level endemism of dung beetles shows a heavy bias towards the Gondwana fragments of Africa, South America, Madagascar and Australia (Table 2.1). There is also evidence that the Afro-Eurasian distributed genera and geographically widespread tribes have an Afrotropical origin (Davis *et al.*, 2002).

Additionally, the basal lineages are African in most phylogenies, and these in turn gave rise to all lineages found throughout the world. Hence all tribes originated in Africa, with the exceptions of the New World Eurysternini, Eucraniini and Phanaeini. Based on Philips *et al.* (2004a), it is likely that the next oldest lineages are those in South America and Australia, while in Monaghan *et al.* (2007), with better taxa representation, there is a mix from various regions of Gondwana.

The three main basal tribes of Dichotomiini, Coprini, and Canthonini may have evolved before the separation of the ancient supercontinent Gondwana within southern Africa. It is possible that lineages of these groups then spread to what is now southern South America and Antarctica, then continued their expansion from Africa, through to perhaps Madagascar and also Australia via Antarctica, while these land masses were connected. If the origin of dung beetles occurred sometime around the Jurassic-Cretaceous boundary (around 145 million years ago), this correlates well as it predates the Mesozoic fragmentation of Gondwana, which may have started 125 million years ago (but see Cattermole (2000) for the beginning of the break up as early as 167 million years ago). Alternatively, with a younger age for the origin of the group, dispersal scenarios are needed to explain present distributions of lineages.

Philips *et al.*, (2004a) and Monaghan *et al.*, (2007) have *Coptorhina* (+ *Sarophorus* in the latter) as the basal lineage. It is likely that the genera *Delopleurus* and *Frankenbergerius* are also part of this most ancient clade of dung beetles, based on Vaz-de-Mello (2007a). The second oldest lineage in Monaghan *et al.* (2007) is *Dicranocara* + *Odontoloma*. Evidence from Sole & Scholtz (2010) includes all of these genera and shows that they compose a single clade that may be the sister to all other dung beetle taxa (but see below for two additional genera).

A slightly different view for the oldest dung beetle lineage is seen in a canthonine phylogeny based on 51 canthonine genera and three dichotomiines including *Coptorhina* (C.A. Medina, unpublished). This topology shows *Coptorhina* as the second oldest lineage. Three taxa forming the most basal clade are the canthonines *Byrrhidium*, *Dicranocara* and *Namakwanus*, a result close to that in Monaghan *et al.* (2007). The study of Sole & Scholtz (2010) supports both hypotheses: *Byrrhidium*, *Dicranocara* and *Namakwanus* form a clade that is part of a larger lineage that includes three other canthonines (*Endroedyolus*, *Peckolus* and *Outenikwanus*). This full clade is one of two that together form the sister clade to the basal lineage, composed of the dichotomiines *Coptorhina*, *Delopleurus*, *Sarophorus*, *Frankenbergerius* and the canthonine *Odontoloma*.

Support for some of the oldest lineages is also found in the tree of Vaz-de Mello (2007a). Excluding the most basal clade of *Eurysternus* plus two genera of oniticellines, the oldest lineage is composed of four dichotomiine genera (*Coptorhina*, *Delopleurus*, *Frankenbergerius*, and *Sarophorus*) and their sister clade of dichotomiines *Onychothecus* and *Paraphytus* plus the canthonines *Byrrhidium* and *Dicranocara*. Therefore, either one of these clades or both as sister lineages may well be arguably the most ancient clade of dung beetles and sister to all other scarabaeines, although the data of Sole & Scholtz (2010) suggests that the former clade (together with *Odontoloma*) is the most likely.

One might suggest that a clade of canthonines cannot be the oldest lineage, as these have the derived rolling morphology and behaviour compared to the dichotomiines *sensu lato* (i.e. the dichotomiines and coprines). However, a great deal of morphological divergence within this potentially basal 'canthonine' clade undoubtedly has occurred since its origin, and perhaps quite recently. Hence, the common ancestor morphologically had the features and behaviour of a more typical dichotomiine but evolved at least some of the characteristics found in rolling dung beetles (supporting their placement in the canthonines). Further, none of the species in these genera (with the possible exception of species of *Byrrhidium* with unknown biology) arc what one would call true rollers; pellets are carried and/or dragged while walking backwards to burrows for burial (Deschodt *et al.*, 2007), a relatively unique behaviour that may be derived from tunnelling.

An examination of the distribution of the potentially basal dichotomiine and canthonine genera may be a clue to the geographical origin of dung beetles. Of the perhaps seven valid *Coptorhina* species (Davis *et al.*, 2008b), two species are found in South Africa, while others are widely distributed in various parts of Africa (Angola, Sierra Leone, Cameroon, Democratic Republic of Congo (DRC) and Somalia). The seven known *Frankenbergerius* species are all found in South Africa. *Delopleurus* includes only three species known from southern Africa, the DRC and the Orient. The ten *Sarophorus* species occur mainly in South Africa, with three of these found as far north as Tanzania. For the canthonines, nearly all the *Odontoloma* species are southern African in distribution (Scholtz & Howden, 1987), while the species of *Byrrhidium, Dicranocara*, and *Namakwanus* are also found in north-western South Africa and southern Namibia. Lastly *Endroedyolus*, *Peckolus*, and *Outenikwanus* are all southern African, too.

The present distribution of a species may not be indicative of where it evolved or reflect past distributions. However, if these groups truly compose the oldest lineage or lineages of dung beetles, the high species diversity in southern Africa may indicate that this is where the Scarabaeinae originated.

2.7 Evolution of activity period

Many species of dung beetles have a particular period during the day when they forage (e.g. see Gill, 1991; Caveney *et al.*, 1995). Flight periods may have evolved to correlate with defecation patterns of mammals. Other species have activity periods based in part on body size. Hence, large-bodied dung beetles, all of which are found in the more derived clades and probably selected for improved competitive ability, can raise and regulate their body temperatures to fly during cooler periods (see Chapter 10). For smaller-bodied species, in contrast, relatively low temperatures may especially constrain their ability to fly at night and dawn (Caveney *et al.*, 1995). There is evidence that some small-bodied forest dwelling species have evolved behaviours to thermoregulate during the day (Gill, 1981).

Within nearly every tribe, there is a large range of activity periods. The oldest lineages of dichotomiines and coprines are typically nocturnally active, but there are many genera of dichotomiines, including a few *Copris*, that are daytime-active (e.g. Kingston & Coe, 1977). Many of the canthonines are active at night (e.g. Hernández, 2002), although this does not appear to be the case in the African fauna (Davis et al., 2009). Most of the more derived rollers in the tribes Gymnopleurini and Scarabaeini are active during the day but, at least within the Scarabaeini, there are both diurnal and nocturnal species within a genus (e.g. *Kheper* and *Scarabaeolus*). The four nocturnal species within *Scarabaeus* included in the phylogeny of Forgie *et al.* (2006) are apical and share a common ancestor, implying that diurnal activity is ancestral and nocturnal activity only evolved once. While most species of Onitini are nocturnal, there is a wide range of activity periods and many genera are diurnal. In 11 species of *Onitis* and *Heteronitis castelnaui*, each, according to Caveney *et al.* (1995), evolved one of five distinct strategies for flight behaviours: dusk crepuscular; dusk/dawn crepuscular; dusk/dawn crepuscular; and nocturnal, late afternoon-dusk and dawn-early morning, or diurnal activity. A wide range of flight duration has also evolved in dung beetles. Vulinec *et al.* (2003), in a study of *Phanaeus* reports daily flight time for one species of less than 10 minutes at dawn, whereas others will fly for up to several hours during one part of the day. Lastly, daytime activity in some taxa has also resulted in the selection for brighter body colour (Hernández, 2002).

2.8 Evolution of feeding habits

Both larvae and adults feed predominantly on herbivorous or omnivorous mammal dung. This resource represents a rich source of nutrients based in large part on the bacterial content, sometimes high nitrogen content, abundant complex carbohydrates and also vitamins and minerals. One would reasonably conclude that the ancestral food source was dung, but this may not be the case; as previously discussed, detritus may have been the ancestral food. Feeding on humus of some form (along with fungus and other microorganisms) is still retained by the larval and adult aegialiine and aulonocnemine aphodiines, both possible sister groups. Moreover, a transition from detritus (or fungus) to dung does not seem to be that difficult an evolutionary step, as coprophagy can be considered a specialized type of saprophagy.

Evidence for detritus feeding as the ancestral feeding behaviour in dung beetles is seen in the tree of Vaz-de-Mello (2007a). This phylogeny shows the saproxylophage taxa of *Paraphytus* and *Onychothecus* as a basal lineage (together with *Dicranocara* and *Byrrhidium*) as a sister clade to *Sarophorus* and the fungal-feeding specialists *Coptorhina*, *Delopleurus* and *Frankenbergerius*. Further, the Montreuil (1998) phylogeny, when correctly rooted, shows one of the older and somewhat isolated clades is that composed of *Paraphytus* (Ethiopian and Oriental) and *Onychothecus* (Southeast Asia). In contrast, the relatively ancestral clade of *Coptorhina*, *Delopleurus* and *Frankenbergerius* feed on fungal fruiting bodies, supporting fungus as the ancestral food in the Scarabaeinae.

In regards to the evolution of dung feeding, a group of canthonines that may be a basal lineage are composed of the genera *Byrrhidium, Dicranocara* and *Namakwanus*, which have species utilizing mainly dung pellets from the dassie or hyrax (*Procavia capensis*) (Deschodt *et al*., 2007). A second clade, *Odontoloma*, has species recorded on a variety of dung types as well as carrion and rotting fungus (Howden & Scholtz, 1987). This is some evidence that dung feeding, even if it were not the ancestral food, did evolve very early in dung beetle evolution.

To conclude, although dung feeding may be the ancestral condition, there is evidence that, ancestrally, scarabaeines were either saprophages or fungivores. Although it is not yet clear (and it may never be so), it is possible that dung feeding evolved two or more times, an intriguing idea for the group known as the true dung beetles!

2.9 Evolution of derived alternative lifestyles

Most dung beetles are generalists to various degrees, and relatively few species have only been recorded on just one food type (Table 2.3). For example, specialists include taxa that utilize only a certain type of food, such as ruminant dung (either pelleted or not), millipede carrion, fungi or elephant dung (e.g. Forgie *et al*., 2002; Larsen *et al*., 2006; Davis & Dewhurst, 1993). Regardless, it is important to distinguish food alternatives used by adults as an alternative source of energy and not for larval brood food. One trend noted is that the New World dung beetles appear to have many more food alternatives compared to Old World species (see below). Nevertheless, the variety of food types does indicate the flexible strategies that have evolved in order for adults to survive and reproduce under various conditions, and is further evidence of the amazing evolutionary shifts that have occurred within dung beetles.

Table 2.3 Number of genera of Scarabaeinae within each region that include likely specialist species on alternative foods.

Tribe	*Saprophagy*	*Frugivory*	*Necrophagy*	*Millipedes*	*Fungivory*	*Inquiliny*	*Carnivory*	*Phoresy*	*Vert. nests*	**TOTALS**
Dichotomiini OW	2	1	–	–	3	–	–	–	–	**6**
Dichotomiini NW	2	6	3	–	2	6	–	3	1	**23**
Dichotomiini AUST	–	–	–	–	–	–	–	–	–	**0**
Canthonini OW	–	–	–	–	–	–	–	–	–	**0**
Canthonini NW	–	3	7	–	–	3	2	1	–	**16**
Canthonini AUST	1	–	–	–	–	–	–	–	–	**1**
Phanaeini	–	4	2	–	1	4	–	–	1	**12**
Eucraniini	1	–	–	–	–	–	–	–	–	**1**
Scarabaeini	1	1	–	1	–	–	–	–	–	**3**
Gymnopleurini	–	1	–	–	–	–	–	–	–	**1**
Eurysternini	–	1	–	–	–	–	–	–	–	**1**
Sisyphini	–	–	–	–	–	–	–	–	–	**0**
Onitini	–	–	–	–	–	–	–	–	–	**0**
Helictopleurini	–	–	–	–	–	–	–	–	–	**0**
Oniticellini	–	–	–	–	1	1	–	–	–	**2**
Onthophagini OW	–	3	2	1	–	14	–	–	–	**20**
Onthophagini NW	–	1	–	–	–	1	–	–	1	**3**
Onthophagini AUST	–	1	–	–	1	–	–	1	–	**3**
TOTALS	**7**	**22**	**14**	**2**	**8**	**29**	**2**	**5**	**3**	

Old World excluding Australian region = OW, New World = NW, Australian region = AUST. Dichotomiini includes the Coprini.

Saprophagy is rare and, as discussed above, may be the ancestral feeding behaviour. Additionally, some species both in *Pachysoma* (Scarabaeini) and *Anomiopsoides* (Eucraniini) use dried plant material, a derived saprophagous behaviour that evolved convergently in southern Africa and Argentina, respectively, possibly to cope with limited amounts of dung in hyper-arid regions (Ocampo & Philips, 2005). Derived detritus feeders also include many species associated with social insects, such as the attine ant associate oniticelline *Attavicinus monstrosus* (see Philips & Bell, 2008). There are at least 20 additional genera of dung beetles in four more tribes (dichotomiines, canthonines, phanaeines, and onthophagines) associated with ants and termites that also probably feed on nest detritus. More may exist, as evidenced from a lack of biological information on many rarely collected taxa, resulting from their lack of attraction to typical baits (e.g. see Larsen *et al.*, 2006; Gillett *et al.*, 2009). Within the 14 genera of onthophagines and the four genera of phanaeines, evidence suggests this association evolved only twice – once in each tribe (Philips, 2004b; 2005) – in contrast to the New World canthonines and dichotomiines, where it probably arose several times within both groups.

All frugivorous taxa were ancestrally dung feeders. Frugivory seems to be particularly prevalent in the Neotropics, with four times as many species recorded using this resource compared to other regions (Halffter & Halffter, 2009). Additionally, at least half of the major dung beetle clades have evolved specialists that feed on fruit (generally that in some state of decay) or seeds. This includes 35 species of the onthophagine *Onthophagus* (e.g. Kohlmann & Solís, 2001), the New World genera *Canthidium* and *Dichotomius*, each with 14 species, and *Canthon* with 13 species reported on fruit.

Necrophagy has evolved fairly frequently, although many of the species reported to feed on carrion are also coprophagous. Regardless of the degree of specialization, it appears that all necrophagous species ancestrally were dung feeders. One of the best known genera that use carrion exclusively is the dead millipede specialist *Sceliages*, a member of the Scarabaeini (Forgie *et al.*, 2002). Moreover, there are several species within *Onthophagus* that are also millipede specialists in Africa (Krell *et al.*, 1997), Southeast Asia (Brühl & Krell, 2003) and perhaps also in other regions where large millipedes exist.

What are the selective pressures that may have lead to the evolution of carrion (and fruit) feeding behaviour in dung beetles? In the New World, Halffter & Halffter (2009) suggested that a shift to these food sources may have been caused by the extinction of many large mammals during the Pleistocene. In the Old World, beetles that feed on millipedes have the advantage of being nearly exclusive users of these carcasses and therefore avoid higher levels of competition at other types of carrion and dung (Krell *et al.*, 1997). It may also be a lack of other necrophagous species, such as the Silphidae in some habitats like the Neotropics, or a result of a relative abundance of carrion in the Afrotropics.

Fungal specialists have evolved in dung beetles at least five times. The African dichotomiines *Coptorhina*, *Delopleurus* and *Frankenbergerius* represent an ancient lineage that are specialists on basidiomycote fungi. One species of phanaeine (*Phanaeus halffterorum* Edmonds) may also be exclusively mycetophagous (Price & May, 2009), as is a species of Australian *Onthophagus* (Bornemissza, 1971). The New World *Liatongus rhinocerulus* (Bates) uses decomposing fungus as food (Halffter & Edmonds, 1982), although it can be collected occasionally in dung traps. Larsen *et al.* (2006) also record a few Neotropical dichotomiines that also may be fungal specialists.

Predation is similar to necrophagy, except that the adult dung beetle actually attacks and kills the food source. This recently derived, and very uncommon, behaviour has evolved at least twice in the Canthonini. *Deltochilum* attacks millipedes, while *Canthon* kills reproductive leaf-cutter ants (Lichti, 1937; Navajas, 1950; Pereira & Martínez, 1956; Larsen *et al.*, 2009).

Several taxa are known to hitchhike on mammals and they have evolved modified tarsal claws for grasping fur around the genital or anal region of their host – a wonderful example of convergence among several different scarabaeine genera. In Australia and New Guinea, phoresy probably evolved once in *Onthophagus*, with several species phoretic on kangaroos and wallabies (Halffter & Matthews, 1966; Matthews, 1972). In contrast, Neotropical phoretics evolved at least five times within each of the following genera: *Uroxys* and *Pedaridium* (formerly *Trichillum*), found on sloths (Ratcliffe, 1980); *Canthon* (*Glaphyrocanthon*) (Pereira & Martinez, 1956); *Canthidium* (Herrera *et al.*, 2002); and *Canthon* (Jacobs *et al.*, 2008), with species found on monkeys.

A second and very unusual case of phoresy evolved in the genus *Zonocopris*. Individuals of the two known species can be found on the exposed mantle of giant land snails in South America, where they apparently feed upon mucus as adults. In studies by Vaz-de-Mello (2007b), use of snail faeces was never observed and he speculates that the beetles may breed in snail carrion after the death of their companion.

A few species of dung beetles have evolved an association with vertebrate nests. Some species of onthophagines (all *Onthophagus*) and several coprines (all *Copris*) are found exclusively in nests, indicating at least two separate origins of this behaviour. One species of phanaeine has also been reported as a possible inquiline in rodent nests (Price & May, 2009). Many additional taxa of dung beetles can also be found in nests, but not exclusively. Nests used include those of various species of rodents and gopher tortoises (Halffter & Matthews, 1966; Ziani & Gharakhloo, 2010).

2.10 Evolution of nidification: dung manipulation strategies

Based on phylogenetic evidence, paracoprid or tunnelling behaviour in some form is the ancestral mode of food relocation. This relatively simple behaviour entails digging beside or underneath the food source. Once the burrow is created, bits of food are carved off the source, perhaps held between the head and forelegs, and they are then carried, dragged, or pushed down into the burrow for feeding by adults and larvae (see Chapters 1 and 8). This pleisiotypic behaviour is widespread in dung beetles and found in the basal as well as the most derived lineages of dung beetles, including members of the dichotomiines, coprines, phanaeines, onitines, oniticellines and onthophagines, as well as one eucraniine.

From ancestral tunnellers, dung beetles known as telecoprids or true rollers evolved the ability to construct a dung ball and roll it over the surface (Chapter 1). It is thought that rolling behaviour first evolved in order to avoid the tremendous competition for this sometimes ephemeral food source that can occur directly on, within and even under the dung pile (see Figure 1.4). It has been suggested that rollers first evolved in open habitats (Halffter & Matthews, 1966), but the presence of rollers in Old World tropical forests indicates that open habitats are not necessary to evolve or maintain a rolling fauna (Krell, 2006). There is no strong phylogenetic evidence yet for where rolling generally evolves, so it seems more reasonable to surmize that, regardless of habitat, rolling behaviour evolved wherever high levels of competition occurred. True rollers are known in the canthonines, scarabaeines, gymnopleurines, and sisyphines.

Many alternatives to 'rolling' have evolved, as dung can be moved overland without the creation of a ball. Most of these behaviours are present in tunnellers to move dung inside burrows, or perhaps in rollers whose food requires no ball construction and, for example, single pellets are translocated. In others, however, the behaviours truly appear to be evolutionary novelties. One unique method, where the beetle grasps and lifts dung with its forelegs and moves forward, is seen in most of the eucraniines (Ocampo, 2004; 2005; Ocampo & Philips, 2005), a group once classified as one of the rollers. A second unique method is known in only two lineages; species in the south-west African *Pachysoma* and the New World *Canthon obliquus* drag food by using their hind legs while moving forward (Scholtz, 1989; Halffter & Halffter, 1989). Thus, this behaviour has evolved convergently from true rolling behaviour in two separate clades on separate continents.

One should be aware that we are still lacking evidence of dung manipulation behaviour in many of the 'rolling' canthonines. For example, species of *Epirinus* (Medina & Scholtz, 2005) that are typically collected in the leaf litter may not be rollers. Also, some south-western African species, such as in the genera *Dicranocara* and *Namakwanus* (Deschodt *et al.*, 2007), are above-ground dung movers that may never have been rollers ancestrally. There are some Australian canthonines that do not make balls nor roll dung pellets or any other bits of dung, but alternatively show tunnelling habits; known as the mentophilines, these taxa are characterized by non-rolling habits and hence have been hypothesized to be of very old origin (Matthews, 1974). However, there is no evidence for their great antiquity and there are some mentophiline taxa, such as the African *Anachalcos*, that do roll balls.

A third strategy of dung manipulation is dwelling or endocoprid behaviour that is only known within the Oniticellini, the Eurysternini and, in a simpler form, in some Dichotomiini. Females lay eggs either within the dung mass in a formed ball or, alternatively, on the surface of the dung, without any attempt at relocating the food (Chapter 1). This is a relatively rare form of behaviour derived from tunnelling behaviour. At least in some oniticellines and eurysternines, this strategy seems to be associated with breeding when fewer species are active and/or they are using older and drier dung (typically late season when rains are diminishing or absent) that is no longer attractive to the majority of species. This behaviour evolved at least once within each tribe.

One last behaviour is kleptoparasitism, a strategy derived from tunnelling species that is known in some small-bodied Dichotomiini and Onthophagini, and which has numerous separate origins (see below). It is found in two behavioural guilds (Cambefort, 1991b). One group digs or enters existing tunnels beneath the dung mass and waits for larger tunnellers to bury dung, which they then steal. Those that parasitize rollers are actually more attracted to the dung balls then to the original source of dung, and they burrow into the ball during its construction or while it is being buried.

In some habitats, kleptoparasites are considered to be the dominant guild of beetles (Krell *et al.*, 2003) and may comprise around ten per cent of the species (Cambefort, 1991b). Kleptoparasitic behaviour may not have appeared until after the large tunnellers and rollers evolved. It also appears to be behaviourally quite labile; while some species of *Pedaria* are obligate thieves, others, such as *Onthophagus acuminatus*, are facultative (Gill, 1991), behaving as a typical tunneller at small droppings, but at large dung piles burrowing into the soil and waiting for large tunneller activity. Cambefort (1991b) suggests that this behaviour may have evolved in species that were accidentally buried by larger dung beetles. Interestingly, this lifestyle may have even led to selection for smaller body size.

2.11 Evolution of nidification: nesting behaviour and subsocial care

Complex nesting behaviour, including a great degree of post-ovipositional care, is a characteristic of many dung beetles, either by the female or, occasionally, involving cooperation by both sexes (e.g. Klemperer, 1983). Subsocial care can be defined as parental care after oviposition that increases the survival and growth of the offspring (Tallamy & Wood, 1986). This is seen minimally in all dung beetles in the provisioning of food, and it continues in some taxa with the protection of eggs, larvae, and even pupae within compound nests. Greater parental care is correlated with lower reproductive output, and perhaps reaches the extreme in *Circellium bacchus*, in which females may only produce a single offspring each year (Kryger *et al.*, 2006).

Parental care may also be critical for reducing pathogens of larvae. When the female is eliminated from the brood chamber, fungus proliferates on the brood balls (Kingston & Coe, 1977; Favila, 1993; Chapter 1). Nearly all species nest below the surface of the ground, and the advantages of doing so are that dung is protected to various degrees from desiccation and use by other individuals (Lumaret, 1995) and it also protects both the adult(s) and brood from surface predators (e.g. Denholm-Young, 1978). Nesting behaviours have been classified into seven patterns by Halffter & Edmonds (1982), including two primitive and five advanced behaviours. Their system is used in the following discussion:

- In the ancestral behaviour (Pattern 1), a burrow is excavated beneath or adjacent to the dung. The tunnel is then packed with dung, an egg is laid, the opening sealed off and the process repeated. Additional branching burrows may be constructed off the initial tunnel for additional brood masses. No further care is supplied by the female (e.g. Cabrera-Walsh & Gandolfo, 1996). This strategy is seen in some of the dichotomiines and basal phanaeines as well as the onitines, onthophagines (see Chapter 8) and most of the oniticellines. Although present in what can be considered derived tribes (oniticellines and onthophagines), this is the ancestral behaviour of all dung beetles. Kleptoparasitic behaviour is derived from this behaviour and has evolved at least eight times within various genera of dichotomiines and onthophagines.

- Pattern 2 is known in some of the dichotomiines, the phanaeines and the coprines (*Catharsius* and *Metacatharsius*). As in Pattern 1, each larval provision is within its own chamber but, in contrast, the dung is formed into a ball and covered with a layer of soil. No further care is provided. It is likely that this behaviour arose at least twice from Pattern 1 within dung beetles, based on phanaeines as derived dichotomiines and its presence in at least some coprines.
- A further modification is the construction of compound nests (Pattern 3). This is where a large dung mass is accumulated and then divided up into several individual brood balls within a large chamber. Male-female cooperation is considerable and brood care continues after the egg is laid. This behaviour is known in some of the coprines (*Copris*) and dichotomiines (*Synapsis* and *Heliocopris*), as well as one Australian canthonine (*Cephalodesmius*). The last taxon has a high level of male-female cooperation and foraging is done mainly by the male. Possibly exceptional in dung beetles is that males and females are monogamous and pair bond for life, and there is also post-hatching provisioning of the food ball (Monteith & Storey, 1981). This behaviour evolved at least twice, assuming a single evolution in the dichotomiines *sensu lato*. Within at least the dichotomiines, it is probably derived from Pattern 2 behaviour.
- Pattern 4 nesters are the typical rollers, where a dung ball is buried, remodelled and used for a single larva. No further care is given in most taxa. In contrast, some species divide the dung mass into individual pear-shaped masses to create a compound nest and there is a high level of parental care (e.g. *Kheper* – see Sato & Imamori, 1988). The presence of rolling in the Gymnopleurini, Scarabaeini and Sisyphini clearly indicates three separate evolutionary events. Notably, the Scarabaeini contains the only dung beetle taxon, *Pachylomerus*, that may have reverted to a more typical tunnelling behaviour (Bernon, 1981) but, as it appears as basal or near-basal, the evidence is equivocal. Regardless of this, based on five independent clades of rolling canthonines, creating and rolling balls has evolved at least eight times.
- Pattern 5 is similar to Pattern 4, but nests are always compound and there is more complex brood care. This form of nesting is known in a species of New World *Canthon*. It may be present in other New World canthonines and possibly the Old World *Anachalcos* (Halffter & Edmonds, 1982). This form has evolved at least twice in the New World from a modified Pattern 4 behaviour.
- Patterns 6 and 7 nesting are forms of endocoprid nesting behaviour and both evolved from the basic tunnelling strategy seen in pattern 1. Pattern 6 is only known in the Eurysternini and is one of the more complex nesting strategies known, with provisional nests and compound nests (Huerta *et al.*, 2005). Additionally, within the genus *Eurysternus*, some species of have evolved more 'r' type strategies, with higher numbers of nests and greater numbers of brood balls but no brood care, while other species with 'k' strategies exhibit high levels of post-ovipositional care. This range of behaviour among species of *Eurysternus* illustrates the incredible flexibility and wide variety of nesting behaviours that have evolved within a single genus of dung beetle. Moreover, *Eurysternus* is the only taxon known where infanticide is not associated with lack

of food; it may be associated with continued vitellogenesis, as seen in species with ancestral nesting behaviours (Halffter *et al.*, 1980). Of the nine *Eurysternus* species that have been studied, eight are dwellers and one (*E. foedus*) actually does relocate dung as a tunneller, with the latter likely a retention of the ancestral behaviour.

- Pattern 7 nesters with various forms of endocoprid behaviour are known in four genera of oniticellines, including *Cyptochirus*, *Oniticellus*, *Drepanoplatynus* and *Tragiscus* (Davis, 1977; 1989; Cambefort, 1982; Cambefort & Lumaret, 1984). This form of nesting may also occur in *Paroniticellus* (see Philips & Bell, 2008) and can be applied to the basal *Paraphytus* (and likely *Onychothecus*) that breed in poorly defined cavities in rotten tree trunks (Cambefort & Walter, 1985). Endocoprid nests within the oniticellines are constructed either immediately below the dung pile in a style that has been called soil surface nesting, or within the dung mass (Cambefort, 1982). In some cases, the entire brood mass (composed of individual sub masses) is covered with a layer of soil. This behaviour evolved from basic tunnelling within the Oniticellini at least twice (based on Philips (2005), who shows *Oniticellus* plus *Tragiscus* and *Cyptochirus* as clades) and a third time within the dichotomiines.

A behaviour that can also be considered dwelling but without brood ball creation is seen in many species of small dichotomiines, such as those in the genera *Trichillum* and *Pedaridium* (Alarcón *et al.*, 2009). Females lay eggs in the dung but make no nest or brood ball and lack subsocial behaviour. One species is even known to move as a prepupal larva to the soil, in contrast to the true endocoprids. This behaviour evolved at least once from ancestral tunnelling behaviour. It has been suggested that this endocoprid behaviour of laying eggs on or in the dung, so similar to many of the aphodiines, may be an ancestral behaviour (Verdú & Galante, 2001), but it seems clear that it is a derived form of tunnelling based on the location of these taxa in existing phylogenies. Further evidence is seen in *Trichillum* and *Pedaridium*, where larvae have a dorsal hump and a flattened caudal end, characteristics thought to have evolved for life within a confined space of either a brood mass or ball (Halffter & Edmonds, 1982). Therefore it is most likely that this group did not retain ancestral behaviours (or morphologies) but has evolved to breed in a simple manner, unlike the relatively complex nesting behaviours seen in closely related clades.

2.12 Conclusions

We now have a reasonable picture of dung beetle evolution, and it is apparent that our current classification poorly reflects their evolutionary history. At least twice as many tribes as are currently defined will probably be recognized in the future. We know that the sister group is probably a lineage of the Aphodiinae, and the split from the common ancestor likely occurred sometime within the Mesozoic, or at least no later than the beginning of the Tertiary. All of the oldest lineages are African, and most of these species are found in southern Africa, the region where the first dung beetle may have appeared. Regardless of age of origin, it is apparent that most of the amazing diversity of dung beetles arose in concert with the diversification of mammals and their dung types in the Cenozoic period, within Gondwanan fragments.

It remains uncertain whether dung feeding truly is the ancestral behaviour for the group. Saprophagy or fungivory may be the first food source, although both may be derived traits. Nevertheless, there is no doubt that dung feeding, if not the ancestral behaviour, evolved early in the history of the group. Activity periods are very labile and vary a great deal even within a genus, and we have no evidence that nocturnal activity is ancestral. The utilization of most if not all alternative food sources, such as frugivory, necrophagy, carnivory and myrmecophily, are derived behaviours and have evolved numerous times within different clades.

Tunnelling underneath or beside the dung deposit is the ancestral behaviour, but many alternative behaviours have evolved to relocate dung horizontally, including ball construction and rolling that has evolved perhaps eight or more times. Nesting behaviour is also very diverse and can range from the simple and ancestral behaviour of laying eggs on a buried and unmodified dung mass, to the more complex and derived behaviours of creating a large brood chamber with individual larval food balls and continued care by the female. The most complex forms of parental care have evolved at least eight times within the main categories of dung manipulation of tunnelling, rolling, and dwelling. Based on what we do know, a large range and variation in habits can occur even within well-defined genera and closely related species – evidence again for the amazing evolutionary lability present in dung beetles. Together with the wide variety and form of morphologies found within the Scarabaeinae, the evidence indicates that an extraordinary degree of evolutionary change has occurred during the history of the group.

2.13 Future work/gaps in knowledge

There is no doubt that we still have a great deal of work to do in order to understand the evolution of various characteristics of this fascinating group of beetles. This is particularly true for phylogenetic hypotheses on the two or three relatively basal but unnatural tribes. At least eleven new tribes should probably be defined, both to make sure that each is monophyletic and, in some cases, to recognize unique ecological habits. We also need to discover the biology of many species, as there is still a great deal that we do not know for many taxa in such aspects as activity periods and level of parental care. The lack of resources for biodiversity studies continues to be an issue, but we are fortunate that most species are relatively easy to sample, and hence we have a reasonable knowledge of their diversity. Perhaps optimistically, we have described 70-plus per cent of the extant species.

A robust phylogeny with many more taxa will be needed. With workers on the Scarabaeinae distributed globally and coordinated via the Scarabaeine Research Network, the ability to amass phylogenetic characters from nearly all taxa on the generic and subgeneric level should be relatively easy. It is critical to increase the number of molecular characters, such as that from nuclear genes, to obtain more robust and, hence, more trusted phylogenetic inferences. Information on amino acids, and genes and gene switches and their control in dung beetles, has only recently begun to be considered (see Chapter 7), yet these would provide useful data.

Morphological data should not be neglected, and has at least two advantages over molecular data. Rare taxa lacking preserved DNA can be included, and it also gives us an important level of understanding into how morphological character diversity arose. This phylogenetic task should be completed before any changes to the current tribal classifications are proposed or accepted. That said, it is apparent that there is overall agreement about the general picture of dung beetle evolution, whether one is using morphology or molecules as a source of data. Hence, we can be optimistic that a robust and stable picture of the evolution of this splendid group of animals will be developed within our lifetimes.

Acknowledgments

Adrian Davis, Frank Krell, and two anonymous reviewers are appreciated for their thoughtful comments. My gratitude to John Andersland for creating figures and to Linda Gerofsky for a final edit.

3

Male Contest Competition and the Evolution of Weapons

Robert Knell

School of Biological and Chemical Sciences, Queen Mary, University of London, London, UK

3.1 Introduction

The spectacular horns found on many species of beetle must rate as some of the most extraordinary structures found in the animal kingdom, both for their extravagance and their diversity. Horned species are found in many families of the Coleoptera, but the majority of them are in the Scarabaeidae, four sub-families of which have significant numbers of horned species: the Dynastinae, Cetoniinae, Geotrupinae and Scarabaeinae. The latter two are the dung-feeding scarabs with which we are presently concerned, and in these two families there is not only a huge number of horned species but also an extraordinary variety of horn morphologies. These range from short single or double horns on the head (Figure 3.1C) to the extravagant structures carried by species such as *Heliocopris andersoni* (Figure 3.1F) and *Onthophagus sexcornutus* (Figure 3.1A), which have numbers of large, complex horns arising from both the head and the pronotum. The variability in these horns, even amongst closely related species, has been recognized since the mid-nineteenth century, with Darwin relaying observations from Bates on the matter (Darwin, 1871):

> 'In the several sub-divisions of the family, the differences in structure of the horns do not run parallel, as I am informed by Mr. Bates, with their more important and characteristic differences; thus within the same natural section of the genus Onthophagus, there are species which have either a single cephalic horn, or two distinct horns.'

Recent phylogenetic studies have confirmed Bates's hunch that that these horns exhibit substantial evolutionary lability. Emlen *et al.* (2005b) found that within a

Ecology and Evolution of Dung Beetles, First Edition. Edited by Leigh W. Simmons and T. James Ridsdill-Smith. Published 2011 by Blackwell Publishing Ltd.

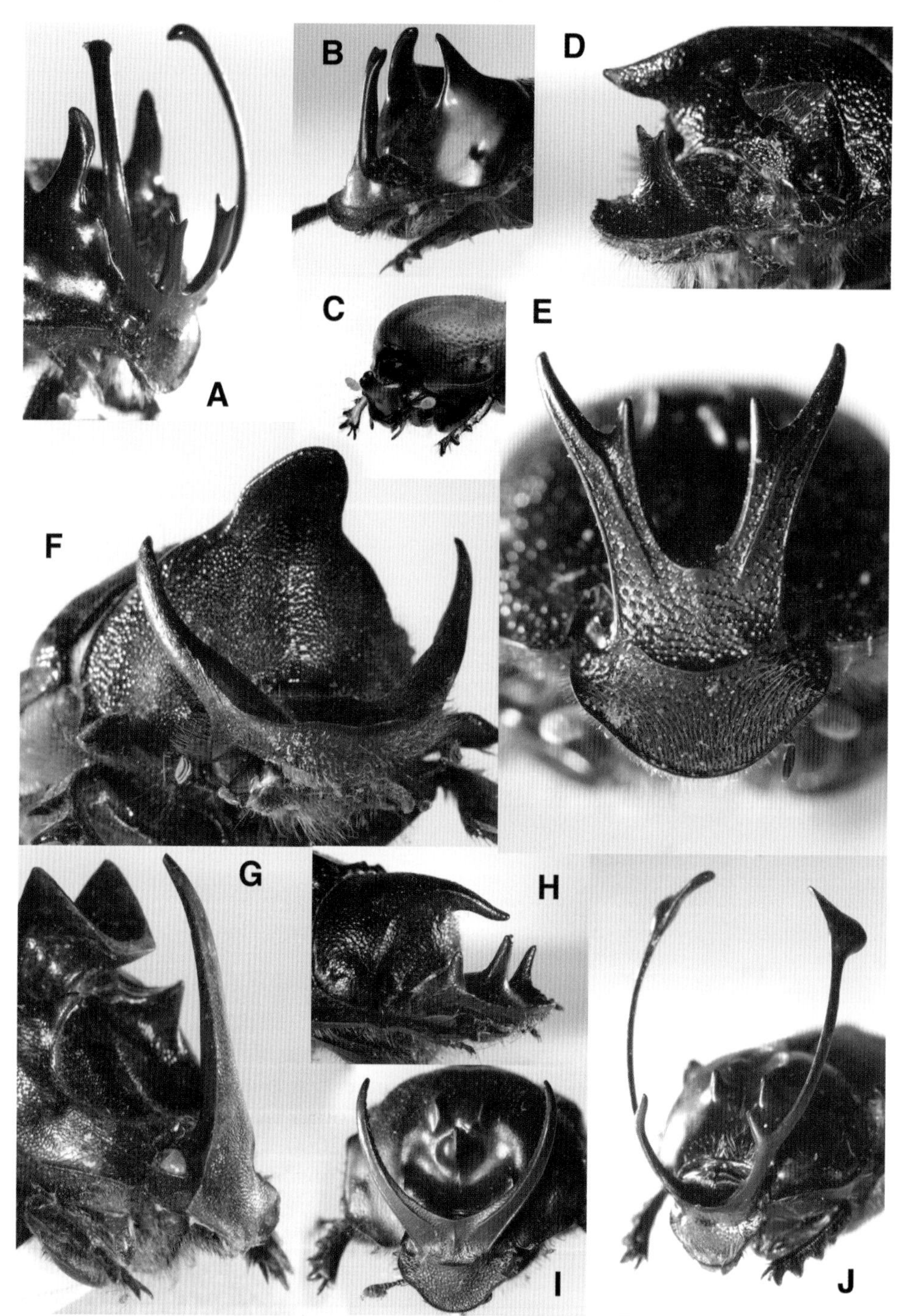

Fig. 3.1 Diversity in horn morphology in dung beetles. A: *Onthophagus sexcornutus*. B: *Oxysternon palaemon*. C: *Euoniticellus intermedius*. D: *Heliocopris hunteri*. E: *Onthophagus imperator*. F: *Heliocopris andersoni*. G: *Coprophanaeus bonariensis*. H: *Heliocopris hunteri*. I: *Onthophagus watanabei*. J: *Onthophagus rangifer*.

phylogeny of 48 species of *Onthophagus* there had been more than 25 evolutionary gains or losses of five types of horn. It seems that the ancestral condition in this particular group of beetles was the possession of a single horn on the head, and that horns have repeatedly been lost, have been gained, and have diversified from this original form.

Not all taxa of dung beetles are horned, however, with many important genera such as *Sisyphus* being entirely hornless. This tremendous morphological variability is clearly in need of an explanation, as are the patterns in the presence and absence of horns. In this chapter I will first focus on the function of these horns and how horn morphology and size are related to the fitness of the bearer. I will then consider how these patterns of diversity in horn presence and morphology might arise, with particular emphasis on the roles of breeding biology, population density and sex ratio.

3.2 Dung beetle horns as weapons

Early workers on beetle horns were not sure of their function. Darwin (1871) discusses them at length in *The Descent of Man and Selection in Relation to Sex*, and concludes that there is little evidence that they are used as weaponry, and that they must therefore be ornaments for attracting females. Although Beebe (1947) described the use of horns in combat between males of a dynastid beetle, it was only towards the end of the 20th century that empirical evidence started to appear that demonstrated a similar role for the horns of dung beetles. This long delay in establishing the function of these remarkable and well-described structures can be attributed to the fact that horned dung beetles usually fight in tunnels (Emlen & Philips, 2006), making observation difficult.

Palmer (1978) was the first to realize that tunnelling beetles will readily excavate in soil between sheets of glass, and used this to observe contests between males of a Geotrupine beetle, *Typhaeus typhoeus,* during which they used their horns as weapons to push each other. The same technique has since been used for species from the Scarabaeinae, and the use of horns as weapons in fights between males has been observed in *Phanaeus difformis* (Rasmussen, 1994), *Onthophagus acuminatus* (Emlen, 1997a), *O. taurus* (Moczek & Emlen, 2000), *Euoniticellus intermedius* (Pomfret & Knell, 2006b), *O. nigriventris* (Madewell & Moczek, 2006) and anecdotally in several more species. These studies, combined with observations of the use of horns as weapons in a number of other beetle species (Brown & Bartalon, 1986; Eberhard, 1979; Eberhard *et al.* 2000; Otte & Stayman, 1979; Siva-Jothy, 1987), have led to a broad consensus among biologists that the horns of beetles are used as weapons during fights, usually between males competing for access to females.

By contrast, there is no evidence of female choice for horns in either *O. taurus*, *O. australis* (Kotiaho, 2002), or *E. intermedius* (Pomfret, 2004). In *O. binodis* there is some evidence for an association between horn length and mating success in the absence of rival males, in that long-horned 'major' males experience higher mating success, but this is attributed by the author to a body size effect rather than a horn effect, with large beetles having higher courtship rates (Kotiaho, 2002). In general, female *Onthophagus* appear to choose their mates based on courtship rate rather than horn morphology (see Chapter 4 of this volume). There is therefore little reason to believe that female choice has played a role in the evolution of beetle horns. However, it cannot be ruled out in every case; possible signalling roles for the horns of Phanaeini, and for the horns carried by some of the more extravagantly ornamented *Onthophagus*, are discussed in the next section.

3.3 Functional morphology of horns

As we have seen, dung beetle horns are extraordinarily diverse, yet we only have detailed descriptions of horn use from a few of these species. This lack of knowledge means that a good understanding of the functional morphology of these structures is still some way away, but a series of studies over the last twenty years have given us an understanding of how the horns are used in some systems. Here I shall relate the form of dung beetle horns to their function, with an emphasis on these systems, and discuss the possible use of horns in some less well-known systems in the light of this knowledge.

Probably the most common horn type is the long, gently curved cephalic horn, often coupled with pronotal sculpturing, as found on most males in the Phanaeini, all male *Copris*, and also on males in many other taxa (Figure 3.1G). These are reminiscent of the cephalic horns carried by many Dynastinae, such as *Oryctes rhinoceros* and *Golofa porteri,* some of which are known to fight by inserting their cephalic horns underneath their opponents. Once this has been achieved and the opponent's grip on the substrate is broken, the defeated opponent can either be flipped onto his back or lifted and held between the cephalic horn and the pronotal horns or sculpturing, allowing the victor to throw his rival to the ground or off the stem where the fight is taking place (Beebe, 1947; Eberhard, 1977; 1979).

Rasmussen (1994) describes similar contests between male *Phanaeus difformis*, with males inserting their cephalic horns beneath opponents and turning them over, and notes that in one case a large male lifted a rival and pinched him against his pronotum using his horn. Rasmussen also reports that *P. difformis* males only fight in this way on the ground at burrow entrances; when males encounter each other in tunnels, the contests are restricted to pushing contests, presumably because the confined space in the tunnels does not allow rivals to be turned over.

Many species of *Phanaeus* will facultatively roll dung some distance prior to burying it, and fights have been reported between males attempting to accompany females rolling dung across the ground (Price & May, 2009). It is tempting to suggest that the ubiquity of long, curved cephalic horns in this genus is a consequence of this habit of fighting on the ground surface, which allows males to lift and flip their opponents. Otronen (1988), however, describes male *Coprophanaeus ensifer*, which also carry long curved horns, as inserting their horns underneath each other in tunnels to allow them to lift and push their opponents, indicating that the use of these horns can vary between taxa.

Many species of dung beetle carry horns that seem to be adapted for pushing, rather than lifting, opponents. Males of the Minotaur beetle, *Typhaeus typhoeus* (Geotrupinae) carry three forward-facing pronotal horns, and Palmer (1978) described their use in detail. Unescalated fights are simple head-on horn-to-horn pushes but, if neither opponent backs down, then one beetle will invert himself in the tunnel so that the two large outer horns engage on the rival's pronotum. The beetles will then engage in a contest of strength that can last up to 75 minutes. A third tactic was described by Palmer as a 'defensive block', whereby a defending beetle edges himself in a tunnel side-on with his back to the aggressor. In these cases, the aggressor uses his horns to lever the defender via the lower edges of the elytra.

Major males of *O. nigriventris* also fight with one male inverted in relation to the other (Figure 3.2), which allows the small posterior horns to engage with the hollow in the cuticle between the anterior and posterior horns, while the longer anterior

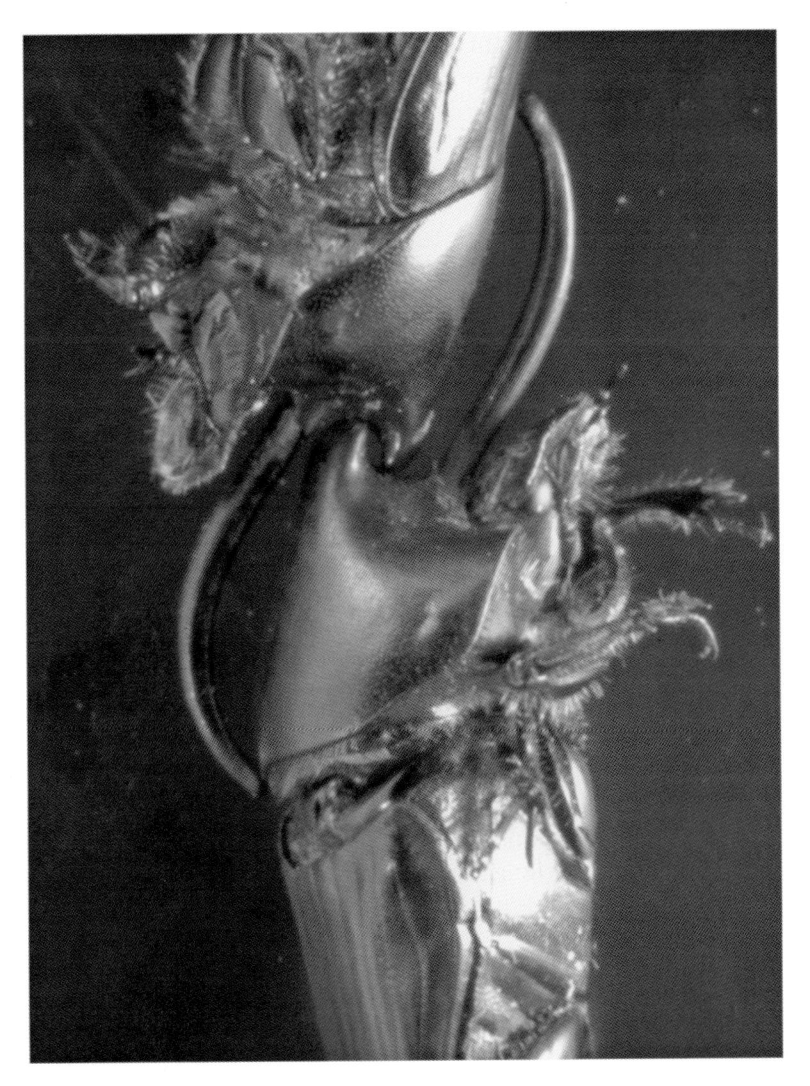

Fig. 3.2 The typical fighting position for *Onthophagus nigriventris*, illustrated using dead specimens (originally published in Madewell & Moczek, 2006).

horn is placed along the curved top of the opponent's pronotum and, in some cases, engages with the gap in the exoskeleton between the pronotum and the elytra (Madewell & Moczek, 2006). While locked together in this position, the beetles will push each other in contests lasting roughly nine minutes, until one is expelled from the tunnel.

Many other beetles have similar horn morphologies, with one or more horns projecting forwards that will engage with an opponent's head or pronotum, and it seems reasonable to suggest that they are likely to be used in a similar way. As an example, consider the three *Heliocopris* species shown in Figures 3.1D, F and H. While the morphologies of these beetles' horns are all different, the overall effect is similar in each species, with a forward-pointing pronotal horn or horns combined with one, two or three upward-pointing cephalic horns. With the head lowered and the cephalic horns pointing forwards, the beetle will present a thicket of pointed weaponry to its opponents. Smaller opponents with less well-developed horns will find their rival's horns fully engaged against their pronotum and head, while they struggle to gain purchase because their own shorter horns are unable to engage their rivals to the same extent.

Rather than the multiple horns found in beetles like these *Heliocopris* spp., many beetles carry more modest armament that is also used for pushing, rather than lifting, opponents. As described above, the two curved horns of major *O. taurus* males engage with the rival's pronotum during contests (Moczek & Emlen, 2000) and the short single horn of *E. intermedius* (Figure 3.1C) is used to pry and push at an opponent's head (Pomfret & Knell, 2006b). It is likely that many of the short, straight or slightly bent cephalic horns carried by other beetles are used in the same way.

The horns of other beetles are likely to be used to engage rival's horns directly rather than the pronotum or head. Consider the horns of *O. imperator* (Figure 3.1E); these animals might lower their heads and engage opponents with the points of the horns, but the shape and location of the horns, coupled with the head extending downwards and parallel to the plane of the horns, suggests the possibility that they are held vertically to block tunnels. Alternatively, the forked ends of the horns would engage with the small pronotal horns, were the beetles to fight with one inverted in relation to the other.

The function of some of the more elaborate horns is harder to understand and has been little studied. In some cases at least, the morphology of the horns might reflect specific details of the beetles' mating systems or the nature of the contests. This is known to be the case in bovids and cervids, where both the overall size and the morphology of the horns or antlers is correlated with factors such as group size, territoriality (Brø-Jørgensen, 2007) and the way that the animals use their weapons in contests (Caro *et al.*, 2003). In the absence of detailed studies of the mating systems of large numbers of dung beetle species, it is difficult to carry out similar studies at present, but this is certainly an area of research that is likely to be fruitful as our knowledge of these animals improves.

Looking at specific details of some of these species with very exaggerated horns, it is possible that the long, curved outer horns carried by species such as *O. sexcornutus* (Figure 3.1A), *O. elgoni* and *O. panoply* function in a similar manner to those of *O. taurus*. However, their great length begs the question of how the bearer manages to bring them forwards in the confined space of a tunnel. Some of these animals carry horns extending upwards from the pronotum, and these might be important during fighting if the males brace themselves within tunnels by pushing up with their legs and pressing the pronotum against the top of the tunnel, as is known to happen in *O. taurus* (Moczek & Emlen, 2000) and *E. intermedius* (Knell, *personal observations*).

The question of how the horns are used is even more acute in the case of *O. rangifer*, which carries horns that are almost the same length as its body (Figure 3.1J), and which are normally carried folded back along the animal's back. Lowering the head raises the horns to the upright position seen in Figure 3.1J, something that would be impossible in most beetle tunnels, which are only a little larger in diameter than the excavator. The horns must therefore either be used in wider tunnels, in the open, or remain parallel to the animal's body when used.

Finally, the females of some dung beetle species carry horns. These are either reduced versions of the male horn (e.g. *Phanaeus difformis* (Rasmussen, 1994)) or different structures that appear to have independent evolutionary origins from male horns (e.g. *Onthophagus sagittarius* (Simmons & Emlen, 2008)). In the case of *P. difformis,* females are reported to fight with other females that attempt to steal dung or take over burrows, but whether the horns are important in these contests is not clear (Rasmussen, 1994).

On the other hand, female *O. sagittarius* use their horns in fights with other females in contests over limited supplies of dung (Simmons & Emlen, 2008; Watson & Simmons, 2010b). Like the males of *O. nigriventris,* the horned females of *O. sagittarius* fight with one individual inverted in relation to the other, such that the cephalic horn engages in the area between the pronotal horn and cephalic horn of the opponent (Watson & Simmons, 2010b).

3.4 Horns as predictors of victory

It is now clear that not only are horns used in fights between (usually) male beetles, but that horn length is an important predictor of victory in these fights. Horn length co-varies with body size, which could be an important predictor of fighting ability, so experimenters have controlled for body size by staging contests between pairs of males matched for size but not for horn length. This technique has demonstrated that males with longer horns are much more likely to win fights in *P. difformis*, (17 out of 20 contests won by the male with the longer horn (Rasmussen, 1994)), *O. acuminatus* (14 out of 16 contests won by the longer horned male when the difference in horn length was $\geq$0.2 mm (Emlen, 1997a)) and *O. taurus* (22 out of 27 contests won by the longer horned male (Moczek & Emlen, 2000)).

Both Emlen (1997a) and Moczek & Emlen (2000) also demonstrated that the probability of winning was related to the magnitude of the difference in horn length. In the case of *O. taurus* (Moczek & Emlen, 2000), 15 out of 15 fights were won by the male with the longer horns when the difference in length was greater than 1 mm, whereas 4 out of 12 fights between beetles with horns that differed by less than 1 mm were won by the animals with the shorter horns (in this species, horns grow up to around 4.5 mm long).

The technique of staging contests between pairs of beetles that are matched for size is useful, but it does not tell us about the relative importance of body size and horn length in determining the outcome of fights. This can be investigated by staging fights between beetles varying in size and horn length and by designating one beetle in each pair as the 'focal male'. The outcome of the fight is then coded as 1 or 0 for a win or loss by the focal male, and a generalized linear model is fitted to the data, with the differences in horn size and body size between the focal male and his rival as predictor variables (Hardy & Field, 1998; Pomfret & Knell, 2006b). To date, this approach has only been used with one species of dung beetle – *E. intermedius*. In this species, both body size and horn size differences were significant predictors of victory when small beetles fought each other but, when fights occurred between large beetles, only horn size predicted victory (Figure 3.3).

Looking beyond the dung-feeding Scarabaeidae, a similar analysis of fights between males of the dynastine beetle *Trypoxylus (Allomyrina) dichotoma* also found that horn length, but not body size, predicted victory. In this case, the authors confirmed this result by staging contests between animals matched for horn length but not for body size (Karino *et al.*, 2005). These results contradict the conventional wisdom that size is the most important factor in contests between animals; future work on the use of horns in contests should clarify whether this is a general pattern.

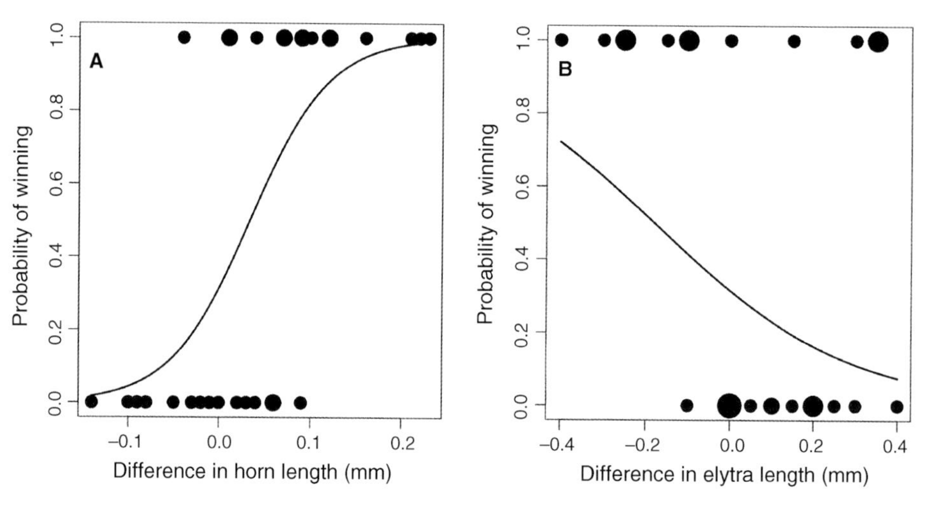

Fig. 3.3 Horn length predicts victory in contests between large males of *Euoniticellus intermedius*. **A**: The x-axis gives the difference in horn length between two male beetles and the y-axis shows the fitted probability of victory from a generalized linear model. The data points show the outcomes of experimental contests, with a zero indicting a loss for the focal male and a one indicating a victory. **B**: The relationship between the difference in elytra size for the same set of experimental contests and the probability of victory. The x-axis gives the difference in elytra length and the y-axis shows the fitted probability of victory. Note that the slope of the line in A is highly statistically significant ($p < 0.0002$), but that in B is not. The size of data points indicates the number of contests corresponding to each point, with the largest representing four contests and the smallest representing one. Figure redrawn using data originally published in Pomfret & Knell, 2006b.

3.5 Are beetle horns simply tools?

Beetle horns are used as weapons in contests, usually between males. There is strong evidence that animals with larger horns tend to win contests and, in some species at least, horn size is more important than body size in determining the outcome of fights. These facts lead us to ask whether beetles with longer horns win their fights because their larger horns are tools that in some way enable them to do so, or whether they win because of some other aspect of their biology that the horns are correlated with. In the latter case, the horns might be functioning to transmit information about the bearer's fighting ability to opponents rather than enabling the bearer to beat an opponent by mechanical means.

I would suggest that beetle horns carry out both functions. Observations of fights between horned beetles, and a consideration of the functional morphology of beetle horns, can lead to little doubt that the horns of many of these animals are used actively in contests to push, pry and lift opponents. In the case of beetles such as *P. difformis*, which uses its cephalic horn to lift opponents (Rasmussen, 1994), a longer horn will enable a male to get his horn into position beneath the body of an opponent, while the shorter-horned opponent is unable to do so, and this may also allow greater leverage to be applied while the opponent is lifted. When considering beetles with multiple horns that fight in tunnels, long pronotal horns can hold a less well-endowed opponent at a distance and allow a cephalic horn to be used against an opponent who is unable to retaliate.

Many dung beetle horns clearly function as tools, therefore, but it is questionable whether all of them do. The horns carried by major males of *O. taurus*, for example, are used in combat and engage with the opponent's pronotum, but whether a longer set of horns provides much mechanical advantage to the carrier is questionable. The males are in contact not only through the horns but also through the head, and it is likely that this is where the majority of the force used to push the opponent is transmitted, rather than through the slender and somewhat flexible horns. Similar questions can be raised about the horns of many other beetles, an obvious example being *O. rangifer* (Fig 3.1J). It is hard to imagine how such long and delicate structures would make effective weapons. In this case, at least, the use of the horns more as signalling structures than weapons has to be considered a possibility.

Evidence is starting to accumulate that horn length is correlated with other traits that will influence fighting ability. In *E. intermedius*, horn length is a better predictor of maximal strength and endurance than body size – both traits that have an obvious connection to fighting ability (Lailvaux *et al.*, 2005). In *O. taurus,* maximal strength is also correlated with horn length once body size has been controlled for, although this relationship is dependent on the animal's condition, with males in poor condition having low strength no matter what their horn length (Knell & Simmons, 2010).

Furthermore, horn length in *E. intermedius* is also correlated, independent of body size, with immunity (Pomfret & Knell, 2006a – see also Cotter *et al.*, 2007, for a study of immunity and morph in *O. taurus*) and, interestingly, with weight gain following eclosion (M. Head & R. Knell, in prep.). The weight gain result is particularly interesting because horn length is determined during metamorphosis, before the maturation feeding period. A possible explanation of these data is that beetles differ physiologically in their ability to assimilate food both as larvae and adults. Horn length is influenced by larval feeding or digestive efficiency, which co-varies with adult assimilation efficiency, so horn length co-varies with the animal's weight gain during maturation.

Finally, it has been suggested that in one case at least, horns might act as visual signals. Most horned dung beetles interact in dark tunnels, so their horns are unlikely to act as visual signals, but, as we have seen, beetles from the Phanaeini frequently interact above the ground and are diurnal (Price & May, 2009). These animals are often brightly coloured and iridescent, and the horn is often a darker colour than the bright pronotum behind it.

Vulinec (1997) demonstrated strong ultraviolet (UV) reflectance from the pronotum in frequencies visible to insects, and suggested that the bright pronotum silhouettes the dark horn, creating a powerful visual signal that could potentially be important in both intrasexual contests and in mate choice. As we will see in Chapter 9 of this volume, dung beetles have acute visual sensitivity and there is little reason to reject outright the idea that this sensitivity might not be brought to bear on the problem of mate and/or competitor assessment. Neither of the detailed descriptions of intra- and intersexual interactions between such beetles includes any behaviour that could be a visual display (Otronen, 1988; Rasmussen, 1994) but, in both cases, the majority of observations were made of animals interacting in tunnels.

3.6 The evolution of horns: rollers vs. tunnellers

Some taxa of dung beetles, such as the Sysiphinae, carry no horns. In some taxa, such as the genus *Phanaeus*, all of the males are horned, while in other taxa, such as the genus *Onthophagus*, there is variation between species, with males of some species being horned, other closely related species having hornless males, and still other species having dimorphic males, some having horns and others not (see Chapters 4, 6 and 7 of this volume).

This variation does not simply reflect variation in male behaviour. Males of many hornless species frequently fight with each other. For example, male *Kheper nigroaeneus* make very large, smooth brood balls that are likely to play a role in attracting females, and they frequently fight with other males for possession of these brood balls (Ybarrondo & Heinrich, 1996); male *K. platynotus* fight to defend females while mate guarding (Sato & Hiramatsu, 1993). So why have these beetles not evolved horns? To answer this question, we have to consider the evolutionary costs and benefits that these structures bring: horns should only evolve when the fitness gains from their possession are greater than the costs.

Growing horns is known to impose a cost on the bearer because resources that could be used in the growth of other body parts are required to build the horn (Emlen, 2001; Moczek & Nijhout, 2004; Simmons & Emlen, 2006). Emlen (2001) compared three species of *Onthophagus* with horns arising from different locations and showed that large horns were associated with reduced sizes of organs close to the horns. Thus, in *O. sharpi*, which has a horn located on the front of the clypeus, males with large horns have relatively small antennae, and in an unidentified species of *Onthophagus* from Ecuador that has horns at the rear of the head, males with large horns have relatively small eyes.

More recent experimental work has shown that these trade-offs are not restricted to organs close to the horns. *O. taurus* males which had the precursor cells that would grow into genitalia ablated while they were still larvae were found to grow larger horns (Moczek & Nijhout, 2004), while *O. nigriventris* males which were similarly prevented from developing horns grew to a larger size and developed larger testes (Simmons & Emlen, 2006). Chapters 4 and 7 of this volume provide detailed discussions of such resource allocation trade-offs.

The possession of horns can also reduce the speed or manoeuvrability of the owner in tunnels (Madewell & Moczek, 2006; Moczek & Emlen, 2000), although not in every case (Pomfret & Knell, 2006b). It is likely that large horns have adverse effects on other aspects of the bearer's biology, such as flight ability.

These costs will be similar across all species, but the benefits arising from the possession of horns will vary between species, depending on the details of each species's breeding and feeding biology. One of the most important variables determining the benefits of horns appears to be whether the contests between males occur in tunnels or in the open.

Most modern dung beetles use either a 'rolling' or a 'tunnelling' strategy to reduce the intense competition for resources that occurs in dung (Hanski and Cambefort, 1991; see Chapters 1 and 2 of this volume). The tunnellers excavate burrows directly beneath the dung and then drag dung down, whereas rollers carve pieces from dung pats, shape them more or less into balls and roll them away. Early workers thought that these two strategies had only evolved once, and that the dichotomy between rollers and tunnellers was a fundamental division within the phylogeny of the Scarabaeinae. However, recent phylogenetic work indicates that tunnelling was the ancestral behaviour in these animals and that ball-rolling behaviour has evolved independently several times (see Chapter 2 of this volume).

Emlen & Philips (2006) mapped the presence or absence of horns onto the phylogeny used for this work and tested for correlated evolution of horns and the behaviour used to sequester dung (tunnelling or rolling). The analysis indicated that horns have evolved eight times within this phylogeny of 46 species from 45 genera, and each gain of horns occurred within a tunnelling, rather than a ball-rolling, lineage (Figure 3.4). This indicates that the method used by beetles to sequester dung for food and breeding has an important influence on the evolution of horns, with horns apparently evolving only in tunnelling beetles.

The reason why this should be is probably that the tunnelling habit has the effect of making resources more defendable (Emlen & Philips, 2006). Male beetles defending tunnels will encounter opponents one at a time and will be able to completely exclude weaker beetles. This will bring greater fitness benefits to beetles carrying horns than would be the case for male beetles such as *Kheper*, which guard resources such as brood balls above ground for a period before burying them (see Chapter 5 of this volume). These will find it much harder to exclude rivals and they will also be open to challenges from more than one challenger at a time.

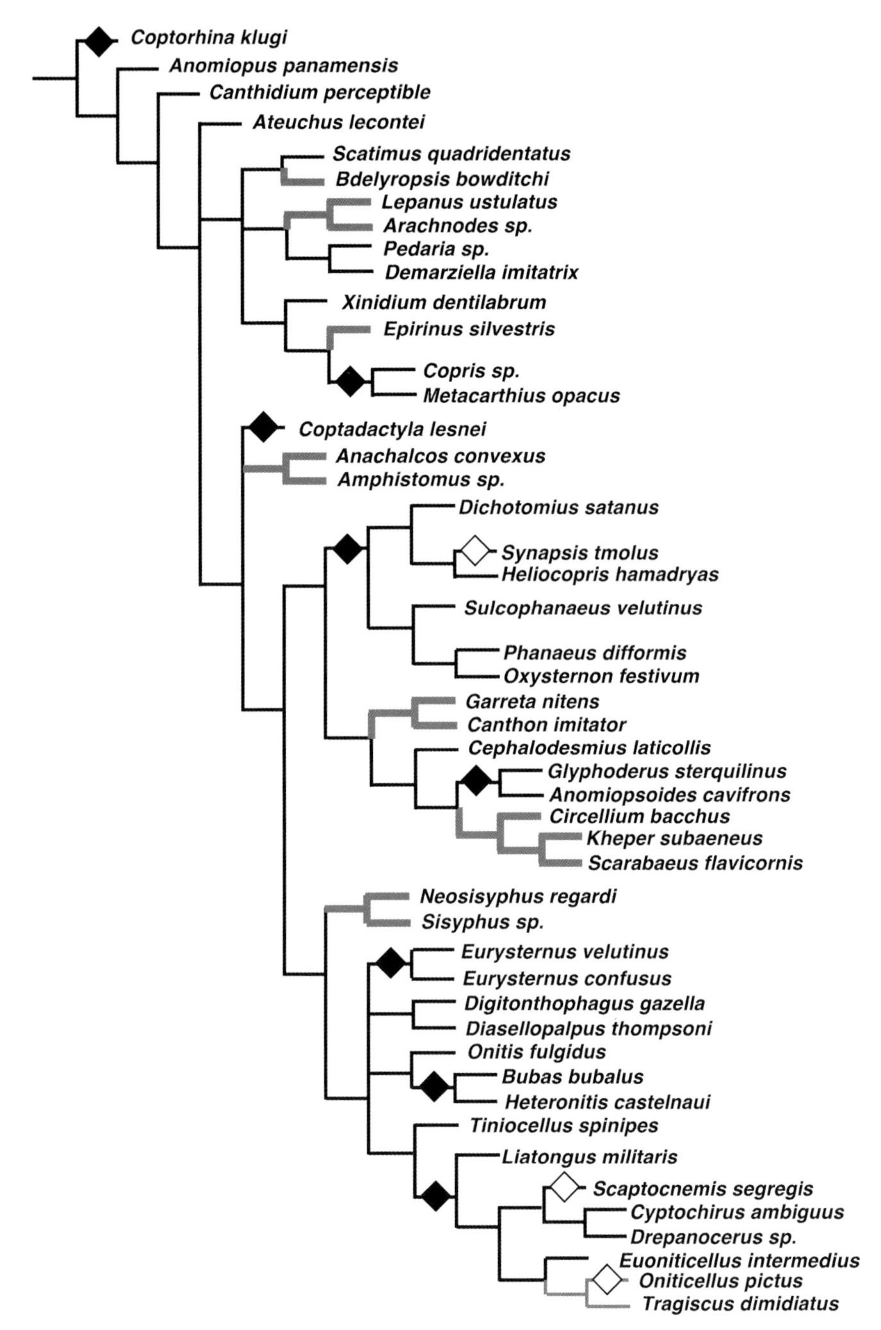

Fig. 3.4 Phylogeny of 45 species of dung beetles. Tunnelling is the ancestral behaviour, indicated by narrow black branches. Non-tunnellers either breed directly within a dung pat ('dwellers'), indicated by narrow grey lines, or roll balls of dung (wide grey branches) Evolutionary losses and gains of horns are indicated by open and closed diamonds respectively. All eight gains of horns occurred on 'tunnelling' branches and one of the three losses of horns occurred on a non-tunnelling branch. Redrawn from Emlen & Philips (2006) with permission.

3.7 The evolution of horns: population density

The dichotomy between tunnellers and rollers seems to explain the larger-scale patterns in the occurrence of horns within the Scarabaeinae, but the smaller-scale patterns remain to be explained. Within some tunnelling lineages there are hornless species; five of the 48 species of *Onthophagus* considered in the phylogeny described in Emlen *et al.* (2005b) were reported as hornless, and five of the 14 species of *Onthophagini* studied by Pomfret & Knell (2008) were hornless. To explain these patterns, we must look to other ecological factors, a number of which have been put forward as potentially being important in determining the strength and nature of sexual selection (Emlen & Oring, 1977; Hamilton, 1979). These include the spatial distribution of resources; the temporal distribution of receptive females; the operational sex ratio (OSR); and population density (Emlen & Oring, 1977). Of these, population density and OSR have been studied specifically in dung beetles.

If males are competing among themselves for access to females, then, as population density increases, the strength of sexual selection should also increase. This is because contact rates between and within sexes will increase, potentially leading to increased reproductive skew within the male population as high-quality males have greater opportunities to monopolize access to females (Emlen & Oring, 1977; Knell, 2009b; Kokko & Rankin, 2006).

This might lead us to expect aggression to increase with population density, so that individuals carrying weaponry would experience increased fitness as density increases. However, empirical studies of male fitness and density have reported both increased and reduced fitness of aggressive males at high densities from different systems, indicating that the relationship between selection for aggression and density is in fact likely to be more complex than a simple increase with density (Knell, 2009b and references therein).

When a species occurs at low densities, males will find it difficult to locate mates and, when they do, they are unlikely to encounter rivals when they make contact. Investment into adaptations to aid in movement and the location of mates will thus bring greater fitness benefits than will investment in weaponry. As density increases, however, it will become easier to locate mates and the probability of encountering a rival male will increase, so aggressive males that invest in weaponry are expected to have increased fitness. As density increases further, however, aggressive males that guard females will be forced to spend an increasing amount of time and energy engaging in costly fights, and they will be more likely to encounter a superior competitor who will beat them and take over the resource.

A further cost to aggression will arise because the risk of sperm competition will increase with density. A given female will be more likely to have already mated before she is encountered by a particular male, and is more likely to re-mate with a rival male relatively quickly. Those resources that are invested in adaptations to increase the probability of winning fights, such as muscles and weapons, will not be available for traits that improve fitness under sperm competition, such as large testes (Knell, 2009b). Studies of dung beetles have elegantly illustrated this latter point: male *O. nigriventris* that were manipulated to stop them growing horns grew relatively larger testes, indicating a trade-off between the resources available for these traits (Simmons & Emlen, 2006; see Chapter 4 of this volume). Thus, the costs of aggression will increase for several different reasons with density:

- aggressive males will pay higher costs in terms of energy, time and the risk of injury because they will be fighting more often;
- they will lose more contests, simply because they will be engaging in more of them;
- they will not perform well in sperm competition.

The costs of aggression, therefore, will increase as density gets higher, but the benefits might not – especially in a system where males guard only one female at a time, thereby limiting the degree of reproductive skew possible. At a high enough density, the costs of aggression will outweigh the benefits. Hence, males that use 'scramble' tactics, whereby they simply try to find unguarded females to mate with, or that use 'sneak' tactics, trying to acquire matings with females who are being guarded by somehow bypassing the guarding male, will have a higher fitness than aggressive males that guard females (Knell, 2009b).

This has yet to be demonstrated by direct behavioural observations in dung beetles, but a study of the forked fungus beetle (*Bolitotherus cornutus*), a horned tenebrionid that is found on polypore shelf fungi in the Eastern part of North America (Connor, 1989), showed that longer-horned males gained a greater fitness advantage in low-density populations than in high-density ones. Horn length was positively correlated with the number of mating attempts per hour in the low density populations only. Connor (1989) notes that in the low-density populations there is usually only one male with long horns per fungus, suggesting that these males are able to monopolize resource patches at low densities but not at high densities.

Evidence for a role of population density in the evolution of dung beetle horns comes from both inter- and intraspecific studies. The latter have made use of the dimorphisms that are well known in many species of *Onthophagus*, with 'minor' males that express reduced or no horns employing 'sneak' tactics, and horned 'major' males aggressively guarding females (Eberhard & Gutierrez, 1991; Emlen, 1997a; see Chapter 6 of this volume). Within a population of males, the smaller males tend to develop into minors and the larger ones into majors. These dimorphisms are believed to evolve via a process of 'status-dependent selection' (Tomkins & Hazel, 2007), whereby small (i.e. low-status) males benefit little from competing aggressively and instead pursue alternative tactics that gain them higher fitness (Hunt & Simmons, 2001).

The proportion of the population developing into each morph is known to respond to selection (Emlen, 1996), and in field populations we can draw inferences about the relative fitness benefits of aggressive (majors) versus non-aggressive (minors) tactics from this proportion. If the majority of the male population develop into majors, for example, this implies that aggressive tactics are relatively beneficial; smaller males that aggressively guard females have been selected over similar-sized ones that did not. The reverse situation, with only a few males developing into majors, indicates the opposite; the relative fitness benefits of aggression are small, and only the largest males have historically been able to acquire higher fitness by the use of these tactics.

One of the best studied of these dimorphic beetles is *O. taurus*. Originally found in the Mediterranean region, in the 1960s and 1970s the beetle was introduced to both Eastern (EA) and Western Australia (WA) and to the Eastern United States, and there are now established populations in all three areas. Moczek *et al.* (2002) measured the relationship between horn length and body size in beetles from the latter two regions, and found that proportionally fewer male beetles from the WA population develop into majors, with some intermediate-sized males that would develop into majors in the Eastern US population developing instead into minors in the WA population. This difference persisted even when beetles were reared in the laboratory under identical conditions for several generations, suggesting that these populations had diverged genetically (see Figure 7.3 in Chapter 7 of this volume).

A further study compared beetles from all three populations (Moczek, 2003) and found that the EA population produced a proportion of major males intermediate between the two other populations. Moczek discussed a variety of possible explanations for the differences between these populations, including differences in density of conspecifics, differences in the density of competitors from other species, differences in body size and differences in sex ratio. Of these, only the density of conspecifics followed the pattern that would be predicted if it were the cause of the differences in the proportion of majors; the density of Eastern US populations was substantially less than that of the EA populations, which were themselves considerably less dense than the WA populations. This is not in itself strong evidence that high population density selects against aggressive strategists, and therefore against horned males. Moczek (2003) points out that these data are from three populations only, and that the relationship shown is correlational, but nonetheless it is certainly suggestive.

For further evidence for a role of population density, we must look to interspecific studies. Firstly, Emlen *et al.* (2005b) scored beetle species as abundant, rare or intermediate in their study of 48 species of *Onthophagus*. Abundant beetles were those known to occur at high densities, that are found in most dung pats and that museum collections often have many specimens of; rare species were those that are only rarely encountered in the field and that are usually poorly represented in museum collections. Both horn length and horn number proved to be significantly correlated with increases in population density, apparently contradicting the conclusion from Moczek's work.

Additional analysis showed that the increase in horn number with density was due to an increased probability of gaining thoracic horns in the most abundant lineages. Emlen *et al.* (2005b) suggested that this might arise because of resource allocation trade-offs between horns and nearby organs and structures, occurring in the individual animal during metamorphosis. It might be the case that thoracic horns trade-off against wings more than other parts of the animal because these two structures are physically close (Emlen, 2001). If this so, then because male beetles in these abundant lineages are likely to have to fly less in order to find mates, they might gain less fitness from large wings than they do from thoracic horns.

Emlen *et al.*'s study compared beetles from all over the world and from a variety of habitats. Pomfret & Knell (2008) studied a single community of Onthophagine beetles in a single savannah habitat in South Africa over two years. Rather than a simple measure of population density, they calculated Lloyd's mean crowding

(Lloyd, 1967) for each species of beetle, a measure that takes into account not only numbers but also aggregation, to give an estimate of the number of conspecifics that an individual is likely to encounter at a particular patch.

Of 14 Onthophagine species that were present in both years, five were hornless. Analysis using a generalized least-squares model, incorporating a phylogeny derived from sequences of the mitochondrial COX1 gene, indicated that both mean crowding and OSR were correlated with the presence or absence of horns. Males from beetle species that had female-biased sex ratios were less likely to carry horns, while males from species that had lower estimates for mean crowding were more likely to carry horns (Figure 3.5).

These two interspecific studies that have considered the role of density or crowding in the evolution of beetle horns have thus returned contradictory results. As discussed above, it has been proposed that the selective advantage of weaponry should first increase and then decrease as population density increases (Knell, 2009b). One possibility therefore, is that these two studies have captured two different parts of the overall picture, with the evolutionary gains of horns associated with increasing density described by Emlen *et al.* (2007) being a reflection of increases from low to moderate densities, and the losses of horns at high densities described by Pomfret & Knell (2008) reflecting increases from moderate to high densities. This simple explanation can, however, be discounted, because most of the relevant gains of horns in the Emlen study were associated with increases from moderate to high population densities rather than with increases from low to moderate.

It must also be remembered that these are studies operating at very different scales: the Pomfret & Knell study considered animals coexisting in a single habitat and used a direct measure of population density, whereas the Emlen study used animals from a range of habitats and used a rather indirect measure of density.

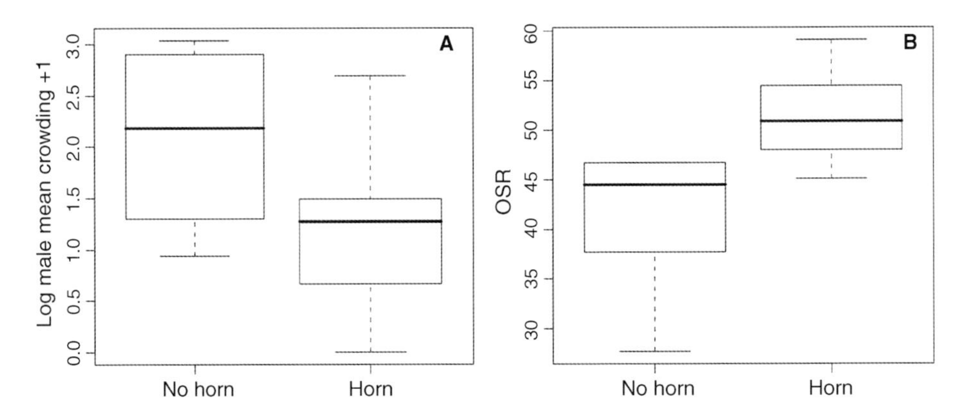

Fig. 3.5 Density and operational sex ratio (OSR) compared between horned (9 species) and hornless (5 species) of Onthophagini sampled at one location in South Africa. **A**: Log mean crowding + 1 of males, a measure of population density. **B**: OSR. For both plots, the bold line is the median, the box is the interquartile range, and the whiskers extend to the furthest data point less than 1.5 times the interquartile range from the box. Originally published in Pomfret & Knell (2008).

A further point to note is that the gains of horns that Emlen *et al.* found to be associated with high densities were almost all gains of thoracic horns, while none of the species included in the South African study carried horns of any size on the thorax. This makes direct comparison of the results difficult.

For the moment, it is probably best to conclude that interspecific studies support an important role for population density in the evolutionary gains and losses of horns in this genus, but that this role might be complex and dependent on other environmental variables as well.

3.8 The evolution of horns: sex ratio

The operational sex ratio (OSR), calculated by dividing the number of sexually active males by the sum of the number of sexually active males and the number of receptive females, is well known as an important determinant of the strength of sexual selection and has been shown to be influential in mating system evolution in a variety of taxa (Emlen & Oring, 1977; Kvarnemo & Ahnesjö, 1996; 2002). This is because a skewed OSR will increase competition for mates among the more common sex. Female fitness is not strongly correlated with the number of matings achieved but male fitness is; therefore, a female-biased OSR is expected to lead to a general relaxation of sexual selection unless males are extremely rare (Jiggins *et al.*, 2000) but a male-biased OSR will lead to increased competition between males and an increase in the strength of sexual selection.

In dung beetles, sex ratio has been examined in two studies. First, Moczek (2003) measured the sex ratio of the three populations of *Onthophagus taurus* mentioned in the discussion of population density. The population was found to be most male-biased in Western Australia, which has the lowest proportion of major males, and most female-biased in the Eastern US population, which has the highest proportion of majors. This might at first seem to go against the accepted wisdom that male-biased sex ratios lead to stronger sexual selection but, as with population density, this can be resolved if we consider that aggression might not be the optimal strategy when competition is fierce. If horned males are unable to defend females effectively in the presence of large numbers of competitors, then it is possible that, in the case of *O. taurus*, sex ratio is acting in concert with population density to select for males that are less likely to develop into majors in the Western Australian populations.

Pomfret & Knell (2008) measured OSR as well as crowding for the community of South African beetles discussed earlier. OSR, rather than the simple sex ratio, was estimated by excluding beetles that were 'callow' and therefore undergoing maturation feeding. In this study, OSR was found to be an important predictor of the presence or absence of horns, with female-biased sex ratios being associated with the hornless condition (Figure 3.5). In this case, therefore, horned species were more likely to have low levels of crowding in even or slightly male-biased sex ratios, whereas hornless species were likely to have high levels of crowding and female-biased sex ratios. These results are at odds with the intraspecific study of Moczek (2003), so clearly more work is necessary to disentangle the effects of population density and sex ratio on the evolution of dung beetle horns.

3.9 Future work

Research on horned beetles has made important contributions to our understanding of the evolution of weaponry in the animal kingdom. We have moved from asking what is the function of the horns to questions regarding their costs and benefits and their extraordinary diversity. In this review, the need for further work on questions such as the role of ecological factors in horn evolution, whether some horns are used as signals rather than weapons, and how horn size is linked to aspects of male quality, has already been highlighted. In the final section of this chapter, I will call attention to some further questions that are of interest but that have not so far received much attention from researchers.

First, what drives the evolution of elaborate horn morphology? The question of whether some of the more slender horns are used as weapons or as signals has already been raised. A separate question is how the elaborate forms of some of the more robust horns, for example those carried by male *O. imperator* (Figure 3.1E), are used and why they have evolved. Some other animal taxa, such as the cervids, carry weapons of similar diversity and complexity (Emlen, 2008), and researchers working on these groups have shown that some of this diversity can be ascribed to differences in mating systems and to the way the animals fight (Brø-Jørgensen, 2007; Caro *et al.*, 2003).

It has also been suggested, however, that an advantage to novel structures in contests might also be a driving force leading to weapon diversity (Emlen, 2008; West-Eberhard, 1983). If the extra tines and notches that are often found on the horns of dung beetles give the beetle some mechanical advantage during fights with opponents who do not have them then, as pointed out by Emlen (2008), this could lead for selection for novelty, which would lead to evolution down species-specific arbitrary pathways. This is an attractive theory that could explain much about the patterns of diversity seen in animal weaponry, but it is yet to be tested. Horned dung beetles would appear to be ideal model organisms with which to investigate this issue further.

A second, related question is why horn morphology in some taxa is so variable, while in others it is not. In *Heliocopris* and *Onthophagus*, for example, the horns are highly variable in number, morphology and location (Figures 3.1D, F and H, and Figures 3.1A, E, I, and J respectively). In genera such as *Oxysternon*, *Phanaeus* and *Copris,* by contrast, all horned males carry a single, curved cephalic horn (although pronotal structures can vary between species). Why have the horns of these latter taxa not diversified morphologically? As discussed earlier in the chapter, we have a number of good descriptions of the use of these horns, which are inserted beneath a rival and used to lift him (Beebe, 1947; Eberhard, 1977; 1979; Otronen, 1988; Rasmussen, 1994). The advantage that might be gained by the addition of novel parts to a weapon that was speculated about in the previous paragraph does not apply to this particular model of horn; the long, slender and slightly curved form could be the best design for this function, so any additions to it might detract from its usefulness.

Finally, why do beetles in some taxa tend to lose their horns while other taxa are all horned? This is exemplified by comparing *Onthophagus* with *Phanaeus* and *Copris*. As has already been noted, all male *Phanaeus* and *Copris* beetles carry a single cephalic horn, whereas in *Onthophagus* a substantial minority of species are hornless. Why, then, do some *Onthophagus* species lose their horns, while *Phanaeus* or *Copris* species may develop smaller horns but do not seem to lose them altogether?

A possible answer may lie in their breeding biology. Both of these genera have low fecundity and high investment per offspring, possibly even more so than other scarab genera such as *Onthophagus*. *Phanaeus* males can show long periods of pre-copulatory mate guarding and will also cooperate with females to build nesting burrows and construct brood balls (Halffter & Edmonds, 1982; Price & May, 2009), and *Copris* females construct nests, often with male assistance, containing only a few brood balls and care for them until adult emergence (Halffter & Edmonds, 1982). Given that these males need to make a substantial investment in time before gaining a mating, it could be that the costs of losing a fight and allowing another male to mate with a female shortly before oviposition are high in comparison with genera like *Onthophagus*.

4

Sexual Selection After Mating: The Evolutionary Consequences of Sperm Competition and Cryptic Female Choice in Onthophagines

Leigh W. Simmons

Centre for Evolutionary Biology, School of Animal Biology, The University of Western Australia, Crawley, Western Australia

4.1 Introduction

Darwin (1871) attributed the evolution of exaggerated secondary sexual characters to the advantages that they gave certain individuals over others of the same sex and species in reproduction. Secondary sexual characters, such as the antlers of deer or the horns of dung beetles, were proposed to arise in males because they served as weapons with which to monopolize females and/or signals to attract and persuade females to copulate. Successful reproduction thus becomes skewed toward males with the most exaggerated traits, resulting in directional selection for continued secondary sexual trait evolution.

Despite initial resistance to these ideas, research over the last three decades has provided considerable support for sexual selection theory (Andersson, 1994; Andersson & Simmons, 2006). For example, the role of male contest competition in the evolutionary divergence of dung beetle horns is now well established (e.g. Eberhard, 1979; Emlen *et al.*, 2005b; Emlen, 1997a; Emlen *et al.*, 2005a; Siva-Jothy, 1987; Pomfret & Knell, 2006b; Hunt & Simmons, 2001; Moczek & Emlen, 2000) (see Chapters 1 & 3). Sexual selection is also widely recognized as a powerful engine of speciation (Panhuis *et al.*, 2001; Seddon *et al.*, 2008).

A major advance in sexual selection theory was the realization by Parker in the early 1970s that sexual selection could continue after copulation (Parker, 1970). When females mate with more than one male, there will be competition among the sperm of different males to fertilize available ova. Parker recognized that sexual selection would favour opposing adaptations in males that both facilitate

Ecology and Evolution of Dung Beetles, First Edition. Edited by Leigh W. Simmons and T. James Ridsdill-Smith.

the pre-emption of sperm present from a female's previous mate and prevent the mating male's sperm from being pre-empted by future males. In the 40 years since Parker's insight into what has become known as post-copulatory sexual selection, the body of empirical evidence supporting a role of sperm competition in the evolution of male reproductive biology has grown almost exponentially (Smith, 1984; Birkhead & Møller, 1998; Simmons, 2001).

Parker highlighted the evolutionary consequences of sperm competition for male reproductive physiology, morphology and behaviour. For example, in a taxonomically wide range of insect species, including dung beetles, accessory gland products have been identified that significantly influence female reproductive physiology (Martinez and Magdalena, 1999; Cruz & Martinez, 1998; Kotiaho *et al.*, 2003, Simmons, 2001). Perhaps best studied in fruit flies of the genus *Drosophila*, accessory gland products have been identified that suppress female sexual receptivity and elevate the rate of oviposition immediately following mating (Wolfner, 1997; Chapman & Davies, 2004).

The adaptive significance of such physiological manipulation of females is clear: females will produce more offspring during a period when they are unreceptive to further mating, and this will enhance the reproductive success of the male transferring the accessory gland product (Fricke *et al.*, 2009). While clearly beneficial for males, male accessory gland products appear detrimental to the long-term fitness of *Drosophila* females, significantly reducing female lifespan (Chapman *et al.*, 1995), and generating sexual selection via sexual conflict (Rice, 1996; Arnqvist & Rowe, 2005).

The best-studied morphological adaptations to sperm competition are the genitalia of male dragonflies and damselflies, which entrap sperm from previous males on the spines that cover their surface so that these sperm are removed on withdrawal of the penis and/or repositioned within the female's sperm storage organs where they cannot gain access to fertilizable ova (Waage, 1979; Simmons, 2001). In this way, the female's most recent mate is assured of gaining almost complete paternity of offspring when eggs are fertilized and laid (e.g. Siva-Jothy & Tsubaki, 1989). Adaptations for sperm pre-emption mean that the last male to mate with a female has a selective advantage over her previous partners. It also means, of course, that males are vulnerable to losing fertilizations if they abandon their mates before they have produced offspring.

Thus we see behavioural adaptations arising for the avoidance of sperm pre-emption by future rivals (Alcock, 1994; Simmons, 2001). In the dung beetle *Onthophagus taurus*, for example, males increase the amount of time spent guarding their mates when other males are present (Moczek, 1999) and will engage their mate in insurance copulations if they encounter another male searching for mates within their tunnels (Hunt & Simmons, 2002c). Such phenotypic plasticity in mate guarding is common in insects, where males avoid the costs of guarding when the risk of sperm competition is low (Simmons, 2001).

It seems clear that adaptations arising from sperm competition have been interpreted as an extension of male contest competition. However, Darwin (1871) envisaged two processes of sexual selection: male contest competition and female choice. Over the last two decades, the focus of research has switched to examining alternative mechanisms for the evolution of traits such as ejaculate composition or mate guarding, considering the role that females might play in determining paternity (Eberhard, 1990; 1996).

While sperm competition is the post-copulatory equivalent of male contest competition, the post-copulatory equivalent of female choice has become known as cryptic female choice (Thornhill, 1983). This is so termed because, unlike pre-copulatory female choice, we are often unable to observe female preferences for particular males; rather, they are manifest as biases in paternity that are determined by females. For example, Eberhard (1985; 2009) argued that the widespread morphological complexity in male genitalia could arise under cryptic female choice if females were more likely to store and use sperm from males able to provide the appropriate genital stimulation during copulation. Some of the best evidence for sexual selection acting on male genital morphology comes from studies of onthophagine dung beetles (Simmons *et al.*, 2009; House & Simmons, 2003; 2005). Thus, it is often possible to raise alternative scenarios to male contest competition to explain traits such as genital morphology that have previously been thought to have arisen via sperm competition.

Distinguishing between selection via sperm competition and cryptic female choice has proved a considerable empirical challenge to researchers in this area (e.g. Birkhead, 1998; Pitnick & Brown, 2000). Nonetheless it is important to do so, because the net action of sexual selection acting on males depends very much on the magnitude and direction of individual episodes (Andersson & Simmons, 2006; Hunt *et al.*, 2009). For example, males who win in sperm competition may not always be the best fathers from the female's perspective (Bilde *et al.*, 2009), so that cryptic female choice might in theory act to counteract the selective advantages expected from sperm competition. Such sexual conflict between males and females over fertilization has been identified as a powerful mechanism of speciation (Parker & Partridge, 1998; Gavrilets, 2000; Gavrilets & Waxman, 2002).

Dung beetles in the genus *Onthophagus* have proved a remarkably useful model system with which to test predictions arising from theoretical models of post-copulatory sexual selection based on both sperm competition and cryptic female choice. In this chapter, I review this literature, showing how sperm competition can drive the evolution of male expenditure on sperm production and how this selection can affect the evolutionary diversification of horns used in pre-copulatory contest competition. Moreover, research on these beetles has revealed how female reproductive fitness can be enhanced via the promotion of sperm competition, and how cryptic female choice and sperm competition can act synergistically in increasing male and female reproductive fitness.

4.2 Sperm competition theory

Parker and his colleagues have developed a series of game theoretical models with which to predict the effect of sexual selection through sperm competition on male ejaculation strategies (Parker & Ball, 2005; Parker, 1998). In sperm competition games, as the models are called, reproductive expenditure is partitioned between expenditure on gaining matings (e.g. searching for females, signalling to attract females or competing with other males for access to females) and expenditure on the ejaculate. Expenditure on the ejaculate is assumed to increase the gain in terms of fertilization success with a given female, but it reduces the number of possible matings, so that net male fitness is the product of the number of matings obtained and the fertilization success obtained per mating.

Thus, there are two assumptions behind Parker's sperm competition games: first, that there is a trade-off between expenditure on the ejaculate and other reproductive activities; and second, that fertilization success conforms to what Parker calls a 'raffle' – the fertilization success of a given male will depend on the number of sperm he has at the site of fertilization relative to other males. The expenditure adopted by a male should depend, then, on what other males are doing in the population – hence the game theoretical approach to modelling ejaculate evolution. Variation in sperm competition risk, or the probability that a female will mate with more than one male, has important implications for the evolutionarily stable ejaculation strategy, because expenditure on the ejaculate will determine the fitness return obtained when males compete for fertilizations (Parker, 1998).

The predictions that arise from Parker's models make intuitive sense. In Figure 4.1A, the predicted evolutionary stable ejaculation strategies are calculated and plotted against the probability that a female mates with two different males. Across species, the models predict that as the risk of sperm competition increases, so male expenditure on the ejaculate should increase. Slopes are plotted for a variety of values of r, which represents a loading coefficient for the raffle. When r is 1, two males who transfer the same number of sperm have an equal chance of fertilizing ova. In other words, sperm competition conforms to a fair raffle. As r becomes smaller, fertilization success is increasingly biased toward the sperm of one male, so that at its extreme there is no longer any sperm competition because one male's sperm are not counted at the time of fertilization. Here, the effect of double mating on ejaculate expenditure will be negligible. Such a situation might arise if paternity were controlled completely by cryptic female choice, so that the numbers of sperm transferred by each male had no influence on whether a female chose to use those sperm or not. Thus, with paternity determined entirely by cryptic female choice, the risk of a double mating by females should have no impact on male ejaculation strategies. With females acting as 'passive vessels' for male competition, the risk of sperm competition should have its maximal impact on male ejaculation strategies.

Within-species predictions are somewhat more complex, because they depend critically on what Parker (1990a) termed 'roles'. A role could be a male's timing of copulation relative to the female's fertile period, his sequence position (first or second to mate), or it could be a male preferred by females during processes of cryptic female choice. Males can thus be in the favoured role – and therefore more likely to gain fertilizations per unit of ejaculate – or in the disfavoured role, being less likely to gain fertilizations for the same unit of ejaculate. If these roles are assigned randomly, then all males should have the same ejaculation strategy because, averaged over all matings, each will have the same mean fertilization success. In contrast, if roles are assigned non-randomly, so that some males are consistently in the favoured role, they should decrease their expenditure on the ejaculate so that disfavoured males have the greater expenditure (Parker, 1990a).

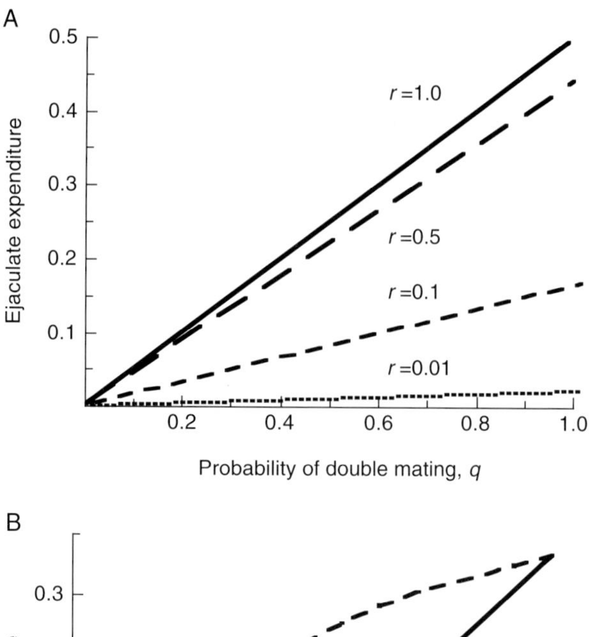

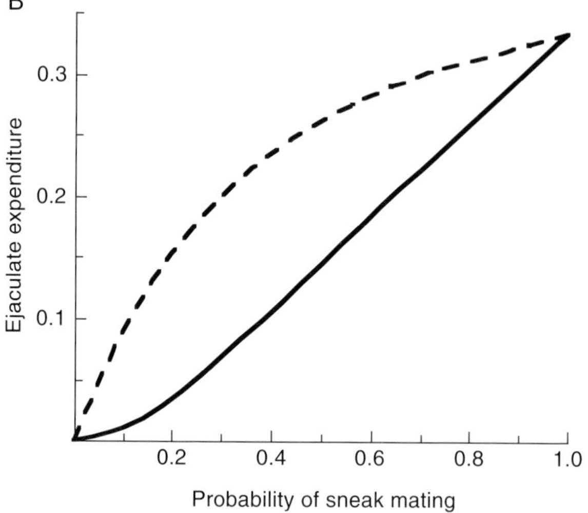

Fig. 4.1 Sperm competition theory. **A:** The evolutionarily stable ejaculation strategy (ESS) is for males to increase their expenditure on the ejaculate as the risk of a double mating by females, q, increases. The strength of this effect is dependent on the loading of the raffle, r. When two males have equal probability of fertilization ($r = 1$; solid line) variation in risk should have its maximum effect on the ESS, but as $r \approx 0$, risk should have no effect on the ESS (various dotted lines; redrawn from Parker *et al*., 1997). **B:** When males adopt alternative tactics of guarding females (solid line) or sneaking copulations (dotted line), there will be an asymmetry in sperm competition risk between tactics, and males subject to the greatest sperm competition should have the higher ejaculate expenditure. When the risk of a sneak mating ≈ 1, all males will be subject to sperm competition and should have equal expenditure on the ejaculate (redrawn from Parker, 1990b).

One situation in which roles will be non-random occurs when males adopt alternative mating tactics. Often, those males unsuccessful in competition will sneak copulations with females, for example when guarding males are occupied in disputes with other males. By the nature of their alternative tactic, sneaks are always subject to sperm competition. Guards will be subject to sperm competition with low probability, dependent on the relative frequency of sneaks and guards in the population.

Figure 4.1B shows a graphical formulation of the predicted evolutionarily stable ejaculation strategy in Parker's sneak-guard model (Parker, 1990b). As the probability of a sneak mating increases, so the ejaculate expenditure of both sneaks and guards should increase, the familiar among-species prediction shown in figure 4.1A. However, for a species with moderate risk of a sneak mating – say, 20 per cent – guards should invest less because they will be free of sperm competition with high probability. The disparity in expenditure between sneaks and guards should decrease with increasing risk of a sneak mating because, at high risk, guards, like sneaks, will be subject to sperm competition with high probability. Both should thus invest maximally in their ejaculates.

4.3 Evolution of ejaculate expenditure in the genus *Onthophagus*

Dung beetles in the genus *Onthophagus* arrive at fresh droppings shortly after they have been deposited, and excavate tunnels in the ground beneath the dung. There they build brood masses from dung that is brought from the surface (Halffter & Edmonds, 1982) (see Chapters 1 and 8). Each brood mass represents the resources available to a single offspring for its entire development from hatching until adult emergence. In some species, males exhibit a suite of morphological and behavioural traits that characterize alternative mating tactics (Emlen *et al.*, 2005a) (see Chapters 6 and 7).

Some males, referred to as majors, develop horns, either on the head or pronotum (or in some species, both) and compete for and guard females (Emlen, 1997a; Cook, 1990; Moczek & Emlen, 2000). Major males also assist females by collecting and carrying dung from the surface to the brood mass (Hunt & Simmons, 2000; 2002b). Minor males remain hornless, resembling females, and they do not assist females in brood provisioning. Rather, they excavate side tunnels and sneak into breeding tunnels in order to copulate with guarded females. Females can thus mate with several males, both within and between breeding events.

On average, sperm competition conforms to a fair raffle, in which paternity is distributed among males in proportion to the number of matings they have achieved relative to their competitors (Tomkins & Simmons, 2000; Hunt & Simmons, 2002c; Simmons *et al.*, 2004). By the nature of their alternative mating tactic, minor males are always subject to sperm competition, while major males will be subject to sperm competition with lower probability, depending on the frequency of the male population that adopt the sneak tactic. The onthophagines thus offer an ideal model system with which to test sperm competition theory, and specifically the sneak-guard models, because among species of *Onthophagus* there is considerable variation in the proportion of the male population that adopt the sneaking tactic (Simmons *et al.*, 2007).

In their test of sperm competition theory, Simmons *et al.* (2007) collected data on body weight, testes weight and the scaling relationship between body size and horn length from 16 species of male dimorphic onthophagines from three geographical regions: Australia, South Africa and North America. They used the methods of

Eberhard and Gutierrez (1991) to determine the threshold body size that distinguished major from minor males (see Chapters 6 and 7) and used this threshold body size to calculate the proportion of the male population that exhibited the minor phenotype, which ranges from 10 per cent in *O. nuchicornis* to as high as 62 per cent in populations of *O. taurus* introduced into Western Australia.

Using a molecular phylogeny of this group of species to control for any phylogenetic inertia, Simmons *et al.* found a positive association between evolutionary changes in testes size and the proportion of males adopting the sneaking tactic that is consistent with Parker's among species prediction (Figure 4.2A). Moreover, as predicted by the sneak-guard model, within 12 of the 16 species examined, minor males invest relatively more in testicular tissue than do major males (Simmons *et al.*, 2007), and at least within *O. binodis*, this tactic-specific variation in testes size is reflected by variation in the volume of ejaculate transferred to the female during copulation (Simmons *et al.*, 1999).

The patterns of testes size variation seen among species of *Onthophagus* suggest that males respond in evolutionary time to selection arising from sperm competition by increasing their physiological investment in sperm production. However, comparative analyses provide evidence of association, rather than cause and effect. Studies of experimental evolution offer a powerful approach for observing responses to selection in male and female reproductive biology. For example, experimental removal of sexual selection by enforced monogamy in the naturally polygamous fly *Drosophila melanogaster* has led to the evolution of ejaculates that are less effective in inhibiting female re-mating, have a reduced competitive fertilization success and are less harmful to females, relative to those produced by males evolving under sustained polygamy (Holland & Rice, 1999; Pitnick *et al.*, 2000, 2001). The technique of experimental evolution has recently been brought to bear on the question of ejaculate evolution in *Onthophagus*.

Simmons & García-González (2008) enforced monogamy within three populations of *O. taurus* and allowed three populations to breed in groups within which females mated with an average of 4–5 males before producing offspring. The females from all six populations of beetles were then allowed to generate offspring in an identical manner, so that the populations differed only in the action of pre- and post-copulatory sexual selection. Within just 11 generations, males from populations subject to sexual selection had increased testes sizes, while males from monogamous populations had decreased testes sizes (Figure 4.2B). Moreover, males from sexual selection lineages had superior competitive fertilization success to males from monogamous lineages.

The evolutionary divergence in ejaculate expenditure within experimental populations of *O. taurus* provides unequivocal evidence that sperm competition contributes to the macroevolutionary patterns in testis size variation seen among species of *Onthophagus* (Figure 4.2A).

4.4 Evolutionary consequences of variation in ejaculate expenditure

Sperm competition theory rests on the fundamental assumption that males are faced with a trade-off between gaining matings and gaining fertilizations (Parker, 1998).

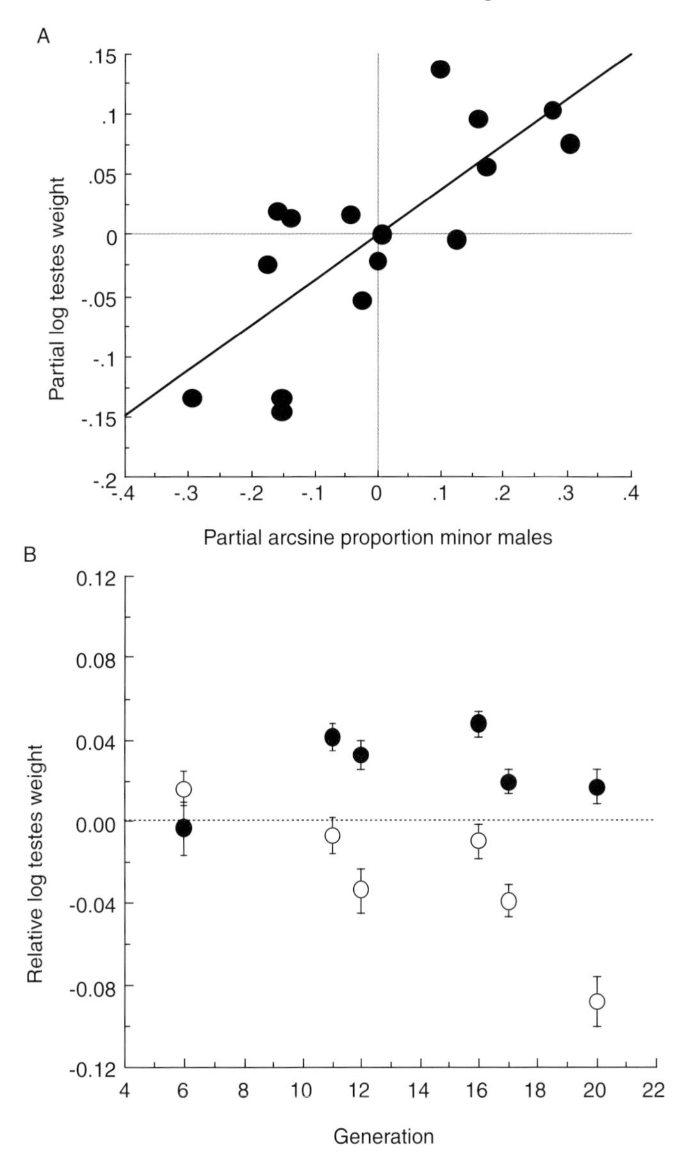

Fig. 4.2 Testes evolution in onthophagines. **A:** Comparative analysis among 16 male dimorphic species of *Onthophagus* demonstrates that after controlling for body size, a species testes size is positively associated with the proportion of the male population that adopts the sneaking tactic. This among-species relationship is robust to control for phylogeny (redrawn from Simmons *et al.*, 2007). **B:** Experimental evolution studies have shown that in *O. taurus*, testes sizes diverged between lineages with enforced monogamy (open symbols) and those with polygamy (solid symbols) after as little as 11 generations of selection (redrawn from Simmons & García-González, 2008).

This underlying assumption has received considerably less attention than have the predictions that arise from the models. Again, the considerable among-species variation in both the strength of selection from sperm competition, and in the expression of horns which are the sexually selected weapons used to obtain mating opportunities (see Chapter 3), make the onthophagines an excellent model for testing the assumptions of sperm competition theory and for exploring the evolutionary consequences of this life-history trade-off for the evolutionary diversification of male morphology.

The production of horns is known to be costly for male onthophagines. During development, males face a resource allocation trade-off such that the allocation of resources to horn growth comes at the expense of the allocation of resources to other body structures (see Chapter 7). Within species, the relative sizes of horns are often negatively associated with the relative sizes of other morphological traits such as eyes, wings and antennae (Emlen, 2001).

Experimental manipulations in which the allocation of resources to one trait are changed, either by artificial selection (Emlen, 1996), or by the ablation of developing tissues (Moczek & Nijhout, 2004), have been shown to influence the growth of adjacent traits. Thus, Simmons & Emlen (2006) cauterized in late instar *O. nigriventris* larvae the area of proliferating cells that would normally give rise to the pronotal horns of unmanipulated beetles. They found that males prevented from allocating resources to horn growth grew larger in body size, and that controlling for body size invested more resources in their testes. Male onthophagines that develop into minors, and so do not allocate resources to horn growth during their development, also develop larger testes. These two observations provide compelling evidence of the trade-off between investment in fighting for females and in gaining fertilizations assumed within sperm competition theory. But this trade-off has broader implications for the diversification of male morphology in this group of dung beetles, and for developmental mechanisms generally.

Developmental trade-offs between morphological traits are most likely due to direct competition for nutrients stored during the larval growth period, or for circulating growth hormones or other signals that are associated with nutrition (see Chapter 7 and Emlen *et al.*, 2006; 2007). Insensitivity to these signals would result in patterns of canalised growth that protect developing structures from resource competition (Shingleton *et al.*, 2008).

In a comparative analysis of 25 species of *Onthophagus*, Simmons & Emlen (2006) found that evolutionary increases in the steepness of horn allometry (i.e. increases in the degree of nutrient-sensitivity of horns) were associated with evolutionary decreases in testes allometry (decreased nutrient-sensitivity or increased canalization). These patterns suggest that one evolutionary consequence of the trade-off between horns and testes in these beetles is that selection for increased horn size under pre-copulatory sexual selection favours the evolution of canalised development in primary sexual traits such as testes.

The second striking pattern to emerge from Simmons and Emlen's comparative analysis was the finding that the evolution of novel horns on the pronotum were restricted to lineages that lack sneaker male phenotypes, while the evolution of novel head horns were independent of the presence of sneak males. Competition for resources becomes more intense as the proximity of horns to adjacent body structures increases (Emlen, 2001). Simmons and Emlen (2006) therefore suggest that the evolutionary origin of pronotal horns is constrained by local resource competition between pronotal horns and testes in lineages subject to sperm competition from sneak males, where intense selection favours the allocation of resources to testes growth and sperm production.

4.5 Theoretical models of female choice

Models of female choice posit that by mating polyandrously and inciting sperm competition, females can ensure that their eggs are fertilized by preferred males. The behaviour of multiple mating is thus the overt expression of female preference, and the male trait that is the focus of selection will be associated with fertilization success. For example, the 'sexy sperm' hypothesis argues that polyandrous females will produce sons that themselves have high fertilization success by virtue of heritable variation in the trait that determines the fertilization success of fathers (Sivinski, 1984; Curtsinger, 1991; Keller & Reeve, 1995; Pizzari & Birkhead, 2002; Evans and Simmons, 2008). This trait could be variation in genital morphology, the number and/or quality of sperm, or seminal fluid composition. The sexy sperm model parallels classic Fisherian sexual selection, in which female preferences for male secondary sexual traits evolve in a runaway fashion, fuelled by a genetic correlation between the preference in females and the trait in males (Andersson & Simmons, 2006).

The 'good sperm' hypothesis, on the other hand, argues that the ability of males to gain high fertilization success is also correlated with their underlying genetic quality, so that males successful in sperm competition sire offspring of generally higher viability and – as with the sexy sperm hypothesis – sons who will have high fertilization success (Yasui, 1997). The good sperm model is thus the post-copulatory equivalent of good genes models for the evolution of male secondary sexual traits (Andersson & Simmons, 2006).

Male secondary sexual traits subject to directional selection from female choice frequently develop condition-dependence, developing in proportion to male condition in such a way as to provide reliable indicators to females of underlying male genetic quality (Rowe & Houle, 1996; Wilkinson & Taper, 1999; Tomkins *et al.*, 2004b). Directional post-copulatory selection on male traits that determine competitive fertilization success, such as testes size and ejaculate quality, should likewise develop condition-dependence (Andersson & Simmons, 2006).

Both sexy and good sperm processes rely critically on the genetic architecture of traits that contribute to competitive fertilization success (Pizzari & Birkhead, 2002). The limited evidence that is available suggests that male traits such as testes size, ejaculate volume and sperm motility can harbour sufficient levels of additive genetic variation to fuel sexy and good sperm processes (Simmons & Moore, 2009; Evans & Simmons, 2008). Although empirical evidence of these processes in action remains scarce (Hosken *et al.*, 2003; Kozielska *et al.*, 2003; Fisher *et al.*, 2006), proof of principle has come from studies of the dung beetle *O. taurus*.

4.6 Quantitative genetics of ejaculate traits

Theoretical models for the evolution of female choice require that traits which contribute to a male's fertilization success exhibit significant levels of additive genetic variation, and, for the good sperm model, that these traits are condition-dependent. Quantitative genetic analyses of sperm competition traits in *O. taurus* provide some of the best evidence of the genetic architecture necessary for the evolution of adaptive cryptic female choice.

Simmons & Kotiaho (2002) used a half-sibling breeding design to examine the magnitude of additive genetic variation in male body size, testes weight, ejaculate volume and sperm length. Testis size is important in sperm competition, varying with mating tactic and sperm competition risk among species of onthophagines and responding to selection imposed via sperm competition (Figure 4.2). In other species, testis weight has been shown to correlate with the size and quality of ejaculates produced (Møller, 1989; Simmons & Moore, 2009).

Sperm length could potentially influence a male's fertilization success. On the one hand, Parker (1982) originally argued that there should be a trade-off between sperm size and number, and that sperm competition should favour increasingly numerous and tiny sperm. Others have argued that the length of sperm tails can influence the flagella forces that facilitate sperm movement and thus contribute to competitive fertilization success (Gomendio & Roldan, 1991; Humphries *et al.*, 2008; Fitzpatrick *et al.*, 2009). Simmons & Kotiaho (2002) also measured condition as the weight of somatic tissue, after controlling for body size. Condition is defined as the pool of resources that are available for allocation to the production or maintenance of traits that enhance fitness (Rowe & Houle, 1996), and residual body weight can provide a good proxy for the availability of stored resources (Kotiaho, 1999).

Testes weight and ejaculate volume were found to have high coefficients of additive genetic variation, and are of a magnitude expected for fitness related traits (Houle, 1992). There appears to be little residual variation in these traits, giving very high levels of heritability (Table 4.1). Despite lower levels of additive genetic variance, heritability was also found to be high for sperm length; indeed, it exceeds the maximum value of 1.0.

The heritability estimates in Table 4.1 were calculated assuming autosomal inheritance. The fact that heritability for sperm length is so high, and that there is no residual variance, strongly suggests that inheritance may be sex-linked to the Y-chromosome. Y-linked inheritance in sperm morphology is not uncommon (Simmons & Kotiaho, 2002; Simmons & Moore, 2009) and provides the genetic architecture most likely to facilitate sexy sperm and good sperm processes (Pizzari & Birkhead, 2002). With sex-linked inheritance, the value for heritability should be halved, making the estimate qualitatively more sensible (Lynch & Walsh, 1998). There is little additive genetic variability for body size (Table 4.1) and strong maternal effects, consistent with what we know about the effects of maternal provisioning of the brood mass on realized adult body size (Hunt & Simmons, 2000). However, condition exhibits relatively high levels of genetic variability, and it is strongly heritable (Table 4.1).

Importantly, Simmons & Kotiaho (2002) found that both sperm length and testis size exhibited strong genetic correlations with condition; across sire families, sperm

Table 4.1 Observational coefficients of variation and narrow-sense heritabilities of body size, sperm competition traits and condition (estimated as the weight of soma after controlling for body size) for the dung beetle *Onthophagus taurus* (modified from Simmons & Kotiaho, 2002).

Trait	*mean*	*SD*	CV_P	CV_A	CV_R	h_S^2	$SE\ h_S^2$
Testis weight (mg)	2.75	0.44	15.84	15.59	2.77	0.97	0.45
Ejaculate volume (mm^3)	3.97	0.71	35.56	22.67	24.40	0.41	0.26
Sperm length (mm)	0.99	0.03	2.64	2.83	0	1.14	0.61
Pronotum width (mm)	5.34	0.27	5.38	0	5.38	0	0
Condition (mg)	82.37	4.01	8.95	8.14	3.70	0.84	0.46

SD: standard deviation. CV: coefficients of phenotypic (P), additive genetic (A), and residual (R) variation. h_S^2: narrow sense heritability due to sires.
SE: standard error.

length was negatively related to condition, yielding a significant genetic correlation of – 0.897, while testes size was positively related to condition, yielding a significant genetic correlation of 0.524. Thus, males of good genetically mediated condition have both larger testes and shorter sperm.

4.7 Empirical evidence for adaptive cryptic female choice in *Onthophagus taurus*

If male *O. taurus* with larger testes and shorter sperm obtain more fertilizations under sperm competition, females could ensure their offspring are sired by males of high genetic quality through multiple matings and the promotion of sperm competition, by virtue of the genetic architecture that links these sperm competition traits to heritable variation in condition (Yasui, 1997).

The contribution of testes size to sperm competitiveness has been established from experimentally evolving lineages of *O. taurus* subject to enforced monogamy or sexual selection (see Section 4.3). García-González & Simmons (2007) examined the role of sperm length in competitive fertilization success, utilizing Amplified Fragment Length Polymorphisms to assign paternity to offspring when females had mated with two males that differed in the length of the sperm they produced. They found that males with shorter sperm obtained more fertilizations than their long sperm producing competitors. Moreover, there is an active role for females in mediating the short sperm fertilization advantage. Females with the largest spermathecae are better able to bias paternity toward males with shorter sperm (Figure 4.3A). As spermatheca size decreases, selection on sperm length becomes increasingly stabilizing for males producing sperm of average length. Thus, spermathecal morphology appears to be the trait responsible for cryptic female choice acting on sperm length in *O. taurus*.

All models of female preference evolution predict that directional selection imposed by female choice will generate linkage disequilibrium between the genes that code for the male trait and the genes that code for the female preference, because females with the preference will produce both daughters with the preference and sons with the preferred trait (Fisher, 1915; O'Donald, 1980; Lande, 1981; Kirkpatrick, 1982; Pomiankowski *et al.*, 1991; Kirkpatrick & Ryan, 1991; Kokko *et al.*, 2006; Jones & Ratterman, 2009). Empirical studies have demonstrated genetic correlations between traits and preferences within the context of precopulatory female choice for male secondary sexual characteristics (Brooks & Couldridge, 1999; Houde & Endler, 1990; Houde, 1994; Bakker, 1993; Gilburn *et al.*, 1993; Wilkinson & Reillo, 1994). Indeed, genetic correlations between male traits and the mechanisms by which females choose these traits provide the evolutionary footprint of female choice (Evans & Simmons, 2008).

In their study of *O. taurus*, Simmons & Kotiaho (2007b) examined the quantitative genetics of spermathecal morphology, finding significant additive genetic variance in the size and shape of the spermatheca. Critically, they found a significant genetic correlation between spermathecal size and sperm length that is only expected from an evolutionary history of cryptic female choice for short sperm (Figure 4.3B).

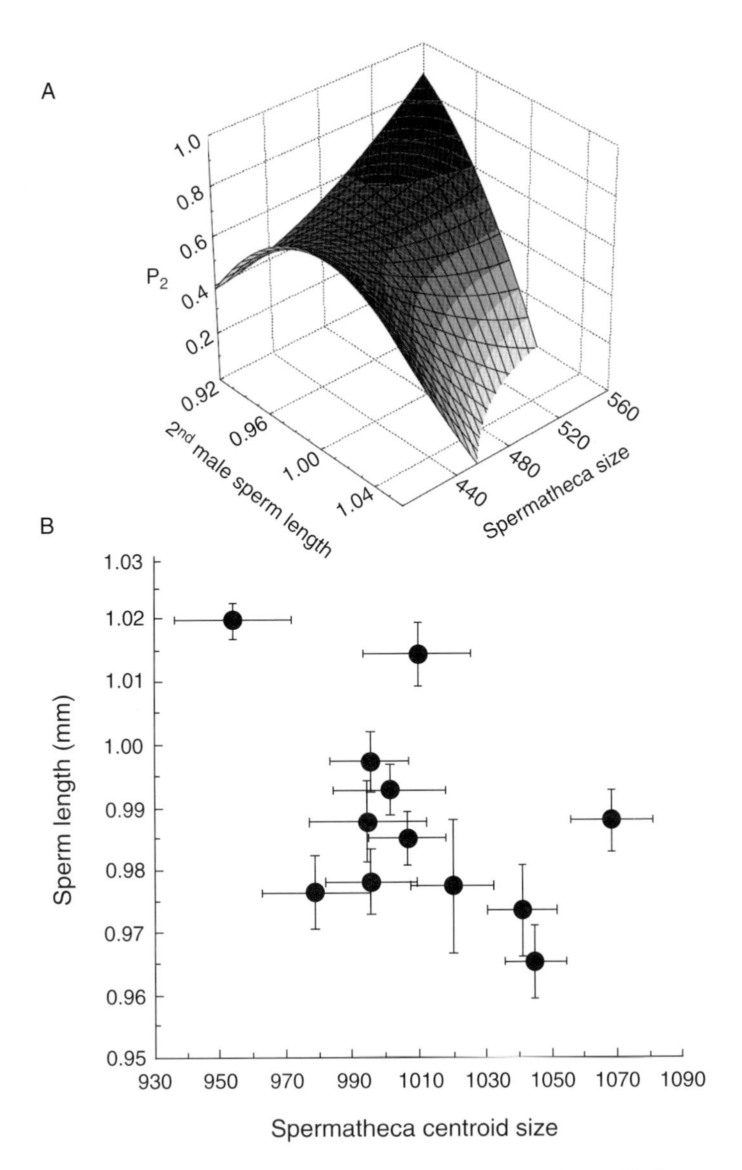

Fig. 4.3 Cryptic female choice in *Onthophagus taurus*. **A:** Males with shorter sperm have a selective advantage at fertilization (P_2) that is mediated by spermatheca size. Females with the largest spermatheca exert directional selection for short sperm, while females with the smallest spermathecae exert stabilizing selection for averaged length sperm (redrawn from García-González & Simmons, 2007). **B:** Sire family mean sperm length ($\pm$SE) plotted against sire family mean spermathecal size ($\pm$SE). The strong genetic correlation between sperm length and spermathecal size (-0.851 ± 0.114) is expected, following a history of selection via cryptic female choice (redrawn from Simmons & Kotiaho, 2007b).

Theoretical models of female preference evolution predict that females should accrue indirect genetic benefits for their offspring in the form of offspring viability (good sperm) and/or competitive fertilization success (sexy sperm). Heritable variation in the traits that promote competitive fertilization success (testes size and sperm length) would thereby support a sexy sperm process in *O. taurus*. But what evidence is there that cryptic female choice can enhance the viability of offspring produced? The genetic correlation between male condition and sperm competition traits should result in females that promote sperm competition through multiple mating producing offspring in better condition.

Simmons and García-González (2008) used their evolving lineages of *O. taurus* to examine the consequences of multiple mating by females for offspring condition. They allowed females from monogamous and polygamous lineages to mate either monogamously or polyandrously and assessed the condition of offspring that those females produced. Polyandrous mating should allow males with larger testes and shorter sperm to sire more offspring and, by virtue of the genetic correlations between condition and these traits, produce offspring of higher condition.

Simmons and García-González found significant effects of selection history and mating regime on offspring condition, and a significant interaction. Females from monogamous lineages had enhanced offspring condition when males competed to father their offspring, but polyandry had little impact on offspring condition for females that had an evolutionary history of polyandrous mating. Thus, the fitness consequences of cryptic female choice for offspring were more pronounced in females with an evolutionary history of enforced monogamy. Such an effect might be expected if enforced monogamy had allowed the accumulation of deleterious mutations in monogamous lineages that were purged by cryptic female choice (Radwan, 2004; Whitlock & Agrawal, 2009).

Maternal effects can have a strong influence on offspring performance (Mousseau & Fox, 1998; Mousseau & Dingle, 1991), making it difficult to ascribe unequivocally variation in offspring performance to paternal genetic effects. For example, females have been shown to increase their reproductive effort after mating with preferred male phenotypes, so that increased offspring fitness could, in theory, be due to differential maternal allocation (Burley, 1988; Sheldon, 2000; Cunningham & Russell, 2000; Qvarnström & Price, 2001). This is a particular problem when assessing paternal genetic effects on offspring performance in species like dung beetles, where maternal provisioning effects on offspring can be very strong (see Chapter 8; Hunt & Simmons, 2000; Kotiaho *et al.*, 2003). Maternal split clutch designs are the best way of controlling for maternal effects, because they allow the assessment of the fitness of offspring sired by different males subject to the same maternal conditions (See Box 4.1).

Using the irradiated male technique to facilitate the identification of a focal male's offspring within a single clutch, it can be shown that offspring sired by males of high condition have a greater probability of surviving to sexual maturity than those sired by males of low condition, even when those offspring share the same maternal genetic and environmental effects (Box 4.1). The available evidence strongly suggests that female *O. taurus* can obtain indirect genetic benefits for their offspring through cryptic female choice for high-condition males. Interestingly, this mechanism of post-copulatory female choice appears to reinforce pre-copulatory female choice in this system.

Kotiaho (2002) found that the females of three species of *Onthophagus* – *O. australis, O. binodis*, and *O. taurus* – all show significant mating preferences for males that deliver high rates of courtship, a female preference also found in a fourth species, *O. sagittarius* (Watson & Simmons, 2010a). In these species, males drum on the female's dorsum and flanks with their forelimbs and head before attempting to engage the female in genital contact. Courtship rate is also dependent upon male condition; males that have been experimentally deprived of nutrition have reduced courtship rates (Kotiaho, 2002), and in *O. taurus* there is a significant additive genetic correlation between male condition and courtship rate (Kotiaho *et al*., 2001). Thus, both pre- and post-copulatory mechanisms of female choice appear to target male condition in *O. taurus*, which, in turn affects offspring viability and the behavioural, physiological and morphological traits that contribute to the mating and competitive fertilization success of male offspring.

A recurrent question in sexual selection research, however, is how genetic variance for fitness can be maintained in the face of strong directional selection (Fisher, 1930; Gustafsson, 1986; Kirkpatrick & Ryan, 1991; Rowe & Houle, 1996; Tomkins *et al*., 2004b; Kotiaho *et al*., 2008). We should expect selection to deplete genetic variance in fitness traits, so that female choice should soon be lost – the so-called 'Lek Paradox'.

Rowe & Houle (1996) proposed that because many traits contribute to the ability of an individual to gather and process resources, condition would present a large target for continued mutational variance that influences the ability of individuals to allocate resources to fitness enhancing traits, and thus for the accumulation and maintenance of genetic variance in condition – so-called genic capture (for a discussion, see Tomkins *et al*., 2004b). Sperm competition traits such as seminal fluid composition and sperm form and function are known to be highly polygenic, offering a very large mutational target for the maintenance of fitness variation (reviewed in Simmons & Moore, 2009).

Box 4.1 Indirect genetic benefits of cryptic female choice in *Onthophagus taurus*

Here I describe a simple experimental approach to control for potential maternal effects when exploring male genetic contributions to offspring fitness. It uses a maternal split clutch design to assess the fitness of a single male's offspring when a female has mated to both high and low quality males.

A sample of 200 male *O. taurus* were weighed and their pronotum widths determined. Body weight was regressed on pronotum width and the residuals saved. The 40 males with the highest positive residuals and the 40 males with the lowest negative residuals were selected as high and low condition males respectively. Then 20 high-condition and 20 low-condition males were exposed to 10 krads of gamma irradiation, sufficient to induce complete embryonic mortality in the offspring they sire (Tomkins & Simmons, 2000; Simmons & García-González, 2008).

Forty females were each mated to two males, one of high condition and one of low condition. For each female, one of her mates was an irradiated male. Females were then established in breeding chambers to construct broods, following standard protocols (Simmons & García-González, 2008).

After one week, breeding chambers were sieved and brood masses containing hatched larvae, and therefore sired by the normal males, were incubated. Emerging adult beetles were housed in single family groups with unlimited access to fresh dung. Three weeks after emergence, when offspring were sexually mature, the numbers of surviving adults per family were counted.

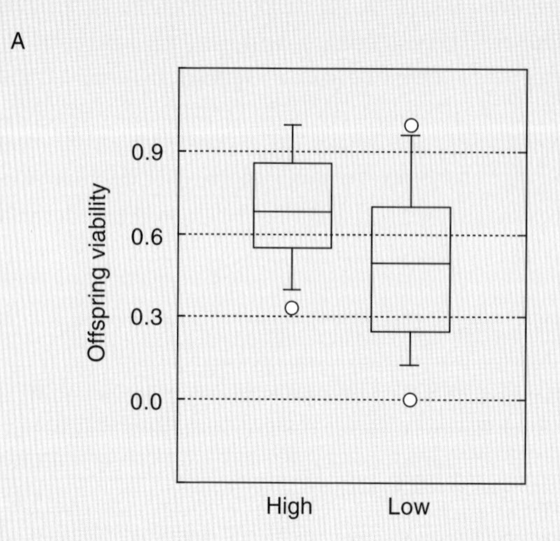

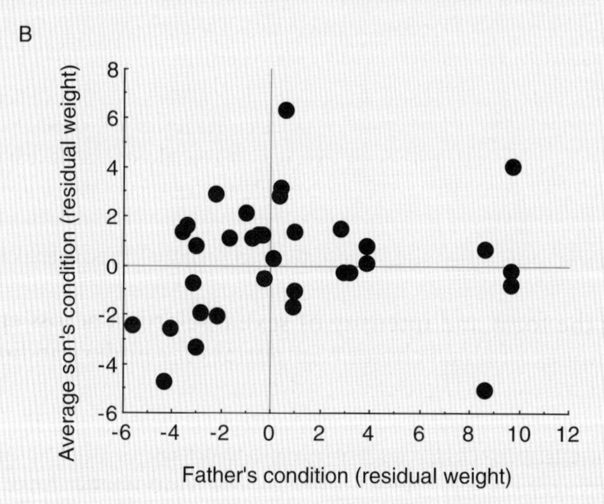

Variation in offspring viability (survival from hatching to sexual maturity) was analysed using a generalized linear model of the number of adults at sexual maturity, using binomial errors, with the total number of hatched larvae as the binomial denominator, and a logit link function. Scaled deviances were compared using F-tests. The predictor variables were the father's sequence position (first or second to mate) and his condition (high or low).

There was a significant effect of male condition ($F_{1,37}=5.73$, $P=0.022$), but no effect of his mating position ($F_{1,37}=0.88$, $P=0.353$). Thus, offspring sired by high-condition males have a greater probability of surviving from hatching to sexual maturity (Figure A). Moreover, for surviving offspring, there was significant non-linear co-variation between the normal males condition and that of his sons (linear effect of fathers condition, $F_{1,\ 32}=4.24$, $P=0.048$; quadratic effect, $F_{1,\ 32}=4.64$, $P=0.039$) (Figure B).

Interestingly, there was no co-variation between the condition of fathers and their surviving daughters ($F_{1,\ 32}=0.12$, $P=0.734$; quadratic effect, $F_{1,\ 32}=0.08$, $P=0.773$) which is consistent with the suggestion of Y-linked genetic effects (see Section 4.6).

4.8 Conclusions and future directions

Parker (1970) recognized that 'females cannot be regarded as an inert environment in and around which [male] adaptation evolved' (Parker, 1970, p.559), anticipating that female processes of post-copulatory sexual selection would have important evolutionary consequences. Nevertheless, much of the 30 years following Parker's insight was spent studying male adaptation to sperm competition, with little consideration of female effects on paternity. A change in focus to the role of females in determining paternity was triggered by Eberhard's (1996) important review of female influences on sperm transfer, storage and utilization, and there was a tendency, as with any new paradigm, for students of post-copulatory sexual selection to reject male influences in favour of female control. However, to reject male influences is as naive as ignoring female influences. Work on onthophagines shows us that both sperm competition and cryptic female choice can both be important selective processes shaping male and female reproductive physiology and behaviour.

Quantifying the form and intensity of sexual selection across episodes of male-male competition and female choice is important for understanding the net action of selection acting on an organism, and thus the evolutionary potential of that selection (Hunt *et al.*, 2009). When episodes of male contest competition and female choice are reinforcing, the evolutionary potential of sexual selection can be enhanced. However, when these episodes are opposing, there can be no net selection and therefore no evolutionary response to either process. Typically, researchers have estimated the form and strength of sexual selection only at single episodes, giving us a fragmented and incomplete understanding of this important evolutionary process.

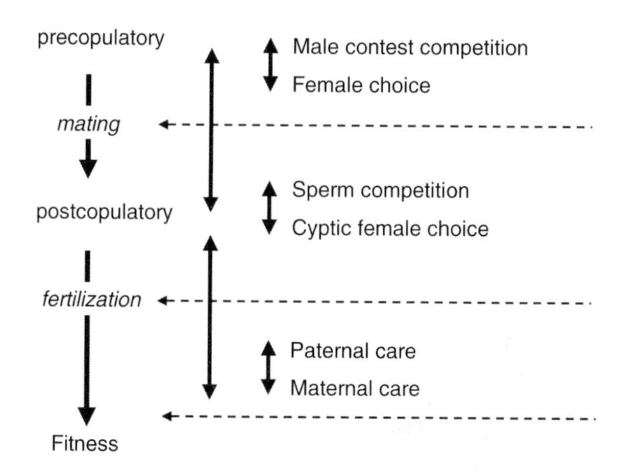

Fig. 4.4 Three potential stages during reproduction at which sexual selection can act. The bi-directional arrows between male and female processes, and between episodes of selection, allow for reinforcing (down) or opposing (up) selection. Opposing selection can generate sexual conflict, the resolution of which will depend on the relative power of each sex to exert their interests over the other sex, and the value to each sex of winning (Parker, 2006). The net form and intensity of sexual selection acting on any particular male or female trait will depend on the interactions between male and female processes within episodes of selection, and on the interactions across episodes of selection.

In reality for most animal mating systems, there are potentially four episodes of sexual selection that need to be quantified: male contest competition; female choice; sperm competition; and cryptic female choice (Figure 4.4). A handful of studies have examined both forms of pre-copulatory sexual selection operating on the same species, but very few studies have extended this into the post-copulatory arena (Hunt *et al*., 2009).

Studies of onthophagine dung beetles suggest that pre- and post-copulatory sexual selection in this group are likely to be reinforcing. Females show a pre-mating preference for males able to deliver high rates of courtship, while, among mated males, those with larger testes have a competitive fertilization advantage. Moreover, females tend to utilize shorter sperm to fertilize their ova. Since all of these traits are genetically linked to condition, pre- and post-copulatory female choice and sperm competition will all select for males of high condition. Condition is also an important determinant of male strength, and will impact a male's ability to compete with other males for access to breeding tunnels and females (see Chapter 3 and Lailvaux *et al*., 2005). The challenge now will be to measure the intensity of selection acting at each of these episodes in order to determine the net selection operating on male condition (Hunt *et al*., 2009).

Both pre- and post-copulatory episodes of selection have been examined in very few other systems. Two studies have identified interactions between pre- and post-copulatory male contest competition. Danielsson (2000) found that large male water striders who were successful in monopolizing access to females in pre-copulatory sexual selection became sperm-depleted because of their greater mating activity, so that small males had a post-copulatory advantage in sperm competition.

Such antagonistic selection might moderate any evolutionary response to sexual selection acting on male body size. A similar phenomenon has been reported in Soay sheep, where dominant rams likewise become sperm-depleted because of their high mating frequency and lose out in sperm competition (Preston *et al.*, 2001).

Some studies have quantified selection acting across episodes of pre-copulatory female choice and post-copulatory sexual selection, though the form of that post-copulatory sexual selection is unclear. In flour beetles, males with more attractive pheromone signals were also more successful in gaining fertilizations when a female mated with more than one male (Lewis & Austad, 1994), while, in fireflies, males with the most attractive bioluminescent displays subsequently showed a lower fertilization success (Demary & Lewis, 2007). In *Drosophila simulans*, males found attractive during courtship also have a higher fertilization success, although the precise traits that are the target of selection in this study are yet unknown (Hosken *et al.*, 2008). In *D. bipectinata*, the male sex comb appears to be the target of both pre- and post-copulatory sexual selection (Polak & Simmons, 2009; Polak *et al.*, 2004).

Finally, in guppies, both pre- and post-copulatory episodes of female choice favour males with more orange colouration on their bodies (Pilastro *et al.*, 2004). Interestingly, even when mechanisms of cryptic female choice are circumvented by artificially inseminating equal numbers of sperm from two males, offspring production is biased toward the more orange sperm donor (Evans *et al.*, 2003). These data suggest that orange males may also have more competitive sperm. Indeed, colourful males do appear to ejaculate faster-swimming and more viable sperm (Locatello *et al.*, 2006), suggesting reinforcing sexual selection across episodes of pre- and post-copulatory female choice and sperm competition.

Male and female interests over when to mate, and which male's sperm should be used for fertilization, will not always coincide, leading to sexual selection via sexual conflict (Arnqvist & Rowe, 2005). Although Parker (1979) defined and analysed the concept of sexual conflict nearly 30 years ago, it is only in the last decade that the full evolutionary consequences of sexual conflict have been recognized and explored, and sexual conflict has become the new paradigm in sexual selection research (see contributions following Tregenza *et al.*, 2006).

It is not uncommon to see broad statements made in the literature concerning the known costs of mating for females. However, such sweeping generalizations are false, ignoring as they do the large literature that shows how mating can in fact be beneficial for females, providing them with direct benefits such as seminal fluid compounds that can enhance fecundity and/or offspring performance (Arnqvist & Nilsson, 2000; Simmons, 2001).

Nevertheless, we know little of the importance of sexual conflict in the mating systems of dung beetles. The fact that in onthophagines, sperm competition and cryptic female choice acting on males is reinforcing rather than opposing, and that females can increase their fitness through the incitement of sperm competition by multiple mating, might suggest that there should be little sexual conflict over mating *per se* (Simmons & García-González, 2008). However, in species with biparental care, there may be significant sexual conflict over offspring provisioning. In *O. taurus*, for example, males will only provision offspring if the females themselves are provisioning, and will not provision at all when the risk of sperm competition is high (see Chapter 8; Hunt & Simmons, 2002b; 2002c). Parental provisioning can have major impacts on offspring size and reproductive performance (Hunt & Simmons, 2000), so that variation in male and female provisioning has the potential to impact the strength of sexual selection both within and between species.

The research summarized in this chapter has taught us much about the mechanisms and evolutionary consequences of sexual selection in just one tribe of dung beetles, the Onthophagines. Future detailed work needs to be conducted on a broader range of taxa. Dung beetles provide an outstanding model system with which to explore the interactions between parental care and pre- and post-copulatory sexual selection. Such work will provide a more holistic understanding of the evolutionary diversification of male secondary sexual traits and parental strategies, both among species of dung beetles and for animal mating systems more generally.

4.9 Dedication and acknowledgement

This chapter is dedicated to Geoff Parker, on the occasion of his retirement from the University of Liverpool. The research described herein was funded by the Australian Research Council.

5

Olfactory Ecology

G.D. Tribe[1] and B.V. Burger[2]

[1]*ARC-Plant Protection Research Institute, Stellenbosch, South Africa*
[2]*Laboratory for Ecological Chemistry, Department of Chemistry, Stellenbosch University, Stellenbosch, South Africa*

5.1 Introduction

As the morning sky over Zululand begins to lighten, and before the intense heat of the sun's rays breaks through the morning mist, the air is filled with the sound of flying dung beetles as they begin to search for dung voided during the night. As the day progresses, waves of dung beetles arrive at the rhino middens, which eventually teem with hundreds of beetles of many species. Out of this mêlée, several dung beetles are seen rolling away balls of dung to which a female clings.

But how do the beetles find ephemeral and patchy dung resources – and, when they arrive at the dung, how do they find conspecific mates? The answer undoubtedly involves olfaction, from the initial location of disparate dung types to the attraction of mates via pheromones that are broadcast using structures whose complexity appears to be linked to nidification behaviour. Semiochemicals clearly influence both the behaviour and ecology of dung beetles and affect nearly every aspect of their existence, even serving as kairomones for phoretic mites.

However, despite this rich research area, unlike other aspects of the ecology and evolution of dung beetles discussed throughout this volume, the olfactory ecology of dung beetles is poorly documented. Here, we provide an overview of the behavioural, morphological, and chemical evidence for the importance of olfaction and chemical communication in the lives of dung beetles, and we highlight important avenues for future research.

5.2 Orientation to dung and other resources

On continents that were once inhabited by vast herds of herbivorous animals, most scarab species adapted to utilize ubiquitous but transient dung as a resource on

Ecology and Evolution of Dung Beetles, First Edition. Edited by Leigh W. Simmons and T. James Ridsdill-Smith. Published 2011 by Blackwell Publishing Ltd.

which to raise their young. The similar size of the antennal clubs of both sexes of most dung beetle species (Halffter & Matthews, 1966) suggests that their primary function is the detection of dung, and that close-range detection of sexual signals is secondary.

The huge number of compounds present in dung is such that little analysis has been done in determining those to which individual dung beetle species respond. Inouchi *et al.* (1988) showed that single antennal olfactory cells of *Geotrupes auratus* (Geotrupidae), which releases food-searching behaviour in response to cow dung odour, responded well to five components of dung-specific odorant, namely 2-butanone (butanone), phenol, *p*-cresol, indole and 3-methylindole (skatole). They found that the dung beetles seem to orient towards food through the R-type I olfactory sensillum cells, responding only to 2-butanone, which is the most volatile among the components, and that R-type II cells may contribute to behavioural function after arrival at the food. The orientation of dung beetles to a dung source was tested by Neuhaus (1983) by allowing *Scarabaeus semipunctatus* to orientate to 1.10^{11} molecules skatole/cm^3 in a wind tunnel. At a wind speed of 5 m/s, the beetles zigzagged over a distance of 30 m within the chemical plume until the source was reached.

Recent studies have provided evidence for clear feeding preferences of scarab beetles colonizing the dung of herbivores, and their ability to discriminate among odours from faeces of various herbivores, based on the numbers attracted to dung-baited traps (see Chapter 2 in this volume). In France, a study using pitfall traps baited with the dung of different herbivore species showed that beetles were more attracted to sheep dung than to cattle, followed by deer and horse dung (Dormont *et al.*, 2007). Eleven of the 27 beetle species collected had significant feeding preferences for one of the four dung types which were confirmed by laboratory olfactometer bioassays where scarab beetles orientated preferentially towards dung volatiles from the dung type preferred in the field. This provided clear evidence that volatile compounds emitted by faeces are involved in the process of resource location and selection.

Dormont *et al.* (2007) also found that each dung type was characterized by a distinct profile of volatiles consisting of more than 20 components, including compounds common to all dung types as well as a few compounds specific to each dung type. Pitfall traps baited with cattle or horse dung revealed similar preferences, but none of the dung beetle species tested seemed linked exclusively to one kind of dung (Dormont *et al.*, 2004).

The distribution of scarabs throughout the world has resulted in them adapting to regions of disparate fauna and flora, as well as to different abiotic conditions. This has given rise to the use of a wide variety of food types other than dung, which they locate through their distinctive odours (see Chapter 2 of this volume). Although most beetles feed on dung, some species feed on carrion, fungi, millipedes and fruit, and beetles will orientate to the odour of these food resources.

This specialization establishes distinct niches for these species. Feeding exclusively on fungi, *Coptorhina klugi* make tunnels below a mushroom into which pieces torn from the parasol by means of their bidentate clypeus are packed, and an egg is deposited in the ball thus formed (Tribe, 1976). The association of fungal fruiting bodies which often grow on drying dung pats may have facilitated this predilection for fungi as food. *Sceliages* and *Onthophagus* species have both been recorded as obligatorily necrophagous only on millipede species that use quinones as defensive secretions (Krell, 1999). Pitfall traps baited with live Diplopoda, or with paper soaked in diplopod defensive secretions, attracted *Onthophagus bicavifrons* in Swaziland by positive chemotaxis (Krell *et al.*, 1998). The most common quinones found in millipedes to which different species of Scarabaeidae orientate are 2-methoxy-3-methyl-1,4-benzoquinone and 2-methyl-1,4-benzoquinone (toluoquinone).

The flightless dung beetle *Thorectes lusitanicus* (Geotrupidae) shows greater attraction to the volatiles of *Quercus suber* acorns than to dung, thus becoming a secondary seed disperser in Mediterranean oak forests, although rabbit dung in particular is needed to raise young (Verdú *et al.*, 2007b). The African dung beetle *Pachylomera femoralis* frequently rolls seeds of the ripe fruit of spineless monkey orange trees (*Strychnos madagascariensis*) (Burger & Petersen, 1991). This behaviour is puzzling, since the beetles are unable to access the seeds without the hard fruit capsule being broken open by monkeys, yet tests showed that the beetles were highly responsive to the volatiles emanating from the seeds. Burger & Petersen (1991) collected volatile organic constituents of the flavour of the fruit in a capillary trap from the headspace gas of the fruit flesh for gas chromatographic-mass spectrometric analyses (GC-MS) and GC-FID/EAD analyses (electroantennographic detection – see Section 5.4.1). The results of a typical GC-FID/EAD analysis of the headspace volatiles of this fruit are depicted in Figure 5.1.

Remarkably sharp and reproducibly strong EAD responses were obtained in all of these analyses. The quantities of the EAD-active constituents released into the atmosphere from the contents of a monkey orange were determined in a wind tunnel at a wind velocity of 0.16 km/h, using a headspace analytical method adapted for this purpose. The results of qualitative and quantitative analyses of the headspace gas of the fruit are given in Table 5.1.

In the field, comparable numbers of *P. femoralis* were found in traps baited with spineless monkey orange and with a mixture of the synthetic compounds (Burger & Petersen, 1991). The individual compounds and mixtures of two or three of them were also attractive to *P. femoralis*.

A series of field trials were conducted to determine the relative attraction of *P. femoralis* to traps baited with horse dung and spineless monkey orange. At the time when these trials were conducted, the spineless monkey orange season had reached its peak and large numbers of the fruit, some of them opened by monkeys and baboons, were lying beneath the trees. In areas with a large population of these trees, the air was permeated with the smell of the fruit and, as anticipated, no dung beetles were caught in traps baited with the fruit, whereas more than 30 *P. femoralis* were found in traps baited with horse dung. In contrast, *P. femoralis* were attracted more strongly to the fruit than to horse dung in areas where there was much animal activity, and where *S. madagascariensis* was scarce or the ripe fruit was not available. In spite of fierce competition for small quantities of horse dung placed in these areas, no dung ball formation or rolling of balls or fragments of dung from these dung sources by *P. femoralis* was observed. In contrast, practically all the *P. femoralis* attracted to the spineless monkey oranges started rolling seeds either immediately on arriving at the fruit, or after inspecting the fruit for a few seconds.

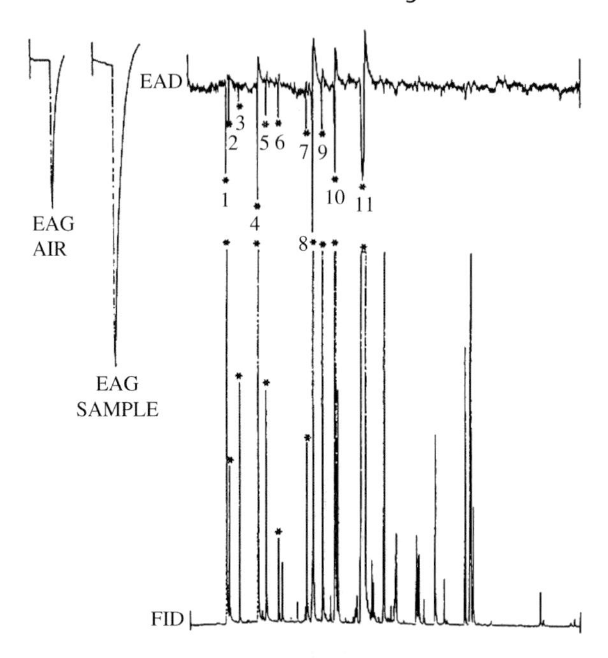

Fig. 5.1 Gas chromatographic analysis of the headspace gas volatiles of the fruit of the yellow monkey orange tree, *Strychnos madagascariensis*, with FID and EAD recording in parallel using an antenna of a female *Pachylomera femoralis* as sensing element. Corresponding peaks in the FID and EAD traces are marked with asterisks and are numbered consecutively. The EAD-active compounds are listed in Table 5.1. (Burger & Petersen, 1991).

It is clear that the attractiveness of the synthetic mixture is at least comparable to that of the fruit flesh of the monkey orange. It is interesting that almost all of these EAD active compounds are C_4 compounds or contain at least one C_4 unit. The attraction of *P. femoralis* to fruit is not a unique phenomenon. Halffter & Matthews

Table 5.1 EAD-Active constituents of the headspace gas of the fruit of the yellow monkey orange tree, *Strychnos madagascariensis* that elicited antennal responses on excised antennae of the dung beetle *Pachylomera femoralis.*

Peak no. in Figure 5.1	*Compound*	*Amount (μg/ml headspace gas)*
1	1-Butanol	0.06
2	Methyl butanoate	0.005
3	Ethyl 2-methylpropanoate	0.02
4	Ethyl butanoate	0.38
5	Butyl ethanoate	0.02
6	Methyl hexanoate	0.01
7	Propyl butanoate	0.04
8	Butyl propanoate	0.06
9	Methyl hexanoate	0.05
10	Butyl 2-methylpropanoate	0.26
11	Butyl butanoate	2.04

(1966) have mentioned the attraction of dung beetles to carrion and rotting fruit and, in Mkuze Game Reserve, a few of the small golden brown dung beetles *Proagoderus aureiceps* were also found in many of the traps baited with spineless monkey orange (Burger & Petersen, 1991).

There are several possible explanations for this puzzling phenomenon. First, it is likely that *P. femoralis* rolls fragments of primate dung, and it is also likely that the fruity aroma of the spineless monkey orange could be present in the dung of primates that feed on these fruit during the early summer months. It is therefore possible that the beetle is misled into believing it is rolling a piece of primate dung. The phenomenon could also represent an evolutionary trick played on the beetles by the spineless monkey orange. Many plants have evolved to exploit insects or other animals to achieve pollination and/or disperse their seeds, and the presence of volatiles in the fruits of monkey orange that are attractive to ball-rolling dung beetles could be an evolutionary adaptation for seed dispersal.

A final possible explanation is supported by circumstantial evidence. In a GC-FID/EAD investigation into the possibility that the abdominal secretion of *Kheper* species could be involved in interspecific or intergeneric communication, a group of 12 volatile compound constituents of an extract of the abdominal secretion of male *K. lamarcki* elicited strong and reproducible EAD responses in the antennae of *P. femoralis*. Eight of these compounds were identified as short-chain esters, from methyl propanoate to ethyl pentanoate, and two of them, – methyl butanoate and ethyl butanoate – attracted *P. femoralis* in bioassays with the synthetic constituents of the aroma of the yellow monkey orange (Burger *et al.*, 1995a). These two compounds contain C_4 moieties, as do all of the EAD-active constituents of the yellow monkey orange aroma. It is therefore possible that *P. femoralis* is attracted to pheromone-secreting *K. lamarcki*, which are likely to be in possession of fresh dung. Similar volatile compounds containing C_4 units were not detected in the abdominal secretions of other *Kheper* spp.

Given that dung beetles live in a world dominated by odour cues, and have evolved acute olfactory senses with which to find scarce and often widely distributed resources, we might expect them to be pre-adapted to use odour cues in mate recognition. In the next section, we provide behavioural and morphological evidence that chemical communication is indeed an important aspect of the reproductive biology of dung beetles.

5.3 Olfactory cues used in mate attraction and mate recognition

The family Scarabaeoidea includes the subfamilies Scarabaeinae (dung beetles) and Melolonthinae (chafers). Within the latter group, it is invariably the females who release the sex pheromone that attracts males (Leal, 1998). In dung beetles, it is often the males that release the sex pheromone, although in some species either sex may do so.

The use by large ball-rolling dung beetles of pheromones to attract the opposite sex was first recorded by Tribe (1976), and the constituents of the pheromones were subsequently analysed by Burger *et al.* (1983). The abdominal glands most involved in production of these pheromones have since then been described from the ball-rolling Scarabaeini, Canthonini, Sisyphini and Gymnopleurini, and the non-ball-rolling Coprini (Pluot-Sigwalt, 1991). The dissemination of pheromones by insects is accomplished in a variety of ways, ranging from the 'brush organs' of Lepidoptera and the setae of cockroaches (Percy & Weatherston, 1974), to scent disseminated on particles as in the butterfly *Danaus* (Pliske & Eisner, 1969; Meinwald *et al.*, 1969). The most complex pheromone dispersal structures described from the Scarabaeinae are in the African ball-rolling genus *Kheper*.

A male *Kheper* arriving at a dung pat first carves out and then moulds a dung ball. Once this ball is complete, the male takes up a headstand position of 45° from the dung surface on top of the ball with his hind legs outstretched (Figure 5.2a). The hind legs are retracted simultaneously inwards towards the sides of the body and are then simultaneously extended rapidly, resulting in a puff of white powder being dispersed into the air above the abdomen. After a short interval of 20 to 30 seconds, the legs are again withdrawn inwards against the sides of the body and again extended rapidly. This behaviour continues until a female arrives, by which time both the hind legs and sides of the abdomen have become coated in the white powder.

A

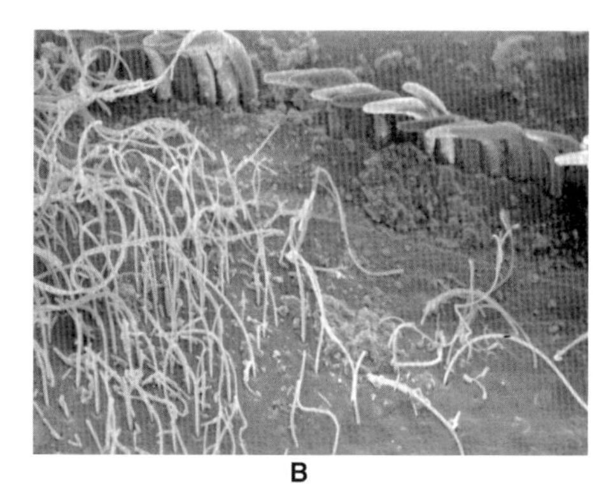

B

Fig. 5.2 **A:** The pheromone release stance adopted by a male *Kheper subaeneus* showing tibial brushes and two rows of abdominal bristles covered in white powder. (Photo: G. Tribe). **B:** Scanning electron micrograph of the proteinaceous carrier impregnated with pheromone emerging from pores in the depression of a male *Kheper nigroaeneus*. X500 (Photo: G. Tribe).

As soon as a beetle of the same species appears within range, the male relinquishes his headstand position and challenges the intruder with his forelegs raised. If the intruder is a male, he will respond in a corresponding way by raising his forelegs to meet the challenge. A fight then ensues, but invariably the beetle which retains the highest position on the pat, usually the one releasing the pheromone, is able to forcefully flip the other beetle off the pat with his forelegs and clypeus. Both residency and relative body sizes of fighters have been found to be important asymmetries influencing contest outcome (Sato & Hiramatsu, 1993). In contrast, a female beetle will lower her forelegs when challenged. The male then approaches her and, after antennal contact – possibly involving the detection of a pheromone – the female is allowed to cling to the dung ball, which the male then rolls rapidly away from the midden.

The carrion-feeding species *Canthon cyanellus cyanellus* (Canthonini) rolls balls of carrion away from a cadaver resource and adopts the same headstand position as the dung-feeding species (Favila & Díaz, 1996). This species also possesses abdominal sternal glands (Pluot-Sigwalt, 1991). The males produce pheromones which only attract conspecific females, and Favila (1988) notes that contact of the male abdomen with the ball chemically marks the food balls they are rolling, and that the attraction is greatest with males associated with a recent nest.

During the later part of the breeding season of telecoprids, the sex ratio often becomes biased in favour of an excess of males, because females remain below the soil with their broods. At this time, a male beetle alters his mate search behaviour; he will roll the brood-ball from the pat, bury it a short distance away and return to the surface of his burrow, where he will release pheromone at the entrance. *Scarabaeus catenatus* males have rarely been observed to release pheromone at the dung pat, but will do so on consecutive days at the entrance to the tunnel in which the ball has been buried if no females have arrived (Sato, 1998a).

It is possibly during the absence of females that the cost of producing copious amounts of volatile pheromone embedded on the proteinaceous carrier becomes most important, because the male must rely less on the dung to bolster attraction. However, from field observations, Edwards & Aschenborn (1988) found male *K. nigroaeneus* never to release pheromone at dung pats, but always to bury brood balls without females and then to release the pheromone at the entrance to the tunnel. Over a period of eight years, while working with more than 1,000 *K. nigroaeneus* in captivity, Burger (unpublished) observed the release of pheromone at the entrance to tunnels where males had buried dung. Edwards & Aschenborn (1988) were able to detect the pheromone odour from a distance of 9 metres.

5.3.1 Morphology of pheromone-producing and -dispersing structures

Female *K. nigroaeneus* possess two types of sparsely distributed pores over the first five abdominal sternites which consist of small direct openings, and dispersed amongst these are tiny depressions in which 24 tiny pores open (Tribe, 1976). The possession of sternal glands in both male and female beetles is widespread within the Scarabaeinae (Pluot-Sigwalt, 1991). While the functions of these glands are unknown, they are likely to be involved in olfactory communication.

Pluot-Sigwalt (1991) suggests that the abdominal glandular system in Scarabaeidae evolved by the elaboration of simple glandular units scattered around the integument, because every evolutionary step towards increasing complexity can be observed within the family. This glandular complexity and the complexity of nesting behaviour appear to be correlated. In tunnelling species, the glandular units increase in density only in some abdominal areas, while in rollers, differentiated glandular patches were found (Pluot-Sigwalt, 1991). In ball-roller species, and in particular those of the genus *Kheper,* these glands are more numerous and are concentrated into distinct patches within depressions, and consequently they have complex dissipating structures. The sternal glands are always sexually dimorphic, and in males they are mainly present in species in which the sexes collaborate closely in nesting behaviour (Pluot-Sigwalt, 1991). This sexual dimorphism strongly implicates sexual selection in their evolution.

The pheromone in the genus *Kheper* is produced from a large interwoven mass of single-celled glands on either side of the first abdominal sternite. These glands open through several hundred minute pores into a depression which resembles a sieve, from which the white flocculent material emerges (Figure 5.2b). Immediately posterior to the depressions, along part of the anterior sections of the second, third and fourth male abdominal sternites, are single rows of short, semi-rigid bristles which curve inwards towards the depressions. Although both sexes have numerous long hairs along the dorsal length of the tibiae of their hind legs, only in the male are the hairs concentrated centrally to form distinct brushes.

The male *Kheper* possess thick brushes on their hind tibia which, as the hind legs are extended, brush the emerging pheromone across the depression and against the rows of abdominal bristles. A puff of powder is seen above the abdomen, which soon, together with the hind-legs, becomes coated with white powder. In the absence of a female, the male will continue to secrete this sex pheromone. Thus it is possible to amputate the tibia of a secreting male, which after an interval, will resume its position and continue secreting. In *Kheper* and some *Scarabaeus* species, the pheromone will emerge like shaving cream from the side where the tibia has been removed and may be collected. The pheromone has a distinct skatole scent, which can be smelled from several metres away. However, the white flocculent material is merely a carrier material impregnated by the volatile pheromone, and it can be periodically collected with forceps.

The pheromone-containing material of *Scarabaeus catenatus,* which also assumes a headstand pheromone release stance, is neither visible nor odorous (Sato, 1998a), whereas in *K. platynotus* and *K. aegyptiorum* it resembles a fine white powder but has no odour (Sato & Imamori, 1986; 1987).

5.3.2 Pheromone-dispersing behaviour

The headstand pheromone release stance (Figure 5.2A) has been documented in ball-rolling species in the genera *Canthon*, *Scarabaeus*, *Kheper*, *Garreta*, *Allogymnopleurus* and *Sisyphus* (Favila & Díaz, 1996; Paschalidis, 1974; Tribe, 1975; 1976). However, the rhythm of brushing behaviour appears to be species-specific. For example, within the genus *Kheper*, one species draws its tibiae once over the pheromone exuding from the depression and then leaves its legs extended for 20–30 seconds before the legs are withdrawn, while another has two or so brushing strokes over the depression in quick succession before the legs are extended (Tribe,

1976). *Garreta nitens*, which assumes the pheromone release stance but possesses no tibial brushes, draws the legs alternately over the abdomen before extending them in the air, while *Scarabaeus* criss-cross their extended tibiae (Tribe, 1976).

The ratio of the length of the brush to that of the tibia within the *Kheper* and *Scarabaeus* genera varies between species, and this appears to be linked to the time of day that they are active. Thus, species such as *K. nigroaeneus*, which fly in the early morning, have the shortest brushes, and the brushes become increasingly longer with increasing crepuscular activity of each species (Tribe, 1976). Presumably this is because diurnal species rely to a greater extent on vision, whereas crepuscular species rely more on olfaction. Variations in the position and shape of brushes occur in the genus *Scarabaeus*. They are situated on the inside of the tibia in *S. goryi*, resulting in a peculiar action when brushing, while long 'crew-cut' brushes are present in the crepuscular *S. zambesianus* (Tribe, 1976)

Various behaviours have evolved to contend with competition when acquiring fresh dung. Bernon (1981) has shown that 90 per cent ($n = 114$ spp.) of the dung beetle species at a dung pat are paracoprids, which construct their brood tunnels directly under the pat, while 10 per cent are telecoprids, which remove dung from the pat and bury it a distance away. Brood space beneath the pat was probably a limiting factor during peak activity (Bernon, 1981), necessitating that ball-rolling species remove dung away from this competition. Although the data are currently rather limited, the available data summarized in Table 5.2 suggests that long-range pheromone signalling may be particularly characteristic of the larger telecoprids.

All Scarabaeinae have some form of gland systems (Table 5.2). Unlike telecoprids, most chemical communication of paracoprids is likely to take place within the confined space of their tunnels, so simpler gland systems are found (Pluot-Sigwalt, 1991). Exocrine glands in the forelegs of dung beetles in the genus *Onitis* were described by Houston (1986). They consist of two groups of exocrine glands on the forelegs, one located on the protrochanters and confined to the male which is present in most *Onitis* species, while the other is located on the procoxae in conjunction with disseminating structures on both the procoxae and profemora, which is present in both sexes and probably occurs throughout the scarabaeine dung beetles. Houston (1986) speculated that these glands may produce pheromones and are involved in close range species recognition and sexual attraction.

5.4 Chemical composition of *Kheper* pheromones

The chemical compounds of *Kheper* display both similarities and differences in their composition. The major volatile constituents of the abdominal secretion of male *K. larmarcki* were analysed as 2,6-dimethyl-5-heptenoic acid, (*E*)-nerolidol and hexadecanoic acid, while skatole is a minor component (Figure 5.3) (Burger *et al.*, 1983). In initial field trials, these compounds were deposited on French chalk as it was considered to be a suitable inert carrier material, although later it was established that the natural carrier consisted of proteins.

Regardless, these field trials proved inconclusive because, in the absence of dung, neither male or female *K. lamarcki* were attracted to the pitfall traps baited with a mixture of the chemical compounds. This is perhaps not surprising, because dung is

Table 5.2 Some general trends in pheromone dispersal structures between Scarabaeine dung beetle tribes.

Tribe	*Abdominal sternal glands[1]*				*Headstand[2]*	*Hind tibial brushes[2]*	*Exocrine glands on fore tibia[3]*		*Nesting behaviour*
	Glands scattered	*Glands grouped*		*Pygidial glands*			*Protochantral*	*Procoxal*	
		Male	*Female*						
Scarabaeini	+	+	+	+	+	+	–	+	Telecoprid
Canthonini	+	+	+	+	+		–	+	Telecoprid
Gymnopleurini	+	+	+	+	+	–	–	+	Telecoprid
Sisyphini	+	–	+	–	+	–	–	+	Telecoprid
Coprini	+	–	–	+	–	–	–	+	Paracoprid
Onitini	+	–	–	–	–	–	+	+	Paracoprid
Onthophagini	+	–	–	–	–	–	–	+	Paracoprid

Data from [1]Pluot – Sigwalt, 1991;
[2]G. Tribe (unpublished);
[3]Houston, 1986. There is considerable variation among species within tribes, and this table only reports on those species for which data are available. *Typhaeus typhoeus* (Geotrupidae) also adopt a headstand position – see Brussaard, 1983 (page 210), but no other information on glandular structures appears to be available.

K. lamarcki	*K. nigroaeneus*	*K. subaeneus*	*K. bonellii*
Hexane-2-one[b]	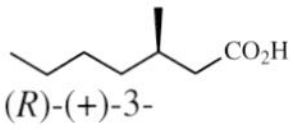 (*R*)-(+)-3-Methylheptanoic acid[c]	Butanoic acid (*n*-butyric acid)	Butanone
3-Hexen-2-one[b]			Propanoic acid
2,6-Dimethyl-5-heptenoic acid[c,d]	Skatole	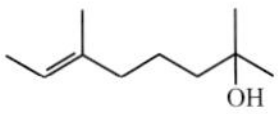 (*E*)-2,6-Dimethyl-6-octen-2-ol	Butanoic acid
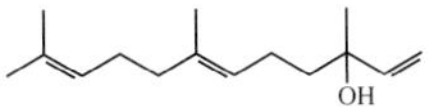 (*E*)-Nerolidol[c]		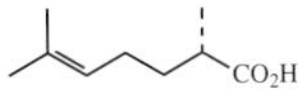 (*S*)-2,6-Dimethyl-5-heptenoic acid[b]	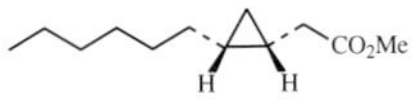 Methyl (1′*R*,2′*R*)-2-2′-hexylcyclopropylacetate
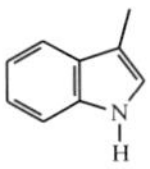 Skatole (3-methylindole)		Skatole	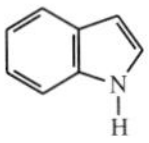 Indole
Hexadecanoic acid (palmitic acid)[b]			Skatole

[a]Constituents listed in order of elution from the columns used for the analysis of the secretions of the respective species.
[b]EAD Responses only in some antennae.
[c]Isolated from abdominal secretion.
[d]Stereochemistry not elucidated.

Fig. 5.3 EAD-Active compounds identified in, or isolated from the male abdominal secretions of *Kheper lamarcki, K. nigroaeneus, K. subaeneus and K. bonellii.*

the primary attractant to which dung beetles respond (Halffter & Matthews, 1966; Tribe, 1975); only secondarily do they respond to the pheromone of a signalling male. It thus seemed unlikely that a bioresponse-guided strategy could lead to the isolation and eventually the identification of the sex attractant of *Kheper* species, so electroantennographic detection (EAD) was used instead to record the response of the beetles to various compounds.

The analysis of pheromones is vastly simplified if the pheromone secretion or material can be trapped from the effluvium of the secreting organism rather than using material extracted from the glandular tissue. A typical sample preparation consists of the extraction of a few milligrams of the secretion with dichloromethane and concentration of the extract. Once it is established that the EAD-active constituents of the secretion can be located in a gas chromatogram by GC-FID/EAD analyses, material containing the sex attractant can then be obtained with much less effort by excising the glands underlying the abdominal pores on both sides of the first abdominal sternite of the male *Kheper* after the guts of the insects have been removed. The glandular tissue can then be extracted with dichloromethane for analysis.

Gas chromatographic analyses (GC) and gas chromatographic-mass spectrometric analyses (GC-MS) revealed that the volatile fractions of the abdominal secretions of the three *Kheper* species were complex mixtures. Qualitatively, the volatile fractions of the secretions were similar, but only a few constituents were common to two or to all three of the species, and the volatile components were present in totally different quantitative ratios (see Section 5.5.2).

The so-called chemical image approach (Albone, 1984) is another method that can be used to find the active constituents in complex semiochemical secretions. However, it requires the comprehensive chemical characterization of the volatile fraction of the secretions of each species and evaluation of the chemical and physical properties of the constituents as potential constituents of sex attractants. The identification and synthesis in pure form of all the constituents of the secretions, including the enantiomers of the long-chain methyl-branched chiral constituents for behavioural tests, would be an almost impossible task. Electroantennography (EAG), or preferably gas chromatography in conjunction with electroantennographic detection (GC-EAD), is the most viable alternative for the detection towards identification of the active component(s) of the sex-attracting secretions of *Kheper* species.

5.4.1 Electroantennographic detection

Electroantennographic experiments with the lamellate antennae of dung beetles are difficult. The insects are too strong to tether, but fortunately the excised antennae of these beetles remain viable for at least 18 hours at moderate temperatures of around 20°C. Contact can be made with the proximal end of an antenna by inserting a micropipette filled with saline solution into the haemolymph of the scapus. Contact with the distal end of the antenna can be made by bringing a saline-filled pipette into contact with the surface of the antenna, but not by inserting it into the antennal club. To prevent the saline solution from draining out of the pipettes at the point of contact with the antennal surface, the viscosity of the saline solution must be increased by the addition of polyvinylpyrrolidone (van der Pers, 1980).

The instrumentation necessary to increase the signal-to-noise ratio of the EAD signal for GC-FID/EAD analyses with the lamellate antennae of dung beetles was first described by Burger & Petersen (1991). GC-FID/EAD analyses with the lamellate antennae of *Anomala cuprea* (Scarabaeidae; Rutelinae) were subsequently reported by Leal *et al.* (1992), with similar success.

5.4.2 Comparison of the responses of beetle species to attractant compounds

Work with *K. lamarcki* produced rather confusing results, largely because CG-FID/EAD analyses mostly produced very weak signals that could not be reproduced. Skatole, 2,6-dimethyl-5-heptenoic acid, (*E*)-nerolidol and hexadecanoic acid were identified in the abdominal secretion of male *K. lamarcki* (Figure 5.3). As noted above, initial attempts at testing these compounds in the field were unsuccessful (Burger *et al.*, 1983).

After instrumentation for GC-FID/EAD analyses became available, further field trials were conducted in Mkuze Game Reserve with the EAD-active constituents, as well as with combinations of the EAD-active constituents, where some constituents gave non-reproducible EAD responses on a few occasions (Munro, 1988). In the absence of dung, no beetles of any species were attracted to traps baited with the natural secretion of *K. lamarcki*, or with any combination of synthetic EAD-active compounds. In combination with dung, large numbers of dung beetles of many species were trapped. Although slightly more *K. lamarcki* females than males were attracted to pitfall traps baited with dung plus these compounds than to traps baited only with dung, this trend was not statistically significant.

Three constituents of the abdominal secretion of male *K. nigroaeneus* gave reproducible EAD responses in GC-FID/FID analyses. One of these compounds is present in such a low concentration in extracts of the abdominal secretion of the males that it was invisible in FID traces, despite relatively large samples having been used for analyses. A similar situation was encountered in the GC-FID/EAD analysis of the volatile fraction of the abdominal secretion of *K. subaeneus* (see below).

Enough of another EAD-active constituent could be isolated from the *K. nigroaeneus* secretion to record its 1H NMR spectrum. Using enantioselective gas chromatography to separate the two enantiomers, this compound was identified as (*R*)-(+)-3-methylheptanoic acid (Figure 5.3). It was synthesized in high enantiomeric purity from (*R*)-(+)-3,7-dimethyl-6-octen-1-ol [(*R*)-(+)-β-citronellol] as starting material (Burger & Petersen, 2002). Skatole was the third EAD-active constituent of the secretion.

Although male and female antennae mostly gave reproducible and qualitatively similar results, a few examples of male and female antennae responding reproducibly to different long-chain compounds on different occasions were encountered. These EAD analyses could not be reproduced with other antennae and the identification of these compounds could thus not be verified by EAD analyses of the synthetic compounds. A pheromone elicits an immediate and constant response, unlike the irregular responses of these compounds. However, the compounds were always long-chain fatty acids such as hexadecanoic acid or methyl or ethyl esters of long-chain acids. This phenomenon has not been investigated further.

Kheper subaeneus occurs in parts of the Hluhluwe Game Reserve, about 50 km from the Mkuze Game Reserve, where *K. nigroaeneus* and *K. lamarcki* are found. A typical GC analysis of the sex attractant of male *K. subaeneus* with FID and EAD recording in parallel, using a female antenna as sensing element, is shown in Figure 5.4.

The first-eluting constituent was identified as butanoic acid (*n*-butyric acid). Another constituent was identified as skatole. Although barely detectable by FID, the third constituent gave a very strong and reproducible EAD response. Male antennae produced similar EAD/FID analyses. The mass spectrum of this constituent contains several ions typically found in the spectra of the monoterpenes. This, and the presence of compounds with terpenoid character in the abdominal secretions of other *Kheper* species, led to the eventual identification of this new tertiary monoterpene alcohol, (*E*)-2,6-dimethyl-6-octen-2-ol (Figure 5.3), which was named (*E*)-subaeneol.

In addition to butanoic acid, (*E*)-2,6-dimethyl-6-octen-2-ol and skatole, which elicited reproducible EAD responses in male and female antennae of *K. subaeneus*, some other constituents of the abdominal secretion also occasionally gave EAD responses. Because of the non-reproducibility of the EAD results, these constituents

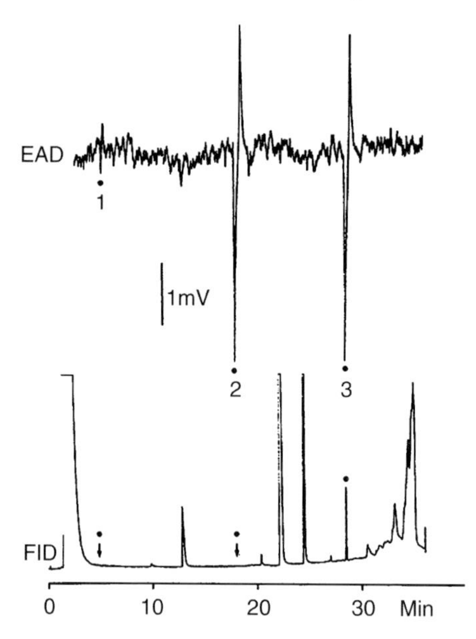

Fig. 5.4 First part of a typical gas chromatographic analysis of the volatile fraction of the male abdominal secretion of *Kheper subaeneus* with FID and EAD recording in parallel, using a female antenna as sensing element. Assignment of EAD-active constituents: 1 = butanoic acid; 2 = (*E*)-2,6-dimethyl-6-octen-2-ol; 3 = skatole. (Burger *et al.*, 2002).

could not be identified. However, the major volatile constituent of the secretion also occasionally gave an EAD response and was identified as the chiral 2,6-dimethyl-5-heptenoic acid, which had previously been identified in *K. lamarcki* (Burger *et al.*, 1983).

By synthesis from an authentic starting material, the EAD-active enantiomer present in the secretion was identified as (*S*)-2,6-dimethyl-5-heptenoic acid (Burger *et al.*, 2002) (Figure5.3). Although the mass spectrum of the synthetic acid is identical to that of the natural compound, and it co-eluted with the natural compound from the enantioselective gas chromatographic column, it did not elicit EAD responses in a large number of male and female antennae that were used in EAD/FID measurements. This is not unexpected, in view of the observation that the natural compound also did not elicit reproducible EAD responses.

The possibility was investigated that the response of an antenna could depend on the gas chromatographic behaviour of carboxylic acids on the gas chromatographic column that was used, and on the concentration of the components flowing over the antenna. However, no clear explanation could be found for the inconsistent results obtained with this compound or with some of the other minor constituents of the secretion. Unfortunately, as in the research on *K. nigroaeneus*, field trials with synthetic materials have so far not produced positive results.

Males of *K. bonellii* hardly ever produced secretion in captivity. The volatile components of the secretion were therefore extracted from abdominal glandular tissue. As with the other *Kheper* species investigated, this extract was also comprised mainly of constituents of relatively high molecular mass such as, for example, branched and unbranched long-chain fatty acids, alcohols and esters, which did not elicit EAD responses in the antennae of *K. bonellii*. The investigation was therefore focused on the more volatile constituents of the glandular extract. Male and female antennae responded reproducibly to the same constituents of the extract, which was not an unexpected result for a sex attractant secreted by a male insect (Leal *et al.*, 1998).

The EAD-active constituents butanone, propanoic acid, butanoic acid, methyl (1′*R*,2′*R*)-2-2′-hexylcyclopropylacetate indole and skatole (Figure 5.3) were identified in the extract from *K. bonellii* (Burger *et al.*, 2008). Male and female *K. bonellii* antennae also responded to butanone. This result is also not unexpected, because 2-hexanone and 3-hexen-2-one, constituents of the abdominal secretion of male *K. lamarcki*, elicited EAD responses in male antennae of that species (Munro, 1988). Although butanone was subsequently found to be present in the solvent as a trace impurity that could have been concentrated during sample preparation for GC-MS analysis, the possibility has to be considered, in the light of the work of Inouchi *et al.* (1988), that it could be a component of the volatile fraction of the insect's abdominal secretion, or that it could have been adsorbed from dung by the proteinaceous carrier material.

Skatole elicited EAD responses in the antennae of the other three *Kheper* species investigated thus far. Methyl cis-2-2'-hexylcyclopropylacetate (absolute configuration not specified) is found elsewhere in nature, and was first identified together with its trans isomer as components of the essential oil of *Croton eluteria* (Euphorbiaceae), a tree-like shrub of the West Indies, by Wilson & Prodan (1976), who named the compound methyl *cis*-cascarillate.

Skatole is an EAD-active component of the male abdominal secretions of all four *Kheper* species investigated. It is most likely sequestered from mammalian dung on which the beetles are feeding. However, this has not been confirmed by studies with isotopically marked skatole or amino acids.

Of all the EAD-active compounds that were identified in the male abdominal secretions, the terpene alcohol (*E*)-2,6-dimethyl-6-octen-2-ol gave the strongest and most reproducible EAD responses. (*S*)-2,6-Dimethyl-5-heptenoic acid is not a terpenoid compound (it lacks one carbon atom), but it has the typical methyl-branching pattern of a monoterpenoid. These two compounds could be sequestered as such from dung, or they could be metabolized by the beetles from terpenoids that are present in plant material. Mono- and sesquiterpenes have been identified as pheromones in many insect species (e.g. Francke & Dettner, 2005), and a few terpenoids that are thought to be involved in semiochemical communication have also been found in the exocrine secretions of mammals (Burger, 2005).

Regarding the unsatisfactory reproducibility of EAD responses, the question is whether this problem could perhaps be ascribed to the fact that the GC-FID/EAD analyses were almost invariably carried out late in the season with antennae of beetles that had spent many months in captivity under often unusual climatic conditions. The observation that somewhat better results were obtained with *K. bonellii*, available within a few kilometres from the laboratory, would seem to support this explanation.

5.4.3 The pheromone-disseminating carrier material

The white flocculent material secreted by *Kheper* species was found to consist of a volatile pheromone component adhering to a proteinaceous carrier. Gas chromatographic analyses (GC) and gas chromatographic-mass spectrometric analyses (GC-MS) revealed that the material extracted from the carrier material and glandular tissue each contained more than 150 constituents that were for the largest part less volatile than the EAD-active constituents discussed in Section 5.3.2. These constituents were mostly long-chain straight- and branched-chain hydrocarbons, alcohols, fatty acids, and methyl and ethyl esters of these and other fatty acids, of which only a limited number were common to the first three *Kheper* species investigated.

Only a few constituents of the material extracted from the abdominal secretions of these species were common to two, or to all three of the species. Quantitatively, the extracts were totally different (Figure 5.5). One explanation for the presence of a large number of long-chain alkanes, alcohols, fatty acids, etc. in the abdominal secretion could be that these compounds augment or modulate the pheromone-disseminating capacity of the proteinaceous carrier material. However, the fact that the antennae of the male and female dung beetles on occasion seemed to respond to some of these compounds could indicate that, depending on circumstances, they could possibly have a semiochemical function. It is also possible that the EAD techniques used in these studies were just not suitable for the detection of such long-chain compounds.

As described above, the protein extruded from the abdominal pores solidifies to produce fibrous material, which looks like cotton wool under a microscope, and which is broken up by the brushes on the tibiae of the secreting male into fine particles that are released into the air (see Section 5.3). The object of this sophisticated pheromone-disseminating mechanism is obviously to deliver the chemical message to recipients with high fidelity.

Quite remarkably, the protein fractions of the abdominal secretions of the three species, *K. lamarcki*, *K. nigroaeneus* and *K. subaeneus*, each consist of only a few proteins, the major proteins having molecular masses of about 15.4 kDa. The amino acid sequences of the proteins of the three species are similar but not identical (Burger *et al.*, 1990). The carrier protein of *K. lamarcki* has a high affinity for 2,6-dimethyl-5-heptenoic acid, a major component of the volatile fraction of the abdominal secretion of this species, and it might therefore be involved in the transport and controlled release of the volatile pheromones. In a comparison of this protein with other common proteins, albumin and the carrier had approximately similar affinities for the acid, whereas trypsin retained only about one-third of the amount of acid that was retained by the carrier protein.

The proteinaceous pheromone carrier of *Kheper* species is physiologically expensive to produce, so it must have an important biological function other than a short-range communication when the sexes meet at a dung pat. Vast amounts of pheromone are released, sometimes for several consecutive days, from the entrance to the burrow in which a male has sequestered dung in the hope of attracting a female. This occurs frequently once most females are below the soil brooding, and the sex ratio is severely biased towards males which remain active. It is possible, in such situations, that the pheromone becomes more important.

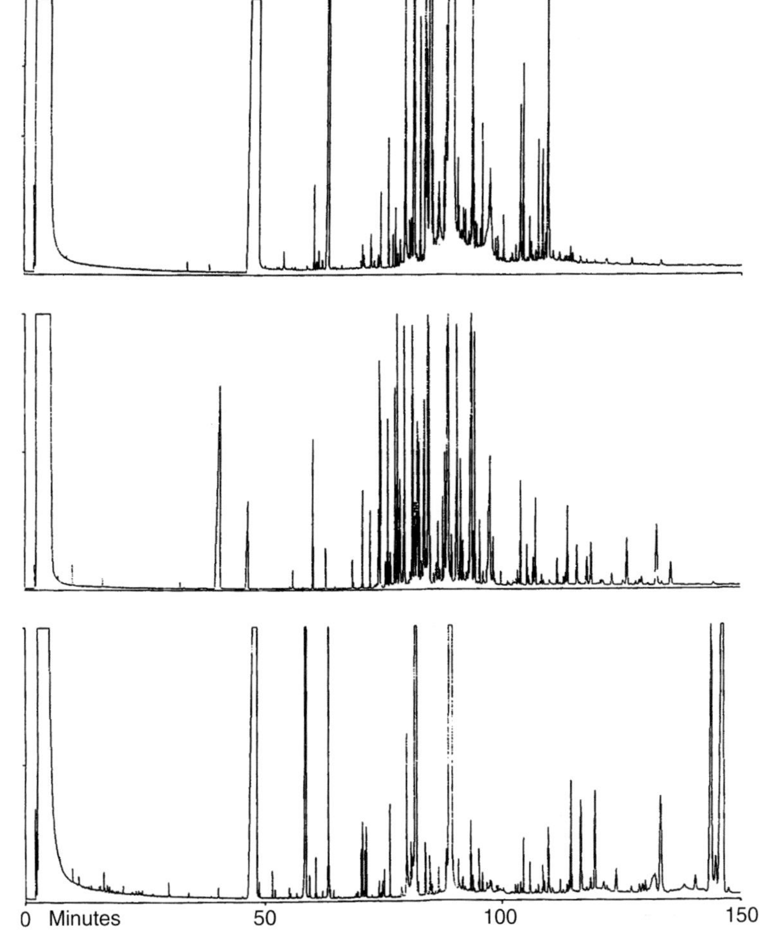

Fig. 5.5 Comparison of the gas chromatographic analyses of the volatile fractions of the male abdominal secretions of *Kheper lamarcki* (top), *K. nigroaeneus* (middle), and *K. subaeneus* (bottom) (from Munro, 1988).

5.5 Kairomones

As with many other species of Coleoptera, dung beetles have several species of phoretic macrochelid mites (Acari: Mesostigmata) associated with them. These wingless mites use dung beetles to carry them to fresh dung, where they feed on a plethora of dung-inhabiting organisms. Dung pats represent an ephemeral resource and, as the pat is shredded and dries out, the mites desert the dung, cling to the hairs on the beetles' bodies and are carried to a fresh source. Only fertilized female mites are phoretic (Costa, 1969).

Using baited pitfall traps, various *Scarabaeus* species were captured in Morocco and the number of *Macrocheles saceri* mites on each beetle was counted. Then, by presenting choice tests using live beetles and cuticular extracts of beetles, Niogret *et al.* (2006) were able to show that the mites were attracted from a distance of a few centimetres to certain beetle species by means of kairomones emanating from the cuticle. *Scarabaeus sacer* was significantly more attractive to *M. saceri* mites when compared with the kairomones of other beetle species that were tested. The cuticular extracts, acting as kairomones, resulted in chemical recognition that determined the phoretic behaviour of the mites but which varied according to the *Scarabaeus* species.

No eye or photosensitive organ has been observed in these mites, and they receive chemical information through chemoreceptive sensilla on anterior palps and the first pair of legs (Niogret *et al.*, 2006). Niogret *et al.* (2006) considered that the various areas to which the mites attach to the beetles may be influenced by the different patterns in the distribution of secreting glands that occur on the sternal parts of the body of Scarabaeidae species (Pluot-Sigwalt, 1991). Thus, the phoretic mites are able to rapidly locate and recognize their beetle hosts through their sternal glands.

Although traditionally thought to be phoretic, macrochelid mites may in some cases be detrimental to their hosts (Polak, 1996; 1998). Kotiaho & Simmons (2001) found that there was a significant deleterious effect of *Macrocheles* on longevity of *Onthophagus binodis* males, where males in a mites-added treatment died on average 15 days earlier than males in a mites-removed treatment. The possibility exists, therefore, that some life-history stages of these mites may turn out to be parasitic.

Regardless of the precise relationship between mites and their hosts, Niogret *et al.*'s (2006) work suggests that dung beetle species have species specific chemical profiles associated with their cuticles. It is likely that such species-specific cues could be utilized by the beetles themselves in mate recognition.

5.6 Defensive secretions

The only defensive secretion recorded for dung beetles is in one of two southern African endocoprid species. Brood-balls of endocoprids are constructed within intact dung pats during winter, when intraspecific competition for dung has ceased. No elaborate pheromone-disseminating structures exist within the endocoprid species, but surprisingly an elaborate chemical defence and associated escape behaviour is found in *Oniticellus egregius* (Davis, 1989; Burger *et al.*, 1995b). When *O. egregius* is disturbed, it flips into the air and becomes cryptic, where it lands amongst the shredded dung while exhibiting thanatosis. It simultaneously releases a brown fluid from the lateral edge of the anterior abdominal segments just posterior to the hind legs (Davis, 1977). This secretion, reminiscent of oil of wintergreen, has been identified as methyl salicylate (oil of wintergreen) and 1,4-benzoquinone (Burger *et al.*, 1995b), a relatively common defensive secretion of the Coleoptera which acts as a repellent (e.g. Blum, 1981). Ants and termites frequently inhabit dung pats in winter, where they would encounter *O. egregius* and would likely be repelled by the wintergreen odour.

5.7 Conclusions and future directions

It is clear from the data reviewed in this chapter that dung beetles have an acute olfactory sensitivity that allows them to locate ephemeral and often patchily distributed resources required for reproduction, and that semiochemicals play an important role in communication. Given their role in mate attraction, semiochemicals are expected to be subject to considerable selection, both as mate recognition signals among species and as cues to individual recognition and mate selection within species (Johansson & Jones, 2007; Thomas, 2010).

As we have seen in Chapters 3 and 4, much is known about other modalities of communication, and the action of selection operating on male and female sexual traits, so it is perhaps surprising that so little is known about selection acting on olfactory signals. Dung beetles are not unique in this regard. Despite the fact that olfactory communication is probably the most ancient and widespread form of sexual communication, our general understanding of the evolutionary pressures acting on olfactory signals is limited (Wyatt, 2003). Perhaps this reflects our own sensory biases as researchers; the exaggerated horns of male dung beetles are both visually striking and easy to study (see Chapter 3). However, chemical signals produced by males in attracting females can be equally complex, as our analyses of *Kheper* pheromones attests. We hope that our chapter will encourage further exploration of this potentially rich area of research, and we certainly believe that dung beetles offer an ideal model system with which to advance our understanding of chemical ecology generally.

Dung beetles have been broadly classified as tunnellers and rollers, and these two breeding systems do appear to be subject to very different modes of selection. Exaggerated male weaponry appears to be more likely to evolve in the context of tunnel defence in paracoprid species, which meet and fight for females in the tunnels beneath the soil surface (see Chapter 3). While the data are too limited for firm conclusions to be drawn, our prospective survey suggests that pheromone signalling to attract females may be restricted to telecoprid species. Clearly, more studies of a greater variety of dung beetles with different breeding biologies will allow a formal evolutionary analysis of this pattern, and we hope our chapter will stimulate such an approach.

There is also a growing awareness that cuticular hydrocarbons (CHCs) play important roles as semiochemicals in insects (Blomquist & Bagneres, 2010). Insect CHCs have been shown to be sexually dimorphic (reviewed in Thomas & Simmons, 2008) and can be subject both to natural selection as species mate recognition cues, and sexual selection as cues to individual mate quality (Blows, 2002; Higgie *et al.*, 2000). Behavioural observations of *Kheper* suggest that males need to make physical contact with females for sex recognition, suggesting that contact semiochemicals may play a role here, and chemical analyses of *Kheper* species have identified several long-chain hydrocarbons that might have a semiochemical function. Machrochelid mites also appear to use cues emanating from the cuticle to recognize their hosts, suggesting that cuticular compounds have some species specificity.

While pheromone dispersal may be characteristic of telecoprids contact semiochemicals might be a more important mode of communication for tunnelling species. As we will see in Chapter 8 of this volume, male onthophagines cooperate with females to provide significant parental provisioning, but they suffer a high risk of being cuckolded by sneaking males that invade their breeding tunnels. Semiochemicals may provide the necessary cues for males to recognize their mates and to allocate provisions based on their confidence of paternity.

Although the biparentally caring *Nicrophorus* burying beetles (Silphidae) are not closely related to the dung beetles, CHCs within them are individually distinct. They provide males with cues regarding whether a given female is a familiar versus a novel mate (Steiger *et al.*, 2008) and also to her breeding state (Steiger *et al.*, 2007). Insect cuticular hydrocarbons can also provide cues to male dominance (Thomas & Simmons, 2009) and to the risk and intensity of sperm competition (Thomas, 2010). Cuticular hydrocarbons have never been studied in dung beetles, but we predict that they will prove to be a rich area for future research.

6

Explaining Phenotypic Diversity: The Conditional Strategy and Threshold Trait Expression

Joseph Tomkins[1] and Wade Hazel[2]

[1]*Centre for Evolutionary Biology, School of Animal Biology, University of Western Australia, Crawley, Western Australia*
[2]*Department of Biology, DePauw University, Greencastle, IN, USA*

6.1 Introduction

Variation in phenotypic expression accounts for intraspecific diversity in an array of traits and behaviours across the animal kingdom, including caste, trophic, seasonal, dispersal and predator-induced polyphenisms, sex ratio investment (all reviewed in West-Eberhard, 2003; Roff, 1996), adaptations to sperm competition (Parker, 1990b) and the often spectacular alternative size-dependent reproductive phenotypes of males (Oliveira *et al.*, 2008). For nearly a century, the existence of such discrete phenotypic variation in natural populations has been of interest to ecological geneticists because it suggests a balance of selective forces at work (Sheppard, 1975). Specifically, it raises the intriguing question of why, over evolutionary time, one alternative phenotype does not drive the other extinct.

Most of the classical ecological genetic studies of variation in nature – the coexistence of discrete differences in shell banding and colour in snails (Cain & Sheppard, 1954) and different mimetic forms characteristic of Batesian mimicry in butterflies (Clark & Sheppard, 1960) – have been aimed at understanding the kinds of selective forces that maintain discrete variation. Most of these studies concerned phenotypic differences that are due to the relatively straightforward expression of simple genetic differences.

In the framework of evolutionary game theory, phenotypic differences that are linked to one or a few loci, and inherited in accordance with Mendelian ratios, have been termed alternative strategies (Gross, 1996; Maynard Smith, 1982). In such cases, negative frequency-dependent selection, where rare phenotypes have higher fitness than common ones, lies behind the maintenance of the phenotypic diversity seen within the population (Gross, 1996; Maynard Smith, 1982). For example, it

Ecology and Evolution of Dung Beetles, First Edition. Edited by Leigh W. Simmons and T. James Ridsdill-Smith. Published 2011 by Blackwell Publishing Ltd.

has been shown recently that the survival of male colour morphs in wild populations of guppy (*Poecilia reticulata*) is negatively frequency-dependent (Olendorf *et al.*, 2006), the selective agent being a predatory fish.

There is a second source of discrete phenotypic variation that arises in large part from environmentally-cued differences in morphology and behaviour (Sheppard, 1975), and this source of phenotypic variation has received significant attention from ecologists and evolutionary biologists only recently. In evolutionary game theory, such strategies are termed conditional strategies. The word 'conditional' is used not because the strategy depends on the organism's condition (although this is often true), but because the strategy contains a conditional clause such as 'if large, do X; if small, do Y'.

Because one genotype is capable of producing different phenotypes depending on the environment, the conditional strategy is a type of phenotypic plasticity (Bradshaw, 1965). However, unlike the many cases where phenotypic variation in response to environmental conditions is depicted as a continuous reaction norm of phenotypes (West-Eberhard, 2003), in the conditional strategy it is discrete phenotypic states that are triggered in response to variation in the environment (Tomkins & Hazel, 2007). One of the shortcomings of the game theoretical approach is the neglect of genetic realism in the modelling of how selection operates on phenotypes, and this has been particularly apparent in some treatments of the conditional strategy (Tomkins & Hazel, 2007).

One way to introduce a genetic framework to conditional strategies is to model them as environmentally-cued threshold traits, where what varies among individuals is the point in the continuum of environmental variation where development switches from one alternative to another (i.e. the switchpoint). It is this idea that is the basis for the environmental threshold (ET) model for conditional strategies.

The example, 'if large, do X; if small, do Y,' can be used to illustrate the idea of a switchpoint and a threshold. Since 'large' and 'small' have to be delimited, the individual's decision rule becomes, 'if larger than *s*, do X; if smaller than *s*, do Y', where *s* is the body size 'switchpoint' or threshold that must be exceeded for phenotype X to be expressed. Where individuals vary genetically in switchpoint, selection can shift the average switchpoint in a population, depending on the fitness of the alternative phenotypes X and Y (more of which, below).

Under the conditional strategy, the fitness of each phenotype can be plotted as a function of an environmental cue (Figure 6.1). That fitness, switchpoints, and cues can all be scaled together makes it possible to model the effect of selection on switchpoints. The fitness functions of the alternative phenotypes have different slopes and elevations, and therefore intersect, and the ET model seeks an evolutionarily stable switchpoint for the population.

The conditional strategy has been widely invoked in insects to explain a diversity of morphological and behavioural phenotypes, including alternative morphologies such as dispersal or resident polyphenisms (Roff, 1986), seasonal polyphenisms (Shapiro, 1976), alternative phenotypes associated with defence (Hazel, 2002) and alternative reproductive behaviours such as sneaking matings, guarding females and the provision of parental care (Emlen, 1997a; Alcock, 1979; 1996; Hunt & Simmons, 1998a; Moczek, 1999).

For example, in many species of dung beetles, males exhibit what appear to be discrete alternative morphologies, most notably in the length and shape of the head

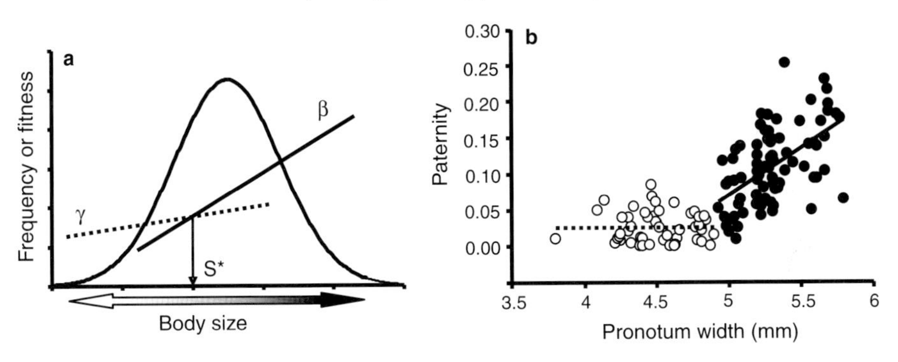

Fig. 6.1 Hypothetical and empirical examples of fitness functions underlying the conditional expression of status/size dependent alternative phenotypes. **a** Hypothetical fitness functions for the alternative phenotypes γ and β. The fitness of the phenotypes changes at different rates in relation to an environmental gradient or cue. In the case of horn polyphenism in dung beetles, this gradient arises from the variation in body size of the beetle (i.e. status), which is in turn determined by the amount of dung provisioned in the brood mass. The tone of the x-axis arrow indicates how the phenotypes express change with changes in body size. The S^* is the point at which selection for an increase in switchpoint is balanced by selection for a decrease in switchpoint; S^* does not always fall where the fitness functions intersect. **b** Status-dependent fitness functions for minor (open symbols) and major (closed symbols) male *Onthophagus taurus* (Hunt and Simmons, 2001). Paternity was assessed by using the sterile male technique and using focal irradiated males among a background of fertile males as well as focal fertile males against a background of sterile males. Data provided by J. Hunt.

and pronotal horns (Chapters 1 and 7), but also in the behaviours associated with these alternative morphologies (Chapters 4 and 8). Larger males, known a majors, tend to develop horns; these males typically attempt to guard females with which they mate, while smaller males, known as minors, develop smaller horns or no horn at all and attempt to sneak copulations with females (Emlen, 1997a; Hunt *et al.*, 1999; Hunt & Simmons, 1998a; 2002a; Moczek, 1999). Whether a dung beetle is a major or minor is usually conditional on its body size, which confers status. Therefore, variation in horn length is often described as a conditional strategy with status-dependent alternative tactics of being a major or a minor.

In many of the studies of dung beetle horns, the environmental threshold model has been explicitly (and sometimes implicitly) assumed as the genetic basis for the production of these alternative phenotypes. Here we take advantage of the diversity of patterns of dimorphism, trimorphism and behavioural plasticity, and the increasing interest in dung beetles as model species for evolutionary and ecological research to review the usefulness of the environmental threshold model as a quantitative genetic framework for understanding conditional strategies.

6.2 The environmental threshold model

The environmental threshold model owes it origins to Wright (1920), who sought to understand all-or-none phenotypes that were not explicable using a simple

Mendelian model. Subsequent papers by Dempster & Lerner (1950), Falconer (1965) and Roff (1996) built on Wright's original work. The idea continues to be an important tool for the analysis of all-or-none traits in animal breeding and agriculture (David *et al.*, 2009).

The model assumes an underlying, normally distributed heritable phenotypic variable, 'liability' (Falconer & Mackay, 1996). The expression of the all-or-none alternative phenotypes depends on whether or not an individual's liability exceeds some fixed threshold value, such that individuals with liabilities less than the threshold always produce phenotype Y, while those with liabilities greater than the threshold experience a developmental reprogramming event (see Chapter 7) and produce phenotype X (Falconer & Mackay, 1996). Evolutionary changes in liability relative to the fixed threshold alter the frequencies of the alternative phenotypes.

The liability model is easily adapted to conditional strategies by making liabilities equivalent to switchpoints. As a result, an individual will express one or the other alternative phenotype, depending, for example, on whether their switchpoint is greater or less than some cue associated with their internal or external environment, condition or status.

The environmental threshold model (ET) was originally proposed as the genetic basis of the environmentally-cued production of cryptic green and brown pupae in butterflies (Hazel, 1977; Hazel & West, 1982), and was formalized mathematically in 1990 and 2004 (Hazel *et al.*, 1990; 2004). In the ET model, the proportion of individuals that express each alternative phenotype depends on both the distribution of switchpoints and the distribution of cues experienced by the population (Tomkins & Hazel, 2007).

In the short term, the frequencies of alternative phenotypes can change solely as a result of changes in cue distribution. For example, the frequencies of major and minor dung beetles can vary due to dietary effects on body size (Emlen, 1994; Hunt & Simmons, 1997; Moczek & Emlen, 1999; Tomkins, 1999; Aubin-Horth & Dodson, 2004 – see also Chapter 7 in this volume), ecological conditions (Emlen, 1997b) or changing social conditions (Forsyth & Alcock, 1990). However, evolutionary changes in the frequencies of alternative phenotypes can only arise through changes in the distribution of switchpoints (i.e. its mean relative to the cue, its variance, or both) (Emlen, 1996; Moczek & Nijhout, 2003; Moczek & Nijhout, 2004; Tomkins & Brown, 2004; Unrug *et al.*, 2004; Aubin-Horth & Dodson, 2004).

6.2.1 Does the development of a horn dimorphism in male dung beetles occur in a manner consistent with the assumptions of the ET model?

The most obvious threshold trait in many species of dung beetle is the dimorphism seen in male horns (Emlen *et al.*, 2005a; see Figure 1.2d in Chapter 1). Here we use the extensive research that has been carried out on these traits as the basis for our discussion of the ET model.

Phenotype frequency and the distribution of cues

Under the environmental threshold model, environmental cues relative to switch-point determine the developmental pathway adopted and the phenotype expressed. The horn dimorphism in dung beetles like *Onthophagus taurus* illustrates the environmental dependence of horn length on an environmental 'cue', but the illustration is not without its complexities. Clearly horn length is tightly related to body size, and body size (or something strongly correlated with body size) can therefore be thought of as the 'cue' that individual beetles use.

At first, perhaps, body size does not seem to be environmental. In reality, however, body size is largely determined by the nutritional environment provided to the larval dung beetle in the form of the brood ball. This is evidenced by the typically low levels of additive genetic variance for body size in dung beetles (Moczek & Emlen, 1999; Kotiaho *et al.*, 2003) and experimental manipulations which show that males who develop in large brood balls tend to be large and bear large horns, while those that develop in small brood balls tend to be small, with small horns (Hunt & Simmons, 1997; Emlen, 1997b; and see Chapter 7 in this volume for a detailed discussion on nutrient-mediated phenotypic plasticity in these beetles). Therefore, short-term changes in the distribution of the environmental cue (brood mass and its consequence for body size) will affect the frequency of tactics in the population (Hunt & Simmons, 1997). Thus, we would expect temporal and spatial variation in morph frequencies simply due to variation in the amount of dung beetle females' provision to their offspring.

Conditional fitness of alternative phenotypes

For the production of alternative phenotypes to be evolutionarily stable under the conditional strategy and the ET model, the fitness functions of the alternative phenotypes must intersect at some point in the range of cue values experienced by the population. For example, if there is a cue that triggers the switch between the alternative morphological and/or behavioural phenotypes, then there is some cue value above which one tactic has greater fitness, and below which the other tactic has greater fitness (Figure 6.1a; Gross, 1996; Hazel *et al.*, 1990). In dung beetles, the prediction is that minor males and major males will show these intersecting fitness functions in relation to changes in body size.

This prediction has been met in a study of *O. taurus* (Hunt & Simmons, 2001; Figure 6.1b). The experiment used irradiation as a means of identifying an individual male's paternity relative to others in the population. In a satisfying confirmation of theory, the fitness functions of major and minor males behaved much as expected, with the fitness functions of the smaller minor and larger major males having different slopes and intercepts, suggesting that small majors would have lower fitness than equivalently small minors, and that larger majors would have greater fitness than equivalently large minors. The greater fitness in major males was also a function of horn length.

This study is one of the few examples where the fitness of alternative phenotypes has been quantified, and we discuss this example further below. Perhaps the only shortcoming of the experiment is that it was not possible to directly compare the fitness of major and minor males of the same size, because it is difficult to coerce small males to express horns and large males to suppress them.

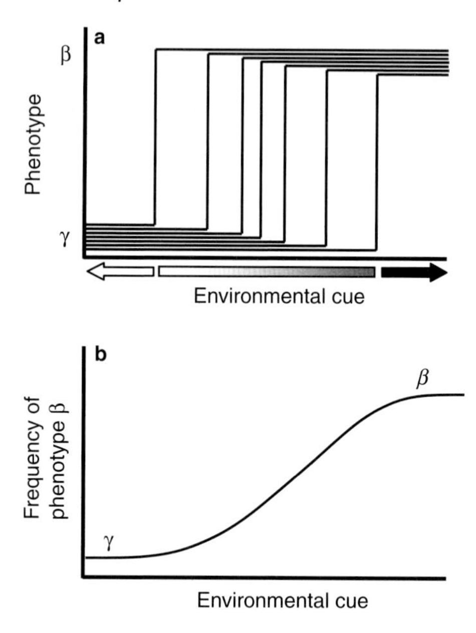

Fig. 6.2 Environmental thresholds of genotypes and populations. **a** The reaction norms of seven genotypes that switch from the gamma to beta phenotype in response to an environmental cue. Each genotype has a different sensitivity to the environmental cue, reflecting genetic variation in switchpoint. **b** The sum effect of individual variation in switchpoint is a cumulative normal response curve that is symmetrical about the mean switchpoint of the population, which is estimated as the cue value at which the two phenotypes are produced with equal probability.

Genetic variation in the switchpoint distribution

For switchpoints to evolve under the ET model, variation in switchpoints must be heritable. In other words, with respect to their response to the environmental cue, some genotypes should more readily switch to producing horns than others (Figure 6.2a). The cumulative effect of switchpoint variation is a logistic shaped population reaction norm of morph frequency on the environmental cue (e.g. body size) (Figure 6.2b).

Perhaps the best intrapopulational evidence for genetic variation in switchpoints in a dung beetle species comes from Emlen's (1996) artificial selection experiment in *O. accuminatus*. In this experiment, male beetles with either relatively long horns or relatively short horns, given their body size, were selected in a replicated experimental design. The result was a divergence in the predicted directions, with lines selected for relatively long horns shifting in switchpoint to produce horns at a smaller body size. This divergence demonstrated significant amounts of genetic variation for the switchpoint in the population – a key premise of the ET model.

Further evidence comes from studies that have shown populational divergence in the average switchpoint. Populations of *O. taurus* have been introduced from Europe to both Western Australia and North America (North Carolina). These exotic populations show divergent switchpoints, both when collected from the field and when the populations have been reared in common garden conditions for two generations, providing strong evidence that the difference in average switchpoints between the populations is genetic (Moczek *et al.*, 2002). The switchpoint of the North Carolina population is at a smaller body size, such that there is a greater proportion of the body size distribution over which major males are expressed (see Figure 7.2b).

Under the general terms of the conditional strategy and the quantitative genetic framework provided by the ET model, populations are expected to vary in a predictable way in response to a factor that influences the fitness difference between the two male morphs (Tomkins *et al.*, in review). Moczek (2003) proposed that differences in population density have been the primary driver of switchpoint divergence, and that higher densities favour the minor male tactic. In support of the density hypothesis, differences in density do co-vary with the switchpoint mean across the North American and Western Australian populations (Moczek, 2003), and the difference in mean switchpoint alters the ratio of male phenotypes in the population (Moczek, 2003).

Density has both proximate and ultimate effects on switchpoint in other arthropod species. For example, mites reared at high density yield fewer males of the 'fighter' morph (Tomkins *et al.*, 2004a) and earwig populations have evolved lower average switchpoints such that more long-forceped 'macrolabic' males are produced in populations with a history of high density (Tomkins & Brown, 2004). Thus there is evidence from dung beetles and other arthropods that intra-population genetic variation exists in switchpoints, and that switchpoints evolve to produce different frequencies of alternative male morphs through evolutionary changes in the switchpoint distribution.

Conceptualizing the interaction between the environment and the genotype

Understanding the evolution of environmentally cued threshold traits revolves around the interaction between genes and the environment. In the horn length dimorphisms of dung beetles, there is clearly a large environmental component that determines male size. However, there are genetic effects on body size that are contributed indirectly through the parents and the environment that they provide for their offspring. For example, genetic variation among females in the size of the brood masses they provision, and genetic variation in the propensity of males to either help provision or influence female provisioning through their ejaculates (Hunt & Simmons, 2000; 2002a; Kotiaho *et al.*, 2003) are all indirect genetic effects on the body size of male offspring, and therefore the morph that they adopt as adults (Chapter 8).

Does it matter for the functioning of the ET model that these genes influence male size? We think not: if there is a set of variable genes in the parents or offspring that influence body size, and if these genes are recombining every generation, then, from the point of view of the genes influencing switchpoint, the body size in which they find themselves is unpredictable. The only way to adaptively match horn size with body size is to have horn development respond to cues that correlate with body size (Emlen & Allen, 2004). The punctuated nature of insect growth makes the response to body size with altered horn growth possible.

Another complicating factor in understanding the interaction between genes and their environment in threshold traits is the existence of multiple cues. The horn dimorphism in dung beetles is sensitive primarily to the amount of dung available to the larva, which determines the size of the beetle (Emlen, 1997b; Hunt & Simmons, 1997; Moczek & Emlen, 1999). However, the quality of the dung also influences horn expression. Beetle larvae reared on dung with a higher nutritional value tend to produce adults with a larger mean body size and produce long horns at larger body sizes than do larvae that have been reared on lower quality dung (Emlen, 1997b; House & Simmons, 2007; Moczek & Emlen, 1999). These differences suggest that an increase in dung quality in the diet acts in opposition to the body size cue for the production of horns. In the context of secondary sexual traits that function to signal male condition, we would expect more exaggeration in the group reared on a higher quality diet but, in the case of dung beetles, the horns are less exaggerated.

The effect of dung quality on horn expression is consistent with the hypothesis that negative frequency-dependent selection on the alternative phenotypes positions the distribution of switchpoints, such that the ratio of morphs in the population remains constant when mean body size varies in response to diet quality (Emlen, 1997b). However, the environmental threshold model can explain the effect of dung quality without invoking frequency dependence. If relative body size (in contrast to absolute body size) is most important in determining alternative phenotype fitness, then the phenotype fitness functions will track changes in body size; variation in dung quality would be a second predictive cue for such shifts.

The existence of multiple cues, such as diet quality and quantity, can be easily included in the ET model and can be visualized as an n-dimensional surface relating the probability of morph expression to n-1 environmental cues (Figures 6.3a and b).

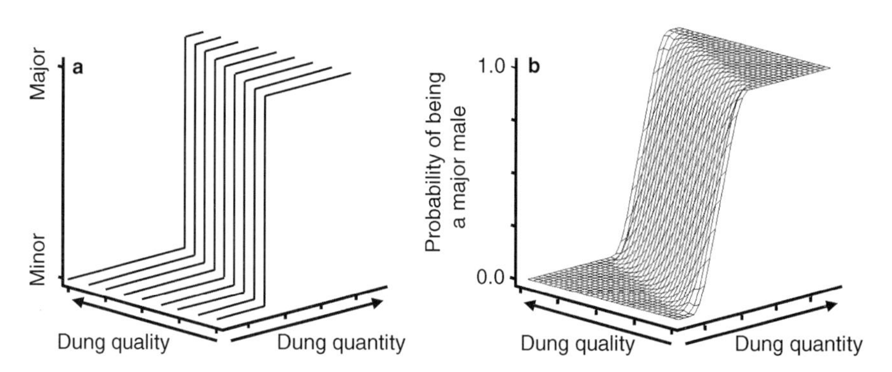

Fig. 6.3 The effect of two environmental cues on a threshold trait. The amount of dung provisioned in a brood mass affects body size and the probability of a male dung beetle becoming a major. The switchpoint in this relationship is affected by the quality of the dung. When dung is of high quality, all individuals in the population are larger and the body size at the switchpoint is greater. **a** Response of a single genotype to variation in nine different levels of dung quality and a range of dung quantities. Individuals either adopt the minor or major phenotype. **b** The population can be seen to respond to the two cues as a smooth curve that passes through the of the environmental cue space; this compares with Figure 2b, where there is only one cue.

Moczek (2002), however, has suggested a non-adaptive developmental explanation for the shifts in threshold associated with changes in diet quality.

Questions about development: threshold traits versus positive allometry

The scaling of organs in relation to body size, the study of allometry, provides a useful way of understanding the relative investment made by individuals of different size. Where organs scale such that large and small individuals invest identically, the organ in question is said to scale isometrically. Frequently, organs scale such that larger individuals invest relatively less in their organs, and such 'negative allometry' is the case for brain size in mammals (Martin, 1981).

The contrasting situation is 'positive allometry', where larger individuals invest relatively more than smaller ones – a pattern particularly common in secondary sexual traits (Bonduriansky, 2007; Bonduriansky & Day, 2003; Kodric-Brown *et al.*, 2006; Tomkins *et al.*, 2010). When plotted against body size and viewed on a linear scale, positively allometric traits are characterized by an upward curve. The greater the allometric exponent (i.e. the slope), the more concave the curve, the lower the intercept of the slope and the longer the curve appears to remain flat before it curves upward (Tomkins *et al.*, 2005; 2006; Tomkins & Moczek, 2009).

Tomkins *et al.* (2005; 2006) proposed that the sigmoid shape of the horn length/body size curve for *O. taurus* was simply a very steep allometric curve in which the physiological limits of horn production created an asymptote, with the result being the strongly bimodal distribution of horn length. In support of their hypothesis, the horn allometries of minor males of *O. taurus* and *O. binodis* (males typically described as hornless) are very steep (Tomkins *et al.*, 2005; 2006), a pattern that is inconsistent with minors being either hornless or developmentally feminized.

In contrast the emergent phenotypes of classical threshold traits occur as two discrete phenotypic states. For example, Wright (1920) proposed the concept of threshold traits as a way to explain the inheritance of piebald patterning, an all-or-none trait in guinea pigs. Although beetle horns appear as two relatively distinct classes, horn length within each class tends to be related to body size and it does seem that, in some species, male horn lengths can be explained as a curvilinear reaction norm of horn length on body size (Tomkins *et al.*, 2005; 2006; Tomkins & Moczek, 2009).

The existence of two morphs that are generated by positive allometry rather than by a sudden developmental reprogramming event is an example of different developmental pathways converging on a single phenotypic solution, the solution being the production of bimodal horn lengths with the modes associated with different body sizes. The environmental threshold model for beetle horn variation is built around the notion that there is a distribution of switchpoints in the population that triggers the developmental reprogramming necessary for horn growth, depending on the strength of some cue (or cues) associated with ultimate body size (see also Chapter 7). If male morphs arise as the consequence of positive allometry, what does this mean for the utility of the ET model?

One of the key differences between the positive allometry model and the threshold model is that, in the latter case, selection can alter switchpoints only where there is variance in the expression of the alternative phenotypes. This contrasts with the positive allometry model, where horn length is a consequence of a continuous reaction norm. In this case, selection anywhere along the reaction norm can have an influence (Tomkins *et al.*, 2005; 2006 Tomkins & Moczek, 2009).

This is a major difference between the two developmental mechanisms by which the different modes in the distribution of horn lengths could be produced, and it has the potential to alter much about their evolution. Under positive allometry, evolutionary change in either the intercept or in the allometric exponent can lead to changes in the frequency of the alternative phenotypes (Tomkins *et al.*, 2005). It seems logical, therefore, that in species where horn length variation is due to positive allometry, selection on phenotype frequencies could be usefully estimated as the sum of selection on the residual variation around the allometric reaction-norm – something we explore further below. The notion is that selection favouring relatively long horns would shift the frequency of alternatives phenotypes towards an increase in major males, while selection favouring relatively short horns would shift the allometry in the opposite direction, generating more minor males.

In support of this idea is the response to selection achieved by Emlen (1996), who, by selecting on relative horn length in *O. acuminatus*, was able to shift relative horn length in opposite directions. Therefore, the distribution of residual horn lengths might respond to selection in a manner similar to the distribution of switchpoints in the ET model. However, detailed mathematical treatment of this possibility is required to put the allometric model in the same context of the current ET model.

Trimorphism – a simple model

Recently, Rowland & Emlen (2009) reported evidence for trimorphic variation in a number of insect taxa, including dung beetles. These trimorphisms are expressed sequentially (gamma, beta and alpha) with increasing body size. At the transitions from one morph to the next, there is a range of body size where both morphs are produced. The transitions are hypothesized to reflect changes in the size-dependent fitness of the three phenotypes (Figure 6.4a). If these trimorphisms are, indeed, the outcome of threshold inheritance cued by body size, then their expression must rely on two distributions of switchpoints in order for the three morphs to be produced (Rowland & Emlen, 2009).

We explored the switchpoint variation in *Phaneaus triangularis* by coding the data for the morphs (alpha, beta and gamma) as dichotomous variables. We then used a logistic regression to estimate the probability distribution of switching from gamma to beta and from beta to alpha males. These probability distributions can be used (see below) to estimate the switchpoint distributions for the morphs and where they lie in relation to the distribution of body sizes (Figure 6.4b). If the cues to switch from gamma to beta are the same (i.e. size) as those influencing the switch from beta to alpha, then it is possible that the neuroendocrine architecture underlying each switchpoint distribution would be similar, and that the genetic correlation between the probability of switching from gamma to beta and from beta

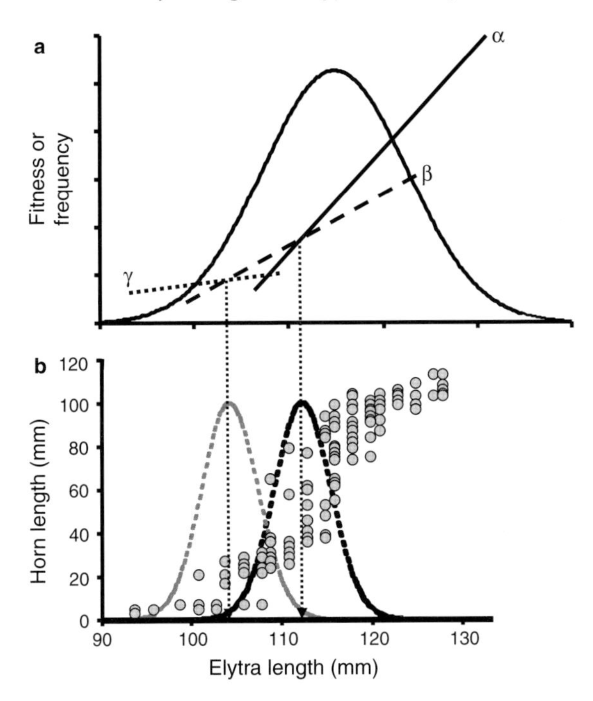

Fig. 6.4 Understanding trimorphic variation. **a** Hypothetical fitness functions of the trimorphic dung beetle *Phanaeus triangularis*, superimposed onto the distribution of body sizes for the data reported in Rowland and Emlen (2009). **b** Scatterplot of the trimorphic variation in *P. triangularis*, alongside the distributions of phenotypic variation in switchpoints for the change from gamma to beta phenotype and from beta to alpha.

to alpha would be high. This would allow the switchpoint distributions to overlap as they do, because no genotypes would be more sensitive to the second switchpoint than they were to the first (potentially causing a reversal in the sequence of morph expression).

However, if the switch mechanisms are genetically correlated, the frequencies of the three morphs in the populations will be constrained, since an evolutionary shift in the beta to alpha switchpoint distribution (e.g. due to an increase in alpha fitness relative to beta fitness) will affect the gamma to beta switchpoint distribution and increase the frequencies of beta males at the expense of alpha males. This would make most Evolutionarily Stable Strategy (ESS) frequencies impossible to attain.

Therefore, it seems more likely to us that if the three morphs are produced by two underlying switchpoint distributions, the two switchpoint distributions are probably uncorrelated. This would mean that changes in the means and variances of both distributions will be able to produce the morph frequencies expected from theory. Finally, the theories (both ET and ESS) require that at some range of intermediate body size, beta fitness is greater than the fitness of both alpha and gamma males. This range of body sizes is quite small indeed (Figure 6.4a).

The evolutionary stability of threshold trimorphisms would be an interesting topic for further research.

6.3 Applying the threshold model

6.3.1 Predicting the mean switchpoint of a population

Using standard quantitative genetics techniques, the ET model can be used to identify the mean switchpoint that would yield an overall selection differential of zero. Therefore, if the observed mean switchpoint (i.e. the body size at which the probability of each alternative phenotype being produced is 50 per cent) is similar to this expected value, it is likely that the mean switchpoint is being stabilized close to its equilibrium value by selection.

It is easiest to understand how the selection differential on variation in switchpoint is calculated, by imagining there are only two body sizes, large and small, occurring with frequencies p and q. In the small individuals, selection favours switchpoints that result in hornless beetles; in the large individuals, selection favours switchpoints that result in horn growth. Hence, selection on variation in switchpoint in the two sizes is in opposition, so the overall direction and magnitude of selection on switchpoint depends on the frequencies of large and small individuals and the fitnesses of horned and hornless beetles of the two body sizes.

To quantify the overall selection differential, we would sum the weighted selection differentials in each of the two body sizes, with the weights being the frequency of those body sizes. In contrast, when body size varies continuously, the direction and intensity of selection at each body size is estimated by the fitness functions for the alternative phenotypes over all body sizes. Because the fitness functions, cue values and switchpoints are continuous, integration is used for the weighting. Now, the overall selection differential, $S(\mu)$, on switchpoint is:

$$S(\mu) = \left(\frac{\sigma}{D(\mu)}\right) \int_{-\infty}^{\infty} [w_X(t) - w_Y(t)]g(t) \quad f(t, \mu)dt \tag{6.1}$$

with fitness functions $w_X(t)$ and $w_Y(t)$ for the alternative phenotypes X and Y, body size distribution $g(t)$ and switchpoint distribution $f(t)$ with variance σ^2 and mean μ. The parameter D normalizes the selection differential (Hazel *et al.*, 1990) and is always positive (see Hazel & Smock, 1993 for the derivation).

The fitness functions can be estimated by regressing phenotype fitness on body size. Ideally, the frequency distribution of body sizes, $g(t)$ should be estimated in the field. The variance (σ^2) and mean (μ) of the distribution of switchpoints are estimated by squaring the standard deviation in switchpoints. Logistic regression of alternative phenotypes (horned or hornless) on body size can be used to estimate the observed switchpoint mean (the body size at which the probability of the alternative phenotypes being produced is 50 per cent). The standard deviation in switchpoint is the body size that lies between where the probability of the alternative phenotypes is 50 per cent and body size where the probability is 84 per cent or 16 per cent. Using all of these estimates (except the observed switchpoint mean), one can estimate the mean switchpoint that would result in a selection differential of zero by plotting $S(\mu)$ for different values of the mean switchpoint (μ) (Figure 6.5a).

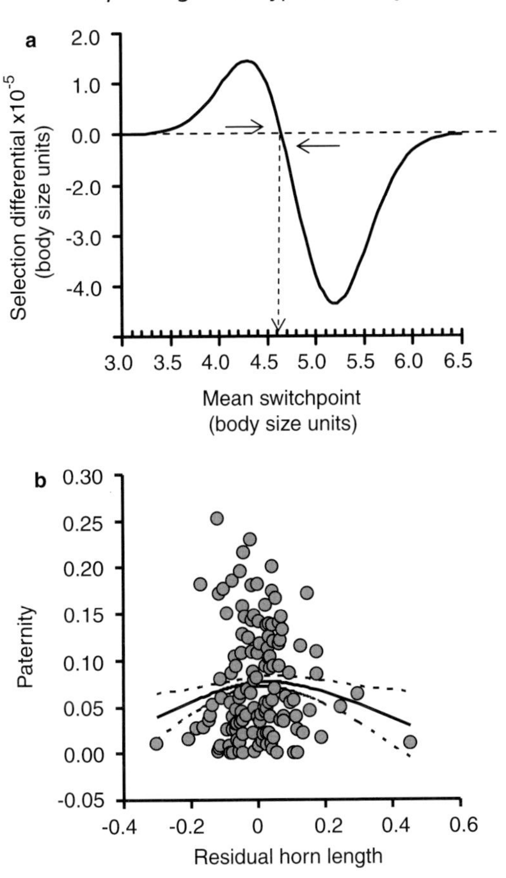

Fig. 6.5 Estimating selection on threshold traits. **a** Graphical solution to the equilibrium mean switchpoint, using the distribution of switchpoints of *O. taurus* taken from the literature (Moczek, 2003) and an experimental estimate of the fitness of the alternative phenotypes from the experiment of Hunt and Simmons (2001). The solid arrows indicate the direction of shift in the mean phenotype switchpoint towards a stable equilibrium (dashed arrow), where the selection differential equals zero. Using 4.84 mm as the mean body size, the selection differential will be zero when the switchpoint mean is 4.66 mm (the observed was 4.92 mm). **b** If we interpret the horn polyphenism in *O. taurus* as a continuous allometry, we can estimate the selection acting on relative horn length using the same data from Hunt and Simmons (2001) experiment. This analysis suggests that selection was weakly stabilizing.

This is the value of mean switchpoint to which the population in question should evolve, provided variation in switchpoint is heritable.

6.3.2 Estimating the selection on thresholds using the ET model

With respect to dung beetle morphs, the selection differential is a function of the differences in slopes of the fitness functions for the two male morphs (i.e. the slopes of fitness plotted on body size for each morph), the frequency distribution of body size (the assumed environmental cue) and the variance in the distribution of switchpoints in the population (from the logistic curve of morph frequency on body size). If the switchpoint distribution of a population is near its evolutionary equilibrium, then, when the observed variables are used in the above formula, the selection differential ($S(\mu)$) on the switchpoint should be close to zero.

Only one study has generated data on the fitness of alternative male phenotypes in male dung beetles that can lend itself to analysis with the ET model. Hunt & Simmons (2001) generated fitness functions for the male morphs of *O. taurus* (Figure 6.1b). The functions describe the change in fitness in relation to body size; these regression equations represent $w_x(t)$ (y = 0.039 ± 0.003 × pronotum width for minor males) and $w_Y(t)$ (y = − 0.586 + 0.131 × pronotum width for major males) in the ET model. The other parameters needed in order to calculate selection that was operating on the switchpoint in their experiment are the distribution of male body sizes ($g(t)$) (all males mean = 4.84 mm, SD = 0.43 mm; focal males mean = 4.97 mm, SD = 0.44 mm) and the mean (estimated as 4.92 mm) and variance (σ^2) of the switchpoint distribution $f(t)$.

In the experiment of Hunt & Simmons (2001), there was no variance in the distribution of switchpoints, since the body size classes of minors and majors as coded by the authors did not overlap. However, we were able to estimate the average switchpoint variance from four Western Australian field populations collected by Moczek (2003). To do so, we calculated the mean of the switchpoint distribution by coding morphs as 0 (minors with a horn less than 1 mm) and 1 (majors with a horn longer than 1 mm) and used the predicted probabilities of being a major from a logistic regression of morph on body size to estimate the mean switchpoint as the body size at which 50 per cent of the males are predicted to be majors. We estimated the standard deviation in switchpoint as the difference between the body size at which 16 per cent of males were predicted to be majors and the body size at which 50 per cent of the males are predicted to be majors (i.e. our estimate of the mean switchpoint).

The four populations collected by Moczek (2003) have an average switchpoint SD of 0.086 mm. Using this estimate for the switchpoint SD to obtain the switchpoint variance, and data from the Hunt & Simmons study to estimate the fitness functions and mean and variance in body size distribution, we used the formula for the selection differential (equation (6.1) above) to estimate the mean body size switchpoint to which the population should evolve (i.e. the switchpoint at which the selection differential will be zero). This analysis indicates that when the switchpoint mean is 4.66 mm, the selection differential on switchpoint variation will be zero (Figure 6.5a). This compares with the observed phenotypic switchpoint of 4.92 mm. There was slight selection towards a lower value for the average switchpoint under the experimental conditions, so, therefore, lines maintained under these conditions might be expected to evolve towards having a higher frequency of major males.

6.3.3 Estimating selection under positive allometry

The lack of switchpoint variance is an unfortunate characteristic of the Eberhard & Gutiérrez (1991) model for the identification of male phenotypes. However, the switchpoint variance can be derived using the Kotiaho & Tomkins (2001) method in combination with logistic regression analysis. This method splits the sample into putative morphs on the basis of the length of the dimorphic character, and a logistic curve against body size can be generated (Tomkins & Brown, 2004). The recent modelling of Knell (2009a) has also increased the statistical repertoire available to researchers by estimating the switchpoint and the variance around the switchpoint, and it is essential reading for the analysis of these traits.

The current debate over the mechanism of morph determination (threshold or positive allometry) (Tomkins *et al.*, 2005; 2006; Tomkins & Moczek, 2009) feeds into the question of how switchpoints are analysed. Only where there is developmental reprogramming (Chapter 7) can the ET model apply in its current form. Where a continuous reaction-norm underlies the expression of alternative phenotypes, the discrimination of alternative morphologies becomes irrelevant and the modality in horn length is simply a consequence of the shape of the allometric curve.

Although the details of how selection acts to change the proportions of alternative phenotypes that arise from reaction norm polyphenisms have been hypothesized, and tested in one case (Tomkins & Moczek, 2009), it requires further detailed experimental investigation. Interestingly, in those cases where alternative phenotypes seem rather clearly to be due to an underlying switchpoint distribution, the estimated variance in switchpoints is large relative to the variation in body size (Table 6.1). However, in those cases where a distinct separation of alternative phenotypes is less obvious, possibly because the alternatives are produced via a continuous reaction norm, the variance in the postulated switchpoint distribution is relatively small (Table 6.1). In the latter case, this might be expected simply because selection can operate on the underlying genetic variation much more readily, so reducing genetic variation.

As outlined above, one way in which selection on continuous allometries can yield different proportions of alternative tactics is if relative trait size is under selection. Emlen (1996) demonstrated this with artificial selection of relative horn length in *O. acuminatus* that yielded divergent switchpoints. Hence, directional selection on relatively large horns (i.e. larger than expected for a given body size) can increase the number of major males as the allometry shifts towards larger body sizes, while selection on relatively short horns will have the opposite effect (Emlen, 1996).

Using data from Hunt & Simmons (2001), we calculated the residuals of horn length of all males using a 4th order polynomial curve in the program '*Curve fit*'. We then used these residuals in an analysis of the paternity data from Hunt & Simmons' (2001) experiment. There was no evidence for linear directional selection on residual horn length ($F_{1,138} = 0.005$, $b = 0.004$, $P = 0.9$), but there was a suggestion of stabilizing selection on residual horn length ($F_{2,138} = 1.76, P = 0.176$, $b_{reshorn} = 0.04 + 0.057$, $t = 0.709$, $P = 0.479$, $b^2{}_{reshorn} = -0.464 + 0.247$, $t = 1.875, P = 0.063$). This trend concurs with the shape of the cubic spline (Schluter, 1988) fitted to the data (Figure 6.5b). Clearly, though, these data are only weakly suggestive. We include them only to show how selection on alternative phenotypes generated from continuous allometries might be quantified.

One caveat is that in major males there is likely to be a greater variance in residual horn length simply because horns are longer; hence, major males might be more exposed to selection on relative horn length than minor males. Another point of

Table 6.1 Switchpoint variation as a proportion of the body size distribution for a range of dimorphic arthropods.

Taxon	*Species*	*Source population*	*Reaction norm or threshold: R/T*	*Switchpoint SD as % of body size SD*
Acarid	*Rhizoglyphus echinopus*[a]	Australia (Lab)	T	236.79
Acarid	*Sancassania berlesei*[b]	Germany (Lab)	T	87.67
Opilionid	*Acutisoma proximum*[c]	Brazil (Field)	T	60.22
Dermapteran	*Forficula auricularia*[d]	UK (Bass Rock, Field)	T	90.26
Dermapteran	*F. auricularia*[e]	Poland, (Krakow, Field)	T	90.6
Dermapteran	*F. auricularia*[e]	Australia (Pemberton, Field)	T	104.72
Dermapteran	*Oreasiobia stoliczkae*[f]	India (Field)	T	76.61
Dermapteran	*Timomenus aeris*[f]	Taiwan (Field)	T	130.13
Hymenopteran	*Synagris cornuta*[g]	West Africa (Field)	T	33.51
Coleopteran	*Phanaeus vindex*[h]	USA (Field)	T	25.94
Coleopteran	*Phanaeus triangularis* $\beta\gamma$[h]	USA (Field)	T	41.11
Coleopteran	*Phanaeus triangularis* $\alpha\beta$[h]	USA (Field)	T	30.3
Coleopteran	*Onthophagus binodis*[i]	Australia (Walpole, Lab)	R	25.6
Coleopteran	*Onthophagus taurus*[j]	Australia Pinjarra (Field)	R	35.98
Coleopteran	*O. taurus*[j]	Australia Narrikup (Field)	R	24.84
Coleopteran	*O. taurus*[j]	Australia Harvey (Field)	R	25.95
Coleopteran	*O. taurus*[j]	Australia Witchcliffe (Field)	R	58.05
Coleopteran	*O. taurus*[k]	Australia (Lab F2)	R	18.54

[a]J.L.T. (unpublished),[b]Tomkins *et al.* (2004a),[c]Buzatto (personal communication),[d]Tomkins & Brown (2004),[e]Tomkins & Moczek (2009),[f]Tomkins & Simmons (1996),[g]Robert Longair (personal communication),[h]Rowland & Emlen (2009),[i]Tomkins *et al.* (2005),[j]Moczek (2003),[k]Moczek & Nijhout (2003).

interest is that allometries can change by selection acting in opposite directions in majors and minors. For example, selection for relatively short horns in minors and relatively long horns in majors will steepen the allometry, while selection in the opposite directions will lower the allometry. Such effects can be analysed by looking for an interaction between male morph and residual horn length. This interaction was non-significant in the Hunt & Simmons data set ($P = 0.69$).

6.4 Future directions

In the same way that horn length varies with body size, so there is a correlation between the propensity of males to help females in their provisioning of the brood masses that is associated with the male phenotypes (Emlen, 1997a; Hunt & Simmons, 1998a; Moczek, 1999). Hunt & Simmons' (2000) data revealed how males below a threshold size do not contribute to the provisioning of the brood masses constructed by females with which they are paired. In contrast, males larger than the threshold size evidently assisted females in the provisioning of brood masses, since brood mass weight exceeds that produced by females working alone (see Figure 8.1b in Chapter 8). This behavioural trait might also be investigated using the threshold model. We would expect there to be genetic variation in the distribution of underlying switchpoints for this trait, and intersecting fitness functions for the different paternal care phenotypes.

Whether the same genes control the expression of the parental care behaviour and the expression of the horn morphology has not been addressed in a quantitative way. If the same genes control horn expression and male parental care, genotypes that switch morphology at small body sizes should also be those that provide care at smaller than usual body sizes, and vice versa with respect to males that switch to the major phenotype at a large size. An intriguing experiment would therefore be to select on male help in animals close to the threshold size and look for changes in the mean switchpoint of horn expression.

Mating behaviour has rarely been considered as a threshold trait, but does lend itself to this analysis. First, a number of studies show that male courtship rate is likely to be an environmental cue which females can use in their decision as to whether to mate or not, with females more readily mating with more actively courting males (Kotiaho *et al.*, 2001; Kotiaho, 2002). To date, there has been no quantification of the fitness of females that reject males with low courtship rates, compared with those that accept them. Obviously, all females need to mate, so the fitness cost would be manifest as the cost of waiting for a better male, rather than the zero fitness derived from not mating at all. Given that to accept a mating or not is an environmentally cued switch, the ET model can be applied and the genetic variance in the switchpoint distribution could be quantified and compared between populations and related to variation in courtship rates.

Hunt *et al.* (1999) documented variation in the residence time of different male morphs in artificially created dung pads. Males leave a dung pad when conditions become unfavourable, and environmental variation in such things as the age of the dung, sex ratio and density are all likely cues that contribute to leaving. For an understanding of the ultimate mechanisms involved in pad residence, an optimality approach could be undertaken. Combining this with an approach that models pad residence as a threshold trait would shed light on the degree to which genetic constraints influence leaving.

Before we can properly understand the threshold mechanism in species with three morphs, the underlying developmental and genetic architecture needs to be more fully understood. One of the first goals should be to determine whether the dual distributions of switchpoints that are assumed to underlie the three morphs are genetically correlated. This is important, since it has implications for whether trimorphic species can evolve to an evolutionary stable frequency. The simplest approach for these species would be a split family design, in which offspring were reared in conditions that either result in them being close in body size to the average gamma/beta switchpoint or close to the average beta/alpha switchpoint. Correlations among families in the proportions of the morphs would be consistent with the switchpoints being correlated. Given that the most likely scenario is no genetic correlation, large sample sizes would be needed to accept the null hypothesis with confidence.

How selection operates to change the ratio of male morphs in a population by altering the mean switchpoint is an area that has new importance. Where there is a reprogramming event that changes the developmental trajectory of individuals that cross a threshold, the ET model describes how this process works. What is less clear is how selection operates to change the phenotype frequencies in species where positive allometry accounts for the phenotypic alternatives. How we quantify such selection and predict responses to it requires modelling effort. Emlen's (1996) artificial selection experiment resulted in a change in the mean switchpoint by selecting for relatively long or relatively short horns across all males. This pattern is expected to shift the continuous reaction norm of horn length on body size, while selecting only on individuals that switch at small or large body sizes is the most powerful way of shifting the position of a true threshold trait.

Separating the predictions of the positive allometry hypothesis from those of the developmental reprogramming hypothesis for polyphenic horn expression is rather difficult, if positive allometry in the horns of minor males is not enough. One difference is that the degree of developmental decoupling between alternative phenotypes should be much greater where reprogramming occurs than where alternative phenotypes are the extremes of a positive allometric continuum (Tomkins & Moczek, 2009).

One way of looking at this question is the genetic correlation between morphs for traits like relative horn length. For species in which the positive allometry hypothesis provides the best fit to the data, the genetic correlations are expected to be high relative to species in which the threshold/reprogramming hypothesis provides the best fit to the data. Such a study would shed light on the degree to which one morph can evolve independently of the other.

Additional studies that quantify the fitness functions of alternative male phenotypes are needed. Unfortunately, these experiments require large sample sizes, and the experimental arenas will have some influence on whether the effects detected relate directly to the field. Perhaps, as genetic techniques become cheaper and more widely available, these experiments will be done in field enclosures, where a closer approximation to fitness in the wild could be assumed. Not all dung beetle species have generation times short enough to make experimental evolution a viable proposition, but even in the rather long-lived *O. taurus*, this has been done (Simmons & Garcia-Gonzalez, 2008). Experimental evolution of thresholds would be an exciting way to explore how changing the fitness of male morphs alters the average switchpoint of a population (Tomkins *et al.*, in review).

The evidence to date is that dung beetle mating systems provide a superb model system for understanding threshold traits. Linking the quantitative genetics of threshold traits with the increasing depth of knowledge of the developmental genetics of polyphenisms in these species (Chapter 7), the evolutionary lability between species (Chapter 2) and their ecology (Chapters 10–13) offers exciting opportunities for research, with broader implications than simply the biology of the insects themselves.

Acknowledgements

We are very grateful to Bruno Buzatto, Douglas Emlen, John Hunt, Rob Longair, Armin Moczek, Mark Rowland and Leigh Simmons for providing data sets that were used in this chapter. We also thank John Alcock, Armin Mokzec, and the editors of this volume for their constructive comments on earlier drafts of this chapter. Marissa A. Penrose helped with the production of numerous figures. This research was supported directly by an Australian Research Council Fellowship to JLT, and a Discovery Grant that employed MAP. WNH was funded by the Fisher Fund for Faculty Development at DePauw and the US National Science Foundation (DEB 0223089).

7

Evolution and Development: *Onthophagus* Beetles and the Evolutionary Developmental Genetics of Innovation, Allometry and Plasticity

Armin Moczek

Department of Biology, Indiana University, Bloomington, IN, USA

7.1 Introduction

Over the past decade, horned beetles in the genus *Onthophagus* have emerged as a promising model system in evolutionary developmental biology (evo-devo) and ecological development (eco-devo). Specifically, *Onthophagus* beetles have attracted the attention of researchers due to:

1. the expression of horns, exaggerated and diverse secondary sexual traits lacking obvious homology to other insect traits;
2. rich phenotypic diversity including morphological, behavioural, and physiological traits;
3. the significance of environmental factors, especially nutrition, in guiding phenotype determination;
4. the fact that *Onthophagus* beetles stand out as the animal kingdom's most speciose genus.

Onthophagus beetles therefore offer a promising microcosm for integrating the developmental genetic underpinnings of phenotypic diversity with the physiological, behavioural and ecological mechanisms that shape this diversity in natural populations.

I begin this chapter by briefly summarizing recent methodological advances in *Onthophagus* research, including the development of gene expression assays, RNAinterference-mediated gene function analysis and genomic/proteomic tools,

Ecology and Evolution of Dung Beetles, First Edition. Edited by Leigh W. Simmons and T. James Ridsdill-Smith. Published 2011 by Blackwell Publishing Ltd.

and by highlighting the power and limitations of each of these approaches. I then explore three frontiers in current evo-devo and eco-devo research and review how the use of genetic and developmental techniques for the study of *Onthophagus* beetles have helped advance these frontiers. Specifically, I explore the developmental genetic regulation of beetle horns as an evolutionary novelty at the heart of one of the most dramatic radiations of animal secondary sexual traits.

Next, I discuss what *Onthophagus* beetles have taught us so far about the developmental regulation of scaling, allometry and form, and some of the forces that shape these mechanisms in nature. And lastly, I review what is known to date about the developmental and genetic underpinnings of plasticity during *Onthophagus* development and its consequences for developmental evolution and diversification, including the origin of novel phenotypes.

Onthophagus beetles emerge as a powerful system with which to integrate micro- and macroevolutionary perspectives of development, and to explore the interplay between environmental, ecological and genetic factors in guiding morphological and behavioural evolution.

7.2 Evo-devo and eco-devo – a brief introduction

Evolutionary developmental biology – abbreviated commonly, for better or worse, as evo-devo – has a long, diverse and convoluted history (Raff, 1996). While the study of ontogeny as a means to gain insight into evolutionary history is as ancient as evolutionary biology itself, it was arguably only recently that evolutionary developmental biology has emerged as a coherent discipline with its own textbooks, journals, funding panels and philosophical debates (e.g. Carroll *et al.*, 2005; Samsom & Brandon, 2007).

In a nutshell, evolutionary developmental biology seeks to understand how developmental processes have originated and been shaped by evolution, and in turn how evolutionary outcomes have been influenced by the properties of development. Evo-devo clearly interfaces with both traditional evolutionary biology and genetics on one side, and with developmental biology on the other, but it cannot be subsumed within them. Specifically, evo-devo asks questions that neither contributing discipline is equipped to address by itself, such as:

- How do novel traits arise from the confines of homology and ancestral variation?
- Is macroevolution merely accumulated microevolution, or are both fundamentally different?
- How does development constrain or bias evolutionary trajectories?

Ecological developmental biology ('eco-devo'), on the other hand, is a discipline that arguably is just now being born, with one foot firmly in 'traditional' evo-devo but with another clearly stepping in directions where no discipline has gone before (Gilbert & Epel, 2009). Specifically, eco-devo seeks to identify the nature of interactions between ecological conditions and developmental processes and their consequences for the developmental biology, ecology and evolution of organisms.

With its focus on environmental factors and phenomena such as phenotypic plasticity, eco-devo builds on, but does not merely repeat, previous work on the evolutionary and quantitative genetics of genotype *x* environment interactions (Schlichting & Pigliucci, 1998; DeWitt & Scheiner, 2003). Instead, while the latter treated development largely as a black box, eco-devo explicitly focuses on filling this box with biological reality.

Specifically, eco-devo addresses questions such as:

- How does the integration of ecological and genetic inputs during ontogeny shape development during the life time of an individual organism?
- How do interactions between ecological and developmental mechanisms affect the amount and type of phenotypic and genetic variation visible to selection?

Evo-devo and eco-devo therefore differ increasingly in their domains and foci. Both share, however, an immense relevance for our understanding of human health issues such as the evolution of diseases and the human body's ability and limits to mount effective defences, or the role of environmental factors in development and inheritance, such as the epigenetic effects of stress, endocrine disruptors or drugs (Gilbert & Epel, 2009).

Insect models have played a critical role in advancing basic evo-devo and eco-devo (Heming, 2003; Carroll *et al.*, 2005). Among them, and as introduced in the next section, dung beetles in the genus *Onthophagus* have emerged as particularly promising models for exploring the interplay between environmental, ecological and genetic factors in guiding morphological, developmental and behavioural evolution.

7.3 *Onthophagus* beetles as an emerging model system in evo-devo and eco-devo

Onthophagus beetles have attracted the attention of evolutionary developmental biologists for a number or reasons (reviewed in Moczek, 2006a). First, a large number of *Onthophagus* species express horns – novel and highly diverse exaggerated secondary sexual traits used as weapons in male combat over breeding opportunities (Emlen, 2000; see Chapter 3 of this volume).

Second, onthophagine diversity is not merely restricted to horns. It extends to many other morphological, behavioural and physiological traits (e.g. resource specificity, nest construction, thermoregulation, sperm competition; see Chapter 1) in a variety of contexts, such as species-differences, sexual dimorphisms or the expression of alternative morphs within sexes (e.g. Simmons *et al.*, 1999; Moczek & Emlen, 2000; Shepherd *et al.*, 2008; see Chapter 6). The latter is particularly significant because, in many cases, the 'same' phenotypic differences, such as the presence or absence of horns, can be caused by different proximate factors, permitting researchers to juxtapose environmental regulation of phenotype determination with traditional genetic or allelic determination (Emlen, 1994).

Lastly, *Onthophagus* beetles stand out as the animal kingdom's most species-rich genus (Arrow, 1951; Balthasar, 1963b; Emlen *et al.*, 2007). At least one subset of these species is now widely distributed and relatively easy to rear and maintain in the laboratory. *Onthophagus* beetles therefore offer a promising and accessible microcosm for integrating the developmental and genetic underpinnings of phenotypic diversity with the physiological, behavioural and ecological mechanisms that shape this diversity in natural populations.

More recently, this effort has been greatly advanced through the development of key genetic and genomic techniques and resources (Moczek *et al.*, 2007). Box 7.1 briefly summarizes the most relevant techniques, the information they can provide and some of their limitations. The sections that follow will discuss how the application of these techniques in *Onthophagus* beetles has begun to advance current frontiers in evo-devo and eco-devo research. Specifically, I will introduce three interrelated focal areas of current research in evo-devo and eco-devo and review the respective advances made possible through the study of dung beetles. Highlighted throughout are some of the most significant remaining gaps in our knowledge, alongside proposed avenues by which future work with dung beetles may be able to fill those gaps. I begin with a focal area as old as evolutionary biology itself: the origin of novelty.

Box 7.1 Developmental genetic tools available in *Onthophagus* beetles: utility and limitations

Overview

Genes and their messenger-RNA and protein products play a pivotal role in instructing developmental processes. Moreover, changes in when and where a given gene is expressed generate important avenues for changing aspects of phenotype expression. Visualizing, or otherwise documenting, if and to what degree a gene is expressed in a given context is therefore key to investigating the actions of developmental pathways and their potential contributions to developmental evolution. This can be achieved through a variety of techniques, several of which are now routinely utilized in *Onthophagus* beetles.

In situ hybridization

In situ hybridization relies on a labelled RNA strand complementary – and thus specific – to the mRNA of a given gene of interest to localize those tissue regions where this gene of interest is transcribed (Wilkinson, 1998). *In situ* hybridizations require that at least parts of the gene of interest have been cloned, sequenced, and are sufficiently unique to the gene of interest to exclude the possibility of false positives. Strictly speaking, *in situ* hybridizations can only document whether or not a gene of interest is transcribed. However, much gene regulation occurs post-transcriptionally, and it is commonplace for genes to be transcribed but for their protein coding regions to not be translated (Gilbert, 2006). *In situ* expression data are therefore appropriate for surveying potential candidate genes given a certain phenotype, but extrapolations toward protein expression and gene function require caution.

Immunohistochemistry

An immunohistochemical approach (antibody staining) can safeguard against some of the limitations of *in situ* hybridization because it relies on protein-specific antibodies to localize protein products, and thus it only detects those genes that have been successfully transcribed and whose protein coding regions have been translated. For researchers working with non-model organisms, the *de-novo* production of an antibody is still often too time-consuming and costly. However, a huge arsenal of antibodies have been developed for model organisms such as *Drosophila melanogaster*, and a subset of these (albeit a small one) can be used across insect orders or, in extreme cases, across phyla (e.g. Panganiban *et al*., 1997).

While immunohistochemical approaches reduce false inferences that may arise due to post-transcriptional regulation, they are, of course, not without limitations. Post-translational modification of proteins and their functions are also ubiquitous during development (Gilbert, 2006), so a protein expressed at a certain time or place is no guarantee that the protein is actually carrying out the hypothesized function, or in fact any function at all. More generally, and similar to *in situ* hybridizations, immunohistochemistry permits investigations only one gene at a time, and only of genes whose sequence and function are at least in part known from other organisms. Both constrains are lessened in another expression method used now across a growing range of organisms – microarray hybridizations.

Microarray hybridizations

Many different types of microarray currently exist (reviewed in Gibson & Muse, 2009). What they all share is that they consist of thousands of microscopic spots, each of which in turn is made up of relatively short DNA molecules. If the sequence of these short DNA molecules is specific for different genes, each spot can act as a probe for a different gene. As such, microarrays permit the simultaneous examination of expression levels of thousands of transcripts. A common route to microarray development is through the development of a normalized cDNA library, i.e. a collection of the mRNA population (converted to cDNA) present in a particular tissue at a particular stage of development. Clones from such a library can then be used to manufacture a corresponding array.

In most cases, arrays are used to detect relative differences in transcript abundance by competitively hybridizing RNA samples obtained from two different tissues or treatments. Samples are labelled with different fluorescent markers and reflectance can be used as a proxy to determine relative expression differences across tissues or treatments. Similar to the methods discussed above, microarray hybridizations typically do not allow strong inferences toward gene function, but they represent a powerful tool for quickly surveying gene expression differences across large numbers of genes, or even on a genome-wide scale.

Two types of arrays currently exist for *Onthophagus* beetles: a spotted cDNA array consisting of 4,000 spots representing approximately 2,700 genes (Kijimoto *et al.*, 2010); and a NimbleGen High Density Array consisting of 138,000 spots representing approximately 14,000 genes (Kijimoto, Moczek & Andrews, unpublished data).

Next-generation sequencing

Next-generation sequencing refers to several innovative sequencing techniques that are capable of economically generating large amounts of sequence data. Their main limitation is that individual sequence reads are relatively short (presently 70–500 base pairs, depending on the specific approach used). However, given the extreme volume of reads generated in a single sequencing run, much overlap exists between reads which can therefore be assembled bioinformatically to larger contiguous sequences (contigs).

Next-generation sequencing is a highly effective and sensitive technique to characterize transcript diversity and to measure transcript abundance, using the number of overlapping reads of the same transcript as a proxy for expression intensity. It is most effective if the genome of the organism to be investigated has already been fully sequenced and can serve as a reference to place reads. If this is not the case, next-generation sequencing itself can be used to generate and assemble a reference transcriptome or genome.

As of the writing of this chapter, the first next-generation sequencing effort for dung beetles has been completed using Roche/454 GS-FLX Titanium platform sequencing. This effort generated >1.3 million sequence reads with >580 megabases of sequence information, which assembled into approximately 50,000 contigs and singletons. Sequences matched >14,000 genes from other organisms already described in public databases, and they also included a large number of protein coding sequences that lack obvious homology to known genes (Choi *et al.*, in press).

2D Differential Gel Electrophoresis (DIGE)

A similar approach to microarrays, albeit on the protein level, involves 2D Differential Gel Electrophoresis (DIGE). Here, multiple protein samples are labelled fluorescently, combined and then separated according to their isoelectric focusing point and size on a two-dimensional gel (Unlu *et al.*, 1997). Proteins shared by samples migrate together, whereas differentially expressed or post-transcriptionally modified proteins occupy unique spots on the gel. Proteins of interest can then be extracted, and their mass and their amino acid sequence can be determined via mass spectrophotometry and blasted against available databases.

RNAinterference

All methods listed above have the power to implicate genes and their products in processes determining phenotype expression. However, all of them ultimately rely on correlations between gene expression and phenotype expression. To examine gene function, gene expression must be perturbed experimentally. Experimental over-expression of candidate genes is commonplace in genetic and developmental model systems, but it is currently beyond the reach of most non-traditional model systems. Experimental down-regulation of gene expression, on the other hand, is feasible via RNA interference (RNAi), a relatively straightforward technique with incredible power (Novina & Sharp, 2004).

Specifically, as an experimental approach, RNAi involves injections of double-stranded RNA fragments, thereby activating a cellular response mechanism that digests exogenous RNA. Digestion fragments are then used to detect other matching RNA molecules, including those made by the organism itself, which are subsequently targeted for digestion as well. Experimental injection of dsRNA that matches a selected target gene made by the organism itself allows researchers to use the organism's own RNAi machinery against itself to deplete transcript abundance of its own genes, including those of interest to researchers.

RNAi has revolutionized experimental evolutionary developmental genetics due to its applicability across a wide range of organisms (Novina & Sharp, 2004) including *Onthophagus* beetles (Moczek & Rose, 2009). It does require pre-existing knowledge of candidate genes (such as sequence and expression data) and is relatively work-intensive, but it is extremely powerful when it comes to identifying and comparing gene function.

7.4 The origin and diversification of novel traits

The origin of novelty in evolution has captivated evolutionary biologists ever since the inception of the discipline (Raff, 1996; Wilkins, 2002; West-Eberhard, 2003). What has to come together – genetically, developmentally or ecologically – for complex novel traits to arise and diversify in nature? Is the origin of novel traits underlain by processes fundamentally different from those that facilitate quantitative changes in pre-existing traits, or is innovation merely an extrapolation of diversification over time (Erwin, 2000; Davidson & Erwin, 2006; Moczek, 2008)? Furthermore, how is innovation initiated within the confines of homology and descent with modification? Where exactly does ancestral variation end and novelty begin?

For instance, one of the strictest definitions of novelty, advocated by Müller and Wagner (1991), is the absence of homology or homonomy (serial homology). Here, novelty begins where homology ends (Moczek, 2008), but exactly where homology ends has become more and more difficult to define, primarily for two reasons (Brigandt, 2002): On the one hand, we have come to understand that a tremendous amount of morphological diversity, including clearly non-homologous traits, is made possible through the use of a relatively small toolkit of ancient, conserved and homologous developmental pathways (Carroll *et al.*, 2005; Shubin *et al.*, 2009). Clearly, homology of development need not imply homology of form. On the other hand, the opposite is also true: unambiguously homologous traits often diverge dramatically with respect to the developmental genetic mechanisms that regulate their expression during development (True & Haag, 2001; Palmer, 2004), a phenomenon also known as developmental systems drift (True & Haag, 2001) or phenogenetic drift (Weiss & Fullerton, 2000).

Homology of form, apparently, need not imply homology of development either. Therefore, where homology ends and novelty begins is murkier than ever. It is precisely at this intersection that research on dung beetles has made several important contributions, focusing on the developmental genetic regulation of horns and horn diversity in the dung beetle genus *Onthophagus*.

7.4.1 Dung beetle horns as novel traits

Beetle horns lend themselves well to the study of innovation in evolution, primarily for the following reasons. Beetle horns are often large, solid, three-dimensional outgrowths that function as weapons in male competition over breeding opportunities (Emlen, 1997a; Moczek & Emlen, 2000; see Chapter 3 of this volume). Beetle horns thus shape in many ways both the morphology and the behavioural ecology of their bearers. At the same time, beetle horns differ dramatically in size, shape, number and location of expression, and much of this variation can be found not only between species but also between sexes, and frequently within sexes, creating much opportunity for comparative approaches.

Lastly, and most importantly, beetle horns are unique structures in a sense that they lack clear homology to other traits in insects. Horns are not modified versions of traditional appendages such as mouth parts, antennae or legs, but instead exist alongside these structures in body regions in which insects normally do not produce outgrowths (Moczek, 2005; 2006a). Horns can therefore be considered an example of an evolutionary novelty that horned beetles invented at some point during their history, giving rise to one of the most impressive radiations of secondary sexual traits in the animal kingdom (Arrow, 1951; Balthasar, 1963a; Emlen *et al.*, 2007; Emlen, 2008).

7.4.2 How horns develop

The horns of beetles become first detectable during the last larval stage as the animal nears the larval-to-pupal moult (Figure 7.1; reviewed in Moczek, 2006a; Moczek and Rose, 2009). At this stage, certain epidermal regions detach from the larval cuticle and proliferate. The resulting tissue folds as it is trapped underneath the larval cuticle, then expands once the animal moults to the pupal stage. It is at this stage that horns become externally visible for the first time.

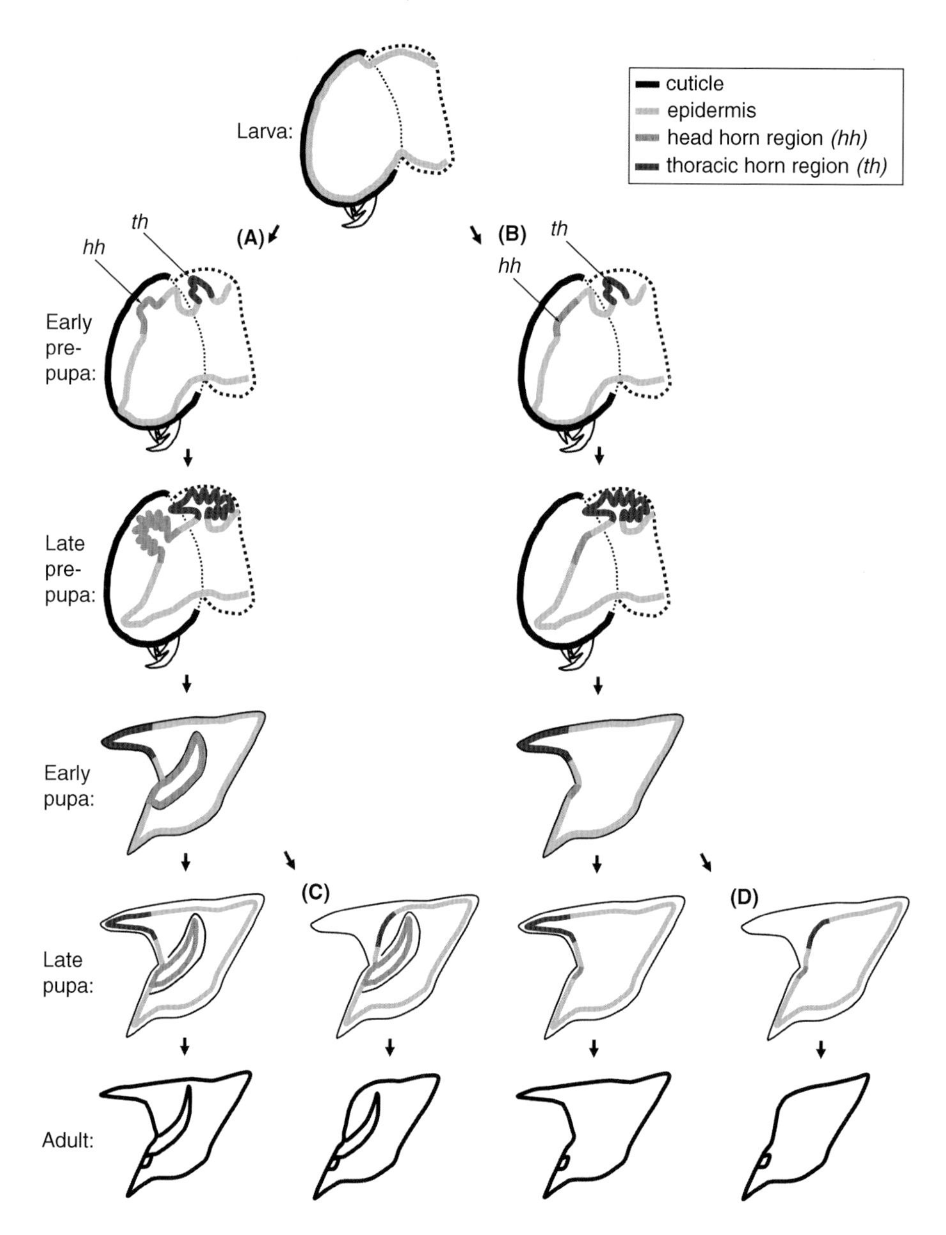

This period of prepupal horn growth is then followed by a period of pupal remodelling of horn primordia. During the pupal stage, horns undergo at times substantial remodelling in both size and shape – and, in some cases, complete resorption prior to the adult moult. After the pupal-to-adult moult is complete, horns have then attained their final adult size and shape.

The horns of adult beetles therefore develop, at least in part, in a similar manner to traditional appendages in other holometabolous insects (Svácha, 1992) and may thus be thought of, at least in developmental terms, as highly simplified appendages. Recent studies now show that many important components of the developmental machinery employed in the making of traditional appendages have, indeed, been recruited into the development and evolution of horns.

7.4.3 The developmental genetics of horn growth

Expression studies using immunohistochemical approaches and *in situ* hybridization (Box 7.1) were the first to implicate several important appendage-patterning genes in the making of horns. Specifically, several transcription factors known to play important roles in establishing the proximo-distal axis of insect appendages (*Distal-less* (*Dll*), *aristaless* (*al*), *dachshund* (*dac*) *homothorax* (*hth*) and *extradenticle* (*exd*)) were also found to be expressed during the formation of horns in the prepupal stage (Moczek & Nagy, 2005; Moczek *et al.*, 2006a). Moreover, all of them but *dac* were expressed in regions of the future horn that were at least consistent with a conservation of gene function.

Fig. 7.1 Schematic development of **(A)** horns and **(B-D)** horn dimorphisms in *Onthophagus* beetles. **(A)** During the last larval instar, the larval epidermis (light gray) fully lines the larval cuticle (black). At the onset of the prepupal stage, the larval epidermis detaches from the cuticle (apolysis) and selected regions (shown here for a head horn (*hh*) and thoracic horn (*th*)) undergo rapid cell proliferation. The resulting extra tissue folds up underneath the larval cuticle. The epidermis subsequently secretes the future pupal cuticle, which, upon the moult to the pupal stage, forms the outermost layer of the pupa, lined once again by a layer of epidermal cells. During this pupal moult, horn primordia are able to expand and unfold and are now visible externally. During the second half of the pupal stage, epidermal cells detach once more. This time, however, no significant growth of horn tissue follows detachment. Instead, epidermal cells secrete one last cuticle and the pupa undergoes one last moult to the final adult stage. **(B)** Development of horn dimorphisms through differential proliferation of prepupal horn tissue (illustrated here for head horns (*hh*) only). During the prepupal stage, presumptive horn tissue proliferates little or not at all, resulting in the absence of external horns in pupae and the resulting adults. This mechanism is used to generate sexual dimorphisms as well as alternative male morphologies for head horns in many species. **(C)** and **(D)** Development of horn dimorphisms through differential loss of pupal horn tissue (illustrated here for thoracic horns (*th*) only). Pupal horn epidermis is resorbed prior to the secretion of the final adult cuticle, most likely via programmed cell death. In many cases, resorption of pupal horn tissue can completely erase the former presence of a thoracic horn. This mechanism contributes to sexual dimorphisms for thoracic horns in many species, and can occur in the presence or absence of (differential) head horn development. Modified after Moczek, 2006a.

Recent gene function analyses (Box 7.1) using RNAinterference (RNAi) mediated transcript depletion clarified these inferences substantially, and in the process they underscored the limitations arising from inferring gene function purely from comparative gene expression data. Specifically, Moczek & Rose (2009) examined the function of three of these genes – *dac*, *hth*, and *Dll* – during horn development in two species of *Onthophagus* beetles. Irrespective of any involvement in horn development, larval RNAi-mediated transcript depletion of all three genes generated phenotypic effects identical or similar to those documented by previous studies in other taxa (e.g. Prpic *et al.*, 2001; Angelini & Kaufman, 2004; Kojima, 2004). This observation was important because it documented that all three patterning genes exhibited conservation of function with respect to the patterning of traditional appendages, as well as the general feasibility of larval RNAi in *Onthophagus* beetles.

In addition, however, this study yielded many surprising insights into the functional regulation of horn development. For example, *dac* did not appear to play any obvious role in the regulation of size, shape or identity of horns, even though it is expressed widely throughout prepupal horn primordia (Moczek *et al.*, 2006a). Thus, even though *dac*RNAi individuals expressed severe *dac* knockdown phenotypes in their legs and antennae, thoracic and head horn expression was completely unaffected. In contrast, *hth*RNAi had a dramatic effect on horn expression, but only affected thoracic, not head, horns in the same individuals, even though *hth* is expressed during the development of both horn types.

The results of *Dll*RNAi made matters even more complicated. While *Dll*RNAi affected the expression of both head and thoracic horns, it did not do so in the same individuals or even the same species. In *Onthophagus taurus*, head horn expression was only affected in large males, whereas horn expression in small and medium-sized males was unaffected. Similarly unaffected was the expression of pupal thoracic horns in both males and females regardless of body size. In *O. binodis*, on the other hand, *Dll*RNAi did affect the expression of thoracic horns in both males and females, though the effect was strongest in larger individuals.

Combined, these results illustrate that horn development evolved via differential recruitment of at least some proximo-distal-axis patterning genes normally involved in the formation of traditional appendages. On the one hand, these results contribute to a by now-common theme in the evolution of novel traits: new morphologies arise through the recruitment of existing developmental mechanisms into new contexts, rather than the evolution of novel genes or pathways (Shubin *et al.*, 2009). On the other hand, they highlighted an unexpected degree of evolutionary lability in the developmental regulation of horns, including the absence of patterning function (*dac*), patterning function in selected horn types (*hth, Dll*) and function in one size class, sex or species but not another (*Dll*).

Different horn types, and even the same horn type in different species, may therefore be regulated at least in part by different pathways. If this is correct, different horn types may thus have experienced distinct, and possibly independent, evolutionary histories. These conclusions receive further confirmation when we take a closer look at the regulation of the second developmental period relevant to adult horn expression: pupal remodelling.

7.4.4 The developmental genetics of pupal remodelling

During the pupal stage, horns are sculpted into their final adult shape. As such, pupal remodelling of horns is not unusual; indeed, all pupal appendages and body regions of holometabolous insects undergo at least some degree of sculpting during the pupal stage (e.g. Nijhout, 1991). What is unusual, however, is the often extreme nature of pupal horn remodelling, especially as it occurs in thoracic horns (Moczek, 2006b). Here, pupal remodelling can result in the complete resorption of pupal horn primordia, causing fully horned pupae to moult into thorax-horn-less adults. Pupal horn resorption occurs in at least one (female or male) or both sexes in all 21 *Onthophagus* species examined to date, suggesting that it is widespread, yet evolutionarily labile, with respect to the sex in which it occurs (Moczek *et al.*, 2006b).

Recent work now implicates programmed cell death (PCD) in the resorption of horn primordial tissue (Kijimoto *et al.*, 2010). PCD is an ancient, highly conserved physiological process employed by all metazoan organisms to remove superfluous cells and their contents during development (Potten & Wilson, 2004). For example, PCD is responsible for removing inter-digit tissue during embryonic development of the human hand; the removal of the tadpole's tail during metamorphosis; and the sculpting of the hind wing projections of swallowtail butterflies (Nijhout, 1991; Gilbert, 2006).

Recent work has now shown that PCD plays an important and dynamic role in the resorption of pupal horn primordia during *Onthophagus* pupal development. Using two bioassays, one for detecting PCD-characteristic DNA breakage, the other for detecting the expression of activated caspases (a class of enzymes used for protein digestion during cell death), Kijimoto *et al.* (2010) showed that most PCD occurred between 24 and 48 hours of pupal life. Most importantly, the same study showed a tight correlation between occurrence of PCD and subsequent resorption of pupal horn primordia.

In *O. taurus*, pupal thoracic horns of both sexes revealed high levels of PCD, matching the sex-uniform, complete resorption of thoracic horn primordia seen in this species. However, very little PCD was detected in the head horns of large male *O. taurus*, which undergo little resorption. In contrast, in *O. binodis*, high levels of PCD were only observed in female, but not male, thoracic horn primordia, this time matching the female-specific thoracic horn resorption characteristic for this species. Thus, the amount of cell death-mediated horn resorption depended strongly on species, sex, and body region, suggesting the existence of regulatory mechanisms that can diversify quickly. Combined, the regulation of pupal remodelling therefore reinforces many of the conclusions reached above for the regulation of prepupal horn growth. As before, a pre-existing developmental machinery, this time PCD, has become recruited into a new developmental context, the sculpting of horns. At the same time, this appears to have permitted the rapid evolution of modifier mechanisms, allowing an ancient developmental process to contribute to species-, sex-, and body region-specific expression of horns.

These insights into the developmental regulation of prepupal growth and pupal remodelling of horns illustrate that regulatory genes whose functions are otherwise highly conserved nevertheless remain able to acquire new functions, and that little phylogenetic distance is necessary for the evolution of sex- and species-specific differences in these functions. Moreover, tracing the diversity of developmental regulation of beetle horns through the phylogeny of horned beetles is beginning to provide surprising insights into the very origins of the first adult horns, as discussed in the next section.

7.4.5 The origin of adult thoracic horns through exaptation

If evolutionary change is, indeed, dominated by descent with modification, everything new has to come from something old (Wake, 1999; 2003). However, tracing this ancestry is often difficult, as ancestral character states may be obscured by long periods of independent evolution of diverging lineages. Alternatively, or in addition, signatures of ancestral character states are often hidden in developmental stages other than the adult. Evolutionary scenarios inferred solely on the basis of adult trait expression are thus bound to overlook these signatures.

While this problem is widely recognized and appreciated, opportunities to integrate developmental perspectives into ancestral character state reconstructions are often limited by the degree to which the ontogenies in question are experimentally accessible. PCD-mediated resorption of pupal thoracic horn tissue, as discussed in the previous section, has provided a good example of how incorporating developmental data can be used to refine evolutionary hypotheses.

PCD-mediated resorption of pupal thoracic horn tissue is ubiquitous among *Onthophagus* species, and all species examined so far express pupal prothoracic horns in both males and females, followed by horn resorption in either one or both sexes in each species (Moczek *et al.*, 2006b). This raises the question as to the adaptive significance, if any, of such transient horn expression. Experimental approaches have now revealed that pupal horns play a crucial role during the larval-to-pupal moult, and especially the removal of the larval head capsule, and do so regardless of whether they are resorbed or converted into an adult structure (Moczek *et al.*, 2006b).

Unlike in larval-to-larval and pupal-to-adult moults, larvae that moult into pupae have absorbed most of their muscle tissue. Instead, they shed old cuticle by means of peristaltic contractions to increase local haemolymph pressure and the swallowing of air to inflate selected body regions. This suffices for the removal of the highly membranous thoracic and abdominal cuticle of larval scarab beetles, but shedding the larval head capsule poses greater challenges, since it is composed of extremely thick and inflexible cuticle. During larval life, this robust cuticle and its inward projections provide important attachment points for the powerful jaw muscles of fibre-feeding scarab larvae, such as *Onthophagus*. Histological studies have now shown that, during *Onthophagus*' prepupal stage, thoracic horn primordia enter into the space between the larval head capsule and corresponding epidermis, fill with haemolymph and then expand. This expansion appears to cause the larval head capsule to fracture along pre-existing lines of weakness. As the larval head moults into a pupal head, the thoracic horn primordium is the first structure to emerge from the head capsule (Moczek *et al.*, 2006b).

Experimental manipulations support a moulting function of pupal thoracic horns. When the precursor cells that would normally give rise to thoracic horn primordia are ablated early in larval development, the resulting pupae do not express a thoracic horn and fail to shed their larval head capsule (Moczek *et al.*, 2006b). Replicating this approach for two *Onthophagus* species yielded similar results, but it failed to elicit any effect in the sister genus *Oniticellus*, i.e. the same surgical manipulation did not impede shedding of the larval head capsule. This suggests that this putative moulting function of thoracic horn primordia may be unique to onthophagine beetles.

Phylogenetic analyses further suggest that the function of horns as a moulting device may have preceded the horns-as-a-weapon function of the adult counterparts and that, ancestrally, all pupal horns were resorbed prior to the adult moult (Moczek *et al.*, 2006b). If correct, this would explain why prepupal thoracic horns appear ubiquitous among *Onthophagus* species, even though only a relatively small subset of species uses them to build a functional structure in the adult.

How could the first adult horns have originated from pupal ancestral structures? The results presented above raise the possibility that the first adult horns could have been mediated by a simple failure to remove a pupal-specific projection via failure to activate PCD at the right developmental time and location. Several anecdotal studies suggest that such events do occur in natural populations, at least occasionally (e.g. Paulian, 1945; Ballerio, 1999; Ziani, 1994; see also Figure 7.2). Clearly, such a failure would have resulted in an adult outgrowth that at first would probably have been small. However, behavioural studies have shown that even very small increases in horn length are sufficient to bring about significant increases in fighting success and fitness (Emlen, 1997a; Moczek & Emlen, 2000; Hunt & Simmons, 2001).

Importantly, the possession of horns is not a prerequisite for fighting, since fighting behaviour is generally widespread among beetles and occurs well outside horned taxa (reviewed in Snell-Rood & Moczek, in press). The first pupal horn that failed to be removed before the pupal-to-adult moult could thus have provided an immediate fitness advantage. Thoracic beetle horns may, therefore, be a good example of a novelty that arose via exaptation (*sensu* Gould and Vrba, 1982) from traits originally selected for providing a very different function at an earlier development stage.

How do these observations help revise earlier hypotheses regarding the evolutionary origin of onthophagine horns? Mapping adult morphologies onto a

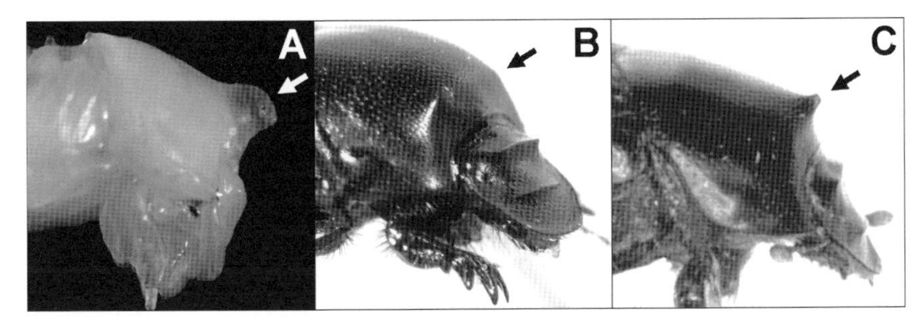

Fig. 7.2 Example of a failure to remove pupal thoracic horns through programmed cell death in *Onthophagus taurus*. Female *O. taurus* express a thoracic horn as **(A)** pupae but remove it via programmed cell death prior to moulting to an adult **(B)**. **(C)** A female *O. taurus* obtained from L.W. Simmons' laboratory culture failed to fully remove the pupal thoracic horn primordium and moulted into a horned female adult.

molecular phylogeny, Emlen *et al.* (2005b) concluded that thoracic horns must have originated a minimum of nine independent times in males and seven in females among just 48 *Onthophagus* species to explain present-day patterns of horn expression. This is a staggering number of independent gains over a remarkably short phylogenetic distance, but it is the inescapable conclusion if adult morphological data are the only source for inferring ancestral character states.

Unfortunately, only a subset of the 48 species in this phylogeny have known pupal morphologies. Incorporating the pupal morphologies of those nine species for which pupae have been described into a re-analysis (and treating the remaining 39 species as treated by the original analysis) is sufficient, however, to make the ancestral character state at all nodes either horned or undeterminable (Moczek *et al.*, 2006b). This suggests that the most parsimonious explanation for the origin of thoracic horns in *Onthophagus* may be as simple as a single gain, followed by diversification of adult horn expression in different species via pupal resorption in one or both sexes. If this scenario is correct, this presumed origin of adult thoracic horns from ancestral moulting devices would illustrate well how convoluted developmental evolution can be, and how it can yield evolutionary novelty from well within the confines of ancestral diversity and homology.

The same complexity in the interactions between development, morphology, and ecology emerges when we shift our emphasis away from the origin and diversification of horns *per se* and toward the diversification of shape and scaling in general, as illustrated in the next section.

7.5 The regulation and evolution of scaling

How organisms and their parts 'know' to what size to grow is a fundamental, yet still poorly understood, question in developmental biology (Huxley, 1932; Thompson, 1942; Gilbert, 2006). Similarly, how the mechanisms by which organisms and their parts regulate their growth evolve and contribute to organismal diversification also remains poorly understood (e.g. Stern & Emlen, 1999; Emlen *et al.*, 2007; Shingleton *et al.*, 2007; Frankino *et al.*, 2005; 2008). Scaling relationships, used here synonymously with allometries, can be depicted in their most simple forms as bivariate plots, correlating size of one trait against, typically, some measure of body size. Comparing these so-called 'static allometries' of different traits measured in the same group of individuals can begin to reveal some of the complexities of scaling relationships (Figure 7.3A).

Allometries range from linear and proportional (where larger animals are essentially proportionally enlarged versions of smaller animals, e.g. tibia length in Figure 7.3A) to flat and largely body size independent (large individuals have a trait with the same absolute size as smaller animals, e.g. paramere size in Figure 7.3A) to sigmoidal (a threshold body size separates two different trait sizes, e.g. horn length in Figure 7.3A). Much the same applies to scaling relationships of the same trait measured in different populations (Figure 7.3B) or species (Figure 7.3C). Intriguingly, even though scaling relationships can vary dramatically for different traits, populations or species, the variance around a given allometry is usually rather small. In other words, trait size tends to scale with body size in a highly

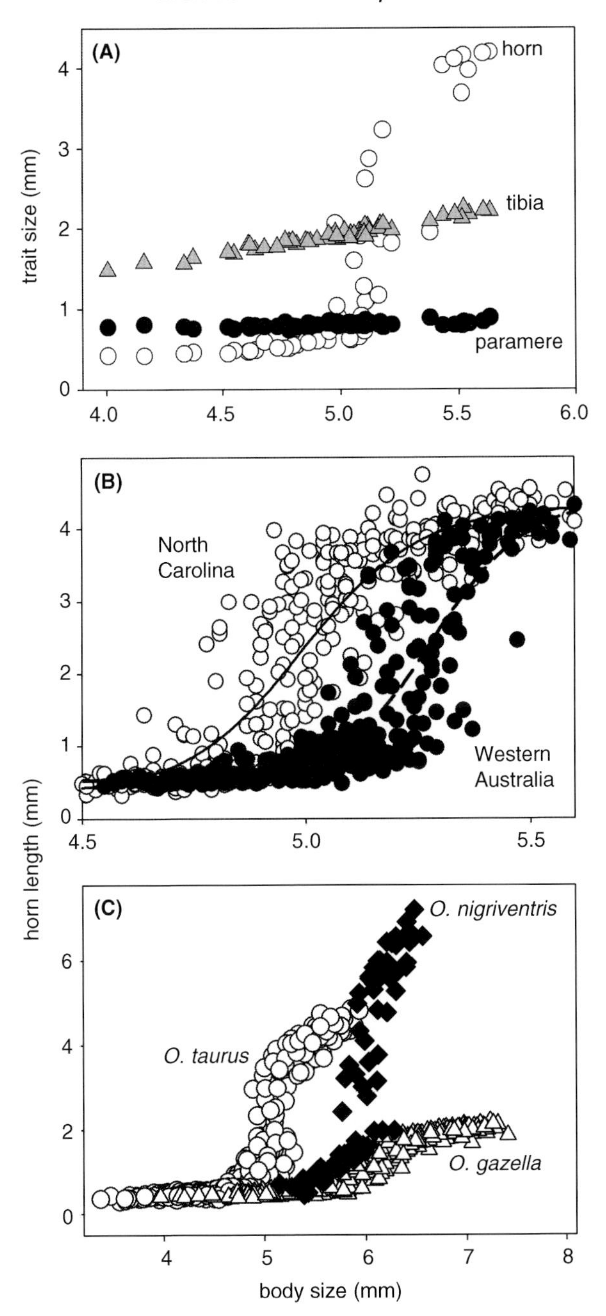

Fig. 7.3 Examples of static allometries and allometric diversity in *Onthophagus* beetles. **(A)** Scaling relationships between body size (x-axis) and the length of the horn, fore tibia and paramere, respectively, in male *O. taurus*. The paramere is part of the male copulatory organ. **(B)** Scaling relationships between body size and horn length in male *O. taurus* collected in North Carolina and Western Australia. **(C)** The same scaling relationship for three *Onthophagus* species. Data in panels (B) and (C) are from Moczek, 2006a.

predictable manner for a given trait and population or species. This begs a number of questions, such as:

- How does each part 'know' to what size to grow?
- How do parts 'know' the size of the remainder of the organism?
- What does it take developmentally, and evolutionarily, to change the relative sizes of parts?

Preliminary answers to these questions are being provided by a growing understanding of the hormonal, genetic and environmental regulators of growth and scaling (e.g. Oldham *et al.*, 2000; Nijhout & Grunert, 2002; Nijhout, 2003a; 2003b; Emlen *et al.*, 2006; Shingleton *et al.*, 2007; 2009). Research on dung beetles, and in particular *Onthophagus*, has made several relevant contributions, which are briefly summarized in the following two sections. The more recent development of genetic and genomic resources now offers the opportunity to deepen substantially our understanding of the regulation and evolution of scaling in dung beetles, as will be discussed subsequently.

7.5.1 Onthophagine scaling relationships: the roles of nutrition and hormones

Larval feeding conditions play a pivotal role in determining the size of adult traits (Emlen, 1994; Moczek & Emlen, 1999). Larvae with access to optimal feeding conditions grow longer, attain larger mass and moult into larger pupae and adults. Moreover, in species in which males are separated by a body size threshold into hornless (minor) and horned (major) morphs (see Chapter 6), nutrition also determines which morph a given male develops into.

Exactly how variation in nutrition is translated into variation in growth is still largely unclear, but several likely important aspects are beginning to emerge (reviewed in Hartfelder & Emlen, 2005; Shelby *et al.*, 2007; Emlen *et al.*, 2007). For instance, juvenile hormone (JH) is an important regulator in insect metamorphosis, but it also appears to play important roles in the determination of alternative morphs such as phase polyphenism in aphids (Hardie & Lees, 1981) and castes in social insects (bees: Rachinsky & Hartfelder, 1990; ants: Wheeler, 1986; 1991; Wheeler & Nijhout, 1983).

In *Onthophagus* beetles, JH applications induce horn development in larvae otherwise fated to develop into small, hornless males, suggesting that JH may also play a role in the nutritional determination of horn expression (Emlen & Nijhout, 1999). Moreover, males from populations that have diverged in the exact body size threshold that separates hornless and horned morphs are differentially sensitive to the same JH manipulation (Moczek & Nijhout, 2002). Similarly, different species and sexes also respond differently to JH perturbations (Shelby *et al.*, 2007). Separate work also provides some evidence that ecdysteroids, a second important class of insect hormones most known for their role in the regulation of moulting, may also regulate aspects of horn expression (Emlen & Nijhout, 1999; 2001; but see Shelby *et al.*, 2007 for a critical evaluation).

Thus far, however, all data available on the endocrine regulation of size and scaling are correlational at best, and mostly derived from relatively crude manipulations. Moreover, only a very limited understanding of natural hormone titre profiles exists for ecdysteroids, and none does for JH. A deeper understanding of the evolutionary endocrinology and its role in the regulation of size and scaling in horned beetles will therefore depend on our ability to document and manipulate hormonal regulation in these organisms in a more quantitative fashion, and to identify the nature of interactions between endocrine mechanisms and their upstream regulators and downstream targets during development.

Toward this end, studies are needed to obtain JH titres as well as titres of the most relevant JH metabolizing enzyme, JH-Esterase, for several *Onthophagus* species. Comparing natural and manipulated titre profiles to transcription profiles compiled through the application of microarrays could then be used to identify putative endocrine-responsive genes, the most promising of which could be analysed functionally via RNAi. A first step in this direction has recently been taken by Kijimoto *et al.* (2010), who first documented programmed cell death (PCD) during early pupal development using standard bioassays, and then used expression profiles for the same developmental stage to identify a number of ecdysteroid-signalling genes that may regulate PCD in a sex-specific manner. This resulted in a first testable model for the developmental genetic regulation of sex-specific PCD in *Onthophagus* beetles.

Another canonical pathway involved in translating variation in nutrition into variation in growth in animals, including insects, is insulin signalling. In an important review, Emlen *et al.* (2006) presented preliminary data implicating differential expression of the insulin receptor as a possible regulator of nutrition-mediated development of alternative morphs in *O. nigriventris*. No validation of these preliminary findings has yet been published, but these results suggest an important and likely pathway as a potential interface between nutritional variation, as it is experienced by larvae, and differential growth of structures as it occurs during the prepupal stage.

Recent microarray studies further support the notion that insulin-signalling genes are differentially expressed in the context of horn development (Snell-Rood *et al.*, in press). Preliminary functional analysis of one such gene, the growth inhibitor *FoxO*, further corroborates this hypothesis (Snell-Rood & Moczek, in review). Insulin signalling may be particularly important in horned beetles, because studies in other insects suggest interesting interactions between insulin signalling and juvenile hormone (Tu *et al.*, 2005).

7.5.2 Onthophagine scaling relationships: the role of trade-offs during development and evolution

All growth requires resources and, if resources are in limited supply, structures that compete for them may find themselves locked in a trade-off (Klingenberg & Nijhout, 1998; Nijhout & Emlen, 1998). In such a situation, enlargement of one structure, whether during the development of an individual or the evolution of a lineage, may only be possible at the expense of another. The notion that resource allocation trade-offs may bias developmental outcomes and evolutionary trajectories is an old one, and a growing number of studies both in the laboratory and in natural populations have now shown that trade-offs are real and potentially widespread (reviewed in Roff & Fairbairn, 2007). However, the underlying mechanisms have remained largely elusive.

Dung beetles have provided some of the most compelling evidence for the power and scope of developmental trade-offs. For example, studying several *Onthophagus* species, Emlen (2001) showed that scaling relationships between body size and traits such as eyes, antennae or wings are affected by the relative amount of horn expression that occurs in their proximity, suggesting that structures that grow in closer proximity to each other may be more likely to engage in a trade-off. Moczek & Nijhout (2004) later expanded this notion by showing that timing of growth may be the main determinant of trade-off intensity. Using experimental manipulations of development in the laboratory, this study showed that structures growing as far apart as head horns and abdominal copulatory organs can still engage in a trade-off, provided they overlap in the exact timing of their respective growth periods. Specifically, males whose copulatory organs were prevented from developing expressed relatively larger horns than untreated or sham-treated males.

Recent comparative studies showed that this developmental trade-off may also bias evolutionary trajectories (Figure 7.4). Studying four recently diverged populations of one species (divergence < 50 years) and an additional ten more distantly diverged species (10,000–38 million years), Parzer & Moczek (2008) showed that in both cases, increased investment into horns was correlated with decreased investment in copulatory organs, and vice versa. A study by Simmons & Emlen (2006) presented at least partly complementary data (see Chapter 4). Experimental inhibition of horn development in *O. nigriventris* resulted in males producing relatively larger testes. While the authors did not find a general relationship between the relative sizes of horns and testes across species, they did observe a negative correlation between the steepness of the body size-horn length allometry and the steepness of the body size-testes size allometry, i.e. species with the steepest horn allometry had the shallowest testes allometry, and vice versa (Simmons & Emlen, 2006). Combined, these data underscore the power of developmental trade-offs over both short and long timescales and the diversity of structures that might engage in trade-offs.

Unanswered still, however, are the developmental mechanisms underlying trade-offs. For instance, it is entirely unclear why the trade-off described by Parzer & Moczek (2008) only extends to horns and copulatory organs, but not legs, even though all three structures overlap in their growth periods (Figure 7.4B). Here again, a combination of more sophisticated endocrine and genetic manipulations, combined with careful quantifications of endocrine and transcription profiles, may provide important hints as to when, where and on what level of biological organization trade-offs may arise. Some of the beginnings of such attempts are described next.

7.5.3 Onthophagine scaling relationships: developmental decoupling versus common developmental programme

The development of alternative phenotypes, such as hornless and horned male morphs in dung beetles, is thought to play important roles in the evolution of organismal diversity, including speciation and the origins of novel traits

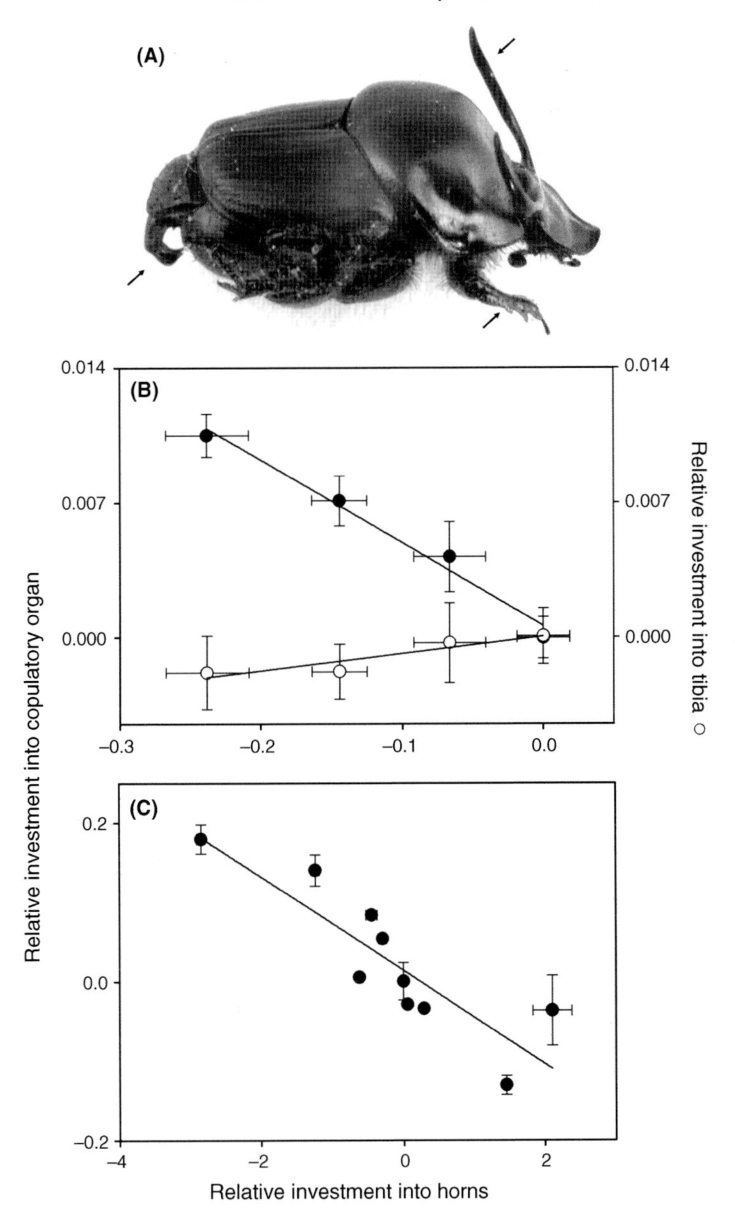

Fig. 7.4 Trade-offs between primary and secondary sexual characters in populations and species of *Onthophagus* beetles. **(A)** Horned male *O. taurus*. Arrows highlight horns, copulatory organ and fore tibia. **(B)** Relative investment into copulatory organ size (left, solid symbols) and fore tibia size (right, open symbols) as a function of relative investment into horn size in four different populations of *O. taurus*. Error bars represent one standard error. **(C)** Relative investment into copulatory organ size as a function of relative investment into horn size in ten different *Onthophagus* species. Data are corrected for differences in body size.Modified after Parzer & Moczek, 2008. Note that in the original figure, residual horn lengths were calculated as residual = expected − observed horn length. The present figure follows the more conventional way of calculating residuals (residual = observed − expected). Main results and conclusions of this study remain the same.

(West-Eberhard, 1989; 2003; Pfennig *et al.*, 2007). One central issue is the notion that alternative phenotypes are discrete developmental products resulting from genetic reprogramming, or decoupling, of development across an environmental threshold. Alternative phenotypes should therefore be able to respond to selection independent of one another, thereby increasing the evolvability of polyphenic, compared to monophenic, lineages.

However, much debate exists about the actual degree of genetic, and by extension evolutionary, independence of alternative morphs. In fact, as detailed in Chapter 6 of this volume, several allometric modelling studies argue against the evolutionary independence of alternative morphs and suggest instead that alternative forms may be the product of a common developmental programme which, through extreme positive allometry (*sensu* Tomkins *et al.*, 2005), may be able to generate discrete morphs without needing a developmental threshold to dissociate alternative developmental pathways (Nijhout & Wheeler, 1996; Tomkins *et al.*, 2005; Tomkins & Moczek, 2009).

It is important to recognize here that much of the evidence in support of genetic re-programming during polyphenic development has been generated using methodologies and taxa that may bias results toward the identification of morph-specific gene expression. For instance, direct tissue hybridizations on microarrays (see Box 7.1) or candidate-gene studies are designed specifically to detect very small differences in gene expression rather than to quantify patterns of shared expression across morphs. Furthermore, the majority of studies of morph-specific gene expression have been conducted in eusocial insects (ants, bees, termites) which may be under unique constraints in the evolution of developmental reprogramming (Snell-Rood *et al*, 2010).

Consequently, while we can be confident that alternative morphs differ in the expression of at least some genes, we know little about the nature, extent and consequences of developmental reprogramming. More generally, we lack the ability to formulate expectations about how much differential expression may be indicative of re-programming. For example, is the development of alternative phenotypes comparable to classic examples of developmental reprogramming such as sex-specific development (Bull, 1983; West-Eberhard, 2003)? Recent transcriptional profiling of alternative male morphs and sexes in two *Onthophagus* species has provided the first tentative answers to these questions.

Contrasting transcription profiles of developing head and thoracic horns, legs and brains in *O. taurus* and *O. nigriventris*, Snell-Rood *et al.* (in press) found that patterns of expression in developing beetle morphs were generally just as divergent as between the sexes. For instance, in the developing head epidermis of *O. taurus*, which gives rise to horns in major males only, overall patterns of gene expression were more similar between females and hornless males than between the two male morphs. In contrast, in the developing brain, patterns of expression in horned males were more similar to those in females rather than the hornless male morph.

It is intriguing to speculate whether the latter similarity may arise from biparental care behaviour shared between females and horned, but not hornless, males. Thus, while differences in gene expression detected between morphs were similar in magnitude to those detected between the sexes, the development of the hornless morph appeared not to be simply due to a 'feminizing' of horned male expression patterns; instead, the nature of differential gene expression across morphs and sexes depended very much on tissue type, body region and species (Snell-Rood *et al.*, in press).

Whether the fraction of differentially expressed genes detected in this study is sufficient to support a genetic decoupling or reprogramming metaphor over a common developmental programme model for beetle horn polyphenisms is a different matter, but one that may be largely semantic rather than biologically meaningful. Alternative phenotypes, no matter how divergent, nevertheless remain similar, and thus a significant degree of shared development, including gene expression, is to be expected. Whether the residual deserves to be taken as evidence for decoupling and re-programming lies largely in the eye of the beholder. The decoupling and re-programming metaphors, however, remain useful if applied to individual genes, their products and their regulators. Genes either do or do not share similar expression levels across morphs, and thus their products do or do not share similar exposure to selection (Demuth & Wade, 2007).

7.5.4 *Onthophagine* scaling relationships: the developmental genetics of size and shape

The diversity of traits, shapes and sizes within and between species, and the importance of both environmental and heritable contributions to phenotype determination, make *Onthophagus* beetles a promising microcosms for exploring development and diversification of size and form. What has stood in the way until recently has been the absence of powerful genetic and developmental tools. As we have seen, this is now changing, and we are beginning to acquire new insights into evolutionary developmental genetics of size and shape. For example, as highlighted above, preliminary data implicate differential expression of the insulin receptor as a possible regulator of nutrition-mediated development of alternative morphs in *O. nigriventris* (Emlen *et al.*, 2006). Additional support for the notion that insulin-signalling genes are differentially expressed in the context of horn development comes from recent microarray studies (Snell-Rood *et al.*, in press) as well as functional analysis of the growth inhibitor *FoxO* (Snell-Rood & Moczek, in review).

Similar progress has been made toward a better understanding of the regulation and diversification of pupal remodelling of horn size and shape, and its contribution to phenotypic diversification. Previous sections have already introduced the role of programmed cell death (PCD) and its putative regulation through ecdysteroid signalling (Kijimoto *et al.*, 2010). Recent work suggests that alternatively, or in addition, *Hox* genes may play a critical role in determining adult horn size and shape through the segment-specific activation of PCD. Specifically, the *Hox* gene *sex combs reduced* (*Scr*), alters the magnitude of sex-specific pronotal horn resorption in *Onthophagus*, suggesting that PCD genes may be among the targets of *Scr* during pronotal horn development (Wasik *et al.*, 2010). This would not be unexpected, as other *Hox* genes are known to regulate PCD in a segment- or organ-specific manner. For example, the *Hox* gene *deformed* (*dfd*) directly controls the expression of *reaper*, an upstream mediator of PCD, during *Drosophila* mouthpart formation (Lohmann *et al.*, 2002).

Most of the developmental genetic regulators of growth and differentiation in dung beetles, and insects in general, remain to be identified and functionally characterized, and the same goes for their interactions as well as their role, if any, in organismal diversification. What is exciting and encouraging, however, is that in *Onthophagus* beetles, the most critical resources that will permit eventually reaching this goal appear to be in place, and that the first preliminary applications of these resources have already yielded many exciting insights. Much of the same can be said for the last focus of this chapter – the evolutionary developmental genetics of plasticity.

7.6 The development, evolution, and consequences of phenotypic plasticity

Phenotypic plasticity, the phenomenon by which a genotype gives rise to different phenotypes in response to changes in environmental conditions, is a ubiquitous property of all organisms (West-Eberhard, 2003). At one extreme, responses to changes in environmental conditions may be thought of as more passive, arising from the many biochemical and biophysical dependencies of biological processes. At the other extreme, such responses may arise through complex, highly choreographed and integrated adjustments of a wide range of traits. Examples of the latter extreme abound among polyphenic insects, such as seasonal or phase polyphenisms, social caste or alternative male reproductive phenotypes (reviewed in Nijhout, 1999; 2003c; Hartfelder & Emlen, 2005; Moczek, 2010).

Phenotypic plasticity has been integrated into quantitative and population genetics as genotype *x* environment interactions and visualized as reaction-norms across environmental gradients (Schlichting & Pigliucci, 1998). An intense debate dominated this field into the 1990s, centring in part around whether or not 'genes for plasticity' are needed to properly model the evolution of plasticity (Via *et al.*, 1995). This was an important issue, because the answer would determine whether plasticity could evolve independent of trait expression within each environment. While both camps vigorously argued their case, neither spent much effort actually exploring the genetic, cellular or developmental basis of plasticity.

This debate is now largely behind us, in part because it has become abundantly clear that many plastic responses to environmental change indeed involve a complex developmental machinery, the genetic basis of which can evolve and diversify on its own, independently of other aspects of trait expression (Moczek, 2009; 2010; Pfennig *et al.*, 2010). Instead, focus has now shifted toward characterizing the exact nature of the genetic basis of plasticity and whether different phenotypic manifestations of plastic responses, from graded to step-wise and threshold dependent phenotype adjustments, share similar genetic underpinnings.

Dung beetles have contributed to this change of focus by providing many examples of plasticity evolution in the context of the expression of alternative male phenotypes (reviewed in Moczek, 2009; and see Chapter 6 of this volume). Specifically, artificial selection experiments (Emlen, 1996) and common garden rearing of divergent populations (Moczek *et al.*, 2002) have provided evidence that body size thresholds separating alternative male morphologies can diverge genetically both in the laboratory and among natural populations. Subsequent developmental studies have offered the first hints that at least some of these divergences may be mediated by heritable changes in juvenile hormone signalling.

Overall, however, the developmental and genetic mechanisms underlying plastic responses remain poorly understood, including in insects generally and dung beetles in particular (Nijhout, 1999, 2003c; Hartfelder & Emlen, 2005; Moczek, 2010). Most importantly, the means by which these mechanisms are able to contribute and bias evolutionary diversification are only now beginning to be grasped. Recent developments in these directions are briefly discussed in the last section of this chapter.

7.6.1 Developmental mechanisms and the evolutionary consequences of plasticity

A growing number of studies have now shown that plasticity, whether polyphenic or otherwise, is often underlain by modularity in gene expression. In other words, different environmental conditions are associated with the expression of different suites of genes (Evans & Wheeler, 1999; 2001a; 2001b; Donnell & Strand, 2006; Hoffman & Goodisman, 2007). In such cases, the frequency of the inducing environment should determine the frequency by which a given suite of genes, or module, becomes expressed in a population within a given generation, and thus becomes visible to selection (Snell-Rood *et al.*, 2010). Rare conditions affecting gene expression only in a subset of individuals within a population should result in genes whose expression is specific to such rare conditions becoming hidden from selection and thus free to accumulate a larger number of mutations relative to genes expressed in every individual and in every generation. In turn, relaxed selection resulting from modularity in gene expression may bring about a fundamental trade-off between mutation accumulation on one side, and the degree of modularity in gene expression underlying plastic responses to environmental changes on the other (Snell-Rood *et al.*, 2010). If this is correct, such a trade-off would have far-reaching implications for defining the costs and limits, as well as the evolutionary consequences, of plasticity.

The notion that restricting gene expression to a subset of the population per generation can result in relaxed selection that may permit mutation accumulation has been examined previously in contexts outside phenotypic plasticity (e.g. evolution of senescence: Charlesworth, 1994; niche conservativism: Holt, 1996). Recent work on maternal effect genes has brought this concept closer to developmental genetics and, in particular, the modularity in gene networks (Cruickshank & Wade, 2008). Maternal effect genes are genes transcribed only by mothers. Mothers then incorporate transcripts, or their protein products, into their eggs. Strict maternal effect genes are only expressed by females and only function during early embryogenesis. The corresponding genes exist, but are not expressed, in fathers. Mutations that occur in paternal copies are therefore passed on to the next generation without being screened by selection.

Assuming equal frequencies of males and females in a population, the strength of selection operating on such genes is half that of the strength of selection operating on comparable genes expressed in every individual, or so-called zygotic genes.

Population-genetic theory consequently predicts that maternal effect genes should accumulate twice the mutation load within populations compared to similar zygotic genes (Wade, 1998). Similarly, theory predicts that, assuming nucleotide substitutions are at least mildly deleterious, maternal effect genes have the potential to diverge many times faster between species than corresponding zygotic genes (Demuth & Wade, 2007).

Both predictions are now matched by empirical data (Barker *et al.*, 2005; Demuth & Wade, 2007; Cruickshank & Wade, 2008). In the most extensive study to date, Cruickshank & Wade (2008) examined sequence variation within and between *Drosophila* species for 39 genes critical for early embryonic development. This list included nine strict maternal effect genes and 30 zygotically expressed genes. Following predictions, they found sequence variation within species to be 2–3 times higher for maternal-effect genes than any other gene class. Similarly, sequence divergences between species (*D. melanogaster* and *D. simulans*) were 2–4 times higher in maternal effect genes than any other gene class. Both findings strongly support the notion that relaxed selection acting on maternal-effect genes causes increased sequence variation within species, which in turn fuels more rapid divergences between species.

Even though this body of work did not explicitly address the consequences of modularity of gene expression, its theoretical foundation and predictions can easily be applied to a developmental plasticity context. Genes whose expression is restricted to individuals experiencing rare environments should exhibit reduced selection and accumulate mutations and, thus, greater sequence diversity within species. This, in turn, should create the potential for more rapid divergence between species, relative to similar genes expressed in every individual in every generation (Snell-Rood *et al.*, 2010). Recent studies on quorum-sensing genes in bacteria, which are induced only in generations exposed to certain population densities, provide support for both predictions (VanDyken & Wade, 2010).

The recent development of genomic resources for *Onthophagus* beetles has now created the opportunity to address these questions also in the context of developmental plasticity underlying the expression of alternative male morphs. Specifically, a series of microarray and DIGE-studies (Box 7.1) has permitted the identification of genes whose expression is specific to horned or hornless male morphs, or shared between morphs or sexes (Kijimoto *et al.*, 2009; Snell-Rood *et al.*, in press). Surveys of sequence variation are now ongoing to determine whether morph-specific genes harbour greater levels of nucleotide diversity within species, and also diverge faster between species, than similar morph-shared genes. Preliminary data on a small number of gene pairs are thus far consistent with both predictions (Snell-Rood & Moczek, unpublished data).

If genes underlying modular plasticity would, indeed, evolve faster than constitutively expressed genes, this could have far-reaching consequences for our understanding of the costs, limits and consequences of modular plasticity. For instance, accelerated mutation accumulation may permit genes underlying modular plasticity to evolve new functions more easily than similar non-plastic genes, and plasticity genes may thus contribute disproportionally to sub- and neo-functionalization events during organismal evolution (Demuth & Wade, 2007; Cruickshank & Wade, 2008). By the same argument, however, accelerated mutation accumulation may cause modular plasticity genes to be more likely to acquire deleterious mutations and devolve into pseudogenes.

The probability of acquiring a deleterious mutation should increase with the rarity by which a given gene is induced and, as such, may place an upper limit on the range of plasticity that can be accommodated through modular plasticity. Modules whose expression occurs very rarely may simply suffer too many mutations to be maintained within populations (Snell-Rood *et al.*, 2010). By the same token, even though relaxed selection may impose constraints on the range of plasticity that can be accommodated through modularity in gene expression, it may pave the way for the evolution of alternative genetic networks for plasticity to evolve, such as integrated networks where the same suites of genes, but via altered types of interactions, contribute to the expression of different phenotypes in different environments. Recent methodological and theoretical advances, including in *Onthophagus* dung beetles, promise that these speculations will soon be followed by empirical evaluation.

7.7 Conclusion

The increasing availability of genetic, developmental and genomic techniques and resources outside classic model organisms has permitted dung beetles to emerge as a promising group for investigating developmental evolution in nature and in the laboratory. In the process, studies on dung beetle evo-devo and eco-devo have begun to contribute to many fundamental and longstanding debates in evolutionary biology. Given the very recent nature of some of these developments, dung beetle evo-devo promises to be an exciting area for future research, with the potential for much discovery and much integration between development, evolution, ecology and behaviour.

Acknowledgements

I would like to thank Leigh Simmons and James Ridsdill-Smith for the opportunity to contribute this chapter. Quinton Hutton, Joseph Tomkins, Leigh Simmons, James Ridsdill-Smith and two anonymous reviewers provided many helpful comments on earlier drafts. Research presented here was supported in part by National Science Foundation grants IOS 0445661 and IOS 0718522.

8

The Evolution of Parental Care in the Onthophagine Dung Beetles

John Hunt and Clarissa House

Centre for Ecology and Conservation, School of Biosciences, The University of Exeter, Penryn, Cornwall, UK

8.1 Introduction

There are few areas of evolutionary biology that have progressed as rapidly as our understanding of the diversity that exists in animal mating systems (Shuster & Wade, 2003). Fundamental to this progress has been our growing appreciation of parental care, as many of the most striking differences in the reproductive behaviour of males and females are intimately linked to their involvement in the care of their young (Clutton-Brock, 1991). Sex differences in parental care play a central role in determining the intensity of sexual selection (Trivers, 1972), which, in turn, is responsible for the evolution of morphology (e.g. Hunt & Simmons, 2000), behaviour (e.g. Radford & Ridley, 2006) and physiology (e.g. Trumbo, 1997).

The evolution of parental care also represents a major breakthrough in the adaptation of animals to their environment. Its effectiveness in neutralizing conditions that are harmful to young is attested by the repeated convergence of parental care strategies in vastly different animal lineages (Tallamy & Wood, 1986; Reynolds *et al.*, 2002). Moreover, the fact that certain parental care strategies are characteristically associated with specific ecological niches (Ridley, 1978; Zeh & Smith, 1985) suggests that parental care is likely to have been an important precursor for range expansion into novel habitats (Ekman & Ericson, 2006), and may even promote speciation (Jones *et al.*, 2003).

Although parental care occurs in most animal taxa (Clutton-Brock, 1991), our understanding of how it evolves has a strong taxonomic bias. By far, the majority of

Ecology and Evolution of Dung Beetles, First Edition. Edited by Leigh W. Simmons and T. James Ridsdill-Smith.

theoretical and empirical research on its evolution has focused on species of fish, birds, and mammals; the evolution of parental care in insects has not attracted the same degree of attention (Clutton-Brock, 1991). Likewise, while our understanding of biparental care systems has increased dramatically over the last decade, most of our limited knowledge still comes almost exclusively from birds (Cockburn, 2006). Consequently, there is a crucial need for more studies on the evolution of parental care across a wider range of animal taxa that exhibit varying parental care strategies (Zeh & Smith, 1985; Clutton-Brock, 1991).

The taxonomic bias in the study of biparental care is surprising, given the similarity that many insect species share with birds in how they provide care to their offspring. Yet, until recently, the evolution of parental care in insects has received relatively little attention. Although caring for eggs or young is generally uncommon in insects (Zeh & Smith, 1985), where it does exist, the forms of parental care are diverse, including the preparation of nests, guarding, brooding or bearing eggs and young, provisioning offspring before and after birth and supporting them to nutritional independence (Zeh & Smith, 1985; Clutton-Brock, 1991). Moreover, although uniparental female care predominates in insects, uniparental male care and biparental care occur at high frequency in certain insect orders (Ridley, 1978; Zeh & Smith, 1985). More importantly, like birds, biparental care in insects is more frequently associated with nest building and the direct provisioning of young (Clutton-Brock, 1991), and it is characteristic of species with intense competition for larval resources (Halffter & Edmonds, 1982; Tallamy, 1984).

However, at present there is far more empirical data on birds than on insects to test between alternate hypotheses. As a result, explanations for the initial evolution and current distribution of parental care strategies in insects rely heavily on attempts to reconstruct the likely selective pressures from conditions that were experienced in the past (Clutton-Brock, 1991). Thus, more rigorous empirical studies are required that examine the costs and benefits to insects of uniparental male and female care, as well as biparental care, and how these relate to environmental conditions (Zeh & Smith, 1985).

The central aim of this chapter is to prove the utility of onthophagine dung beetles as an insect model for studying the evolution of parental care. To address this aim, we have structured the chapter into three major sections. We first provide a general overview of parental care theory, starting with the more conventional view that is commonplace in some prominent behavioural ecology textbooks (e.g. Alcock, 2009), and then move on to explain some of the more recent theoretical developments on this topic. We then illustrate this extensive theoretical base, using empirical studies on the evolution of parental care in onthophagine dung beetles. We will show that parental care in this genus shares many similarities with birds, but that there are a number of important differences that make them particularly well suited for empirical research on the evolution of parental care. We conclude with a brief section outlining our major findings and propose a number of directions for future research on the evolution of parental care in the genus. Our hope is that researchers reading this chapter will give greater consideration to using insects as their organism of choice when studying the evolution of parental care.

8.2 Parental care theory

The evolution of parental care has a rich theoretical base, and the aim of this section is to summarize the major theoretical developments on this topic. We start by presenting the 'conventional' view of the theory that is used to explain the evolution of parental care in some prominent behavioural ecology textbooks (e.g. Alcock, 2009) and undergraduate courses. We then outline some of the more recent theoretical developments that directly challenge these conventional views. Throughout this chapter, we follow Clutton-Brock's (1991) definition of parental care as '*any form of parental behaviour that appears likely to increase the fitness of the parent's offspring*'.

8.2.1 A conventional view of parental care theory

Darwin (1871) noted that it was typically males that were more eager to mate with any female, whereas females were often more discriminating in their choice of a mate. However, it was not until 1948 that a proximate reason was provided to account for this sexual difference. Bateman (1948) demonstrated that variance in the reproductive success of *Drosophila melanogaster* was greater among males than females, because a male's success increased proportionately with the number of matings, while a female's success did not increase substantially after the first mating. He argued that this finding was directly related to the way that males and females allocate their energy to the production of gametes. Since males generally produce far more, smaller gametes than females (known as *anisogamy*), a male's reproductive success is not limited by his ability to produce sperm but by his ability to fertilize eggs, while a female's success is governed by her ability to produce eggs rather than by having them fertilized. Given the widespread occurrence of anisogamy, Bateman (1948) argued that his findings should be applicable to the majority of sexually reproducing organisms. However, anisogamy alone could not explain those instances where the traditional sex roles were reversed (Clutton-Brock, 1991).

The significance of post-gametic parental care was first appreciated by Williams (1966), who noted that females made greater physiological sacrifices in the production of each offspring than did males. However, it was not until six years later that Trivers (1972) formalized this concept within a strict theoretical framework. Trivers advanced the arguments of Bateman (1948) and Williams (1966) in three major ways:

- First, he recognized that offspring production did not always end with the formation of gametes, and emphasized the need for a holistic approach that considered all possible ways that parents contribute to offspring fitness. Consequently, he combined all forms of parental care in a single term - parental investment - that he defined as '*any investment by the parent in an individual offspring that increases the offspring's chance of surviving (and hence reproductive success) at the cost of the parent's ability to invest in other offspring*' (Trivers, 1972).
- Second, he highlighted that all forms of parental investment could be measured by reference to their negative effects on the parent's ability to invest in additional

offspring. Thus, from the outset, Trivers advocated the integration of parental investment and life history theories and recognized that natural selection would operate on this fundamental trade-off to determine how parents allocate their investment to current and future breeding attempts.

- Finally, he stressed that parental investment was an important determinant of sexual selection, with the relative parental investment provided by males and females controlling the intensity of sexual selection operating on each sex. He argued that the sex providing the greatest parental investment will become the limiting resource, and that individuals of the sex investing least will compete amongst themselves to breed with members of the sex investing most. While he recognized that it was typically males that compete for mates, because they usually invest little in offspring, Trivers's formulation also explained the occurrence of role reversal in mating systems (e.g. Simmons, 1992).

Five years later, Maynard Smith (1977) applied the principles of game theory to the evolution of parental care. He identified two possible parental care strategies that could be employed by either sex - provide care or desert - and noted that the fitness of each strategy in a given sex will depend on the strategy adopted by the majority of individuals in the other sex. Maynard Smith demonstrated that four possible evolutionary stable strategies (ESS) could evolve, each demanding a necessary set of conditions to be maintained in the population (Table 8.1). While he demonstrated that the most frequently adopted ESS was for one parent to provide exclusive care, he also noted that biparental care could be the ESS if two parents can achieve twice

Table 8.1 The reproductive payoffs to male (below dotted lines) and female (above dotted line) parents adopting the caring and deserting strategies (modified from Maynard Smith, 1977). In all instances, $P0$, $P1$ and $P2$ equal the probability of offspring surviving when no care is given, care is given by one parent and care is given by both parents, respectively, such that $P2 \geq P1 \geq P0$. A male that deserts has p chance of mating again, a female that deserts produces W offspring, and one that guards produces w offspring, such that $W \geq w$. There are four possible ESS solutions:

1 **Both parents desert:** this exists when $WP0 > wP1$, otherwise the female will care, and when $P0\ (1 + p) > P1$, otherwise the male will care;
2 **Male cares exclusively**: this exists when $WP1 > wP2$, otherwise the female will care, and when $P1 > P0\ (1 + p)$, otherwise the male will desert;
3 **Female cares exclusively**: this exists when $wP1 > WP0$, otherwise the female will desert, and when $P1\ (1 + p) > P2$, otherwise the male will care;
4 **Both parents care**: this exists when $wP2 > WP1$, otherwise the female will desert, and when $P2 > P1\ (1 + p)$, otherwise the male will desert.

		FEMALE	
		Care	Desert
MALE	Care	$wP2$	$WP1$
		$wP2$	$WP1$
	Desert	$wP1$	$WP0$
		$wP1\ (1 + p)$	$WP0\ (1 + p)$

the reproductive success as one and/or if the deserting parent has little or no chance of obtaining additional mates (Table 8.1).

Despite its simplistic nature, Maynard Smith's (1977) game theory model emphasized that both the costs and benefits of providing care are integral components in determining the reproductive payoffs gained from, and thus the intensity of selection operating on, strategies of parental care.

8.2.2 More recent developments in parental care theory

In the previous section, we presented the sequence of events traditionally thought to lead to the evolution of parental care:

1. Anisogamy generates differences in the potential reproductive rates of male and females. All else being equal, the sex with the smaller pre-mating reproductive investment (males) has a higher potential reproductive rate and competes for access to the more heavily investing sex (females), who can afford to be choosy.
2. Differences in potential reproductive rates generate sexual differences in post-mating parental care because the fitness payoffs from deserting are usually larger for males than for females, making male uniparental care less likely to evolve.
3. Differences in post-mating parental investment further elevates male competition for females, which again exaggerates sexual differences in mating behaviour.

This sequence of events leads typically to the correlation between pre-mating and post-mating investment in offspring that predominates in most animal taxa. It is the view of parental care that still predominates in many behavioural ecology textbooks and undergraduate courses, despite several recent theoretical models demonstrating a major flaw in the argument (Queller, 1997; Houston & McNamara, 2002; Wade & Shuster, 2002; reviewed by Kokko & Jennions, 2003a; 2003b).

The problem is that this argument violates the simple fact that average male and female fitness *must* be equal (Wade & Shuster, 2002). The problem with assuming unequal average fitness between the sexes is that if females are spending most of their time caring, how can the fitness gains of a deserting male be higher when it will be much harder for him to find a mate?

Wade & Shuster (2002) highlighted the problem by reanalysing Maynard Smith's (1977) original game theory model. They showed that because his model does not assume equal average fitness for the sexes, it has an internal inconsistency, generated by the fact that deserting males gain their extra offspring by mating with females that do not exist in the population and therefore do not contribute to female fitness (Wade & Shuster, 2002). In the model, males can therefore gain more total paternity than females can produce offspring.

The flaw can be corrected, however, so long as it is explicitly stated where additional offspring come from. For example, Wade & Shuster (2002) model the scenario where males gain a share of paternity in the broods of caring males, or are able to mate with females that have deserted caring males. As soon as the additional paternity of a deserting male comes at a cost to other males in the population (i.e. any paternity gained by one male is considered lost paternity to another male), then the models become self-consistent and the logical flaws disappear (Houston & McNamara, 2002).

Wade & Shuster's model, along with an insightful paper by Queller (1997), have also challenged the long-standing view that parental care generates sexual selection. Instead, they claim that the direction of this causal relationship should be reversed. Queller noted that the costs of parental care *must* be measured in the currency of future offspring production. When viewed this way, the conventional explanations for sex differences in parental care disappear, because there is no inherent bias towards female uniparental care if both parents have the same future prospects for reproduction. This will be true, on average, in populations with equal sex ratios. However, whenever one sex becomes rarer in the population, then caring will be more costly for that sex (Queller, 1997).

Queller pointed out that one way for males to become rarer than females is when sexual selection generates non-random variance in male mating success. If some males are consistently more successful than others in obtaining mates, these males will have a higher average rate of reproduction than females, which elevates their gains from deserting.

A further way that sexual selection can influence the evolution of parental care is through the effect of sperm competition on certainty of parentage (Queller, 1997). Whenever sperm competition exists, males are less certain of their parentage than females and the conventional view is that males should provide less care (Trivers, 1972). However, as Queller (1997) points out, low paternity does not necessarily imply low future reproduction for males, because every future offspring must still have a father. Thus, low paternity does not produce a sex bias in the costs of providing care *per se*. The only important question that remains is whether the benefits of caring are smaller for males than for females. For any given offspring, the benefits for a male of caring are reduced by his lower average parentage, meaning that males should provide less care than females.

Collectively, this work provides compelling evidence that the causal relationship between parental care and sexual selection, as portrayed in more conventional parental care theories, should be critically reassessed. It is important to note, however, that this outcome primarily concerns the initial evolution of sex differences in parental care. It does not ignore the fact there is inevitably feedback between levels of parental care and the intensity of sexual selection (Clutton-Brock, 1991). As females spend more time caring for offspring, the greater the intensity of sexual selection will be on males as they compete to gain access to even fewer receptive females in the population. Thus, although these recent theoretical developments may be at odds with more conventional views, both approaches still recognize that the relative costs and benefits of providing care play a key role in understanding how and why the sexes differ in the amount of care they provide.

8.3 Testing parental care theory using onthophagine dung beetles

Dung beetles provide a rare example of parental care in terrestrial invertebrates, and they have become a useful model for empirically testing parental care theory because of the ease with which their mating system can be manipulated experimentally. Onthophagines, which represent one of the most speciose genera of dung beetles, have been particularly important in this regard. In this section, we start by introducing the mating system of onthophagine dung beetles and highlight the many similarities and differences that this genus shares with birds. We then illustrate how empirical research using this genus of dung beetles has been successful in testing some fundamental predictions of theoretical models. In particular, we will focus on the following five topics:

1. The benefits of care to offspring and the costs to parents
2. The behavioural dynamics that occur between the sexes when provisioning offspring together
3. The role that confidence of paternity has played in the evolution of paternal care strategies
4. How parents decide the optimal amount of care to provide offspring within a heterogeneous environment
5. The implications that parental care has for the evolution of offspring phenotype.

While we focus on these issues, it is important to note that these are by no means the only interesting topics that could potentially be addressed with this genus of dung beetle.

8.3.1 Parental care in onthophagine dung beetles

During reproduction, members of this genus typically remove portions of dung and pack it into the blind end of tunnels excavated beneath the dung pad. A single egg is then deposited into an egg chamber and sealed; the egg and its associated dung provision constitute a brood mass (Halffter & Edmonds, 1982). The dung provision represents the entire food source for the developing larvae, as well as offering protection against harsh environmental conditions and attack from parasites.

Many onthophagines exhibit male dimorphisms, in which large 'major' males develop enlarged horns on their head and/or pronotum and small 'minor' males essentially remain hornless (see Chapter 6 in this volume and also: Cook, 1987; Emlen, 1996; Hunt & Simmons, 1998b). Alternative male phenotypes are also frequently associated with alternative mating tactics, where major males fight for access to females and minor males sneak copulations (Chapter 4; Cook, 1988; Emlen, 1997a; Moczek & Emlen, 2000; Hunt & Simmons, 2001).

In many species, alternative phenotypes also differ in their parental care tactics, with major males assisting the female during construction of the brood mass (Lee & Peng, 1981; Cook, 1988; Sowig, 1996a; Hunt & Simmons, 1998a). Although females are capable of successfully provisioning an offspring alone, assistance by major males has varied positive effects on reproduction in the *Onthophagus* species thus far examined. In *O. binodis* and *O. vacca*, male assistance leads to an increase in the number of brood masses produced but does not alter the average weight of brood masses (Cook, 1988; Sowig, 1996a). However, in *O. gazella* and *O. taurus*,

Table 8.2 Some of the key similarities and differences between onthophagine dung beetles and many species of birds in how and when they provide care to offspring.

Parental Care	*Beetles*	*Birds*
Is the majority of care provided before egg laying?	Yes	No
Is the majority of care provided after egg laying?	No	Yes
Is there biparental care of offspring?	Yes	Yes
Can females provide successful uniparental care?	Yes	Yes
Can males provide successful uniparental care?	No	Yes
Do parents provision more than a single offspring at a time?	No	Yes
Is there direct interaction between parents and offspring?	No	Yes
Do parents forage independently when gathering food for offspring?	No	Yes

male assistance increases the weight of brood masses produced but does not alter the total number of brood masses (Lee & Peng, 1981; Hunt & Simmons, 1998a). Consequently, biparental care is likely to have important fitness consequences for both parents and offspring in this genus of dung beetles.

The onthophagine mating system shares many similarities with that of birds, but also many important differences (see Table 8.2). Onthophagine dung beetles, like many species of birds, commonly exhibit biparental care when provisioning offspring, yet, in both taxa, females are able to provision offspring successfully in the absence of a male. However, unlike many bird species, male dung beetles are unable to provision offspring by themselves. The major reason for this difference is that care in onthophagine dung beetles is provided before egg laying, whereas in birds, the bulk, and often most costly forms of care (e.g. incubation, feeding, nest defence), are provided after egg laying. Thus, without a female present to deposit an egg during the final stages of brood mass construction, a male will not provide care.

This temporal difference in care giving also prevents direct interaction between parents and offspring in onthophagine dung beetles, a key feature of almost all parental care systems in birds (e.g. Wright & Cuthill, 1990). Likewise, because onthophagine dung beetles only provision a single offspring at a time (i.e. each brood mass only has a single offspring), there is no interaction between offspring. This contrasts strongly with most bird species, where chicks compete directly for parental care in the nest (e.g. Lichtenstein & Sealy, 1998).

A final key difference between parental care in onthophagine dung beetles and birds is how they forage for resources when provisioning offspring. During biparental care in birds, males and females often forage independently of the nest. In contrast, because male and female dung beetles share a single tunnel during provisioning, their activities are highly dependent on each other (Hunt & Simmons, 2002b).

We argue that many of the features outlined above make onthophagine dung beetles an extremely useful alternative model to birds for studying the evolution of parental care. First, because offspring are provisioned in discrete brood masses, the amount of care received by offspring can be manipulated directly by adding or subtracting dung from the brood mass (Sections 8.2.2 and 8.2.5) and manipulated indirectly by varying the quality (Section 8.2.5) and/or quantity (Section 8.2.2) of dung made available to parents when provisioning. Manipulations enable both the costs (to parents) and benefits (to offspring) of parental care to be measured experimentally. Moreover, because only one offspring develops per brood mass, the often complicated effects of offspring competition do not exist.

The ability to rear beetles in the controlled conditions of the laboratory also facilitates manipulation of their mating system. For example, contrasting biparental with uniparental female care enables the benefits of paternal care to offspring to be determined (Section 8.2.2), and manipulating the number of minors in experimental populations enables the relationship between confidence of paternity and level of care to be examined (Section 8.2.4). Furthermore, the ability to rear large numbers of beetles under controlled conditions permits powerful quantitative genetic breeding designs to determine the underlying genetic basis of parental care behaviours (Section 8.2.6).

Finally, because males and females share a breeding tunnel and forage together, their parental care behaviours can be easily observed (Section 8.2.3). While this is achievable in many bird species, it is typically much more difficult.

8.3.2 The costs and benefits of parental care in onthophagine dung beetles

The costs and benefits of parental care play a central role in most theoretical models of how parental care evolves (Trivers, 1972; Maynard Smith, 1977), yet surprisingly few studies have quantified these key parameters. We know considerably more about the benefits of parental care to offspring than we do about the costs to parents, and this is particularly true for onthophagine dung beetles.

There is clear evidence for a number of *Onthophagus* species that the amount of dung provisioned by the mother into the brood mass influences offspring size at emergence (Lee & Peng, 1981; Emlen, 1994; Hunt & Simmons, 1997; 2000; 2004; Moczek, 1999; Kishi & Nishida, 2006). In *O. gazella* (Lee & Peng, 1981), *O. acuminatus* (Emlen, 1994), and *O. taurus* (Moczek, 1999; Hunt & Simmons, 2000), correlational studies have shown that larger offspring emerge from heavier brood masses, while manipulations of brood mass weight by adding or subtracting dung have provided direct evidence for this relationship in *O. taurus* (Hunt & Simmons, 1997; 2004). However, a recent correlational study in *O. ater* and *O. fodiens* (Kishi & Nishida, 2009) did not find a relationship between brood mass weight and offspring size, although this study was characterized by small sample sizes. In at least one of these species (*O. taurus*), body size at emergence is an important determinant of both male (Hunt & Simmons, 2001) and female (Hunt & Simmons, 2002c) reproductive success. There is also limited evidence that brood mass weight influences offspring survival in *O. taurus*, with offspring being more likely to survive to emergence when they develop in heavier brood masses (Hunt & Simmons, 1997; 2004).

There is also limited evidence for the benefits of paternal care to offspring from male removal experiments in *O. taurus*. In this species, the amount of care provided by a uniparental female increases linearly with her body size (see Figure 8.1a and also Hunt & Simmons, 2000). In contrast, major but not minor males provide

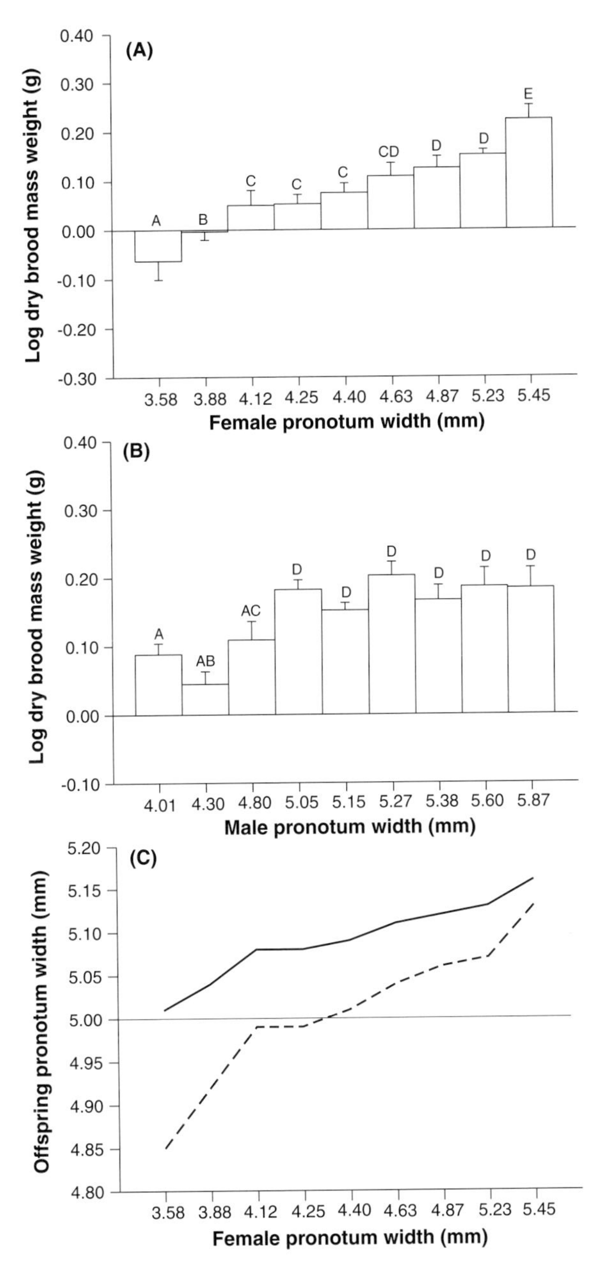

Fig. 8.1 The mean dry weight of brood masses produced by (A) unassisted females (i.e. uniparental care) of different body sizes and (B) male-female pairs (i.e. biparental care) where males (but not females) differ in size. Size categories with different letters are significantly different at $\infty = 0.05$. In (B), the solid horizontal line represents the mean dry brood mass weight that would be produced if females were provisioning unassisted. The broken horizontal lines represent the standard error about this mean. (C) Graphical representation of the relative influence of maternal effects (dashed line) and combined maternal and paternal effects (solid line) on the phenotype of male offspring. The horizontal line represents the pronotum width at which sons produced can be assigned to the major morph. Re-drawn from Hunt & Simmons (2000).

a fixed level of care, irrespective of their body size (Figure 8.1b; Hunt & Simmons, 2000). As offspring size is largely determined by the weight of the brood mass in this species (see Section 8.2.5), the extra care provided by major males has an important effect on the growth and eventual emergence size of offspring. In fact, when females breed alone, only the largest females are capable of provisioning enough resources to produce a major son (Figure 8.1c). However, with paternal assistance, females of all size classes are capable of producing major sons (Figure 8.1c).

In general, we know very little about the costs to parents of providing care. Numerous field studies on birds have used clutch size manipulations to demonstrate that parents often experience a cost as a result of chick rearing (reviewed by Lindén & Møller, 1989). However, such studies have only gone part way in demonstrating a cost of parental care, because not all aspects of reproductive effort are manipulated by varying the number of nestlings (Partridge & Harvey, 1988). For example, a number of empirical studies have demonstrated that both mating (e.g. Stutt & Siva-Jothy, 2001) and the production of eggs can be costly to females (e.g. Partridge *et al.*, 1987). Therefore, to provide direct evidence for a cost of parental care, manipulative experiments are required which control for costs that may be accrued during earlier reproductive stages.

In a series of experiments manipulating the mating system of *O. taurus*, Hunt *et al.* (2002) showed that uniparental female care reduces female survival, and that this cost was independent of egg production. They manipulated both the mating status of females (mated to either a minor or a major male, or unmated) and the amount of dung made available to provision offspring (75 g, 125 g and 300 g). Their findings showed that while virgin females do not produce brood masses, when mated females are provided with more dung they produce more brood masses that are lighter in weight. Not surprisingly, virgin females survived longer than mated females at each dung level, and the survival of mated females decreased as dung availability increased. Consequently, this experiment provided clear evidence for a cost of reproduction to females, but it could not determine whether this cost was due to producing more eggs or providing more care.

To distinguish between these alternatives, Hunt *et al.* (2002) mated females to either a major or a minor male and then let these males reside with the female while provisioning offspring. Females provisioning offspring with assistance from a major male provided relatively lower levels of care than unassisted females (i.e. paired with a minor male) (see Section 8.2.3 in this volume and Hunt & Simmons, 2002). Therefore, although male assistance leads to the production of significantly heavier brood masses, a female's contribution to parental care decreases when assisted.

However, paternal assistance does not affect the number of brood masses produced (Hunt & Simmons, 1998a, 2002b), and hence all females invest the same into egg production. Hunt *et al.* (2002) found that the mean number of brood masses produced by assisted and unassisted females increased with dung availability. Moreover, while mean brood mass weight decreased with dung availability in unassisted females, that produced by cooperative pairs was unaffected by dung availability. Therefore, their finding that female survival is increased when receiving paternal assistance provides definitive evidence that maternal care reduces lifespan in this species.

8.3.3 Behavioural dynamics of the sexes during biparental care

In uniparental care systems, parental care should be affected by the trade-off between the benefits of current investment to offspring fitness and the costs of this investment to future reproduction (Williams, 1966). However, in systems with biparental care, an individual's optimal investment will not only depend on this trade-off but also on the amount of investment provided by the cooperating partner (Williams, 1966; Trivers, 1972). Whenever members of a cooperative pair do not share the common goal of maximizing joint fitness (Trivers, 1972), the investment optima of the sexes will differ, generating sexual conflict over the amount and division of care. Consequently, the net investment of two parents is likely to reflect the outcome of a contest played between the sexes over behavioural or evolutionary time.

A number of theoretical models have attempted to predict how the care provided by one parent should change relative to the care being provided by its partner (Chase, 1980; Houston & Davies, 1985; Winkler, 1987; McNamara *et al.*, 1999; Jones *et al.*, 2002; Johnstone & Hinde, 2006). Early models assumed that each parent provided a fixed level of care that maximizes its fitness based on the care provided by its partner (Chase, 1980; Houston & Davies, 1985; Winkler, 1987). Under most conditions, the amount of care provided by two cooperating individuals is expected to be negatively correlated over evolutionary time, with the shortfall of one partner being compensated for by the other (Chase, 1980; Houston & Davies, 1985; Winkler, 1987).

Later models (McNamara *et al.*, 1999; Jones *et al.*, 2002; Johnstone & Hinde, 2006) permitted parents to negotiate the care they provide in real time (i.e. behavioural as opposed to evolutionary time). McNamara *et al.* (1999) also predicted that if one parent reduces its care, the ESS response by its partner is to increase its own care but not so much that it completely compensates for the lost care. However, Jones *et al.* (2002) and Johnstone & Hinde (2006) showed that it is theoretically possible for this negotiation between parents to lead to partial, full or even no compensation, depending on the shapes of the cost and benefit functions for parental care, and the degree to which each parent is informed of their offspring needs.

Empirical evidence for compensatory responses of parents to changes in the level of care provided by their partner has come almost exclusively from studies on birds with biparental care, and these have yielded conflicting results. A recent meta-analysis across 54 bird studies suggested that the mean response was for one parent to partially compensate for any reduction in care by the other. Although females appeared better than males at compensating, this depended on the type of manipulation used (i.e. parent removal versus handicapping) (Harrison *et al.*, 2009). Despite this average effect, considerable variation still exists in how birds respond to the level of care provided by their partner, with some species exhibiting only partial compensation for their partner's reduction (e.g. Markman *et al.*, 1995), while others exhibit complete compensation (e.g. Wolf *et al.*, 1990) or exhibit no compensatory response at all (e.g. Lozano & Lemon, 1996). Similarly, partner removal studies in insects (e.g. Suzuki & Nagano, 2009), fish (e.g. Itzkowitz *et al.*, 2001) and mammals (e.g. Cantoni & Brown, 1997) have also revealed mixed results, although the number of experiments on non-avian taxa is much smaller (Harrison *et al.*, 2009).

To determine whether parents adjust their level of care relative to their partner's contribution in *O. taurus*, Hunt & Simmons (2002b) observed the behavioural dynamics of parents during biparental care. They showed that *O. taurus* exhibits a clear division of labour between the sexes when providing care. During the provisioning of a brood mass, females allocated approximately 84 per cent of their time towards parental care duties, while males allocated only 48 per cent of their time. This contrasts with a number of bird species, where the male's contribution to offspring feeding either equals (e.g. Smith *et al.*, 1988) or exceeds that of the female (e.g. Sasvàri, 1990).

Furthermore, while both parents were able to perform all caring behaviours, each sex exhibited specific parental roles. A male's primary role was to remove dung from the pad and deliver it to the female in the brood chamber (Figure 8.2a), while the female's primary role was to incorporate and pack this dung into the brood mass (Figure 8.2b). It has been argued for a number of bird species (e.g. Markman *et al.*, 1996) that sex-specific parental roles facilitate the efficiency with which biparental care is provided to offspring.

Contrary to the general theoretical prediction of a negative correlation between the levels of care provided by the two members of a cooperative pair (Chase, 1980; Houston & Davies, 1985; Winkler, 1987; McNamara *et al.*, 1999), Hunt & Simmons (2002b) found that the level of care provided by the sexes was significantly positively correlated. Interestingly, Johnstone & Hinde (2006) showed that if there is a large informational asymmetry between partners over offspring need, the poorly informed parent may match changes in the care provided by its partner. Thus, if females are better informed over offspring needs than males, this may result in males coordinating their care giving with the female rather than acting independently. Alternatively, when parents coordinate the care they provide, they may receive greater fitness returns by exhibiting reciprocal altruism (Trivers, 1971). Under such conditions, it may be more appropriate to view the actions of cooperating parents as an iterated game, where an individual cooperates on its first move and then continues to match its partner's previous move in subsequent actions (Axelrod & Hamilton, 1981). Such 'tit-for-tat' strategies are known to be evolutionary stable if animals play stochastically (e.g. Boyd, 1989) or if parents differ in their quality (e.g. Leimar, 1997).

An interesting outcome of such models is that when the probability of repeated interactions between parents is increased, individuals are predicted to not only cooperate more but develop greater generosity (Lazarus, 1989; Sherratt & Roberts, 1998). In *O. taurus*, the restriction of parents to a breeding tunnel ensures that they interact repeatedly during offspring provisioning, and this may provide the necessary conditions favouring the coordination of caring behaviours between the sexes. Moreover, the absence of direct interactions between parents and offspring in this species may place a greater emphasis on interactions between parents in determining investment optima in this species.

Finally, the work of Hunt and Simmons (2002b) showed that females compensate for a reduction in male care, with uniparental females providing, on average, 15 per cent more care than biparental females. However, despite the increased level of care provided by unassisted females, this compensation is incomplete. The total number of caring behaviours performed by uniparental females was, on average, 14 per cent lower than for biparental females, resulting in the production of brood masses 11 per cent lighter in weight.

Why, then, do unassisted females only partially compensate for the lack of a partner? It may be because unassisted females are physically unable to compensate

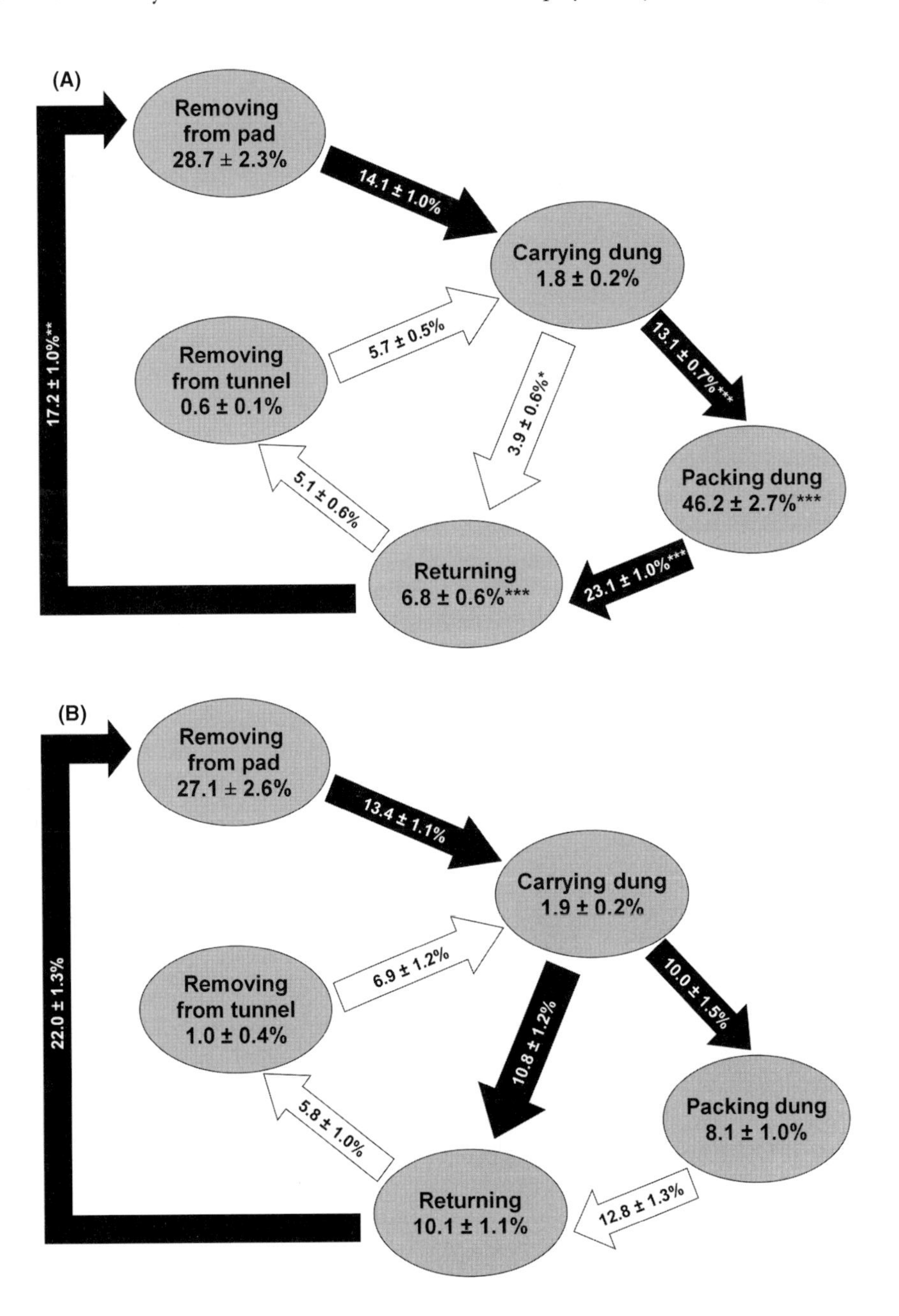

for the loss of their partner. In birds, incomplete compensation is often the result of physical limitations caused by time constraints on the rate at which food can be gathered (e.g. Wright & Cuthill, 1990), or because unassisted females are already working at their physiological limit (e.g. Drent & Daan, 1980). An alternative explanation is that the strong sex-specific parental roles observed in *O. taurus* make it difficult for unassisted females to compensate fully for the absence of a male. Clearly, more work is needed to determine the proximate mechanism involved in incomplete compensation by females.

8.3.4 Confidence of paternity and paternal care

Hamilton's rule ($rb - c > 0$) states that the evolution of parental behaviour should depend exclusively on the costs to the parents (c) or the benefits to the offspring (b) because the relatedness (r) between parent and offspring should be the same for all offspring and for both parents (Hamilton, 1964). However, sperm competition occurs in most animal taxa (Simmons, 2001) and is particularly intense in onthophagine dung beetles (see Chapter 4). This can significantly reduce the relatedness of males to a given cohort of offspring, and Trivers (1972) proposed that this uncertainty of paternity would favour male desertion, because failure to assure paternity prior to investment would put a male at a selective disadvantage when competing against more reproductively selfish individuals. Consequently, variation in a male's confidence of paternity has been proposed to explain the observed inter- and intrasexual differences in the magnitude of paternal care provided (Ridley, 1978).

The relationship between confidence of paternity and paternal care has received considerable theoretical attention. Early models by Maynard Smith (1978) and Grafen (1980) concluded that low confidence of paternity alone is insufficient to select against paternal care because it will, on average, affect all males equally. These models, however, assume that paternity remains constant across all matings, that individuals are unable to assess their own level of paternity, and that the only cost of providing investment is missed opportunities for re-mating. When one or more of

Fig. 8.2 Ethogram showing the parental behaviour pathways for (A) female and (B) male *O. taurus* during offspring provisioning. The behaviours performed by parents during offspring provisioning were broadly categorized as either parental or non-parental. Parental behaviours directly contributed to the production of the brood mass and included removing dung from the pad or tunnel, carrying this dung to the brood chamber and packing it into the brood mass, and then returning to either the pad or tunnel to collect more dung. The values presented under each parental behaviour represent the mean (± SE) percentage of time spent performing that behaviour. The values inside the arrows represent the mean percentage (± SE), out of all possible behavioural transitions, that one parental behaviour was followed by another. The most frequently taken pathway is highlighted with black arrows. Since each parental behaviour could be followed by a non-parental behaviour, these percentages do not necessarily add to 100 per cent. To compare the proportion of time spent performing each parental behaviour and the transitions between each behaviour between the sexes, paired *t*-tests were used. Significant differences between the sexes are presented as asterisks in the female ethogram (** $P \pm \leq 0.001$, *** $P \leq 0.0001$). Re-drawn from Hunt & Simmons (2000).

these assumptions are relaxed, reduced confidence of paternity can, theoretically, select against the evolution of paternal care (Werren *et al.*, 1980; Winkler, 1987; Westneat & Sherman, 1993; Xia, 1992; Whittingham *et al.*, 1992).

Empirical studies of the influence of paternity on paternal care have again come predominantly from studies of birds. Some have reported positive associations between paternity and paternal care (e.g. Sheldon & Ellegren, 1998), while others have found no such relationship (e.g. Whittingham & Lifjeld, 1995). It is difficult to assess the general influence of confidence of paternity on paternal care from such studies because of the diversity of the different experimental manipulations that have been employed (Wright, 1998). Thus, concrete evidence that confidence of paternity can generate facultative variation in patterns of paternal care still remains elusive in birds (Wright, 1998).

More conclusive evidence that paternity can influence the evolution of paternal care comes from experimental studies on fish (Neff & Gross, 2001; Neff, 2003) and insects (Hunt & Simmons, 2002c; García-González *et al.*, 2003). However, these studies are restricted to only a small number of species.

Hunt & Simmons (2002c) examined the relationship between confidence of paternity and paternal care in *O. taurus* by experimentally manipulating the number of minor 'sneak' males in experimental populations containing a single major male and examining how this male adjusted his level of care. They showed that the paternity of a major male declines with the number of sneak males in the population. However, the relationship was non-linear, with paternity declining most rapidly between a frequency of one and three sneaks and stabilizing thereafter at about 50 per cent (Figure 8.3a).

Next, they showed that pairs exposed to sneak males suffered a reduction in brood mass weight, while control pairs (without sneaks) showed no change in brood mass weight (Figure 8.3b). The reduction in brood mass weight was non-linear, with the greatest decline occurring between no sneaks to three sneaks, then stabilizing beyond this frequency (Figure 8.3b). Moreover, when exposed to sneak males, pairs showed a linear reduction in the number of brood masses produced, while control pairs showed an increase in the number of brood masses produced over time (Figure 8.3c).

In a final experiment, Hunt & Simmons (2002c) recorded the behaviour of major males when provisioning without sneak males, and when provisioning with either 1 or 3 sneak males. They found that the number of sneaks had a significant effect on the rate of desertion by the major male: only 17 per cent of major males deserted in the absence of sneak males, while 74 per cent deserted when one sneak was present and 78 per cent when three sneaks were present.

If a major male did not immediately desert when exposed to sneak males, he increased the proportion of time spent guarding females and decreased the proportion of time spent provisioning the brood mass. Conversely, in the absence of sneaks, major males increased the proportion of time spent provisioning and decreased the proportion of time spent guarding. Not surprisingly, the proportion of matings by a major male decreased with increasing sneak frequency (control = 100 per cent, 1 sneak = 70 per cent, 3 sneaks = 55 per cent), in close agreement with the negative relationship between paternity and sneak frequency seen in Figure 8.3a.

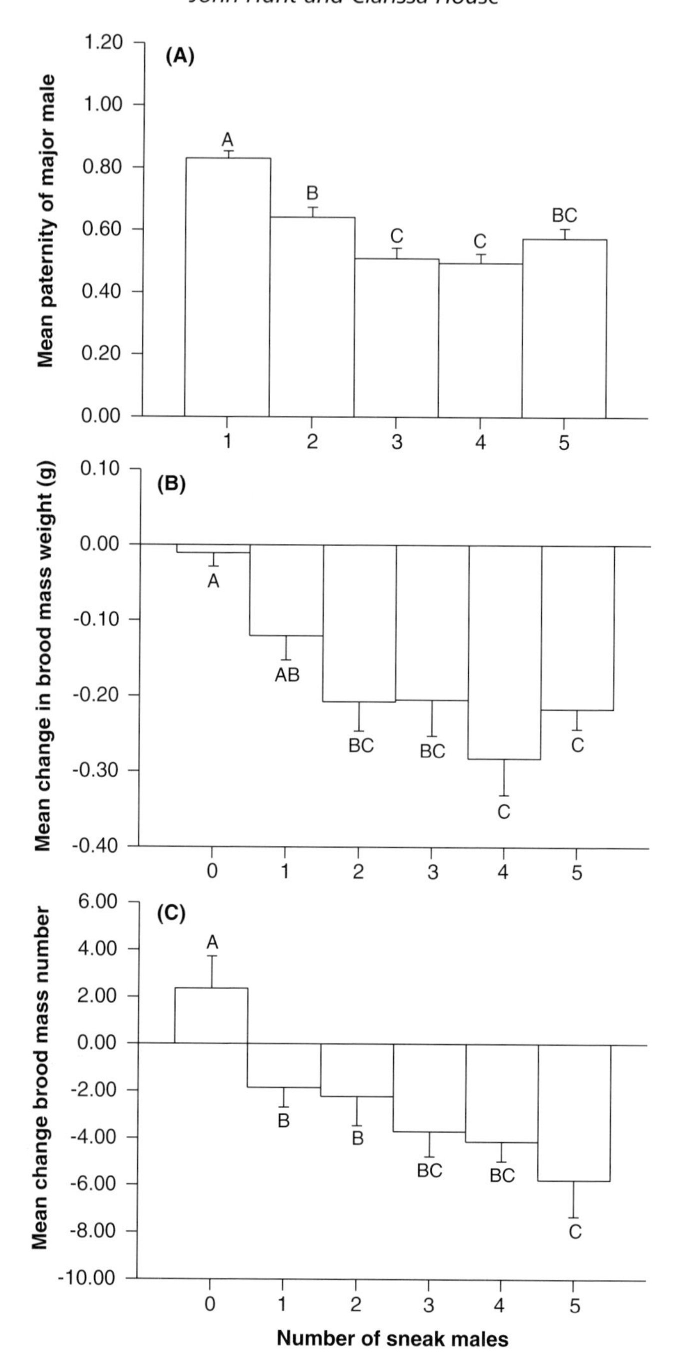

Fig. 8.3 The relationship between sneak frequency and: (A) the proportion of offspring sired by the major male; (B) the change in the mean weight of brood masses produced by a major male-female pair; (C) the number of brood masses produced by a major male-female pair. In each instance, treatments with different letters differ significantly at $P \leq 0.05$. Re-drawn from Hunt & Simmons (2000).

The level of care provided by a major male was not related to the number of matings he attained but, rather, there was a negative relationship between the amount of care he provided and the number of fights he had with sneak males intruding into the breeding tunnel. Thus, by assessing the number of fights with sneak males, a major male appears able to gauge his confidence of paternity and adjust his investment accordingly.

8.3.5 Do parents optimize the care they provide?

Given the importance of parental care to offspring fitness, parents should be expected to maximize their investment in each offspring. However, resources are often finite, so that the more investment in each offspring, the fewer the number of offspring that can be produced. Consequently, parental investment strategies are expected to represent a trade-off between the benefits of the investment to individual offspring and the costs to future reproduction (Trivers, 1972).

Theoretical models examining whether parents should optimize the care they provide have largely examined the trade-off between offspring size and number (Smith & Fretwell, 1974; Parker & Begon, 1986; McGinley *et al.*, 1987; McGinley & Charnov, 1988). Smith & Fretwell (1974) presented a graphical model where they assumed that the relationship between parental investment and offspring fitness showed diminishing returns, and that offspring required a minimum level of parental investment to survive. They concluded that parental fitness should be maximized by a single, optimal level of investment in all offspring. Subsequent extensions of this model incorporating developmental constraints (Parker & Begon, 1986), multiple resources (McGinley & Charnov, 1988) and variation in the availability of resources experienced during investment (McGinley *et al.*, 1987) have also provided support for a single optimal level of investment.

A general criticism of the modelling approach taken by Smith & Fretwell (1974) is that it assumes that the relationship between parental investment and offspring fitness is constant across environments. This is unlikely to be true with factors such as environmental uncertainty, predation pressure and the extent of resource competition all likely to shape this relationship (Parker & Begon, 1986; McGinley *et al.*, 1987). Theoretical models including the effects of environmental heterogeneity on parental care strategies have produced conflicting results. Some models have shown that when the environment is unpredictable, variable parental investment strategies will be favoured (e.g. Kaplan & Cooper, 1984), while others have shown that environmental heterogeneity is unlikely to favour variable investment per offspring (e.g. McGinley *et al.*, 1987). Part of the reason for this inconsistency of predictions is likely due to the vastly different modelling approaches used in these studies. However, these models do highlight the fact that the theoretical debate over how parents should invest in offspring is far from being resolved.

In line with the above theoretical models, empirical studies have also shown that there is great diversity in the way that parental investment strategies vary with environmental conditions (reviewed by Bernardo, 1996). Some studies have shown that the variance in offspring investment increases in unfavourable environments (e.g. Thompson, 1984), while others have shown that only the mean level of investment increases (e.g. Brody & Lawlor, 1984).

This variation in levels of investment provided by parents is also apparent in the onthophagine dung beetles, where many species are known to adjust the size of brood mass with environmental conditions. For example, *O. taurus* (Moczek, 1999; Hunt & Simmons, 2004), *O. atripennis* (Kishi & Nishida, 2006) and *O. ater* (Kishi & Nishida, 2009) all reduce the weight of the brood mass produced as dung quality increases, while still producing larger offspring. Moreover, *O. vacca* (Sowig, 1996b) and *O. taurus* (Hunt & Simmons, 2004) both produce heavier brood masses when provisioning in relatively dry soil versus wet soil, presumably to protect the developing offspring from desiccation. Collectively, these studies suggest that the observed plasticity in levels of parental care may represent an optimal strategy to maximize parental fitness.

A potentially useful model for exploring the adaptive significance of parental care strategies is to examine the costs and benefits of marginal increments in the allocation of resources by a parent. This approach, referred to as the marginal value theorem (MVT), assumes that the fitness gains (W) from provisioning an offspring depend on how long the parent spends performing the activity (t), and that the fitness gains of this investment show exponentially diminishing returns that can be estimated according to Equation 8.1:

$$W(t) = W_{\max}[1 - \exp(-rt)] \tag{8.1}$$

where r is the rate at which W rises to its asymptote, W_{max}.

MVT predicts that the optimal time spent providing care (t^*) will occur when the marginal gains of provisioning a given offspring fall below those received by provisioning a new offspring. The optimum, t^*, can be found graphically by constructing the tangent from the cost, C, associated with provisioning a new offspring, to the gain curve (Figure 8.4). The optimum can be approximated mathematically using equation 8.2:

$$t^* \approx \frac{1}{r}\left[\ln(rC+1) + \frac{\ln(rC+1)}{rC+1}\right] \tag{8.2}$$

$t^*$is only influenced by r and C (a larger C increases t^* while an increase in r reduces t^* (Figure 8.4).

To determine whether the plasticity that female *O. taurus* exhibits when provisioning offspring with dung differing in quality (cow versus horse dung) and in soil moisture (5 per cent versus 10 per cent water content) represents an optimal strategy, Hunt & Simmons (2004) estimated the key parameters necessary for the MVT approach (Table 8.3) and used these to predict the optimal level of care that a female should provide in each environment. They showed that offspring provisioned on high-quality horse dung were characterized by a slightly higher rate of increase in body size per unit of parental investment (r) and a markedly higher asymptotic body size (W_{max}), compared to offspring provisioned on lower-quality cow dung (Table 8.3, Figure 8.5a). However, these parameters did not differ with soil moisture content (Table 8.3, Figure 8.5b).

Next, Hunt & Simmons (2004) showed that the minimum amount of care needed for an offspring to survive in each of the environments (t_{min}) was lower

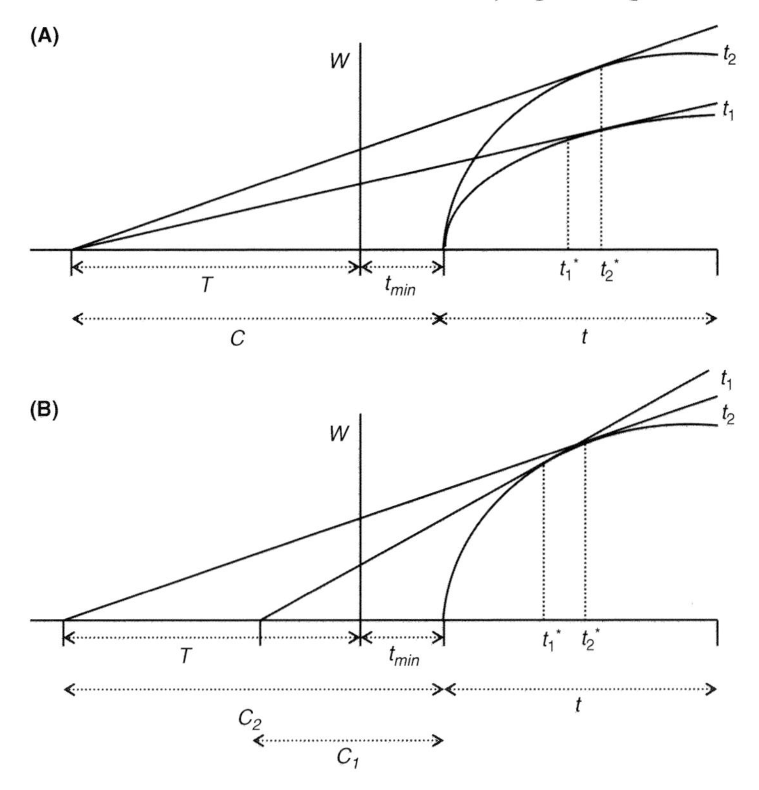

Fig. 8.4 Graphical representation of the Marginal Value Theorem (MVT). (A) Two gain curves are shown. In t_1, the fitness gain (W) rises more slowly with time (t) and has a lower asymptotic fitness (W_{max}) than in t_2. Assuming a fixed total time cost (C), the tangent to gain curve t_1 predicts a lower optimum (t^*) than the tangent to gain curve t_2 because of the greater rate of fitness gain. (B) Increasing C (from C_1 to C_2) will have the effect of increasing t^* for any given gain curve because the tangent of the line will intersect the curve at a higher value of t. It is assumed that a parent has some minimum investment (t_{min}) that must be made in order to produce an offspring and a cost associated with providing investment (T). The total cost, C, is therefore $t + t_{min}$. In Hunt & Simmons's (2004) treatment of female provisioning in *O. taurus*, W was quantified as offspring body size and t was measured as the time taken to provision a brood mass. t_{min} was calculated as the minimum amount of time taken to provision an offspring that yielded 95 per cent likelihood of survival and T was estimated as the time required to construct a breeding tunnel (see text for more details). Re-drawn from Hunt & Simmons (2000).

when provisioning with horse dung than cow dung, and when provisioning in soil with high moisture content versus low moisture content (Table 8.3).

One of the major costs of providing care in onthophagines is the time it takes to construct the breeding tunnel (T), and Hunt and Simmons showed that T did not differ with dung quality but was considerably shorter when females constructed tunnels in moist soil compared to dry soil (Table 8.3). The total cost for providing care to offspring ($C = t_{min} + T$) was therefore higher when provisioning with cow dung versus horse dung, and when provisioning in low moisture soil versus high moisture soil (Table 8.3).

Table 8.3 The key parameters used in Hunt & Simmons (2004) MVT estimation of optimal parental care in *O. taurus* and how they differ with dung quality and soil moisture. Parameters are explained in full detail in the text and in Figure 8.5

	Dung quality		*Soil moisture*	
Parameter	*Horse*	*Cow*	*Low*	*High*
W_{max}	4.90 mm	4.45 mm	5.10 mm	5.10 mm
r	0.0063 mm/min	0.0061 mm/min	0.0034 mm/min	0.0033 mm/min
t_{min}	140.89 min	154.63 min	168.07 min	156.86 min
T	364.44 min	364.44 min	369.23 min	180.11 min
C	505.33 min	519.07 min	537.14 min	336.97 min
Predicted t^*	281.43 min	299.19 min	413.82 min	333.99 min
Observed t^*	736.91 min	861.28 min	914.23 min	866.31 min

Collectively, the estimated MVT parameters set a clear picture of the costs and benefits of providing care in these alternate environments. When provisioning offspring on high-quality horse dung, the marginal gains (r and W_{max}) are higher and the costs (C) are slightly lower, compared to provisioning with lower-quality cow dung. In contrast, when provisioning in soil varying in moisture content, the marginal gains were largely constant but the costs of providing care were much lower on soil of high moisture content.

By substituting these crucial MVT parameters into Equation 8.2, Hunt & Simmons (2004) showed that the optimal amount of care provided by a female should be lower on horse than cow dung, and also lower when in high moisture soil than in low moisture soil (Table 8.3). They then compared these theoretically predicted optima with observed levels of care provided by females in each environment. On both dung types and soil moistures, the observed levels of care provided by females agreed qualitatively with the theoretically predicted optima (Table 8.3). However, in all cases, the observed level of care was over twice the theoretically predicted optima (Table 8.3).

There are a number of possible reasons for the lack of a quantitative fit between observed and theoretically predicted levels of parental care, but most centre on the inaccuracies of estimating the gains curve and/or costs of providing care. The MVT approach taken by Hunt & Simmons (2004) assumes that offspring body size is linearly related to fitness, and that the relationship between parental care and fitness shows diminishing returns. While this is likely to be the case in females, it seems improbable in males, given the non-linear relationship between body size and reproductive success in this species (see Chapter 6 in this volume, and Hunt & Simmons, 2001).

Similarly, it is likely that the estimates of costs are severely underestimated in *O. taurus*. Although the time costs of provisioning a breeding tunnel are undoubtedly important in onthophagine dung beetles, this cost is likely to have been greatly reduced in the laboratory with females provisioning on a homogeneous soil mixture. Furthermore, it is extremely unlikely that this represents the only cost to parents of providing care. It is at least promising that when females are given an active choice between provisioning on horse and cow dung, they show a clear

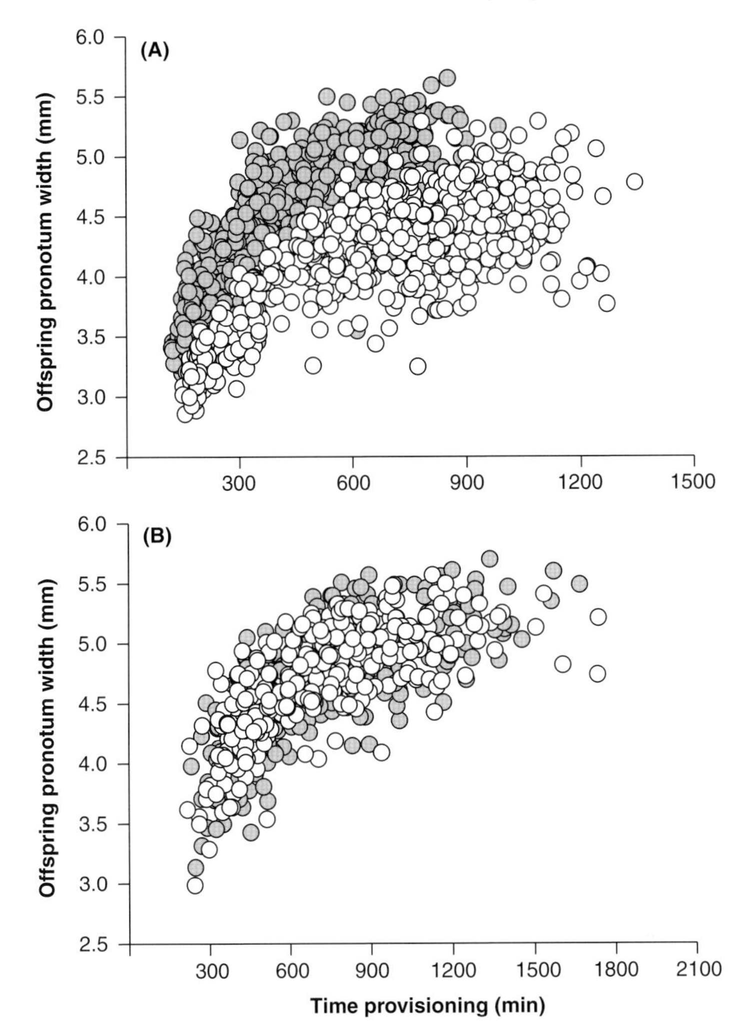

Fig. 8.5 The relationship between the time taken to construct a brood mass (per offspring investment) and offspring pronotum width (offspring fitness) for offspring developing on (A) horse dung (closed symbols) and cow dung (open symbols) and on (B) high soil moisture (open symbols) and low soil moisture (closed symbols). Re-drawn from Hunt & Simmons (2000).

preference for the dung type yielding greater fitness returns, producing 3.5 times more brood masses on horse than cow dung (Hunt & Simmons, 2004).

8.3.6 Evolutionary quantitative genetics of parental care

The majority of empirical research on the evolution of parental care has examined the costs of providing care to parents and the benefits received by offspring. Considerably less attention has been given to studies examining the level of genetic variation that exists in parental care behaviour needed for its evolution (but see

Mavrogenis & Papachristoforou, 2001; Snijders *et al.*, 2001; Rauter & Moore, 2002). In many species, parental care represents one of the most important environments that the offspring encounter during development. If there is genetic variation among parents in the quality of the care they provide, then indirect genetic effects (IGEs) will exist (Wolf *et al.*, 1998). Thus, while parental care represents an environmental effect to the offspring receiving it, if it has a genetic basis, then this environment can be inherited across generations and therefore can contribute to the evolutionary process (Wolf *et al.*, 1998).

Hunt & Simmons (2002a) estimated the relative contributions of genetic and non-genetic effects to the levels of maternal care provided by daughters, and their body size, in *O. taurus*. They showed that while offspring body size exhibited little genetic variance ($h^2 \pm SE = 0.01 \pm 0.07$), the weight of the brood mass produced by females exhibits substantially higher levels of additive genetic variance (0.13 ± 0.09).

Furthermore, they showed that body size exerts a strong maternal influence on the weight of brood masses produced, which, in turn, accounts for 22 per cent of the non-genetic variance in offspring body size. Maternal body size accounted for 5 per cent of the variation in the number of offspring produced, but there was no genetic variance for this trait. Offspring body size and brood mass weight exhibited positive genetic ($r_A = 2.58 \pm 9.74$) and phenotypic correlations ($r_P = 0.40 \pm 0.11$). Together, these findings suggest that both indirect genetic effects, via maternal care, and non-genetic maternal effects, via female size, play important roles in the evolution of offspring phenotype in this species.

Although previous work has concluded that body size is not heritable in this species (Mozcek & Emlen, 1999), Hunt & Simmons (2002a) show that the conditions necessary for the evolution of body size do exist in *O. taurus* (Wolf *et al.*, 1998). Whenever parental care has a genetic basis, there is potential for this care to drive the evolution of phenotypic traits that exhibit little or no additive genetic variance (Wolf *et al.*, 1998). This contradicts more mainstream evolutionary theory, which posits that phenotypic traits must have a genetic basis to evolve (Lynch & Walsh, 1998). Furthermore, indirect genetics effects can, in theory, impede or accelerate responses to selection and can generate large time-lags in the evolutionary response to selection (Kirkpatrick & Lande, 1989; Wolf *et al.*, 1999). In *O. taurus*, the positive genetic correlation between offspring pronotum width and brood mass weight suggests that the indirect genetic effect is likely to promote the evolution of body size across generations; however, considerably more work is needed before this can be proven empirically.

8.4 Conclusions and future directions

The aim of this chapter was to prove the utility of onthophagine dung beetles as a valuable insect model for studying the evolution of parental care, and to present this fascinating genus as a feasible alternative to birds when studying parental care. To this end, we have provided a small taste of what is possible, and have shown how research on this genus has progressed our understanding of how parental care evolves. However, many more questions remain to be answered, and we believe that onthophagine dung beetles will continue to prove to be fruitful research organisms for many years to come. We also note that detailed studies of other taxa, such as the rollers, are likely to be equally valuable, especially given their tendency to provide continued care after oviposition (Chapter 1). While the list of possible research questions is exhaustive, we outline a few in this concluding section that we believe will provide greater depth to our understanding of how parental care evolves in this diverse genus of dung beetles.

There is a crucial need for more research on the costs and benefits of paternal care. While we know that major males of a number of onthophagine species provide assistance, we still know very little about exactly how much care they provide and the consequences this has for offspring fitness. For example, is paternal care always linked to the male dimorphism, and does this relationship differ with environmental and social conditions? What patterns of paternal care are found in male monomorphic species?

We also know very little about other forms of paternal care in onthophagines, such as protection from predators and preventing brood parasitism. The later is particularly likely to be important in those *Onthophagus* species that reach high densities and therefore experience intense competition for dung resources (e.g. Moczek & Cochrane, 2006), or those that are subject to high rates of brood parasitism (González-Megías & Sánchez-Piñero, 2004). Nothing is known about these costs of paternal care.

More research is also needed on whether parents optimize the amount of care they provide to offspring – and, if so, how this is regulated. Only two studies have attempted to quantify this empirically (Hunt & Simmons, 2004; Kishi & Nishida, 2008). Quantitative fits to theoretical predictions were poor in Hunt & Simmons (2004), and this is undoubtedly due to the costs of providing care to offspring being greatly reduced in a laboratory environment. Thus, future studies would benefit immensely by attempting to measure the costs of providing care in a more biologically realistic manner.

We also need more studies that estimate the benefits of parental care. While numerous studies have shown that parental care increases offspring body size at emergence in onthophagines (Moczek, 1999; Hunt & Simmons, 2004; Kishi & Nishida, 2006; 2009), the link between adult body size and fitness is less clear. Hunt & Simmons's (2001) work on *O. taurus* shows that the relationship between male body size and reproductive success is non-linear, but this relationship remains largely unexplored for any other onthophagine species.

Another interesting research direction is the potential for parental care to act as a mechanism of evolutionary change. Hunt & Simmons's (2002a) quantitative genetic work on *O. taurus* demonstrates that maternal care has a genetic basis in this species and therefore represents an indirect genetic effect that can potentially facilitate the evolution of offspring body size, despite having a low level of additive genetic variance. However, more empirical research is desperately needed to test this concept. One possible approach would be to artificially select on both the level of maternal care and body size in this species and to examine the rate at which these traits evolve. If maternal care promotes the evolution of body size, lines selected for increased levels of maternal care should show a greater response in body size, compared to lines selected for increased body size.

Another interesting research direction that deserves more attention is the potential for parental care to facilitate range expansion to novel environments. It is not a coincidence that this genus of dung beetles occupies most of the major habitat types in its worldwide distribution (see Chapters 2, 11 and 13). Could this widespread distribution be the result of the substantial care provided by members of this genus? Fox *et al.*'s (1997) classic work shows that plasticity in egg size has played a major role in host expansion in the seed beetle *Stator limbatus*. We propose that similar plasticity in brood mass weight, particularly in response to environmental factors such as soil moisture (Sowig, 1996b; Hunt & Simmons, 2004), has the potential to play an equally important role in promoting range expansion in onthophagines.

Finally, perhaps the most promising direction of research that has yet to be pursued in onthophagines is to examine parental care from a phylogenetic perspective. Recently, Emlen *et al.* (2005a) constructed a robust molecular phylogeny for 48 species of *Onthophagus* worldwide. Thus far, this phylogeny has only been used to examine the evolution of horn diversification in this genus (Emlen *et al.*, 2005a; Emlen *et al.*, 2005b), leaving immense scope to examine a variety of questions on the evolution of parental care.

Indeed, many of the research directions outlined above can be addressed at the macroevolutionary scale. For example, by mapping paternal care strategies on this phylogeny, we can determine the number of evolutionary gains and losses of paternal care across *Onthophagus* species and whether or not paternal care strategies map onto male dimorphisms (i.e. do only male dimorphic species exhibit paternal care?). The relationship between a male's confidence of paternity and paternal care can also be tested across species, both directly, by relating the average degree of polyandry measured in wild populations to levels of paternal care (Bretman & Tregenza, 2005), or indirectly, by comparing the frequency of sneak males in a species to levels of paternal care.

More detailed evolutionary questions can also be addressed, such as the relationship between parental care and range expansion, and whether levels of parental care drive phenotypic evolution. In the former, the relationship between environmental factors and the level of parental care across species may provide important insight into whether parental care has facilitated range expansion. In the latter, examining the relationship between the level of parental care and phenotypic traits, such as body or horn size, across species will enable researchers to determine if parental care has constrained or facilitated phenotypic evolution in this genus. While such phylogenetic approaches will require a mammoth effort, we believe that they are essential to translate parental care strategies within species to divergence observed among species.

Acknowledgments

We thank two anonymous reviewers for valuable comments on this chapter. John Hunt was funded by the Natural Environment Research Council and by a Royal Society University Fellowship.

9

The Visual Ecology of Dung Beetles

Marcus Byrne[1] and Marie Dacke[2]

[1]*School of Animal, Plant and Environmental Sciences, University of the Witwatersrand, Johannesburg, South Africa*

[2]*Vision Group, Zoology, Sölvegatan 35, Lund, Sweden*

9.1 Introduction

The sun has poured energy, as light, onto the earth for at least four billion years. But our ability, as animals, to use that light for remote sensing of the environment, has existed for a little over half a billion years. Proper vision burst into life during what is known as the Cambrian explosion (Land & Nilsson, 2002). In less than five million years – the 'blink of an eye' in evolutionary time – animals evolved large functional eyes. This set off a dramatic change in their lifestyles, accompanied by limbs, claws and jaws – all the paraphernalia of life as a predator or prey (Parker, 2003).

Light contains a great deal of useful information. Detecting patterns of bright and dark allow for discrimination of objects in the environment without making contact with them. The wavelength (essentially colour) and the plane of polarization of that light communicate additional meaning if they too can be detected. The performance of an eye in deciphering this information can be assessed by its resolution and sensitivity.

Resolution describes the finest details of a given scene that the eye can resolve; it depends both on how closely the photoreceptors of the retina can be packed and on the optics of the eye. Sensitivity is the light-collecting ability of an eye; it describes how many photons of light are required to distinguish between the light and dark points that can be resolved in a visual scene. For a flying dung beetle, avoiding obstacles while foraging by starlight, sensitivity will be every bit as important as resolution. Nevertheless, that beetle will need sufficient resolution to execute a safe landing, while pushing its vision to operate at the lower limits of available light.

The attainment of vision is credited with driving the Cambrian explosion, when the numbers of animal phyla shot from three to 38, overnight in terms of evolution. The sun has risen on those 38 phyla ever since, with one or two exceptions which

Ecology and Evolution of Dung Beetles, First Edition. Edited by Leigh W. Simmons and T. James Ridsdill-Smith. Published 2011 by Blackwell Publishing Ltd.

have become extinct (Gould, 1990; Parker, 2003). What is notable for us, in our exploration of dung beetle vision, is that compound eyes with multiple facets are discernable in those early ancestors of the arthropods. These eyes are found in creatures from the Burgess shales of Canada, the Sirius Passet fauna of Greenland and the Chengjiang fauna of central China, implying that their evolution happened suddenly and simultaneously, all over the world (Land & Nilsson, 2002).

Compound eyes entered the vision arena about 540 million years ago. They now occur widely in the arthropods and are therefore the most common eye design on Earth. They made their first appearance in the crustacea, chelicerata and trilobitomorpha, and then in all insects, which evolved 145 million years later in the Devonian. The compound eye probably arose from a collection of eyespots in a worm-like ancestor of the arthropods, and it may be echoed in the stemmata found in the larvae of holometabolous insects today (Bitsch & Bitsch, 2005).

Compound eyes are usually paired convex structures distributed around the outside of an animal's head, with most insects having a single pair of compound eyes. In some insects, such as the Hymenoptera, Hemiptera and Diptera, the compound eyes are often accompanied by three ocelli. The Coleoptera lack ocelli, which look and behave like simple eyes, but are generally incapable of forming a sharp image. Ocelli are thought to be used for light/dark detection, horizon orientation and, possibly, detection of polarized light (Krapp, 2009). Indeed, it would appear from most investigations that these simple eyes are built in such a way as *not* to form an image. And what is even more surprising is that, given the severe restrictions that we shall see are imposed on vision by the compound eye structure, the ocelli have not evolved into more efficient camera-type eyes like our own. Other arthropods, such as spiders, have managed to evolve effective simple eyes (usually eight, sometimes six), distributed in pairs around the head. Again, the different eyes perform specialized tasks such as movement or polarization detection (Schmid, 1998; Dacke *et al.*, 1999; 2001; Mueller & Labhart, 2010).

However, even though most insects carry only one pair of compound eyes, the restrictions imposed by having a single pair have been overcome in many groups by partitioning the multiple elements of each eye into specialized regions with specific tasks. These can take the form of 'acute zones', where photoreceptors are more closely packed to increase the resolution of potential prey items or prospective mates. For example, the 'love spot' in the eyes of some flies is a high-resolution acute zone found only in males and is used for spotting and tracking females on the wing (Collett & Land, 1974; Land, 1997). In some species, strips of photoreceptors at the top of the eye in the 'dorsal rim area' are adapted to detect polarized light, which ants, among many other insects, use for orientation (Herrling, 1976; Wehner, 1997).

The spectral sensitivity of different eye regions may also be tuned to light of different wavelengths in order to improve performance at specific visual tasks (Stavenga, 1992). This division can clearly be seen in the enormous compound eyes of dragonflies, some of which have bright red or yellow-coloured dorsal regions, underlined by a dark ventral section. The dorsal portion is tuned to short-wavelength light in the UV-blue range for maximum acuity in spotting prey against the blue sky, which is dominated by short wavelengths (Labhart & Nilsson, 1995).

Some insects go even further and have evolved eyes with separate dorsal and ventral units, having distinctly separate tasks. Some of the most bizarre examples are associated with secondary sexual characters of male bibionid flies and male mayflies. These males have massive skyward-pointing eyes, with double-sized facets, purely for seeking out female silhouettes against a bright sky (Zeil, 1983).

Many dung beetles have eyes which are completely divided horizontally, but these are *not* sexually dimorphic and are therefore almost certainly not involved in mate-seeking. The purpose of the division is, however, not yet fully clear. We do know that the dorsal eyes have a role in direction-finding, which appears to be largely restricted to a specialized upper region of the dorsal eye. The dung beetle ventral eye may be involved in measuring optic flow – the movement of a scene across the retina – which is used by some flying insects such as bees to measure speed and distance (Baird *et al.*, 2005; Srinivasan *et al.*, 1996; 2000). We will return to these points later in our discussion, and will also consider how sexual selection exerts its influence on dung beetles eyes, but first the basic structure of the eye must be explained.

9.2 Insect eye structure

9.2.1 The apposition eye

The most obvious outward features of compound eyes are the multiple facets of the corneal lenses that give these eyes their name. Each of these little optical elements focuses light – but in a way that depends on the internal structure of a particular eye. Most diurnal insects, including the majority of beetles, possess apposition eyes in which each lens directs light into a single group of receptors called the rhabdom. The rhabdom consists of retinula cells bearing microvilli, which are filled with light-sensitive pigment. These microvilli are often closely packed together to form a light-sensitive column called a fused rhabdom, which lies directly in the light path of each corneal lens.

The whole unit, made of the optical elements of a corneal lens and an underlying clear light guide called the crystalline cone, along with the photoreceptors and pigments, is called an ommatidium (Figure 9.1a). The compound eyes of ants may have as few as two ommatidia in the ponerine *Concotio concenta*, while its big-eyed formicine relative *Gigantiops destructor* has thousands, which is typical of a visual predator (Hölldobler & Wilson, 1990). Fast-flying, aerial-hunting dragonflies, for example, have up to 30,000 of these elements per eye. Therefore, the structure of the eye is often a very obvious external expression of the lifestyle of an animal. Dung beetles occupy the middle ground, with between 1000 to 12,000 ommatidia per eye, depending on the species and the role their eyes fulfil in their particular habitat (M. Byrne, unpublished data).

9.2.2 The superposition eye

The number and size of the ommatidia affect the overall size of the eye and give important clues as to the visual ecology and behaviour of a particular species (Land, 1997). However, there is an important internal structural difference in the eyes of

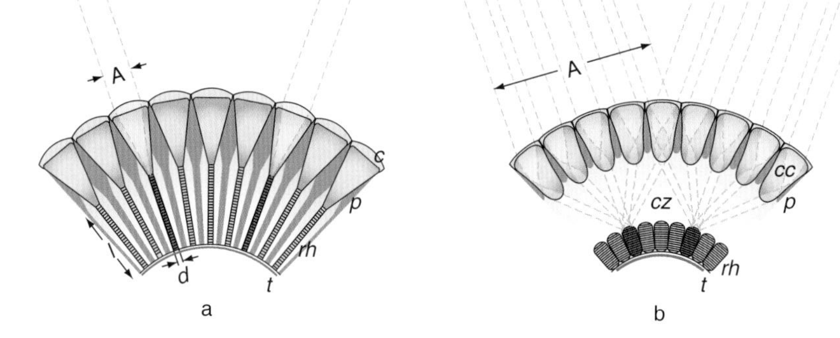

Fig. 9.1 Two types of compound eyes. The paths and fates of parallel light rays, incident on the external eye surface, are indicated for each. The target photoreceptor is shaded dark red. **a**: A focal apposition compound eye. In this eye, light reaches the photoreceptors in a single ommatidium exclusively from the small corneal lens located directly above. This eye is typical of day-active insects. **b**: A refracting superposition compound eye. In this type of eye, a large number of corneal facets and the bullet-shaped crystalline cones collect and focus light across the clear zone towards single photoreceptors. *A* is the diameter of the aperture, *c* is the corneal facet lens, *cc* is the crystalline cone, *cz* is the clear zone, *d* is the diameter of the rhabdom, *p* is the screening pigment, *t* is the tapetum, *l* is the length of the rhabdom, *rh*. Diagram courtesy of Dan-Eric Nilsson (modified).

some insects, which dramatically improves the light-collecting ability of the eye and allows it to see in dim light. This distinction is created by the simple inclusion of a space (the clear zone) between the distal optical elements of the eye and the proximal receptors of the rhabdom (Nilsson, 1989; Figure 9.1b). This creates a superposition eye, where one rhabdom can receive light from more than one facet, immediately increasing the light-gathering capacity of that rhabdom.

As expected, superposition eyes can be found in many crepuscular and nocturnal insects, including dung beetles. In *Copris elephenor*, light is collected from over 500 lenses and directed into a single rhabdom (Warrant & McIntyre, 1996), which improves the sensitivity of the eye 100-fold in dim light, compared to that of an equivalent apposition eye. This optical sleight of hand is achieved by a cunning piece of biological engineering which evaded researchers until the nineteenth century. In 1891, Sigmund Exner calculated that for a lens to have enough light-bending power to redirect that light back across that lens' own axis and onto a rhabdom not physically aligned to that lens, it would need to have a refractive index that differed along the path that the light takes through it (Exner, 1891). This sparkling insight was confirmed when interference microscopes were later invented (Land, 1988). The crystalline cone forms a lens cylinder, which is effectively a rod in which the gradient of the refractive index is highest along its central axis and falls to zero at the rod's perimeter (Land, 1988).

The difficulty of directing large numbers of light ray bundles onto a single rhabdom had been thought to reduce the resolution of superposition eyes and create a blurred image. However, work on *Onitis* dung beetles, whose superposition eyes deliver light from 50–300 facets on to one rhabdom, has shown that their eyes resolve images as well as an equivalent-sized apposition eye (Warrant & Mcintyre, 1990a).

The key features of the superposition eye are the clear zone and the optical structures beneath the cornea. The crystalline cone generally lies just beneath the corneal lens and, together, these optical elements direct light onto the rhabdom (Figure 9.1b). As with every other feature of the eye, these elements vary with eye function and to some extent eye phylogeny.

Stan Caveney (1985) tried to use the corneal lens and crystalline cone structure to unravel the phylogenetic history of eye structure in polyphagan beetles, with particular reference to the Scarabaeidae. Despite the information available on Scarabaeidae eye structure, Caveney concluded that it is only useful as a 'phylogenetic indicator', and added that superposition eyes are not monophyletic and that the diurnal superposition eye of the Scarabaeinae appears to be a derived character.

Molecular analysis has eventually revealed, like so many other aspects of dung beetle biology (including the evolution of ball-rolling and flightlessness (see Chapter 2)), that various eye morphologies may have arisen more than once within different groups of the Scarabaeinae (Forgie *et al.*, 2006; Monaghan *et al.*, 2007), driven by their ecology and behaviour more than their ancestry.

9.3 Eye limitations

Image resolution is ultimately limited by the diffraction of light, which describes the spread and interference of light waves after they pass through an aperture. The image degradation (blur) produced by a lens (aperture) is inversely proportional to the size of that aperture; the smaller the lens being used to create an image, the more blurred that image will become. As a consequence of diffraction, a point light source is imaged by a small lens as a 'blur circle' and an extended image, which can be considered to consist of many point light sources, appears equally blurred.

The blurring effect of diffraction is obviously very pronounced in the compound eyes of all insects, where the small lenses limit the resolution of the eye (26 μm in *Pachysoma striatum*, a large diurnal 'ball roller' (Dacke *et al.*, 2002), 39 μm in *Onitis westermanni*, a medium to large crepuscular tunneller (Warrant & McIntyre, 1990b)). As there is no point in having retinal receptors packed more closely than the blur circles of the image, the diffraction of the aperture also dictates how fine the retinal mosaic of receptors should be (Land & Nilsson, 2002).

Some insects attempt to address the constraints of diffraction by pointing more large facets towards a given point in space, reducing the angle between adjacent ommatidia. This does increase the resolution (sampling frequency), but it flattens the eye by increasing the radius of curvature, which potentially makes the eye bigger and therefore more costly. The diameter of the rhabdom is also influenced by the radius of the eye and the diameter of its lens, and these parameters are usually matched within the eyes of an insect, across a range of families and eye sizes, and are fine-tuned to the light intensity in which the creature lives (Land, 1997).

Nevertheless, the resolution of a compound eye is inevitably poor. Receptors in a bee's eye, for example, sample the world in 1.1° wide pixels, compared to 0.014° in our eye. This results in '*a picture . . . executed in rather coarse wool-work and viewed at a distance of a foot*' (Mallock, 1894). To improve resolution, the compound eye has to increase the size of every lens, with the net result that the eye size increases with the square of resolution, compared to a linear relationship in our simple camera eyes (Land & Nilsson, 2002). As a result, to achieve human eye-like resolution, a compound eye would have to be one metre in diameter – which would be difficult for any animal to carry, let alone fly with! Nilsson (1989) succinctly described the situation: '*evolution is fighting a desperate battle to improve a basically disastrous design*'. This severe phylogenetic constraint on eye size will surface repeatedly as we examine how evolution has shaped the eyes of dung beetles.

Insect compound eyes may have relatively poor spatial resolution, but they compensate to some extent by possessing excellent temporal resolution, meaning that they are fast. A fly's photoreceptors receptor potential can communicate changes in light intensity for frequencies up to 300 Hz, compared to about 75 Hz in humans (Weckström & Laughlin, 1995). If a dung beetle is flying at speed through the African savanna, where it has to avoid branches and tall grass stalks, its low-resolution eyes can compensate to some degree by returning rapidly changing spatial details as high-frequency temporal variation (Anderson & Laughlin, 2000). This reduces the likelihood of collision, even though the scene is only coarsely perceived.

9.4 Dung beetle vision

Eyes are expensive both to build and to maintain (Laughlin, 2001), which leads to the question: what sort of vision do dung beetles need? What tasks do they need vision for, and what quality of vision do they require? Dung beetles locate their food primarily by smell (see Chapter 5). However, vision will be used for obstacle avoidance during foraging flights, and for orientation in those species that sequester dung by rolling balls of it along the earth's surface. Use of vision in underground tunnels has also been raised by Emlen in relation to male beetles defending tunnels (Emlen, 2001). Given the extremely low light levels that probably prevail in the soil under a dung pat, this seems unlikely. However, both diurnal and nocturnal dung beetles are equipped with 'nocturnal' superposition eyes. So, because superposition eyes are considered to be an adaptation to vision in dim light, we will first consider the way dung beetles see at night.

9.4.1 Dim light vision

Spatial summation

Maximizing photon capture is particularly critical for any nocturnal animal. The principal mechanism employed to increase photon capture is to enlarge the eye's size, and this is a simple response to low light conditions seen in all animals. Night vision for dung beetles is solved in the same way as for all other nocturnal animals, namely by increasing the eye's size and opening the effective pupil wider.

However, because the eyes in question are compound and therefore small, there are several serious drawbacks which must be overcome, the first of which are the minute lenses. Increasing the size of a lens will counter the blurring effect of diffraction *and* improve the light gathering ability of its own ommatidium, but will at the same time reduce the resolution of the eye overall. This is because, for an eye to keep the same radius, making the lenses bigger will reduce the total number of lenses it is possible to carry and will consequently increase the angle between neighbouring lenses and their ommatidia. This interommatidial angle relates, in dung beetles and most other insects, directly to the density of the retinal mosaic and therefore to the resolution of the eye. In an eye with large interommatidial angles, fewer receptors will be looking at a given portion of a scene, giving a coarser, low-resolution image. The nocturnal onitine beetle, *O. aygulus*, has an interommatidial angle of 3°, compared to a 2.2° interommatidial angle in the day-flying species *O. belial*; this, nevertheless, still gives the eye reasonable resolution (McIntyre & Caveney, 1998).

Just as the human eye can adapt to changing light conditions so can the superposition compound eye, altering its structure to control the amount of light that will strike the photoreceptors. This is achieved in a manner analogous to the action of the pupil of a simple camera eye. The light-adapted superposition pupil decreases in size by moving pigments into the clear zone, which then restricts the number of lenses through which light can reach a particular photoreceptor. Conversely, by moving these pigment screens out of the clear zone, the dark-adapted eye allows light from many more lenses to reach a particular photoreceptor.

The superposition 'pupil' marks out the area of facets that can focus light onto a single photoreceptor (Figures 9.1b & Figs. 9.2). Pigment granules are contained within specialized pigment cells located distally around the crystalline cones, in the clear zone, and also proximally between the rhabdoms (Warrant & McIntyre, 1990b; Figure 9.1). Differential movement of these pigments away from the clear zone towards both the distal and proximal eye surface allows a progressive increase in light entering the rhabdom. Reversing the process reduces photon capture and causes considerable improvement in image quality by preventing leakage of light between rhabdoms.

As expected, the eyes of a dung beetle from a dim habitat tend to have a much larger superposition aperture than that of its relative living in a bright habitat (McIntyre & Caveney, 1998). The nocturnal dung beetle *Ontis aygulus* has a larger eye aperture ($A = 845\ \mu m$) than its crepuscular congeneric, *O. alexis* ($A = 655\ \mu m$) and the diurnal *O. belial* ($A = 309\ \mu m$) (Figure 9.2). So, the eye aperture not only varies between related species but can also change within an individual of a species by pigment migration, depending on the light conditions it is experiencing.

Temporal summation

Not surprisingly, optical trade-offs between sensitivity and resolution are matched by neural compromises that the dung beetle eye makes to improve vision under dim light conditions. In addition to spatial summation – where the signals from many neighbouring photoreceptors are pooled – the neural system allows the eye to increase its exposure time by temporal summation of photons (Warrant, 1999a). When sacrificing spatial resolution to improve the number of discernable intensity levels in a given scene viewed in dim light, vision unavoidably becomes coarser.

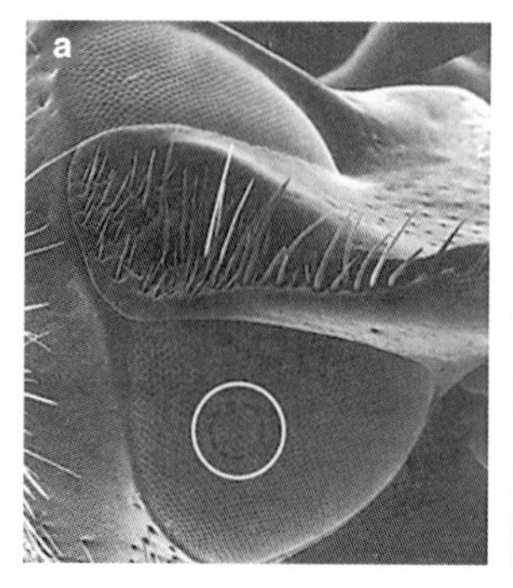

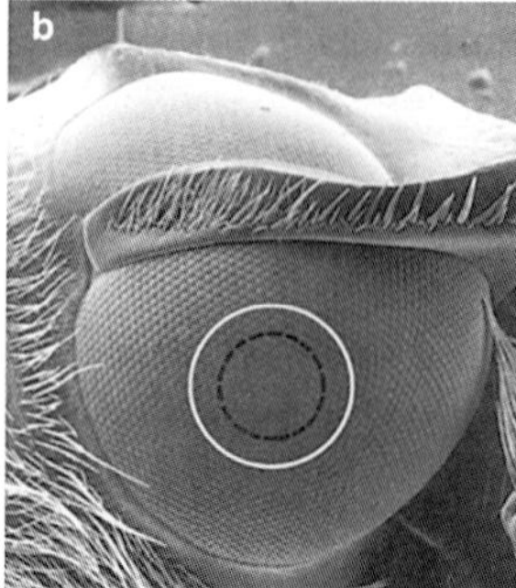

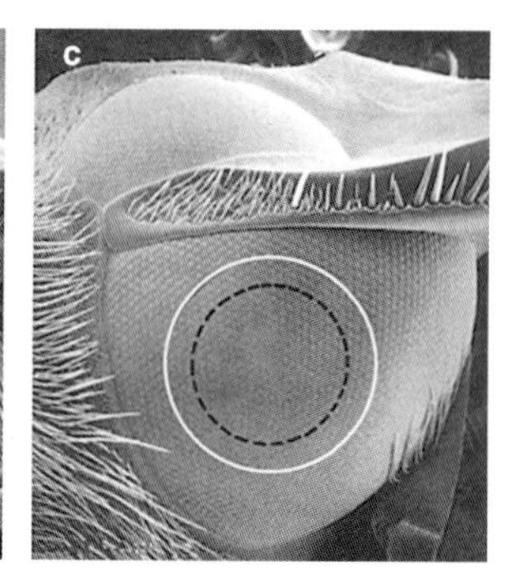

Fig. 9.2 The superposition aperture of the eyes of dung beetles from the genus *Onitis*, which correlates to their diel flight activity. Note the progression from the relatively small aperture (A in Figure 9.1b) of the diurnal (a) *O. belial* (width A = 309 μm), to the larger aperture of the crepuscular (b) *O. alexis* (A = 655 μm) and still larger nocturnal (c) *O. aygulus* (A = 845 μm). The circular superposition aperture is indicated in white on the surface of each eye. The dashed dark circles refer to 'effective apertures', in which each facet contributes light equally. In reality, facets near the edge of the aperture contribute less light than those near the centre. From McIntyre & Caveney, 1998.

The same information arguments apply to the resolution of contrasts in time. If a single photoreceptor views a moving scene, then the photoreceptor's response will rise and fall as brighter and darker details of the scene pass through its visual field. As the scene moves faster, the ability of the photoreceptor to code these contrast details reliably declines, due to the photoreceptor's finite response speed (Srinivasan & Bernard, 1975). The response of the photoreceptor will eventually become too slow to collect sufficient information to support reliable contrast discrimination, relative to the rate of movement of the body which carries that eye through the scene. Eventually the receptors will become too slow to support motion vision, which is crucial for the control of altitude, heading and obstacle avoidance.

Dung beetles are essentially creatures of the soil, but many are also accomplished fliers, able to forage over many kilometres for dung (Paschalidis, 1974). Flying brings obstacles rapidly into view, and the nervous system has to react quickly to generate an appropriate response. Slowing down the photoreceptors has its drawbacks for night-flying beetles, which need to detect low-contrast, rapidly approaching obstacles in their flight path. Temporal summation of photons will thus either cause a decline in flight control performance if flight speed is maintained (Chittka *et al.*, 2009), or force the animal to allow for an increase in response time by slowing down its flight. Night-flying dung beetles should thus either fly relatively slowly, which puts them at risk from aerial predators, or be fast but sloppy fliers. When testing the precision of flight and landing onto dung baits during the night, it was found that the majority of the nocturnal beetles flew close to, but not onto, the bait; they landed some distance from the dung and walked on to it. This suggests that 'quick and dirty' is the better night-time dung-foraging flight tactic (Byrne *et al.*, 2009).

Body size, a theme that we will encounter again, is strongly correlated with eye size, and small species of dung beetle stop foraging when it gets too dark to fly

(Caveney & McIntyre, 1981; Byrne *et al.*, 2009). Conversely, small day-flying species fly accurately onto dung pats. This is presumably both efficient and helpful in evading both aerial and terrestrial predators (Byrne *et al.*, 2009).

9.4.2 The tapetum and enlarged rhabdoms

Nocturnal dung beetle species use two other optical tricks which improve their dim light vision. First, they increase the size of the rhabdoms in the retina. This is seen in several tunnelling genera, including *Copris* and *Onthophagus* (Gokan & Meyer-Rochow, 2000), which increase both the area and often the length of the photon capturing units in the retina. This adaptation is most strikingly seen in nocturnal ball-rolling species, which, having negotiated a flight path through the bush at night, have the additional challenge of terrestrial orientation in rolling a dung ball to an unseen nest-site in the dark. Consequently, these species possess relatively large rhabdoms (Dacke, 2003; Dacke *et al.*, 2003a).

The second optical device also involves the retina, and endeavours to catch every last photon striking it. Placement of a mirror (the tapetum) on the back surface of the retina reflects photons not captured on the inward journey, back out again through the microvilli of the rhabdom, where they get a second chance of being absorbed. The dung beetle tapetum is formed by tracheal tubes, which cause the eye to reflect light shone into it. This gives the eye an eerie 'eye glow', which is a demonic red in nocturnally active dung beetles if the eyes are viewed at night along the beam of a torch. This effect rapidly disappears as the eyes light-adapt by moving pigment distally and proximally along the sides of the ommatidia to restrain photons within a single ommatidium. While it lasts, however, this effect gives the observer some idea of the size of the superposition pupil in the dung beetle eye.

The superposition eye is considered to be the more derived type of arthropod eye and, because of its superior light gathering ability, it is assumed to be adapted for vision in poor light (Warrant, 1999b). Nevertheless, there are many examples of diurnal insects that possess superposition eyes, including Scarabaeidae (McIntyre & Caveney, 1985; Gokan & Meyer-Rochow, 1990) and Lucanidae (Gokan *et al.*, 1998). This should not come as a surprise, given that many dung beetles are day-active, but nonetheless they have retained the superposition eye design. Shielding individual rhabdoms from their neighbours with pigment (Figure 9.1), which migrates back and forth along the sides of the crystalline cone on a daily rhythm, accompanied by lateral migration towards and away from the rhabomere in the retinula cells, maximizes resolution at the expense of sensitivity. Sensitivity is not compromised in daylight, when photons are overabundant, therefore pigment migration is seen in most diurnal superposition eyes (Warrant & McIntyre, 1990a; 1996).

Of the species examined to date (Meyer-Rochow, 1978; Gokan 1990; Meyer-Rochow & Gal, 2004), the day-active dung beetles have modified their superposition eyes by having relatively short focal lengths, smaller apertures (Warrant & McIntyre, 1993; Caveney & McIntyre, 1981) and narrower rhabdoms (Dacke *et al.*, 2003a; M. Dacke, unpublished data). Dung beetles use their 'nocturnal eye' in daylight by moving pigments to restrict the amount of light entering each ommatidium, which in many ways is the most common (and obvious) way to adapt a superposition eye for daytime employment.

9.4.3 The canthus

It should be apparent by now that the overall appearance of a dung beetle's eyes, from external features to internal structure, is a good indicator of the behaviour of their owner:

- Nocturnal flyers have large hemispherical eyes with smooth, glassy, corneal surfaces (Figures 9.3c, f), and the canthus (an incursion of the cuticle into the eye which divides the eyes into dorsal and ventral halves in many ball-rolling species – Figures 9.3a–e) is often almost absent (Figure 9.3f). Internally, the crystalline

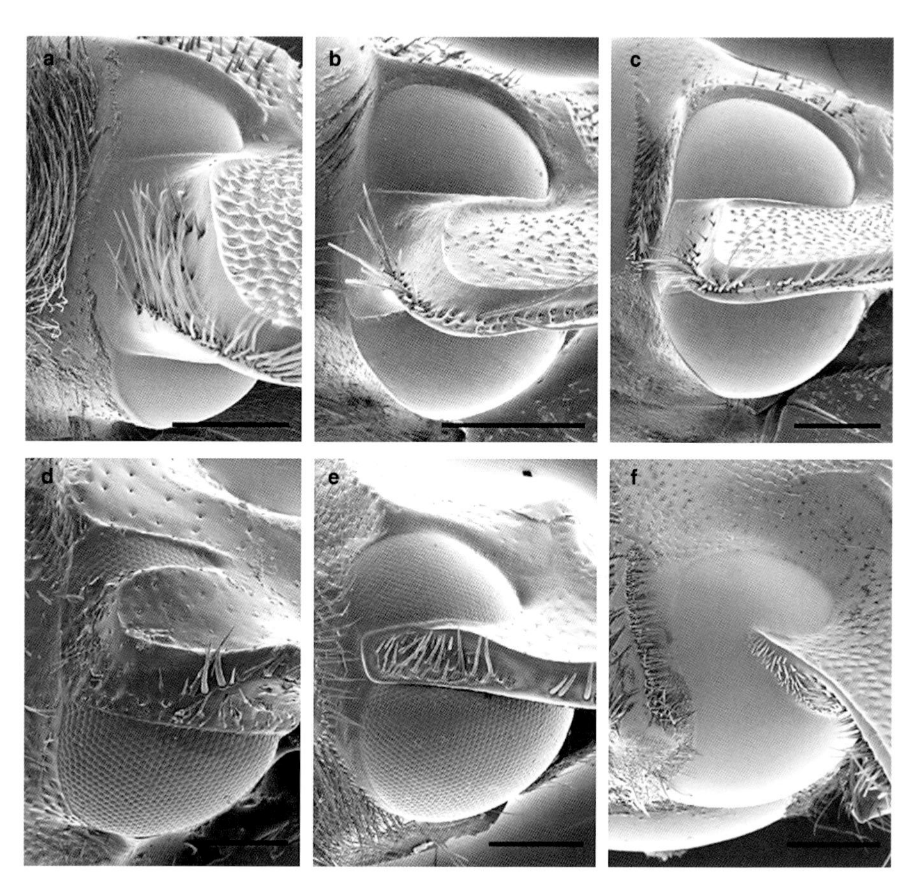

Fig. 9.3 Scanning micrographs of a lateral view of the right, dorsal and ventral eyes of ball-rolling (a–c) and tunnelling (d–e) dung beetle species with different diel activities. **a**: *Kheper nigroaneus*, diurnal. **b**: *Scarabaeus zambesianus*, crepuscular. **c**: *Scarabaeus satyrus*, nocturnal. **d**: *Euoniticellus intermedius*, diurnal. **e**: *Onitis caffer*, crepuscular. **f**: *Catharsius tricornutus*, nocturnal. Scale bars: a-c, f: 1 mm; d: 200 mm, e: 500 mm. SEMs courtesy of M. Dacke.

cones are large and the clear zone is wide (Warrant, 1999b; McIntyre & Caveney, 1985).

- Crepuscular dung beetles also have large eyes, often totally divided by a complete canthus, but the corneal facets are conspicuous, particularly in the ventral eye (Figure 9.3e). The crystalline cone is pinched at its midline to form a 'waist' (*cf.* Figure 9.1, 'bullet-shaped' crystalline cones).
- Day-flying dung beetles generally have smaller eyes, obvious corneal facets (Figure 9.3d), and the cones are tapered and strongly 'waisted' and lined with screening pigment, while the clear zone is relatively narrow. These 'diurnal' features generally reduce the maximum aperture of the eye, which is ultimately dictated by the diameter of the cone waist, limiting the angular acceptance of light from the ommatidial optics (Caveney & McIntyre, 1981). The canthus is varied in diurnal beetles, from complete and massive, to somewhat reduced.

The function of the canthus, however, in all species remains enigmatic. Most commentators assume that it serves to protect the eye surface from abrasion underground, but such completely divided eyes are also seen in aquatic Gyrinidae (whirligig beetles), where clearly the horizontal divide represents a division of labour between the two eyes (Wachmann & Schrörer, 1975). The dorsal region looks at the water surface and sky, while the ventral eye looks down into the water. Some Tenebrionidae carry good examples of a canthus, but none to match the complete division seen in scarabs or whirligig beetles.

The reduction of the canthus in nocturnal tunnelling dung beetle species is an obvious device to increase the available eye size. The nocturnal rolling species have, however, for some reason retained a complete canthus (Figure 9.3c), despite having to orientate in low light conditions. Both groups nevertheless share a beautiful glassy finish to their corneas, which completely obscures the division between the lens facets and could possibly serve to protect the eye from being scratched.

The function of the canthus therefore remains an intriguing subject for future research. The absence of facets also opens another optical question, because a consequence of a smooth eye surface will be that individual lenses will not have an arched outer curvature, which then indicates that much of the refractive power is delivered by refractive index gradients in the lens and the crystalline cone.

9.5 Visual ecology of flight activity

9.5.1 Diel flight activity

Many dung beetles have distinct diel flight activity patterns, supposedly to avoid competition for fresh dung (Hanski & Cambefort, 1991; Caveney *et al.*, 1995; Tribe, 1976; Stickler, 1979) (see Chapter 1). Daily peaks in dung production have also been considered as a driver of dung beetle activity patterns, particularly because dawn and dusk are expected to be most fruitful for finding fresh dung, due to changes in activity by both diurnal and nocturnal mammals at these times (Gill, 1991). Elephant dung specialists, for instance, would be predicted to be diurnal because the ratio of daytime to night-time foraging periods decreases with mammalian body size (Owen-Smith, 1988).

Prompt detection of fresh faeces via airborne odours is important, since dung dropped in the daytime will dry more rapidly. Nevertheless, differences in diel activity will potentially decrease interspecific interference competition. Some beetles fly only in sunlight, others only in the steady darkness of night, and still others choose to fly in the rapidly changing light levels of dusk or dawn. This also is the time of day when predators may have more difficulty tracking a small moving object, as the light intensity swiftly drops down many orders of magnitude. A variety of aerial predators, including birds and robber flies, prey on dung beetles during the day (Gill, 1991), while both bats (Gill, 1991) and owls hunt them at night (Doube, 1991).

The other advantage of flight at dusk could be temperatures low enough to prevent overheating (Bartholomew & Heinrich, 1978), but warm enough to not require a major metabolic effort to elevate body temperature for take-off; and critically, enough light by which to see.

9.5.2 Crepuscular flight activity

The flight activity of crepuscular dung beetles has been more closely studied than that of their diurnal or nocturnal counterparts, and again reveals how beautifully the eye has evolved to fit its task. Twelve different species of dung beetles from the same genus *Onitis* were allowed to select their flight time during exposure to natural daylight. Some species, notably *O. alexis,* were found to have a very narrow window of flight activity which lasted only 30 minutes (Caveney *et al*., 1995). Later analysis of the optics of nine of these species showed that the structure of their eyes largely matched their flight activity period (McIntyre & Caveney, 1998). *Onitis alexis*' flight window is so finely tuned to light intensity that its onset of flight time shifts forward by almost an hour as the seasons progress, moving further away from the summer solstice.

From these studies, it is again obvious that the size of the body limits the size of the eye, as body size is correlated to eye size and nocturnal fliers tend to be large (Cambefort & Walter, 1991; Byrne *et al*., 2008). Large body size of individuals within a single species has even been shown to confer superior visual foraging ability on European bumblebees, *Bombus terrestris* (Spaethe & Chittka, 2003). As we know, a large eye can potentially provide a dung beetle with a large superposition aperture. Along with large rhabdoms, these features contribute to eyes of high sensitivity. For now we should also consider how body size constrains the beetles' ability to raise and maintain an elevated body temperature (see also Chapter 10).

9.5.3 Endothermy and vision

Endothermy is known from both day-flying (Bartholomew & Heinrich, 1978) and dusk- or night-flying dung beetles (Heinrich & Bartholomew, 1979; Caveney *et al*., 1995). The dusk- and night-flying onitine beetles were all found to be able to raise their body temperatures to around 35–40 °C, sometimes 19 °C above the ambient temperature (Caveney *et al*., 1995) This neat physiological ploy is again, like night-flying, restricted to larger beetles but, interestingly, their minimum body mass was found to be just 0.4 g – well below the 2 g theoretical body mass threshold proposed by Bartholomew and Heinrich (1978).

Avoidance of competition was proposed as the reason for the widely separated diel activity patterns of four of these species of *Onitis* (McIntyre & Caveney, 1998), but at least six of the other species of *Onitis* overlapped in their dusk flight times. Three species were often found sharing the same dung pat. This overlap suggests that dusk flight could be a predator-evasion strategy rather than purely driven by avoidance of interspecific competition.

The precise timing of flight could also be coupled to the physiological requirements of raising body temperature above ambient. The thermal offset between the ambient temperature and the body temperature required for flight will also be lower at dusk than at dawn. Even when below the horizon, the rising sun will provide enough light to allow the relatively small eyes of the beetles to work effectively, but the air might be too cold for their owners to invest energy in heating themselves up for flight. Consequently, few beetles fly at dawn (Paschalidis, 1974; Tribe, 1976; Stickler, 1979; Caveney *et al.*, 1995).

A rise in body temperature could possibly also improve vision by heating up the retina of the eye. Day-flying dung beetles show broader peaks of diel flight activity than their crepuscular counterparts, but nevertheless divide the day up into 'preferred' activity periods for some species (Tribe, 1976; Stickler, 1979), even when photons are available in excess. However, specific preferences for early morning or midday flight may again be driven by other factors. The interplay of body temperature and vision in dung beetles awaits further investigation.

9.5.4 Body size and flight activity

Day-flying beetles range from the very large *Pachylomerus femoralis* to the tiny *Onthophagus stellio* (26 mm vs. 2 mm pronotum width respectively). As we have learned from the crepuscular species, these differences in body size will have profound effects on their eyes and vision. The limitations of body size on eye size will even cause very small members of some groups of beetles, generally considered to usually have superposition eyes, to exhibit apposition eye morphology without a discernable clear zone (Meyer-Rochow & Gal, 2004).

The optical geometry of the superposition eye dictates that as the eye gets smaller, the clear zone must shrink, until it reaches a point at which it will disappear. This has been demonstrated in the eye of the Scarabaeine beetle *Paraphytus dentifrons* (Meyer-Rochow & Gal, 2004), which is less than 5 mm in length. This is not an exceptionally small dung beetle, particularly for a day-flying tunnelling species (Hanski & Cambefort, 1991). A small beetle would be expected to carry a small superposition/apposition eye, which is unlikely to gather enough light for reliable vision at night. Not surprisingly, small Scarabaeinae do not to fly at night (Byrne *et al.*, 2009; Caveney & McIntyre, 1981).

The dorsal eyes of small day-flying species tend to be flat, with a disproportionately large radius of curvature (Figure 9.3d). This might be an adaptation to avoid the clear zone loss predicted by Meyer-Rochow & Gal (2004). However, as a result, the visual field will be severely restricted but the resolution improved. This eye design is ideally suited to gather polarization compass information, but its role in orientation is unknown at this stage, as these species tend to be tunnellers rather than rollers and are unlikely to use a polarization compass for terrestrial orientation. So, exactly how this feature would affect the quality of vision from the eye remains to be tested, but nevertheless serves to highlight the variability of eye structure within the subfamily (Scarabaeinae).

Despite the eye being a widespread and largely conserved organ which is essential for survival in our visual world, it remains very plastic in its morphology, because eyes are expensive to build and run (Niven *et al.*, 2007). The eye therefore has evolved through structural changes to maximize its function in different habitats for different lifestyles. But what about the expense of growing an eye? Can the structure itself be traded off within an individual beetle against other features that are of competing importance? The answer is, surprisingly, yes. Another of life's great selection pressures – that of finding a mate and reproducing – is the driving force, trading precious eye size against secondary sexual characteristics in the expectation of getting mated.

9.6 Sexual selection and eyes

Doug Emlen and his co-workers have conducted a long series of elegant experiments on the ecological and behavioural consequences of horn size in tunnelling dung beetles, mainly in the genus *Onthophagus*, where large horned males defend tunnels in which females are building nests. The tunnel entrance represents a discrete defendable patch, and males with proportionately larger horns for a given body size have a competitive advantage in disputes over tunnel ownership (see Chapter 3).

Dung beetle horns (and eyes) grow from imaginal discs which are regional invaginations of the larval epidermis (Chapter 7). These grow during the prepupal phase, when the larva has stopped feeding and is purging its gut in preparation for pupation and metamorphosis. At that stage, the horns look like concentric rings of cuticle which are telescoped inside each other (see Figure 7.1). The resources used for horn development are consumed at the expense of neighbouring structures, such as antennae if the horns are at the front of the head, wings if they are on the thorax, and eyes if the horns form at the back of the head, as they do in many *Onthophagus* spp. (Emlen, 2001).

In the latter species, eye size is negatively correlated with horn size, and the hornless females generally have relatively larger eyes than the males. The decrease in the size of structures (antennae or eyes) associated with horns on the head is substantial – as much as 20–28 per cent (Figure 9.4). For the eyes, this results in a reduction in the number of ommatidia, which probably reduces the visual field and overall resolution and sensitivity of the eye. Therefore, placement of the horns in dung beetles might be driven by the costs of the horns rather than their benefits. Of 161 species of *Onthophagus* reviewed by Emlen (2001), those that fly at night were significantly less likely to bear horns on the base of the head, where they would cause a reduction in the size of the eye.

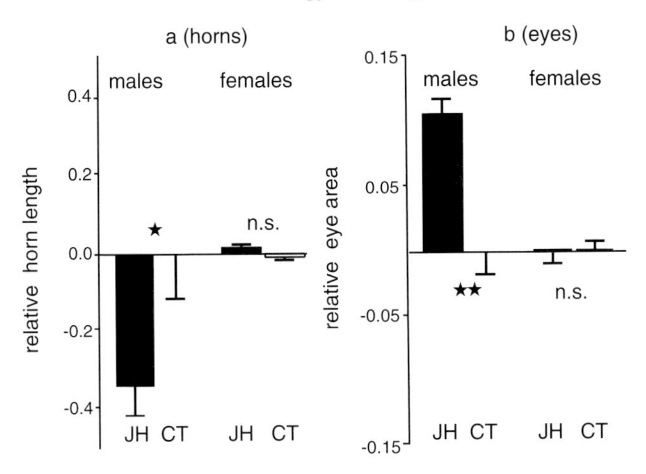

Fig. 9.4 Trade-off between eyes and horns in *Onthophagus taurus*. Animals treated with juvenile hormone (JH) analogue – solid bars; acetone control treatment (CT) – open bars. **a**: Borderline size males were induced by JH to develop significantly shorter horns (relative to body size) than control males. Females never develop horns and were unaffected by the treatments. **b**: Horn reduction was accompanied by a significant increase in eye size. Female eye size was unaffected by the treatments. ★$p < 0.05$; ★★$p < 0.001$. From Nijhout & Emlen, 1998.

Horn size is correlated with body size in male dung beetles, and the relationship is sometimes sigmoidal rather than linear, resulting in a threshold size below which horns don't develop (minor males), and above which they do (major males) (see Chapter 6). Consequently, two different body shapes appear: those with larger horns and subsequently relatively smaller eyes, vs. those with no horn and relatively larger eyes. Both of these are, to some degree, matched to the demands of the alternative reproductive tactics (sneaking or fighting) which males adopt to reproduce.

Topical application of a juvenile hormone analogue to final instar *O. taurus* larvae destined to be horned major males, induced them to form horns that were significantly shorter than those of their untreated counterparts. Treated males showed a significant increase in relative eye size compared to untreated males, confirming the developmental trade-off between horns and eyes (Figure 9.4). Female eye size was unaffected by the treatments (Nijhout & Emlen, 1998). This leads to the prediction that small males and female beetles should perform better at visual tasks than their horned, large male counterparts.

It also suggests that vision in ball-rolling dung beetles (which are not so obviously sexually dimorphic, and never possess horns) could play a role in constraining the evolution of secondary sexual traits on the heads of these beetles. The spurs on the legs of ball-rolling males of *Pachylomerus femoralis* and *Sisyphus mirabilis* are very distant from the head region and relatively small compared to body size, so are therefore unlikely to represent a resource competitor for the eyes. It is thus worth examining the role vision plays in ball-rolling dung beetle species and to ask questions about the constraints that vision places on this foraging behaviour.

9.7 Ball-rolling

Egyptian amulets often depict *Scarabaeus sacer* with the sun, apparently an indication that the beetle was responsible for its movement across the sky. This association of ball-rolling beetles with the sun could be one of our oldest records of animal behaviour. Interestingly, it is also one which has escaped the close attention of contemporary biologists until very recently. This is possibly because it is so obvious what the beetles are doing, but it is not necessarily clear exactly how they do it (Matthews, 1963; Byrne *et al.*, 2003).

9.7.1 Orientation by ball-rolling beetles

Most ball-rolling dung beetles fly several kilometres to locate dung, firstly by its smell, in a place they most probably have never visited before (Chapters 5 and 11). After landing either close to or on the dung, ball-rollers forage rapidly and openly on the surface of fresh dung pats, usually in bright sunlight. Their objective is to sequester a portion of the dung pile in a site some distance from its source, away from competitors or predators, where they can manipulate it safely from interference or risk of dehydration. The ball has one of three fates: either to be eaten by the owner; or employed by a male as an enticing nuptial meal for a female; or used as a brood ball into which a single egg is laid (Tribe, 1975).

Because beetles will fight for and steal dung balls, and also make costly metabolic decisions as to whether to maintain their body temperature above ambient or not, competition is assumed to be fierce (Heinrich & Bartholomew, 1979; Ybarrondo & Heinrich, 1996). To avoid ball theft by larger or hotter beetles, the ball-owner will roll the prize away by walking backwards, away from the dung, along a straight line path. This is both the quickest and safest way to escape competition for dung and to sample the unfamiliar habitat for a good dung burial site.

The beetles use the sun as their principle orientation cue (Byrne *et al.*, 2003). Artificially moving the apparent position of the sun, in a classic experiment using a mirror while blocking the view of the sun with a board (Santschi, 1913) usually causes a correlated reorientation response from the beetle. But otherwise, by making the beetles fall, or by generally putting them through an obstacle course of visual and proprioreceptor tasks, we find that beetles stick steadfastly to the compass bearing they have chosen when they departed from the dung pat (Byrne *et al.*, 2003; Figure 9.5a).

It is unclear what factors, if any, influence the beetles' choice of direction, but it appears that the very act of making a new ball provides the behavioural circumstance under which a new compass bearing is chosen (Baird *et al.*, 2010). However, once the bearing is set, the beetles can keep to their chosen direction with reference to the sun, and this feat is performed by the dorsal eyes. Because the foraging trip is usually one way, not directed to a known goal and fairly rapid (taking no more than 20 minutes), compensation for movement of the sun is unlikely to be necessary.

When the dorsal eyes are painted over, or the beetle's view of the sky is simply obscured by a cap, the orientation response breaks down completely and beetles move erratically around in irregular circles (Figure 9.5b). However, even when their vision is not artificially impeded, the sun may well be hidden by clouds or foliage, so

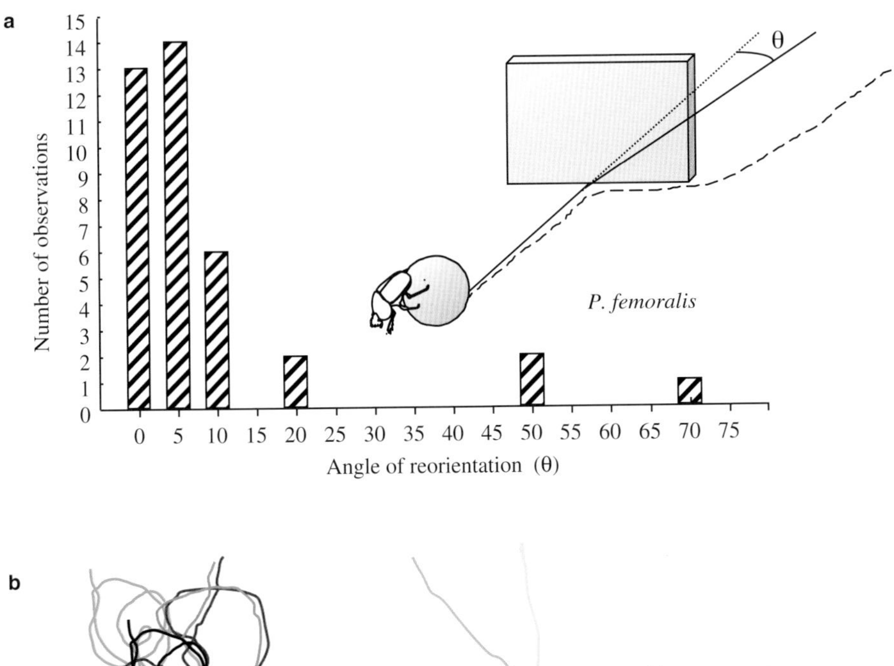

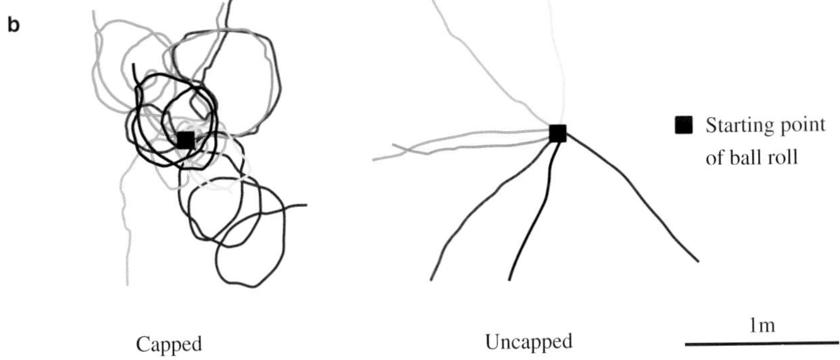

Fig. 9.5 a: The response of ball-rolling dung beetles to a barrier placed in the path of the beetle. All turning angles were treated as positive. Dotted lines: ideal orientation direction (= perfect correction to original rolling direction). Dashed lines: actual orientation direction. θ = angle between actual and ideal orientations. From Byrne et al., 2003. **b**: Tracks taken by the ball-rolling dung beetle *Kheper nigroaneus* when its view of the sky is undisturbed or obscured by a cap glued to the pronotum which blocks both the dorsal and ventral eyes' view of the sky. Tracks recorded from overhead video.

use of an alternative directional cue is anticipated, given that many other orientating insects use a hierarchy of terrestrial and celestial cues to steer a straight-line route between ports of call (Wehner, 1984).

A second hint of an additional direction-finding cue comes from an intriguing component of the beetle's rolling behaviour which involves it climbing on top of the ball to then rotate about the axis of the ball, with its head extended and level to the horizon. This curious behaviour is exaggerated when the beetle cannot see the sky. We believe that this 'rotation dance' is a type of visually mediated orientation mechanism that allows the beetle to determine or maintain a desired course direction (Baird *et al.*, 2008). It could help dung beetles to orientate towards a particular compass bearing using visual cues, which we believe could be either distant landmarks or the pattern of polarized light. Since the elegant experiments of Karl von Frisch (1949), we know that honeybees can navigate using polarized light from just a small piece of blue sky. Subsequently, dozens of other insects in several different orders have been shown to use this celestial cue, among which are the dung beetles (for review, see Wehner & Labhart, 2006).

9.7.2 The polarization compass

One aspect of the world of light, of which we are completely unaware because we cannot see it, is the polarization pattern in the sky. When light from the sun strikes the earth's atmosphere, it becomes scattered and reflected by air molecules. As a result of this, a pattern of polarized light is formed across the sky (Brines & Gould, 1982; Wehner, 2001). This pattern is centred on the sun and varies in a systematic way during the course of the day, both in the plane and degree of polarization, providing a cue with directional properties for many of the animals that can perceive it (Rossel, 1993; Wehner & Labhart, 2006).

Wehner's model animal has been the pharaoh ant, *Cataglyphis bicolour*, which forages for prey in the hot deserts of North Africa, where the saltpans they inhabit provide few reliable landmarks for navigation. However, the sun, or a sky full of polarized light cues, can provide a dependable signal independent of variation in terrestrial landmarks (Wehner, 1997). A dung beetle from the deserts of south-western Africa employs the same tactic to forage safely in a relatively featureless landscape.

Pachysoma striatum is an unusual dung beetle in many respects. It is flightless, which may be a respiratory adaptation to conserving water (Duncan & Byrne, 2005; see Chapter 10). It consequently forages on foot, collecting dry rodent pellets, and establishes a semi-permanent nest from which it departs repeatedly to scavenge over the surface of the sand (Holter *et al.*, 2009). It then returns home, dragging the dung pellets behind it, while walking forwards. Whereas the outward route is a meandering trail through potential dung localities, the homeward path is directly towards the nest (Scholtz, 1989). Moving landmarks or obscuring the disc of the sun does not upset this beetle's capacity to find home (M. Dacke & C. Scholtz, unpublished data). This is because the nest is most likely found by path integration, which requires a combination of compass bearing cues and distance measurements.

The ability to return to a nest is not unique to the Scarabaeinae, as a similar orientation system is also seen in the geotrupid beetle *Lethrus* (Frantsevich *et al.*, 1977). While it remains unclear how *Pachysoma* measures the distance walked – ants count steps (Wittlinger *et al.*, 2006) and flying bees integrate the optic flow experienced by the eyes (Esch & Burns 1995; 1996; Srinivasan *et al.*, 1996) – we do know how it finds the compass bearing of its home direction.

9.7.3 Polarization vision

The ability to measure direction of travel resides first and foremost in the gross and fine structure of the dung beetle eye, which is divided into a small dorsal and a large ventral unit by a huge canthus (Figure 9.6). The internal morphology of the eye is typical of the scarabaeid superposition eye (Dacke *et al.*, 2002), with flower-shaped

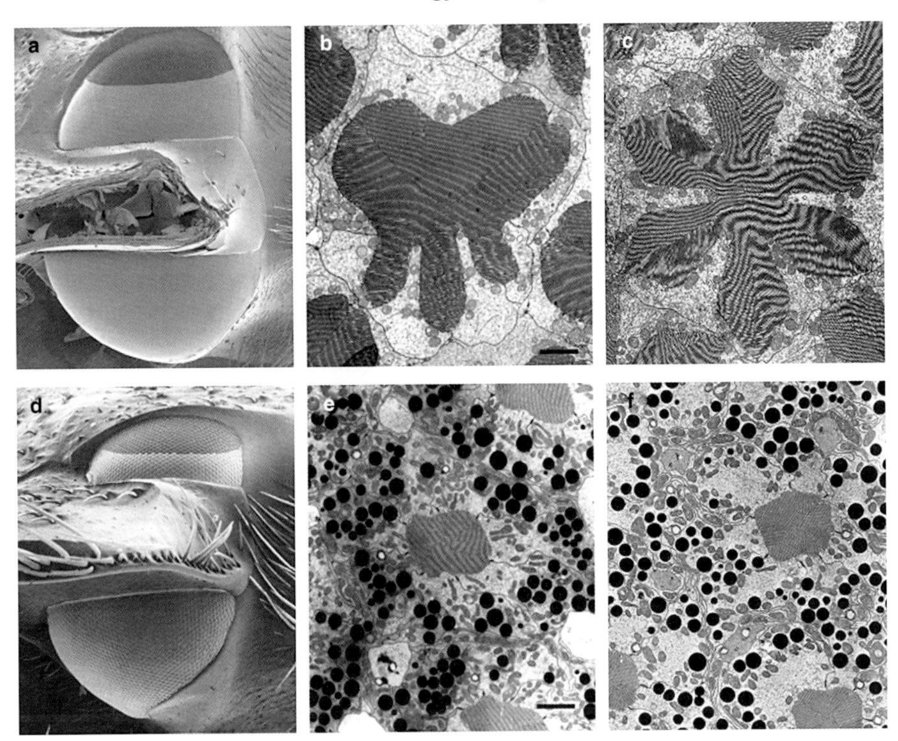

Fig. 9.6 Scanning electron micrographs of the eyes of the crepuscular beetle *Scarabaeus zambesianus* (a) and the diurnal beetle *Pachysoma striatum* (d). The dark grey area in the dorsal eyes marks the extent of the polarization sensitive dorsal rim area. Cross-sections of rhabdoms in the dorsal eyes of *S. zambesianus* (b, c) and *P. striatum* (e, f). The rhabdoms in the dorsal rim area (b, e) have microvilli orientated in only two directions and differ in their structure from those in the lower portion of the dorsal eyes (c, f). Also note differences in the size of the rhabdoms and the amount of pigmentation (black granules) in the retinula cells between the two species, both morphological adaptations to the diel activity of the respective species. Scale bar: 2 μm (b–e). Modified from Dacke et al., 2003a and Dacke et al., 2002.

rhabdoms and an obvious clear zone. However, along the dorsal rim of the dorsal eye, the microvilli in each rhabdom are arranged at right angles (orthogonal) to each other, forming a polarization detector in each ommatidium (Figure 9.6e).

When a microvilli-type photoreceptor is aligned to the axis of skylight polarization, it will be maximally stimulated (Snyder & Laughlin, 1975). This is because the photopigment (rhodopsin) dipole in the cell membrane preferentially absorbs light along its long axis, which, due to physical constraints of space, is aligned along the long axis of the microvillus. When two adjacent sets of microvilli are arranged at right angles to each other, comparison of the signal between the pair creates a polarization detector which is independent of intensity (Nilsson & Warrant, 1999).

An orthogonal arrangement of microvilli can be found also in the eyes of several other dung beetle species (Dacke, 2003; Dacke *et al.*, 2003a; Gokan, 1989a; 1989b; 1989c; 1990; Gokan & Meyer-Rochow, 1990; Meyer-Rochow, 1978), including tunnelling and ball-rolling species. Intracellular recordings of the rhabdom cells in *P. striatum* show them to be 13 times more sensitive to UV light polarized along the axis of the microvilli than at a plane 90° to these structures (Dacke *et al.*, 2002). The dorsal-most ommatidia in the dorsal eyes of the dung beetle *P. striatum* are thus ideally suited for polarized light detection and are equivalent to the polarized light detectors found in many other navigating insects (Labhart & Meyer, 1999). The ommatidia in the ventral eye look like their counterparts in the dorsal eye area outside the dorsal rim, with typical flower-shaped rhabdoms bearing microvilli orientated in more than two planes. Specific functions of these eye regions remain to be investigated further.

9.7.4 Polarization vision in dim light

Utilization of polarized light in the harsh glare of the desert sun is one thing; but can this mechanism be modified to function in the falling light of dusk, when most gracile ball-rolling beetles cease activity and the more stout crepuscular and nocturnal tunnelling guilds move in? The answer is most definitely yes, and two factors make this strategy viable for dim-light, ball-rolling dung beetles.

First, even though light intensity drops dramatically as the day slides into night, the maximal degree of polarization (which is at 90° from the setting sun) is conveniently found in the zenith of the sky. At this time of night, the polarization pattern, because it is always centred on the sun, subsequently runs north to south in a straightforward highway through the stars. This doesn't necessarily guide the beetles, which will orientate in any direction they choose relative to this cue, but it is invaluable to researchers, who, although they cannot see the polarization pattern, know which way it is orientated, and therefore how it can be manipulated.

Second, the right-angled (orthogonal) arrangement of the microvilli in the rhabdom constitute a polarization detector that analyzes the direction of polarization independent from the intensity of the light. This is, of course, only true for as long as this detector is sensitive enough to receive a reliable signal. Insects, however, have been shown to detect the direction of highly polarized light at light intensities which are even lower than that of a clear, moonless night sky (Herzmann & Labhart, 1989).

Given the potential to fill this gap in the diel cycle, which is free of other ball-rolling competitors, *Scarabaeus zambesianus* has adapted its eyes to use the polarized light of dusk for orientation. This reveals once more how diverse dung beetles are, and how perfectly their visual systems are adapted to different ecological opportunities.

Scarabaeus zambesianus appears at fresh dung sometime around sunset. Stretching our eyes to their limits, we can just make out dozens of dark blobs, each about the size of a large coin, frenetically running about on the sandy soil, rapidly making balls and heading off with directional confidence into the gloom. Setting them our standard orientation tasks (as in Byrne *et al.*, 2003), they perform with a precision equivalent to their daytime relatives (Dacke *et al.*, in press). Reorientation of the skylight polarization pattern, by means of a large polarizing filter, makes the beetles realign themselves to the filter as they pass underneath it, and then back to the

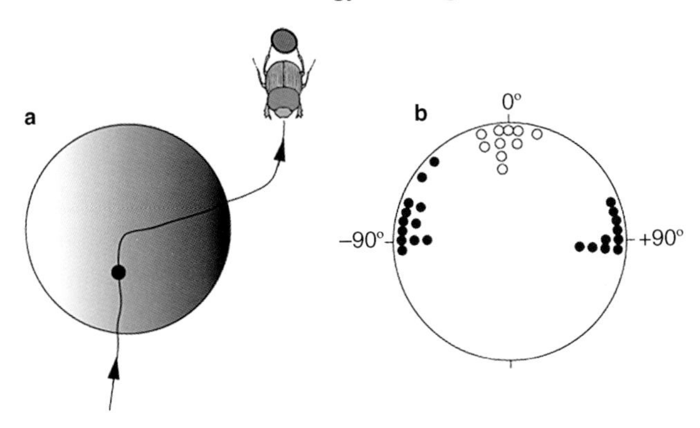

Fig. 9.7 *Scarabaeus zambesianus*' response to polarized light while rolling its ball of dung. **a**: The path taken by a beetle when a UV-transmissive polarizing filter, with its e-vector transmission axis oriented perpendicular to that of skylight, is placed over the beetle at location ● along its path. Arrows mark the direction of movement and the circle the diameter of the filter (42 cm). This beetle turned by 70° under the filter. Note that the beetle reorients back to its former direction of travel when the polarization pattern of the open sky is again visible. **b**: Circular distribution of turns made by 22 beetles in response to the shifted polarization pattern; closed circles. The average turn was 80.6°. Circular distribution of turns made by 10 beetles in response to a polarization pattern aligned with the celestial polarization pattern; open circles. Note that the beetle rolls backwards with its head down. Modified from Dacke *et al.*, 2003b.

skylight pattern as they pass back into its influence (Dacke *et al.*, 2003b; 2003c; Figure 9.7).

Histology of the eye of *S. zambesianus* reveals a dorsal rim area similar to that of *Pachysoma striatum*, but with huge rhabdoms several times longer and wider than its diurnal counterparts (Figure 9.6). A tracheal tapetum at the base of the rhabdom acts like a mirror and reflects light back through the microvillar array of the rhabdom, effectively doubling the length of the rhabdom. These optical differences alone are calculated to make the most dorsal part of the eye of *S. zambesianus* 5.6 times more sensitive to light than its equivalent in *P. striatum* (Dacke *et al.*, 2003a).

However, in reality, this difference in sensitivity is likely to be a great deal larger, because *S. zambesianus* can be expected to have a much wider superposition aperture than the diurnal *P. striatum*. As we know, the superposition eyes of beetles from dim habitats tend to be much larger than those of their relatives living in bright habitats (McIntyre & Caveney, 1998). Moreover, the rhabdoms in the dorsal rim area of *S. zambesianus* are isolated from each other by a reflective tracheal sheath, which constrains photons within the rhabdom until they are absorbed, whereas the rhabdoms of *P. striatum* are isolated by light-absorbing pigments which effectively swallow stray photons (Figures 9.6e, f). This morphological difference will make the sensitivity of S. *zambesianus*' eye even greater still (Warrant & McIntyre, 1991).

By extending its foraging time towards the night, *S. zambesianus* can potentially avoid predators and competition from big ball-rolling species such as *Kheper nigroaneus* and *Pachylomerus femoralis*. But unlike daylight, when there are always photons to spare, some nights are much better than others. The moon may be more fickle than the sun in its manifestations, but it nevertheless is potentially available for 180 days in the year when foraging time could be extended from a limited dusk period into the night. During these nights, when the moon has risen by the time the polarization pattern from the sun has vanished from the sky, it can take over the night-time role of an orientation guide – that is, if the beetles' eyes are sensitive enough to use the moonlight as an orientation cue.

Scarabaeus zambesianus naturally doesn't fail to impress; not only can it use the disc of the moon as an orientation signal for straight-line ball-rolling, it can also use the one-million-times-dimmer polarization pattern of light from the moon to beat a straight-line path away from the dung pile with its ball (Dacke *et al.*, 2003a; 2003b). This is an outstanding illustration of the sensitivity of the beetles' polarization compass, and as yet it remains the only animal shown to be able to perform this feat of nocturnal orientation by polarized light from the moon. However, the structure of the rhabdoms found in the eyes of other nocturnal dung beetles strongly suggests that they, too, should be able to orient to the polarization pattern of the moon (Dacke, 2003; Gokan, 1990).

9.8 Conclusions

In terms of our understanding of their biology, dung beetles aren't quite up there yet with 'that stupid little saprophyte' (William Morton Wheeler, on *Drosophila melanogaster* – from Grimaldi & Engel, 2005), but we do know a great deal about them, as this volume attests, and some of it is about the way in which they perceive their visual world. This knowledge has emerged as a result of careful work from three generations of investigators and, fortunately for the next generation, there is still much to be discovered.

The driving forces behind the gross external morphology of the eye are not understood, and in particular the function underlying the division of the eye into a dorsal and ventral unit (Figures 9.3 and 9.6). The canthus is responsible for this separation and is generally considered to have a protective role, which then does not explain how nocturnal tunnelling species, which have an extremely reduced canthus, manage to avoid destructive eye damage. The ventral eye generally appears similar to its dorsal counterpart (Gokan & Meyer-Rochow, 1990), although *Onitis* species show obvious differences in the size of the facets (McIntyre & Caveney, 1985). But beyond that, little is known about the visual role the ventral eye plays in either flight or orientation.

Systematic analysis of scarab eye structure has been attempted (Holloway, 1969; Caveney & McIntyre, 1981) but has stumbled, probably because the dung beetle's eye is a general-purpose structure which has been modified for day vision as well as night vision. In these roles, it has been shown to provide for orientation on the ground and in the air, during daylight or in dim light, while, at the same time in some species, trading off resources with other body structures such as horns. Inevitably, the eye and its function differ greatly between species, which may well be using their visual systems to assist in avoiding competition with each other for the same ephemeral resource.

Because the dung beetle eye is not specialized for one task, it will be an expensive structure to operate. Strausfeld (1976) estimates that 79 per cent of the 340,000 cells in the brain of a fly are part of the visual system. Vision consumes processing power and energy (Laughlin, 2001; Niven *et al.*, 2007) and evolution generates solutions to these types of pressures, so that division of labour is common in eyes, including our own, where we concentrate high resolution vision in the fovea. Insects do the same by developing specialized zones of visual cells for detecting polarized light, as we have seen in the dorsal rim of *Pachysoma striatum* and *Scarabaeus zambesianus*. Nevertheless, the rate of information flow into the eye is immense and therefore has to be filtered at every processing step.

Photoreceptor dynamics should therefore theoretically be adapted to the speed at which an insect flies (Srinivasan & Bernard, 1975), and this has been borne out by experimentation (O'Carroll *et al.*, 1996; 1997). The dorsal rim area of the polarization detector is probably an example of complementary visual adaptation in dung beetles, but the exact structure and limitations of the so-far-unique nocturnal polarization compass is still not fully understood and is therefore worthy of further investigation.

General purpose eyes are evolutionarily plastic because of the multiple roles they serve, and they are expected to evolve over time in response to a species's needs. The horns of *Onthophagus taurus*, exported from Europe to the USA and Australia, have diverged in response to novel selection pressures in their exotic homes (see Figure 7.2b and Chapter 7). Given the nutrient allocation trade-off between horns and eyes, we might expect eyes also to have diverged in these exotic populations. Moreover, given that more than 100 dung beetle species have been translocated around the world, this type of natural experiment is just bursting with opportunities to focus on eye evolution; and that is in addition to the 5,000 other species which have stayed at home and are visually fine-tuned to their local habits and habitats.

Finally, with the knowledge we have of the Scarabaeinae, and the tractable nature of many of its species for breeding and behavioural observations, we have a model organism with which we can push further into the world of visual ecology and understand how evolution shapes the most popular eye design on our sunlit planet.

10

The Ecological Implications of Physiological Diversity in Dung Beetles

Steven L. Chown[1] and C. Jaco Klok[2]

[1]*Centre for Invasion Biology, Department of Botany and Zoology, Stellenbosch University, South Africa*
[2]*School of Life Sciences, Arizona State University, Tempe, AZ, USA*

> Before the sun becomes too hot, they are there in their hundreds, large and small, of every sort, shape and size, hastening to carve themselves a slice of the common cake.
> *Jean Henri Fabre (1918)* The Sacred Beetle and Others

10.1 Introduction

The extraordinary diversity of dung beetles, and of the scarabaeoids in general, has been a source of general fascination for natural historians and of specific scientific interest to researchers in a wide range of fields. Not only has this interest resulted in documentation of a staggering array of morphologies, behaviours and resource acquisition strategies, it has also led to an increasingly profound understanding of the mechanistic basis and evolutionary history of this diversity (e.g. Halfter & Matthews, 1966; Scholtz, 1989; Doube, 1991; Scholtz & Chown, 1995; Davis *et al.*, 2002a; Pérez-Ramos *et al.*, 2007). Along the way, it has also contributed substantially to theory in ecology and evolutionary biology (Doube, 1987; Hanski *et al.*, 1993; van Rensburg *et al.*, 2000; McGeoch *et al.*, 2002; Emlen *et al.*, 2007), as highlighted by the chapters in this volume.

A recurrent theme emerging from much of this work has been the significance of body size in all aspects of dung beetle ecology and evolution. Indeed, in few other groups has the interplay between size and other traits been as well explored as it has been in these beetles. For example, adult body size not only has a significant effect on resource acquisition (Heinrich & Bartholomew, 1979), but resource acquisition in turn has a profound effect on body size, owing to the influence of larval resource availability on adult size (Emlen & Allen, 2004; see Chapter 8 of this volume).

Ecology and Evolution of Dung Beetles, First Edition. Edited by Leigh W. Simmons and T. James Ridsdill-Smith.

Recognition of this interplay has elucidated many of the often complicated relationships among factors influencing and influenced by size in insects (and other organisms) at the mechanistic, ecological and evolutionary levels (for general reviews, see Chown & Gaston, 2010; Stillwell *et al.*, 2010).

In this chapter, we explore these relationships for ecophysiological traits that are not only thought to have played an important role in the structuring of dung beetle assemblages (Bartholomew & Heinrich, 1978; Krell *et al.*, 2003; Verdú *et al.*, 2007a; Roslin *et al.*, 2009), but which also play parts in determining dung beetle geographical ranges and richness (Davis, 1997; Lobo *et al.*, 2002; Chefaoui *et al.*, 2005; Davis *et al.*, 2008a; Duncan *et al.*, 2009), so ultimately influencing the likely responses of these species and the assemblages they constitute to changing environments (Davis *et al.*, 2002b; Botes *et al.*, 2006; Nichols *et al.*, 2007; see Chapters 11, 12 and 13). We focus on thermoregulation, thermal tolerance, water balance and gas exchange. We show not only how these traits interact with size and various other aspects of dung beetle ecology, but also how this work has contributed to the development and testing of important general concepts in biology. These include Rapoport's rule (Gaston *et al.*, 1998; 2008), the likely role of the sub-elytral chamber for minimizing respiratory water loss (Cloudsley-Thompson, 1964; Duncan & Byrne, 2002) and the evolutionary origins of discontinuous gas exchange in insects (Chown *et al.*, 2006).

10.2 Thermoregulation

The ability of scarabaeoid beetles to elevate thoracic temperatures above ambient has been known since early work on *Geotrupes stercorarius* (Krogh & Zeuthen, 1941). Subsequent studies have established this ability across a wide range of species in the group, with several investigations demonstrating endothermic thermoregulation during flight and some other activities (e.g. Bartholomew & Casey, 1977; Nicolson & Louw, 1980; Morgan & Bartholomew, 1982; Chappell, 1984; Morgan, 1987; Chown & Scholtz, 1993; Saeki *et al.*, 2005; Seymour *et al.*, 2009). Endothermy and thermoregulation have also been demonstrated in dung beetles, mostly during flight, although body (thoracic) temperatures may be elevated above those expected from incident radiation in some other activities, such as ball-rolling (Bartholomew & Heinrich, 1978; Heinrich & Bartholomew, 1979; Caveney *et al.*, 1995; Ybarrondo & Heinrich, 1996; Verdú *et al.*, 2004; 2006).

In the first multi-species comparative study of thermoregulation in the group, it was shown that metathoracic temperatures are related to mass up to about 2.5 g and are then mass-independent (Bartholomew & Heinrich, 1978), with the implication that thermoregulation via active abdominal heat transfer must be taking place in the larger species. Later work on a range of *Onitis* species demonstrated that larger species are, indeed, more capable of thermoregulation than smaller ones, measured as relative independence of thoracic temperature from ambient temperature (Caveney *et al.*, 1995).

This finding was confirmed in a further multi-species study incorporating animals from Spain, Mexico and East Africa (Verdú *et al.*, 2006). Species above a mass of 1.98 g are endothermic thermoregulators during flight (i.e. they are heterothermic),

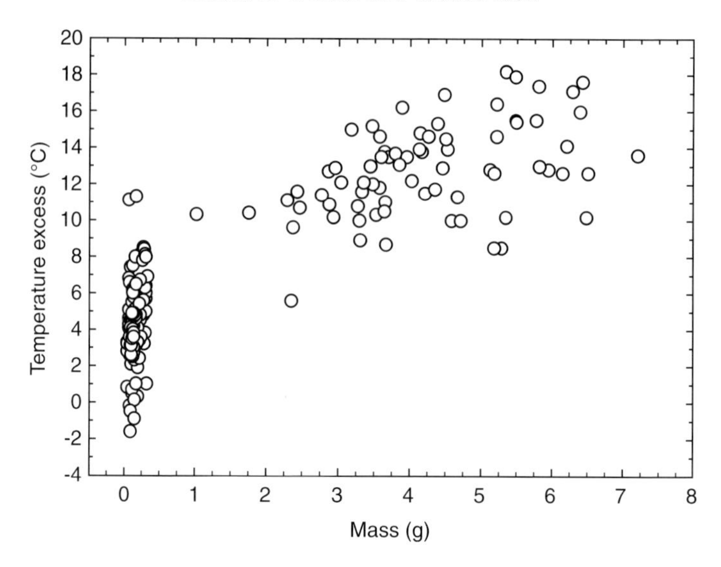

Fig. 10.1 Measurements of temperature excess (T_{ex} – measured as thoracic minus ambient temperatures) vs. mass in individuals of eight dung beetle species varying considerably in mass (mean ± S.E. [n] g: *Pachylomerus femoralis* 4.29 ± 0.21 [56], *Onthophagus* sp. 0.126 ± 0.003 [71], *Sisyphus* sp. 0.089 ± 0.013 [20], *Scarabaeus* sp. 0.186 ± 0.0026 [5], *Garetta* sp. 0.14 [1], *Scarabaeus (Kheper) lamarcki* 3.68 ± 0.21 [19], *Proagoderus* sp. 0.242 ± 0.009 [38], *Oniticellus* sp. 0.08 [1]) made across a range of temperatures at Tembe Elephant Park in northern Kwazulu-Natal South Africa. Piecewise linear regression (as implemented in *Statistica* v.8, Statsoft, Tulsa, Oklahoma) revealed that the break point is in the region of 1.6 g. The relationship below the 1.6 g breakpoint between $\log_{10}$ mass and temperature excess is $T_{ex} = 3.463 \pm 0.792 * \log_{10}$ mass $+ 7.391 \pm 0.696$, $F_{(1,137)} = 19.1$, $p < 0.00001$, $R^2 = 0.116$, SE of estimate = 2.0347.

whereas, in species below this mass, thoracic temperatures (T_{th}) are more closely related to ambient (T_a) conditions. Independent work at other sites bears out the idea that above a mass of about 2 g, thoracic temperatures and temperature excess ($T_{ex} = T_{th} - T_a$) are largely mass independent, while below this mass, they vary with size (Figure 10.1).

In consequence, early suggestions that large beetles are probably thermoregulating have been reasonably well supported by later comparative work. More detailed studies have also confirmed that this is the case and that, in some species, active transfer of heat via the abdomen is involved (Chown *et al.*, 1995; Verdú *et al.*, 2004). Most of this work has employed techniques that have been subject to criticism and are often no longer considered those most suitable in many situations where thermoregulation is being investigated (reviewed in Chown & Nicolson, 2004). However, for free-flying insects, alternatives such as implanted thermocouples and infrared thermography remain problematic (see Seymour *et al.*, 2009).

Not only might the ability to elevate thoracic temperatures enable dung beetles to explore a range of foraging options that otherwise would perhaps not be open to them (Ybarrondo & Heinrich, 1996), it also improves the likelihood of success in contest competition. Both in *Scarabaeus laevistriatus* and in *Scarabaeus* (*Kheper*) *nigroaeneus*, elevated thoracic temperatures following flight provide a competitive advantage over conspecifics and other species, relative to the situation where thoracic temperatures are closer to ambient conditions (Heinrich & Bartholomew, 1979; Ybarrondo & Heinrich, 1996). Large size also provides an advantage, though it appears that thoracic temperature elevation is most significant.

Although thoracic temperatures have a considerable influence on ball-making time and ball-rolling speed (Bartholomew & Heinrich, 1978; Heinrich & Bartholomew, 1979), it does not appear that high thoracic temperatures are routinely maintained during ball-making and -rolling in these species. Rather, other strategies may be used to avoid competition, such as reduction of ball construction time and volume in *S. nigroaeneus* (Ybarrondo & Heinrich, 1996).

The importance of thermoregulation for utilizing different kinds of resources, and for success in contest competition, suggests that differences in thermoregulatory ability may also play a role in niche differentiation in dung beetles. Recent work on Mexican dung beetles (Verdú *et al.*, 2007a) has shown that species with similar thermal niches differ in trophic habitats, while those with different thermal niches tend to have similar trophic habitats. Although it also appears that species with high temperature excesses tend to favour high-altitude, low-temperature environments, the relationship requires additional investigation. A negative relationship between T_a and T_{ex} is expected simply on a statistical basis, because $T_{ex} = T_{th} - T_a$, resulting in the non-independence of the x and y data (see Brett, 2004). Nonetheless, it is clear that substantial variation in timing of activities, resource use and thermoregulatory abilities is characteristic of dung beetles (see also Caveney *et al.*, 1995) and is likely related to the intense competition for resources found in some instances (see, for example, Doube, 1990; Hanski & Cambefort, 1991).

Not all large dung beetles are capable of endothermy or thermoregulation. The absence of endothermy seems to be closely related to flightlessness, which is not unexpected, given the importance of the flight musculature for both warm-up and endothermy during flight (reviewed in Heinrich, 1993). Perhaps the most conspicuous example of such a lack of endothermy is found in the large, flightless *Circellium bacchus* from southern Africa. Early work suggested that the species is not able to maintain body temperatures different from those expected from black bulb temperature (Nicolson, 1987). This was later borne out in a more extensive analysis, which showed that this species benefits from none of the advantages thought to accrue from the ability to elevate thoracic temperatures above ambient (such as improved competitive ability), despite relationships between temperature and activity that would suggest this should be the case (e.g. between thoracic temperature and walking speed – Figure 10.2) (Chown *et al.*, 1995).

On the basis of these and other data, it was suggested that *Circellium bacchus* may have survived in Addo Elephant Park and other sites in the southern and western Cape of South Africa owing to continuous availability of resources and the absence of many other competing large dung beetles. It was also argued that the previous distribution of this species elsewhere in Africa may have been linked to the continuous availability of dung associated with middens of the black rhinoceros, because of *C. bacchus*' lack of competitive ability. However, more recent work

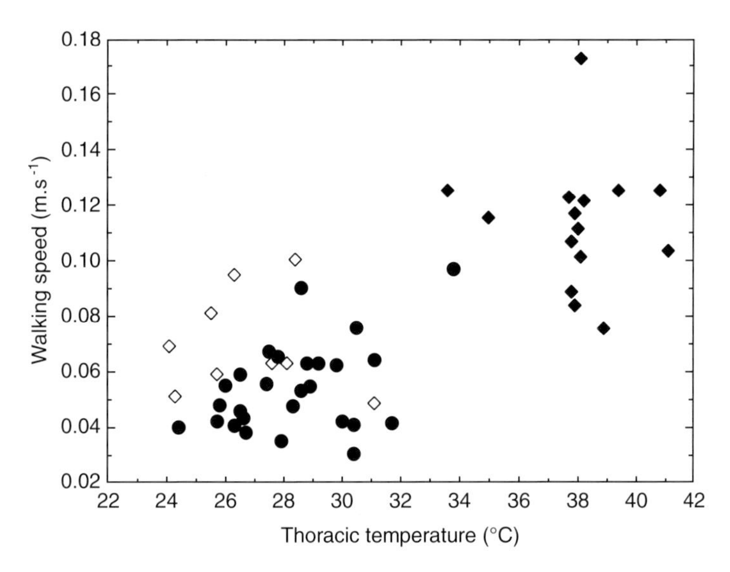

Fig. 10.2 Relationship between thoracic temperature and walking speed in *Circellium bacchus* (filled circles) and in *Pachylomerus femoralis* (open diamonds before flight, closed diamonds after flight). The relationship is significant in the former species, but in the latter only if the two data sets are combined. After flight, but not before, the latter species is substantially faster than the former. (Data and statistical findings from Chown *et al.* (1995)).

suggests that this might not be the case, although compelling evidence to distinguish these ideas remains elusive (for further information on the biology of *C. bacchus* see Kryger *et al.*, 2006).

Another flightless species which is incapable of endothermy is *Pachysoma gariepinum* from the Namib Desert (see Scholtz, 1989; Sole *et al.*, 2005 for discussion of the genus). In this species, thoracic temperature is closely related to operative temperature of models exposed under similar conditions (a mean from three dried beetle models), with the slope of the relationship showing no significant difference from 1 (Figure 10.3A, $t_{(1,24)} = -1.82$, $p > 0.05$), signifying no thermoregulation. Mean temperature of the three operative models is effectively explained ($R^2 > 0.98$) by wind speed, solar radiation, ambient temperature (dry bulb) and vapour pressure deficit (Figure 10.3B), with removal of any one of the variables significantly reducing the Akaike Weight of the model (from $w_i = 0.99$ to $w_i < 0.001$).

These data are similar to those for *P. denticolle*, a species with characteristically orange elytra, where abdominal temperatures range between 26.5 °C and 42.9 °C (Holm & Kirsten, 1979). Like the other dung beetle species studied (see above), walking speed in *P. gariepinum* is closely related to thoracic temperature (Figure 10.3C). However, maximum speeds recorded in *P. gariepinum* are almost double those found in other telecoprids, and this may have to do with the unusual foraging behaviour of the entire genus *Pachysoma* (Scholtz, 1989; Davis *et al.*, 2008b). Although resources may not necessarily be patchy, given that dry dung and/or detritus may be used, depending on the species (Scholtz, 1989; Sole *et al.*, 2005;

Holter *et al.*, 2009), time constraints, given the climate of the area inhabited by most members of the genus, may have selected for considerable foraging speed (see Dunbar *et al.*, 2009 for a general discussion of time constraints). Alternatively, or in combination with time constraints, considerable predation in the dune area where these species are active could explain rapid running speeds and mimicry among species with orange elytra (Holm & Kirsten, 1979).

Although much of the emphasis in the study of dung beetle thermoregulation has been placed on endothermy, various forms of behavioural thermoregulation are also likely, especially in the ectothermic species (e.g. Scholtz, 1989). These aspects have not been especially well explored for the group, though they are undoubtedly important, as timing of activity (Caveney *et al.*, 1995; Verdú *et al.*, 2007a) and the use of thermal refuges (e.g. Nicolson, 1987) suggest. Clines in colour in widespread species, related to environmental temperatures, are also thought perhaps to serve a thermoregulatory function (Davis *et al.*, 2008c). Clearly, this is an aspect of dung beetle physiological ecology that deserves additional attention.

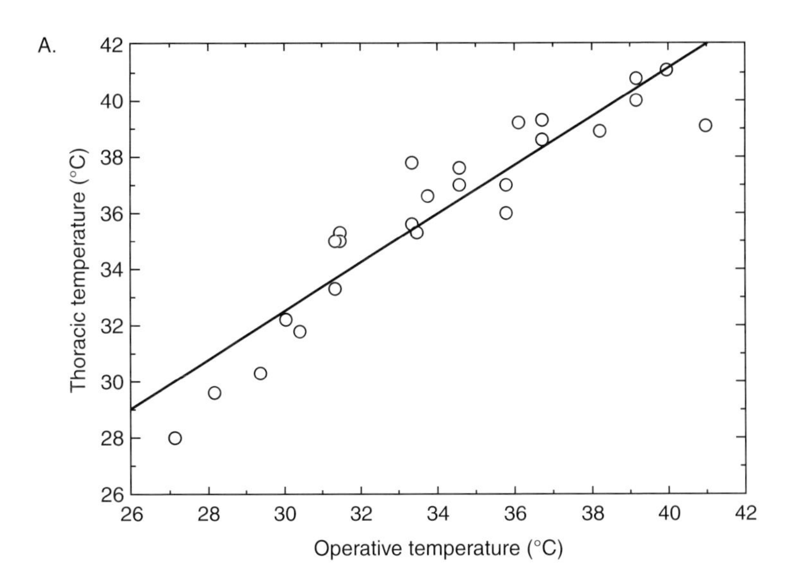

Fig. 10.3 Thermal relations and habitat temperatures for *Pachysoma gariepinum* at Hohenfels in the Namib Desert. **A**: Relationship between thoracic temperature (T_{th}) of free-running individuals, determined by thoracic puncture with needle-embedded 40 gauge thermocouple and operative temperature (T_e) based on the mean of three dried beetle models ($T_{th} = 6.62 \pm 2.35 + T_e{}^{*}\,0.875 \pm 0.069$, $F_{(1,24)} = 162.8$, $p < 0.0001$, $R^2 = 0.866$, SE of estimate = 1.266). **B**: Microclimate data over five full days indicating solar radiation (crosses), operative temperature models in the sun (open squares and diamonds), an operative model in the shade (filled triangle) and shaded dry bulb temperature (filled circles). The table indicates the relationship between the mean of the operative models and the independent variables indicated, which explain c. 99 per cent of the variance in the former. **C**: Relationship between walking speed (measured as in Chown *et al.*, 1995) and thoracic temperature of free-running individuals, determined by thoracic puncture with needle-embedded 40 gauge thermocouple (Walking speed = $-0.238 \pm 0.08 + T_{th} \times 0.0127 \pm 0.0022$, $F_{(1,24)} = 33.1$, $p < 0.0001$, $R^2 = 0.563$, SE of estimate = 0.038).

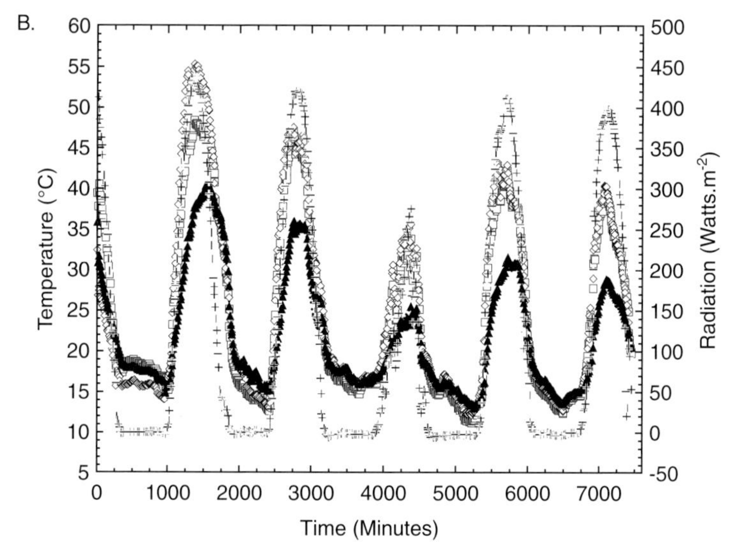

Variable	Estimate	x^2	p
Wind speed (m.s^{-1})	-0.713	299.1	0.00001
Dry bulb temperature (°C)	1.022	1046.8	0.00001
Solar radiation (W.m^{-2})	0.024	889.2	0.00001
Vapour pressure deficit (kPa)	-0.365	22.1	0.00001
Model fit (Deviance/df = 863.2/740 = 1.6)			

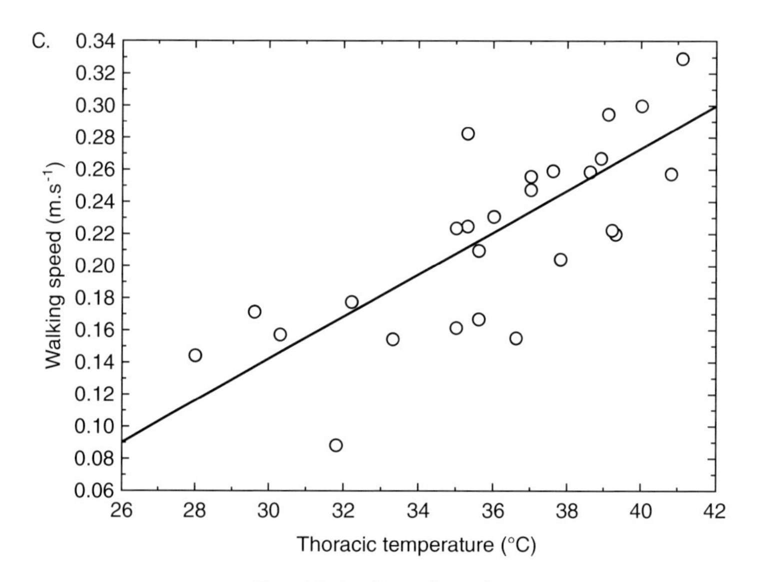

Fig. 10.3 (*Continued*)

10.3 Thermal tolerance

Among their several findings, Verdú *et al.* (2006) noted that the maximum temperature tolerated by dung beetles during flight is 42°C. This value is close to that found for both endothermic and ectothermic dung beetle species in southern and eastern Africa (Bartholomew & Heinrich, 1978; Holm & Kirsten, 1979; Ybarrondo & Heinrich, 1996). It is also borne out by work on *P. femoralis* (Chown *et al.*, 1995) and on several species from northern Kwazulu-Natal (from which data for Figure 10.1 were derived). For *P. femoralis*, it was noted that the highest value of 43.4 °C for T_{th} is probably an overestimate, because beetles close their elytra once caught, so elevating T_{th} (for general discussion see Stone & Willmer, 1989). In consequence, a slight overestimate of sustainable maximum T_{th} likely applies to all of the investigations undertaken to date.

Nonetheless, a maximum sustained T_{th} of 42 °C is close to the short-term critical thermal maximum found in an independent study of 26 dung beetle species across an altitudinal gradient in the Kwazulu-Natal Drakensberg (Gaston & Chown, 1999a). Here, critical thermal maximum (CT_{max}), defined as the onset of thermal-induced spasms following heating at a rate of 0.5 °C min^{-1}, varied between 40 and 52 °C. These values are typically higher than maximum sustained T_{th}, as might be expected from the fact that critical thermal limit values are typically higher than those of upper lethal temperatures measured over longer time periods (Chown & Nicolson, 2004), and especially when CT_{max} values are estimated using rapid heating rates (Chown *et al.*, 2009).

Thus, an upper limit to activity of *c*. 42 °C seems likely, and indeed is more or less in keeping with that found for many other insect species (Addo-Bediako *et al.*, 2000; Hoffmann *et al.*, 2003; Chown & Terblanche, 2007; but see also Calosi *et al.*, 2010). The mechanistic basis of this limit is probably protein and membrane structural damage rather than oxygen limitation of thermal tolerance (see Klok *et al.*, 2004; Chown & Terblanche, 2007).

As is the case in many other insect groups, in those dung beetle species where critical limits have been investigated, critical thermal minimum (CT_{min}) shows much greater variation than does CT_{max} (Gaston & Chown, 1999a; for review of information on insects see Chown, 2001; Chown & Terblanche, 2007). In the specific case of the Drakensberg beetles, this translates to a much greater thermal tolerance range at high than at low altitudes, which is likely to be due to a much greater ambient annual temperature range at higher elevations. In turn, this is associated with a wider mean elevational range for assemblages occurring at increasingly higher altitudes (Gaston & Chown, 1999a).

In combination, these findings provide among the earliest and most comprehensive evidence in support of the climatic variability hypothesis thought to underlie Rapoport's rule or the Rapoport effect – the increase in latitudinal or elevational range of species with latitude or elevation, respectively (Stevens, 1989; Gaston *et al.*, 1998; 2008; Addo-Bediako *et al.*, 2000; see Chapter 11 in this volume). According to Stevens (1989), to survive at higher latitudes (or elevations), individual organisms need to be able to withstand greater temporal variability in climate than those at lower latitudes (or elevations). In consequence, the species to which these individuals belong can attain wide latitudinal or altitudinal extents.

A range of studies of other organisms (including insects) has now investigated and, in some cases, has found support for the climatic variability hypothesis (Addo-Bediako *et al.*, 2000; Pither, 2003; Calosi *et al.*, 2008; 2010), although the significance of the Rapoport effect has been shown perhaps to be more local than global. For recent discussion and data, see Gaston & Chown, 1999b; Orme *et al.*, 2006; Gaston *et al.*, 2008; Beketov, 2009; Šizling *et al.*, 2009.

The relative lack of variation in upper critical limits and considerable variation in lower limits suggest that the latter may play a more significant role in determining species distributional ranges than the former, at least where climate variables are significant (see Davis *et al.*, 2008a; Duncan *et al.*, 2009). However, the effects of temperature on development rate are also likely to be significant. Again, though, this is a relatively poorly researched area (see Edwards, 1986; Walsh *et al.*, 1997) despite much interest in developmental effects on morphology (see Chapter 7) and the effects of agrichemicals on dung beetles during their development (e.g. Lumaret *et al.*, 1993; Wardhaugh *et al.*, 2001; Lumaret & Errouissi, 2002).

10.4 Water balance

Along with some measure of environmental temperature, water availability – usually in the form of rainfall – is consistently identified as a correlate either of dung beetle richness, or assemblage turnover, or the distribution and/or activity of individual species (e.g. Hanski & Cambefort, 1991; Davis, 1996; Lobo *et al.*, 2006; Davis *et al.*, 2008a; see Chapter 13). In consequence, the ability to regulate water balance likely varies considerably among dung beetle species.

Perhaps surprisingly, only a few studies have specifically sought to investigate such variation from a mechanistic rather than a correlative perspective, although more generally in insects it is clear that water balance traits are closely related to variation in water availability (reviewed in Hadley, 1994; Addo-Bediako *et al.*, 2001; Chown & Nicolson, 2004). The most comprehensive studies of dung beetles have concerned the role of patterns of gas exchange or variation in metabolic rate as modulators of respiratory water loss in species from environments that vary in water availability (Duncan & Byrne, 2000; Chown & Davis, 2003; but see also Chown *et al.*, 1995).

However, to begin with the components of water loss (largely respiratory and cuticular – see Hadley, 1994) would mean glossing over the roles of variation in water contents and water loss rates in modulating desiccation resistance, how these interact with survival time to determine desiccation tolerance and how these variables, in turn, are related to size. These are all significant aspects of water balance in insects (see e.g. Hadley, 1994; Gibbs, 2002; Gibbs *et al.*, 2003; Chown & Klok, 2003; Matzkin *et al.*, 2009), but ones which have been poorly investigated for dung beetles. Therefore, we begin by exploring these interactions for 18 species of dung beetles from southern Africa, assessed using a gravimetric approach to water loss, with mass being measured at regular intervals (as a proxy for water loss) following drying over silica gel until death (Table 10.1).

Although water lost before death varies among species, it is strongly related to body mass (and presumably, therefore, water content), with little residual variation

Table 10.1 Mean ± S.E. of starting mass, mass lost before death, rate of mass loss and survival time for the 18 species of dung beetle examined by Klok (1994) using gravimetric methods to assess water loss (i.e. assuming mass loss is a reasonable proxy for water loss). Listing follows the taxonomy in Davis *et al.*, (2008b). The measurements were made at 5 per cent RH, 27 °C, 12:12 L:D cycle by periodic weighing of individuals until death, with the interval before death being recorded as the endpoint of survival (see Chown *et al.*, 1995 for a synopsis of the methods and Klok, 1994 for complete details).

Species	*n*	*Mass (mg)*	*Mass lost (mg)*	*Rate of loss* ($mg.h^{-1}$)	*Time (hours)*
Catharsius tricornutus	20	1314 ± 61.8	392.4 ± 27.8	14.12 ± 0.82	27.6 ± 1.1
Copris amyntor	20	300.6 ± 16.0	115.3 ± 6.4	1.67 ± 0.15	79.2 ± 8.4
Anachalcos convexus	20	1190 ± 59.5	307.8 ± 19.1	5.25 ± 0.35	64.4 ± 6.0
*Circellium bacchus**	14	6488.7 ± 331.9	2626.4 ± 157.8	14.10 ± 1.93	211.6 ± 17.7
Garetta nitens	20	358.1 ± 24.0	89.3 ± 7.1	7.20 ± 0.75	15.8 ± 2.0
Pachylomerus femoralis	16	4266.8 ± 272.5	1267.5 ± 94.5	13.30 ± 1.19	106.2 ± 11.4
Scarabaeus zambesianus	20	1247.7 ± 69.3	320.6 ± 25.7	10.94 ± 0.83	32.8 ± 3.6
Scarabaeus proximus	20	675.3 ± 45.5	230.7 ± 23.1	5.13 ± 0.79	54.6 ± 6.2
Scarabaeus (Kheper) cupreus	20	1790.6 ± 71.9	559.9 ± 25.1	5.80 ± 0.39	101.3 ± 5.2
Scarabaeus (Kheper) nigroaeneus	20	2113.8 ± 122.4	560.8 ± 35.3	9.30 ± 0.62	61.4 ± 1.9
Scarabaeus (Scarabaeolus) rubripennis	20	157.9 ± 7.4	68.3 ± 3.8	6.00 ± 0.43	126.0 ± 7.9
*Pachysoma gariepinum**	20	1744.4 ± 77.6	597.1 ± 61.7	2.20 ± 0.29	392.8 ± 64.2
*Pachysoma striatum**	20	794.9 ± 32.6	304.4 ± 19.8	1.30 ± 0.06	235.3 ± 15.6
Sisyphus impressipennis	20	98.7 ± 3.6	25.9 ± 2.0	1.39 ± 0.12	21.3 ± 2.4
Onitis caffer	20	523.3 ± 27.6	176.7 ± 10.3	9.56 ± 0.53	18.8 ± 0.8
Phalops ardea	20	107.5 ± 3.7	24.2 ± 0.3	3.12 ± 0.28	9.5 ± 1.7
Onthophagus sapphirinus	20	125.2 ± 6.6	34.5 ± 1.7	0.79 ± 0.09	52.0 ± 4.3
Liatongus militaris	40	84.6 ± 1.7	26.4 ± 0.73	1.50 ± 0.04	18 ± 0.5

*=flightless species.

(Figure 10.4A). By contrast, both water loss rate and survival time are also correlated with initial body mass, but substantially greater residual variation is present in both variables after the effects of mass are accounted for (Figures 10.4B and 10.4C). These relationships are apparent when phylogenetic non-independence is accounted for using PDAP (see Garland *et al.*, 1993) and a tree built using the phylogenies provided by Davis *et al.* (2008b). The strong relationships between $\log_{10}$ mass and $\log_{10}$ mass loss, and between $\log_{10}$ mass and $\log_{10}$ rate of water loss, remain

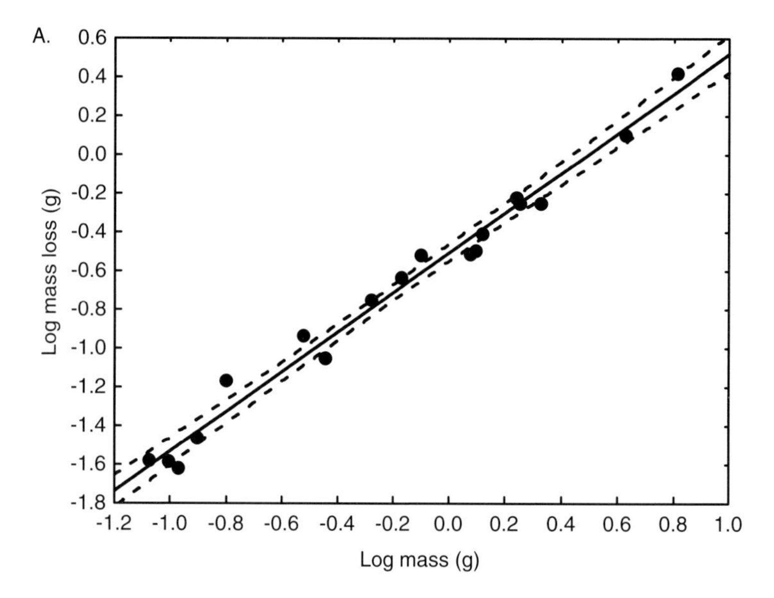

Fig. 10.4 The relationship in 18 species of dung beetles between log body mass and: **A: log water loss before death**. The equation for the fitted line is $\log_{10}$ water loss $= -0.506 \pm 0.021 + 1.026 \pm 0.034 \times$ log body mass, $F_{(1,16)} = 899.1$, $p < 0.0001$, $R^2 = 0.981$, S.E. estimate $= 0.082$. Using independent contrasts, the slope is 0.994, with an R^2 of 0.975 ($F_{(1,15)} = 64.2$, $p < 0.0001$). **B: $\log_{10}$rate of water loss**. The equation for the fitted line is $\log_{10}$ rate of loss $= -2.291 \pm 0.080 + 0.577 \pm 0.134 \times \log_{10}$ body mass, $F_{(1,16)} = 18.67$, $p < 0.0006$, $R^2 = 0.510$, S.E. estimate $= 0.319$. Note that the open square for *Pachysoma gariepinum* and the closed square for *Scarabaeus westwoodi* are data from Davis *et al.* (1999), collected using infrared flow-through methods. Using independent contrasts, the slope is 0.689, with an R^2 of 0.605 ($F_{(1,15)} = 24.5$, $p < 0.001$). **C: survival time**. The equation for the fitted line is $\log_{10}$ survival time $= 1.834 \pm 0.095 + 0.446 \pm 0.159 \times \log_{10}$ body mass, $F_{(1,16)} = 7.87$, $p < 0.0127$, $R^2 = 0.288$, S.E. estimate $= 0.379$. The desert-dwelling *P. gariepinum*, *P. striatum* and *Scarabaeus rubripennis* are indicated. Using independent contrasts, the relationship is no longer significant ($F_{(1,15)} = 3.27$, $p = 0.091$). Flightlessness is not significant as an explanatory variable of residual variation in the mass – mass loss/rate/time relationships (GLZ $p > 0.05$), whereas this is the case for habitat (xeric or mesic) (GLZ $p < 0.05$ in all cases). However, using the Akaike weight, wing status always remains a factor in the best models (Akaike weights are typically 0.1 larger for models including wing status than for other models). Other factors, such as activity time (diurnal or nocturnal) or functional group membership, never achieve significance (generalized linear models including log mass and either activity time or functional group: $0.227 < p < 0.797$).

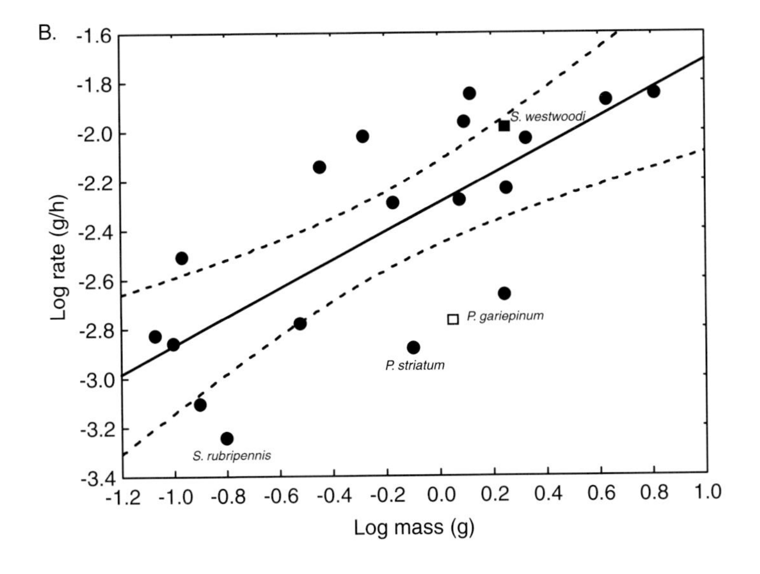

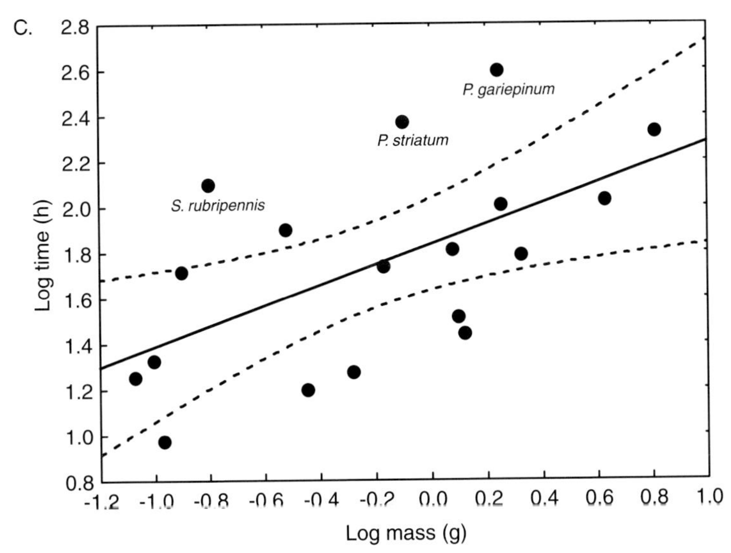

Fig. 10.4 *(Continued)*

when the independent contrasts are examined. However, the relationship between $\log_{10}$ mass and $\log_{10}$ time is no longer significant (see Figure 10.4).

While flightlessness has long been thought of as an adaptation to arid conditions (Cloudsley-Thompson, 1964; Roff, 1990; Scholtz, 2000), it is typically not significant as an explanatory variable of residual variation in the mass – mass loss/rate/time relationships, whereas this is the case for habitat (Figure 10.4 legend). Nonetheless, wing status always remains a factor in the best models, whereas factors such as activity time (diurnal or nocturnal) or functional group membership (see Doube, 1991) never achieve significance (Figure 10.4 legend). Thus, if survival time is taken

Table 10.2 Outcome of a generalized linear model (normal distribution, identity link function) investigating the influence of starting mass, mass lost before death, rate of mass loss, wing status (flying or flightless) and habitat type (xeric or mesic) on survival time of desiccating conditions in 18 species of dung beetles.

Variable	χ^2	*p*	*Estimate*
Log mass	4.56	0.033	0.380 ± 0.167
Log mass loss	11.57	0.0007	0.622 ± 0.154
Log rate	65.73	0.00001	-1.020 ± 0.039
Wing status	2.38	0.12	
Habitat type	0.57	0.45	
Wing status * Habitat type	1.78	0.28	
Model fit deviance/df = 0.019/11 = 0.002			

as a partial metric of fitness (recognizing that fitness includes reproduction), it is clear that variation therein is modulated by initial mass, rate of water loss and water loss tolerated before death (Table 10.2), and that, in turn, these variables are affected by wing status and the type of habitat in which the species are most commonly found. Indeed, it is clear from Figure 10.4C that species from arid areas, including the small, winged *S. rubripennis* and the much larger, flightless *P. gariepinum* and *P. striatum*, all have much longer survival times than their mesic counterparts. However, this is not true of all the arid dwellers, because both *Scarabaeus proximus* (previously *Drepanopodus proximus* – see Forgie *et al.*, 2006) and *Circellium bacchus* are indistinguishable on these grounds.

It seems likely that size is an important component of survival for *C. bacchus* (see also Chown *et al.*, 1995), which might account for smaller size in populations further to the mesic southwest, while behavioural avoidance may be a key factor for *S. proximus*. However, suggestions have also been made that the habitats occupied by *C. bacchus* may be more arid than those of the *Pachysoma* species because of predictable fog in the Namib desert and because of the use of a sub-surface moisture layer for the rehydration of their food (Duncan & Byrne, 2005). Clearly, the availability of water from fog and below the soil surface are important contributors to the water balance of the *Pachysoma* species (Scholtz, 1989; Duncan & Byrne, 2005), but the data presented suggest that they have evolved much lower water loss rates for their size than many other species, while this is not the case for *C. bacchus*.

Although it is reasonable to expect that the gravimetric study provides an overestimate of water loss rates generally (given that energy use must accompany water loss over the trial period), comparison of data from *P. gariepinum* obtained this way and from infrared detection of actual water loss (Chown & Davis, 2003) show that the results are reasonably similar (Figure 10.4B). The good fit to the body mass-water loss rate relationship of data from the mesic *S. westwoodi*, a species examined using infrared methods only, further supports the generality of the gravimetric trials (Figure 10.4B). Thus, flightlessness does seem to be associated with low rates of water loss and extended survival times, although not exclusively so. Moreover, flightlessness might have evolved for a variety of other reasons, such as constant habitat conditions, high food availability and high population densities (Roff, 1990; Scholtz, 2000).

Nonetheless, several ways exist in which flightlessness may contribute to a reduction in water loss. Selection for water saving might either act singly or in concert with other factors (see Chown, 2002) to promote the evolution and retention of this trait.

- First, flightless dung beetles have a strongly rounded shape – much more so than their winged counterparts. Round body forms may well promote a reduction in water loss on purely physical grounds (Chown *et al.*, 1998), although the comparative physiological data to test this idea have not been gathered.
- Second, flightlessness is accompanied by generally lower active and resting metabolic rates in insects (Reinhold, 1999; Addo-Bediako *et al.*, 2002; Chown *et al.*, 2007). If respiratory water loss is a significant component of total water loss, then a reduction in metabolic rate must be an important factor contributing to water savings (Woods & Smith, 2010).
- Third, the sub-elytral chamber might contribute to water savings as a consequence of retrograde airflow through the beetle spiracular system (Cloudsley-Thompson, 1964), so leading to a decline in water loss rate effectively owing to boundary effects and increasing path length (for discussion, see Duncan & Byrne, 2002; Duncan & Dickman, 2009). These latter factors are highly dependent on the extent to which respiratory water loss is a significant contributor to total water loss in beetles.

Much controversy has swirled around the significance of respiratory water loss (RWL) in insects (reviewed by Chown, 2002). The typically low percentage contribution of RWL to total water loss (TWL) was long considered an argument in favour of the unimportance of the former, and led to the predominance of cuticular water loss (CWL) as the most significant water loss avenue. However, comparisons of percentages are misleading, *inter alia* because simultaneous modulation of both CWL and RWL would lead to no change in relative percentage water loss.

In dung beetles, it has been shown by comparison of the measured rates of loss among species from mesic and xeric areas that both CWL and RWL contribute significantly to variation in total water loss rates (Chown & Davis, 2003). Moreover, variation in metabolic rate and duration of particular periods of the discontinuous gas exchange cycle (see below) contribute significantly to variation in RWL rate. Beetles from xeric environments tend to have lower cuticular and respiratory water loss rates, lower metabolic rates, longer periods of complete or partial spiracle closure and shorter periods when spiracles are fully open than their mesic counterparts (Chown & Davis, 2003). Thus, it is clear that modulation of respiratory water loss is important in dung beetles and may well be a factor underlying the evolution of flightlessness.

If this were the case, similar trends should be found in winged species. One indication that this is the case comes from *Scarabaeus spretus* from the Western Cape of South Africa. In this species, acclimation temperature has a significant effect on standard metabolic rate, such that rates are high at low acclimation (rather than measurement) temperatures and low at higher acclimation temperatures. Typically, such data are interpreted in the context of metabolic adjustments to low temperature (reviewed in Clarke, 1993; Chown & Gaston, 1999). However, other interpretations are also plausible, and it may be impossible to distinguish metabolic rate depression following acclimation at high temperatures (for water or energy savings) from metabolic rate elevation after acclimation at low temperatures (e.g. to prime the energy delivery system for immediate functioning) (Davis *et al.*, 2000).

Simultaneous measurement of water loss and metabolic rate in *Scarabaeus spretus* showed that, in this winged species, although metabolic rate varied in the expected manner, much greater control was exercised over water loss rates. Thus, despite high acclimation temperatures, which should lead to higher water loss rates, water loss rate was reduced, especially at high measurement temperatures (Terblanche *et al.*, 2010). Hence, even within species, physiological changes to limit water loss by modulation of gas exchange dynamics are clear.

The role of the sub-elytral chamber in reducing respiratory water loss has turned out to be very different from that originally proposed, and work on several species of dung beetles has shown just how divergent are the original supposition and the empirical findings. In the first species studied, *Circellium bacchus*, little gas exchange at rest takes place via the abdominal spiracles within the sub-elytral chamber. Instead, gas exchange, during the large, open or burst phase of discontinuous gas exchange, is almost entirely dominated by tidal airflow through the large right mesothoracic spiracle (Duncan & Byrne, 2002).

Thus, airflow is typically not retrograde, with water loss being restricted by the humid sub-elytral chamber, as originally thought. Rather, it appears that gas exchange through a single spiracle, equipped with a sieve plate, reduces cross-sectional area over which exchange takes place – thus lowering water loss – and that airflow is either anterograde or tidal. Nonetheless, pressure measurements taken from the tracheal system and sub-elytral chamber (Duncan *et al.*, 2010), and little evidence of water loss from the chamber (Duncan, 2002), suggest that, during interburst periods, oxygen may enter the tracheal system from the sub-elytral chamber, with little water loss owing to michrotrichial sealing of the chamber (Duncan *et al.*, 2010).

Increasing use of the mesothoracic spiracles for gas exchange at rest appears to be a graded response to aridity. In the mesic *Pachylomerus femoralis*, the largest contribution to gas exchange is made by the abdominal spiracles (88 per cent), while in *Pachysoma gariepinum* and *P. striatum*, from more xeric environments, the contribution made by the mesothoracic spiracles increases substantially (20 per cent and 46 per cent, respectively) (Duncan & Byrne, 2005). In *C. bacchus*, the contribution is 79 per cent.

Predominant use of the mesothoracic spiracles during gas exchange at rest has also been demonstrated in three species of flightless desert tenebrionids (Duncan, 2003; Duncan & Dickman, 2009) This has provided generality for the idea that water saving is effected by reduction of the cross-sectional area over which water can be lost (see Kestler, 1985 for the underlying principles, and additional discussion in Schilman *et al.*, 2008). Thus, the role of the sub-elytral chamber is much more complex than previously thought. While it does appear to be partially involved in water retention, though not via retrograde airflow, it may also function as a water storage area by enabling expansion of the abdomen during and after drinking, as suggested by Slobodchikoff & Wismann (1981) (see also Schilman *et al.*, 2008).

Drinking has been recorded in all of the tenebrionid species investigated (Seely *et al.*, 2005; Duncan & Dickmann, 2009) and seems likely in the dung beetles, given the mechanisms of feeding in this group (Holter & Scholtz, 2005; 2007). Irrespective, it appears that a long-held notion about the way the sub-elytral chamber functions has been laid to rest.

10.5 Gas exchange and metabolic rate

One of the main reasons that dung beetles have proven so useful for investigating the components of water loss is that many of them exchange gases discontinuously when at rest, i.e. they show discontinuous gas exchange cycles (DGCs). Indeed, discontinuous gas exchange has now been recorded in *Anachalcos convexus*, *C. bacchus*, *Pachylomerus femoralis*, *Pachysoma hippocrates*, *Pachysoma gariepinum*, *Pachysoma striatum*, *S. flavicornis*, *S. galenus*, *S. rusticus*, *S. spretus*, *S. westwoodi*, and possibly also in *Sisyphus fasciculatus*, though in this species the cycles are less clear (Lighton, 1985; Davis *et al.*, 1999; Duncan & Byrne, 2000; 2005; Duncan *et al.*, 2010; Terblanche *et al.*, 2010).

In essence, discontinuous gas exchange consists of the alternation of three distinct gas exchange periods or phases, during which the spiracles are closed (C phase), flutter (F phase) or are entirely open (O phase). It is during the period of spiracle closure that definitive measurements of water loss through the cuticle can be made, so allowing cuticular and respiratory water loss components to be distinguished (see Chown & Nicolson, 2004).

Discontinuous gas exchange has been well studied in a range of insects and other tracheated arthropods (for reviews and examples, see Lighton, 1994; 1996; Lighton & Fielden, 1996; Klok *et al.*, 2002; Lighton & Joos, 2002; Chown *et al.*, 2006; Quinlan & Gibbs, 2006; Clusella-Trullas & Chown, 2008), although, by comparison with the numbers of insect higher taxa, the extent of the work is still reasonably narrow (Marais *et al.*, 2005). Moreover, many species do not show DGCs (Lighton, 1998; 2002; Lighton & Ottesen, 2005; Schilman *et al.*, 2008; Duncan & Dickman, 2009).

For those animals that do show discontinuous gas exchange at rest, the reasons for the evolutionary origin and maintenance thereof have been controversial, owing to several competing hypotheses, many of which seem to enjoy some support (reviewed in Chown *et al.*, 2006). Although some investigators would continue to disagree (e.g. Lighton & Turner, 2008; Schilman *et al.*, 2008), recent work on dung beetles (see above and Duncan & Byrne, 2000; Chown & Davis, 2003) and on a variety of other insects (Hetz & Bradley, 2005; White *et al.*, 2007a; Schimpf *et al.*, 2009) has settled on the original hypothesis of limitation of water loss (Buck *et al.*, 1953) and the newer idea of limitation of oxidative damage (Bradley, 2006) as the most likely reasons for the evolution and maintenance of this gas exchange pattern.

As was noted above, in addition to gas exchange pattern, metabolic rate may also be modulated to effect a reduction in water loss. Not only has this been demonstrated to be the case for some species of dung beetles (Chown & Davis, 2003), it has also been argued to apply in others (e.g. Duncan & Byrne, 2005 for *C. bacchus*). Because metabolic rate may vary for many reasons, most importantly as a result of size and temperature variation (see Davis *et al.*, 1999 for dung beetles; Chown *et al.*, 2007 and Irlich *et al.*, 2009 for insects; and Brown *et al.*, 2004 for general review), but also for other reasons (reviewed in Chown & Nicolson, 2004), it is important to take these factors into consideration before assessing the likely roles of water conservation in influencing rate variation.

For the dung beetles as a whole (as opposed to specific studies such as the one undertaken by Chown & Davis, 2003), the data are at present perhaps too sparse to undertake such an analysis. Assuming that flow-through methods are more appropriate than closed system approaches to respirometry (see Addo-Bediako *et al.*, 2002), only 12 species of dung beetles (excluding the Aphodiinae) have been investigated. The relationship between mass and metabolic rate has a relatively low coefficient of determination, which is not unexpected, given the small number of species involved. For three species, at least two separate estimates of metabolic rate are available and these vary quite widely (especially in the case of the *Pachysoma* species). However, the different data have only a small effect on the allometric relationship (Figure 10.5).

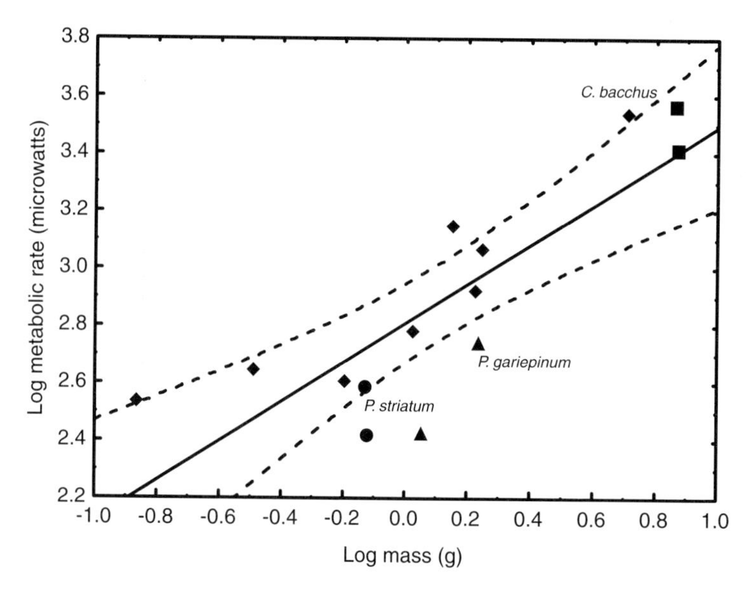

Fig. 10.5 Plot of $\log_{10}$ metabolic rate against $\log_{10}$ body mass for the 11 species of dung beetles investigated using flow-through respirometry. The additional data points (duplicate symbols of the same shape) indicate species for which multiple values are available. The fitted line includes these multiple data points. Excluding them provides a regression of either $\log_{10}$ mass $= 2.823 \pm 0.073 + 0.636 \pm 0.155 \times \log_{10}$ metabolic rate, $F_{(1,9)} = 16.73$, $p < 0.0028$, $R^2 = 0.611$, S.E. estimate $= 0.242$, or of $2.876 \pm 0.057 + 0.655 \pm 0.120 \times \log_{10}$ metabolic rate, $F_{(1,9)} = 30.05$, $p < 0.0004$, $R^2 = 0.744$, S.E. estimate $= 0.186$. These alternatives depend on which metabolic rate data are used where a species has been measured more than once. ■ = *Circellium bacchus*, ♦ = *P. gariepinum* and ● = *P. striatum*.

The data available indicate that the *Pachysoma* species (*P. hippocrates*, investigated by Lighton (1985), was excluded as an outlier in the entire analysis) have metabolic rates that are perhaps lower than expected simply on the grounds of mass, but that *C. bacchus* falls within the confidence intervals, suggesting it has a metabolic rate which might be expected for a beetle of its size (*contra* Duncan & Byrne, 2005). Nonetheless, by comparison with all insects, *C. bacchus* has a metabolic rate lower than expected for such a large size, which is also the case for the *Pachysoma* species.

Overall, the exponent of the metabolic rate allometric relationship (c. 0.636–0.655) is no different statistically to that found for the insects overall (i.e. 0.820, $t = -1.187$ or -1.375, $p > 0.05$, for values of 0.636 or 0.655, respectively – see Figure 10.5) and not significantly different to the theoretically expected values (see Glazier, 2005) of either 0.66 or 0.75 ($p > 0.05$ in both cases), as might be expected given the low coefficients of determination. Thus, not much can be said about the extent to which the metabolic theory of ecology's proposed general scaling exponent of 0.75 (Brown *et al.*, 2004) applies to this group. However, it should be noted that such a universal exponent has now been questioned on both theoretical (Apol *et al.*, 2008) and empirical (Glazier, 2005; White *et al.*, 2007b; Makarieva *et al.*, 2008) grounds, and several alternative ideas exist to explain variation in allometric exponents (e.g. Kozłowski *et al.*, 2003; Glazier, 2008; 2010).

Perhaps one of the most intriguing of these ideas is the way in which cell size and number contribute to increases in size and, consequently, to the likely value of the scaling exponent. Within a species, an increase in size mediated by cell size alone should leave the exponent close to 0.67, whereas an increase in size mediated by an increase in cell number only would suggest a value of 1.0 (Kozłowski *et al.*, 2003), with interspecific values lying somewhere in between.

Some insect groups are ideal to test these ideas, because of considerable intraspecific size variation. So far, however, it has only been done with the ants, providing some support for Kozłowski *et al.*'s (2003) ideas (Chown *et al.*, 2007). Dung beetles are clearly another model group in this respect because of the considerable size variation within many of the species in the group (Emlen & Allen, 2004; Emlen *et al.*, 2007). They also lend themselves to rearing (Walsh *et al.*, 1997) and to investigations of metabolic rate. Moreover, assessments of cell size and number, using the eyes as proxies (see Chown *et al.*, 2007), should also be feasible (see Chapter 9 of this volume).

Although controversial, the metabolic theory of ecology has re-focused attention on the significance of metabolic rate variation as a factor underlying much variation in the ecology and evolution of biodiversity (see also Ricklefs & Wikelski, 2002). Little empirical attention has been given to the interplay between energetics and ecology in dung beetles; however, one study, in West African habitats, suggested that the high energetic requirements of telecoprids likely constrain them to open savanna areas, whereas apparently competitively inferior species, with lower energy demands, may utilize other areas (Krell *et al.*, 2003).

While this may well be the case, given what is known of functional group activity and habitats (see Doube, 1990), the costs of activity and its influence on beetle habitat use, assemblage structure and time of activity have not been comprehensively investigated for any single assemblage. Moreover, changes in the composition (size and species identity) of dung beetle assemblages may also be driven by differences in resource availability and habitat structure (Hanski & Cambefort, 1991; Steenkamp & Chown, 1996; Botes *et al.*, 2006; Nichols *et al.*, 2007; and see Chapters 12 and 13 in this volume). Nonetheless, the work by Krell *et al.* (2003) suggests that investigations of the energetics of dung beetle assemblage structure would repay the effort that assemblage level studies require (see Gaston *et al.*, 2009 for additional discussion).

10.6 Conclusion and prospectus

By comparison with many insect groups, dung beetles have been relatively well investigated from the perspective of the ecological implications of their physiological variation. Moreover, they have proven themselves to be highly tractable models for the kinds of physiological work that provides insight into broader ecological questions, including: the reasons for wider species geographical ranges at higher elevations and latitudes in many groups; the physiological factors that are likely to play a role in determining assemblage membership; and how physiological trait variation influences abundance structure and range limits (Gaston, 2003; 2009).

In the latter case, because much correlative (or climate envelope) modelling has been done on a range of species, it is clear which factors are responsible for regional variation in richness and species occurrence, with climate being most significant at a regional level (Davis *et al.*, 2008a) and perhaps, also, for species introduced to new areas (Duncan *et al.*, 2009). However, these models are limited in several ways (e.g. Chown & Gaston, 2008; Jeschke & Strayer, 2008) and can most usefully be augmented by mechanistic models that enable range distributions to be forecast without using the range data (Kearney & Porter, 2009).

Because dung beetles have proven to be tractable subjects for the kinds of physiological investigations required for such mechanistic models (Edwards, 1986; Chown *et al.*, 1995; Davis *et al.*, 2008c), they may also be a group for which further such investigations can inform this important area spanning the boundary between macroecology and macrophysiology. In our view, this would be a useful area for future work, and particularly also for assessing the dual roles of phylogeny and spatial autocorrelation on the patterns and processes underlying variation in dung beetle assemblages (see Freckleton & Jetz, 2009). The latter have largely been neglected in the investigation of large-scale ecological and physiological patterns and processes in the group (though see Lobo *et al.*, 2002; Chefaoui *et al.*, 2005) and deserve additional attention.

Such work might also contribute to an integrated understanding of the relationships between ecological and physiological variation at the intraspecific, interspecific and assemblage levels, an area in need of investigation (see Gaston *et al.*, 2009). Indeed, in many ways, the dung beetles represent an under-utilized opportunity to understand how physiological variation may contribute to local assemblage structure and regional variation in richness and turnover. Investigation of the latter ecological aspects was stimulated early on, perhaps, by the realization that the group forms an ideal one for these kinds of investigations (see reviews in Hanski &

Cambefort, 1991; and Chapter 13 of this volume) and by the prospects for the use of dung beetles as biocontrol agents of dung and dung flies in Australia (Doube *et al.*, 1991; see Chapter 12). Much excellent ecological work has emerged as a consequence (see reviews in Hanski & Cambefort, 1991; Scholtz *et al.*, 2009; and this volume).

Ongoing realization that dung beetles provide considerable ecosystem services (Nichols *et al.*, 2008) and are under threat from habitat change (Davis *et al.*, 2004; Nichols *et al.*, 2007) and presumably several other global change drivers (Davis *et al.*, 2008c) has led to renewed attention on the group, especially in efforts to understand these impacts (see Chapter 13). Not only is it now becoming clearer what major factors might affect regional variation in richness and turnover, but how these global change drivers may continue to influence dung beetle assemblages into the future is also being increasingly investigated (Verdú & Galante, 2002; Davis *et al.*, 2008a). Investigation of the physiological basis of these interactions would offer additional insight.

Thus, excellent opportunities exist for the integration of mechanistic physiological work and assemblage-level ecological studies in dung beetles. If such a research agenda were to be pursued, much would be achieved, not only in understanding the group, but also in contributing to ecology and evolutionary biology more generally. In this way, a long-standing tradition in the study of these organisms would be continued.

Acknowledgments

We thank the editors for inviting this contribution and for their useful comments on the chapter. Clarke Scholtz deserves special thanks for introducing us to the delights of scarab biology in the first place, and for his good company and assistance on many field trips. His findings about and work on the Scarabaeoidea continue to be a source of inspiration. Erik Holm also assisted us in the collection of beetles in arid areas and made many useful suggestions for what we should read and think about working on. Berndt Janse van Rensburg assisted in Tembe Elephant Park, and Anel Garthwaite assisted with literature searches. Ted Garland made PDAP available to SLC. Two anonymous referees are thanked for helpful comments on a previous version of the manuscript. Our work on dung beetles has been supported by the National Research Foundation of South Africa through grants both past and current, and by Pretoria, Stellenbosch and Arizona State Universities.

11

Dung Beetle Populations: Structure and Consequences

Tomas Roslin[1] and Heidi Viljanen[2]

[1]*Department of Agricultural Sciences, University of Helsinki, Finland*
[2]*Metapopulation Research Group, Department of Biological and Environmental Sciences, University of Helsinki, Finland*

11.1 Introduction

One of the most topical questions of modern ecology is what factors structure the composition of local species assemblages in a landscape context? What forces generate spatial variation in the local abundance and regional distribution of species (Leibold *et al.*, 2004)? During the last few decades, metapopulations (Hanski, 1999; Hanski & Gaggiotti, 2004) and metacommunities (Holyoak *et al.*, 2005) have become paradigms central to our understanding of spatial patterns and dynamics. Compared to a previous focus on local processes such as competition and density dependence in birth and death rates, the realization that spatial processes of emigration and immigration may critically affect what species occur where, and at what abundances, offers a new starting ground for models of both population dynamics (Thomas & Kunin, 1999) and of community structure (e.g. Holt, 1993). Yet, the extent to which the structure of local communities reflects the imprint of random forces (Bell, 2001; Hubbell, 2001), of interspecific differences in habitat requirements (Chave *et al.*, 2002; Urban, 2004; Freestone & Inouye, 2006) or dispersal abilities (Leibold *et al.*, 2004), or of actual interaction among species (Tack *et al.*, 2009) remains the topic of an ongoing debate (Cottenie, 2005; Driscoll, 2008; Pandit *et al.*, 2009).

In the context of this debate, theory roams wild but empirical data are scarce. So far, key features of spatial population structure have been resolved for relatively few

Ecology and Evolution of Dung Beetles, First Edition. Edited by Leigh W. Simmons and T. James Ridsdill-Smith. Published 2011 by Blackwell Publishing Ltd.

taxa (Jacobson & Peres-Neto, 2010). For insects, processes such as dispersal rates and population turnover have been quantified largely for butterflies, which have quickly become a model system for studies in spatial ecology (e.g. Hanski, 1999; Ehrlich & Hanski, 2004). In the absence of general evidence, it has frequently been assumed that metacommunities are formed by populations of multiple species which all perceive the landscape in the same general way. Yet such a simplifying assumption may not be warranted – or should at least be tested (Hanski & Simberloff, 1997; Roslin, 2001a). How species-specific differences in dispersal capacity, habitat preferences and resource use translate into patterns in community structure is a factor which ecologists have only recently started to explore (e.g. Pandit *et al.*, 2009).

In this context, dung beetle communities offer a promising model system. Given their amenability to both observational and experimental approaches (e.g. Finn & Giller, 2000; Finn, 2001; Rosenlew & Roslin, 2008), the insects in this group have been used widely as model systems to test theories on spatial ecology (e.g. Hanski & Cambefort, 1991; Roslin, 2001b) and, more pragmatically, to develop conservation planning and strategy (see Chapter 13).

In this chapter, we explore how the spatial structure of dung beetle populations affect local community composition. For two widely differing regions of the world, we examine what we know about where dung beetles occur in the landscape, how dispersal knits local populations together and how that affects their evolution and dynamics. We argue that species differ fundamentally in the way that they perceive and use the landscape, and that understanding this variation is a key to understanding the community of dung beetles found in a given dropping. Ultimately, this understanding should be built on tracing the roots of present-day variation over long periods of time and over large spatial scales.

11.2 Study systems

It is a warm and sunny day. Two droppings simultaneously hit the ground. In a Finnish pasture, a cow raises its tail and two kilos of wet dung splash onto the grass. In the rain forest of Madagascar, a lemur sits tranquilly on its branch. Fifteen metres below, a small pellet plops quietly to the steep hillside below. In both cases, only minutes elapse before the first few beetles arrive on this resource. Flying back and forth against the wind, the beetles cue in on their quarry. Soon, the dung is teeming with life as tens, or even hundreds, of individuals have landed on the dropping.

Scenes like these are repeated almost everywhere on the globe (save Antarctica and Greenland; Hanski & Cambefort, 1991). To illustrate fundamental factors shaping dung beetle communities worldwide, we focus specifically on the dung beetles colonizing North European cattle dung and Malagasy lemur dung. These systems differ strongly in terms of both past and present setting, hence illustrating both the diversity and the similarities which we expect to find across dung beetle populations across the globe. Let us start by taking a more in-depth look at each system. For consistency, we will adopt the taxonomy of Hanski & Cambefort (1991, p. 34).

11.2.1 The Finnish cow pat

Any dung pat deposited in Finland will be surrounded by a landscape of distinct spatial structure. In a still much-forested landscape, the pastures form relatively small patches of habitat (Figure 11.1), immersed in a matrix of forest. Of the total land area, forests account for 86 per cent (Finnish Forest Research Institute, 2005),

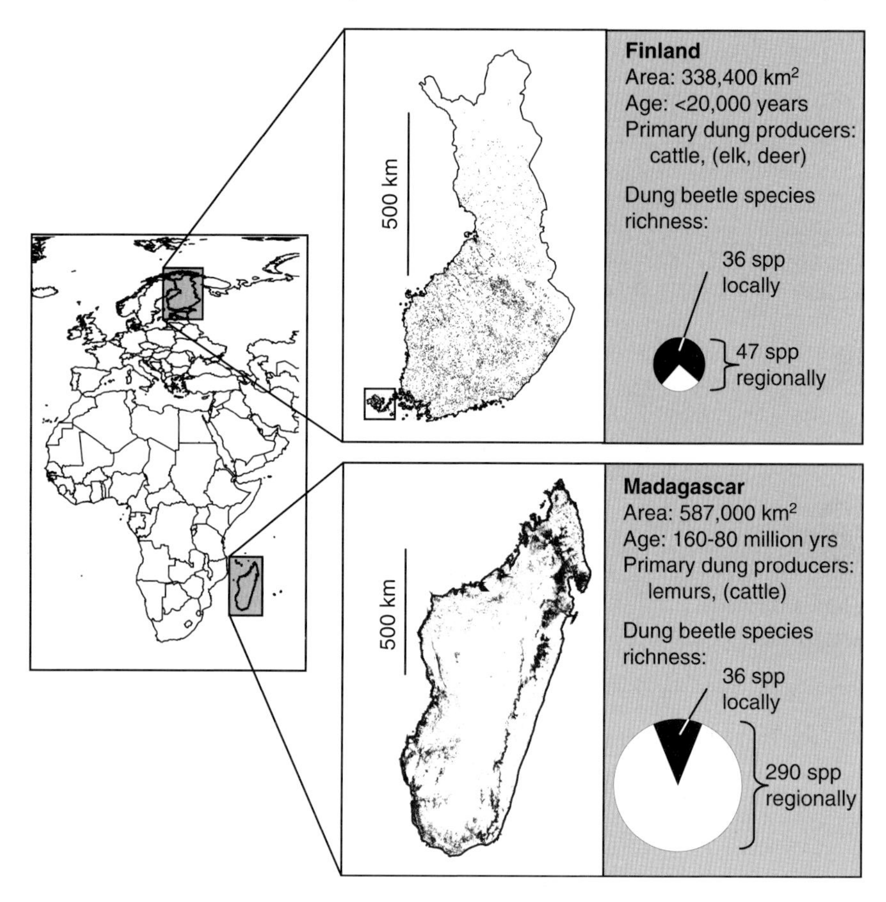

Fig. 11.1 Description of the two study systems: Finland and Madagascar. In Finland, the prime resource for dung beetles is cattle dung, and hence we map the outline of all pastures. For Madagascar, most endemic dung beetle species inhabit forests, and hence we map the extent of forest remaining in year 2005[i]. Figures in the grey boxes show key statistics on the area and the dung beetle fauna. Here, the area of the pie chart shows the size of total national species pool, the black fraction of which can be found coexisting locally[ii], within an area of *c.* 450 km^2. Note that the absolute number of species coexisting locally is exactly the same (36 species), despite a six-fold difference in the global species pool. In the map of Finland, the black box identifies the islands of Åland, referred to on p. 235.

[i] Data source for Finland: Anonymous (2008) and Information Centre of the Ministry of Agriculture and Forestry (personal communication); for Madagascar: MEFT *et al.* (2009). [ii] For Madagascar, this figure refers to the intensively studied Ranomafana National Park (Viljanen *et al.*, 2010b), for Finland to areas of comparable size within the species-rich southernmost Finland (recalculated from Roslin and Heliövaara, 2007).

whereas pastures account for only 0.4 per cent (Information Centre of the Ministry of Agriculture and Forestry, 2008, personal communication).

The current dung supply is principally dependent on domestic cattle, introduced over the last 3,000–4,000 years ago (Ukkonen, 1993, 1996, personal communication). The current herd of around one million cattle will graze mostly open pastures, together with 96,000 sheep, 6,600 goats and 25,000 horses. Forests offer a more limited supply of droppings from roughly 210,000 individuals of reindeer (*Rangifer tarandus*), 100,000 European elk (*Alces alces*) and 45,000 individuals of other deer species (mainly white-tailed deer, *Odocoileus virginianus* and roe deer, *Capreolus capreolus*).

To understand the set of species colonizing this dung supply, we need to understand how they got here. Importantly, Finnish dung beetle communities have not been around for long. Today, the nation spans an area of 338,400 km^2, connected in the east to a large land mass. However, as recently as 18,000 years ago, all of Fennoscandia was covered with ice, as were large tracts of surrounding Europe (Donner, 1995). Hence, all of the dung beetle species that we find here at present have spread here from somewhere else. Some of these species evolved in cool forests or mountains; some in warmer, open habitats. Within their range lived a broad range of large-bodied herbivores, from horses to deer, wild cattle and mammoths (Kahlke, 1994). That dung beetles and other insects simply retreated to more suitable climate zones when the ice came, and later reinvaded Europe as the climate became more suitable, is shown by fossil remains (Hanski, 1986, 1991; with references therein).

The dung beetle fauna of Finland currently consists of 47 species (Figure 11.1). As is typical of most north temperate dung beetle communities (Hanski, 1991), most of these species belong to the family Aphodiidae, with 36 species in the genus *Aphodius*, two in the genus *Aegialia* and one each in *Heptaulacus*, *Oxyomus* and *Psammodius*. In addition, there are three species of *Onthophagus* (Scarabaeidae, tribe Onthophagini), and two large dor beetle species in the family Geotrupidae (Silfverberg, 2004; Roslin & Heliövaara, 2007). None of these taxa are endemic to the region. Even so, the range sizes of different species vary substantially, both in terms of range within Finland (Figure 11.2a) and distribution throughout surrounding areas (*cf.* Section 11.3).

11.2.2 The Malagasy lemur pellet

For dung beetles in Madagascar, the landscape is essentially the mirror image of that encountered by Finnish beetles. In Madagascar, the majority of native dung beetle species occur in forests (Koivulehto, 2004; Wirta *et al.*, 2008), which, in the currently much-deforested landscape, are surrounded by a matrix of open habitats (Figure 11.1).

Contrary to the situation in Finland, the association between Malagasy dung beetles and dung producers reflects a long local history. Madagascar is the world's fourth largest island (587,000 km^2, i.e. 1.7 times the area of Finland), and it was separated from the African continent and from India about 160 and 80 million years BP, respectively (Briggs, 2003; de Witt, 2003). This geographical history has resulted in highly distinct communities of both dung producers and dung beetles (*cf.* Vences *et al.*, 2009). Apart from three species of hippopotamus, there have never been native ungulates (*sensu* Artiodactyla, Perissodactyla and Proboscidea) on the island. Rather, the dung beetles have radiated in parallel with another source of dung – the lemurs, an endemic group of primates (Wirta *et al.*, 2008). While most of the current species are rather small, the lemurs were formerly more diverse. Before the arrival of humans some 2,300 years ago, the fauna included at least 17 large-bodied (> 10 kg) species (*cf.* Burney *et al.*, 2004), providing an assortment of dung types.

Since those times, the dung supply has changed substantially. Cattle were introduced on the island more than 1,000 years ago (Burney *et al.*, 2004) and now offer an abundant resource. There are currently some 7 million cattle across the area, reinforced by 1.5 million goats and sheep, as well as the odd horse and donkey (Rahagalala *et al.*, 2009 and references therein). Nevertheless, as we shall see, the dung supply from domestic animals remains vastly underused by the local pool of beetle species.

The long history of isolation of Madagascar, and interacting patterns of evolution among local taxa, has created a distinct dung beetle fauna. Taxonomically, the fauna is relatively well known, due to the pioneering work of Lebis (1953); Paulian & Lebis (1960); Paulian (1975); Bordat *et al.* (1990) and multiple recent revisions by Montreuil (2003a; 2003b; 2004; 2005a; 2005b; 2005c; 2006; 2007; 2008a; and Montreuil & Viljanen (2007). From phylogenetic, biogeographical and ecological perspectives, the dung beetles of Madagascar have recently been studied by the group of Hanski (Hanski *et al.*, 2007; 2008; Wirta *et al.*, 2008; Wirta & Montreuil, 2008; Viljanen, 2009a; 2009b; Rahagalala *et al.*, 2009; Wirta, 2009; Viljanen *et al.*, 2010a; 2010b). Together, this body of work has revealed a unique species composition: not only are most dung beetle species endemic to this region, but so are higher taxa.

Two tribes of Scarabaeidae strongly dominate the Malagasy dung beetle fauna – the Canthonini and the sub-tribe Helictopleurina, which belongs to the tribe Oniticellini (Paulian & Lebis, 1960; Wirta *et al.*, 2008). Of these, Helictopleurina is completely endemic to the island and consists of just one genus, *Helictopleurus*, with 66 described species and subspecies (Paulian & Lebis, 1960; Wirta *et al.*, 2008). Canthonini has seven endemic genera and 191 described species (Paulian, 1975; Montreuil, 2006). Other tribes of Scarabaeidae in Madagascar include one genus of Scarabaeini with three endemic species (*cf.* Forgie *et al.*, 2005) and six species of *Onthophagus* (Onthophagini), which are thought to be introduced (Paulian & Lebis, 1960; Davis & Scholtz, 2001; Davis *et al.*, 2002). In addition, there are records of some 30 species of Aphodiini and Didactyliini (Aphodiidae; Bordat *et al.*, 1990; Mate, 2007). Hence, the total species pool amounts to some 300 species or six times more than that of Finland, with different phyla dominating the scene.

11.3 Range size

With our two study systems now outlined, we will turn to what we know about how dung beetle populations are structured. We start by examining distribution patterns at a broad spatial scale (across Finland and across Madagascar), then turn to patterns in the fine-scale distribution of species within landscapes. After examining where the species occur, we look at how they disperse between sites. Finally, we turn to the consequences of dung beetle population structures – to how patterns and processes are reflected in patterns of genetic variation, and how dung beetle populations are likely to respond to landscape change.

We start by addressing the large-scale distribution of species, i.e. the general size of the distribution area of species within the two regions examined. In both the Finnish and the Malagasy landscape, the basic resource (dung of different types) is distributed everywhere, across the full nation, yet droppings deposited in different parts of the area will attract a different species complement.

Interestingly, the evolutionary history of dung beetles within Madagascar has resulted in complex patterns in species' distribution. The endemic forest dung beetles typically have small ranges (Figure 11.2b; Viljanen, 2009a; 2009b; Viljanen *et al.*, 2010a), as have many other taxa in Madagascar (Wilmé *et al.*, 2006). Hence, despite overall high species richness, local assemblages are relatively species-poor (less than 40 species; Figure 11.1; Viljanen, 2009a; Viljanen *et al.* 2010a; 2010b) and the relationship between local species richness and latitude is comparatively weak (Figure 11.2d).

At least part of this pattern may be attributed to competitive exclusion, an inference supported by patterns of low local species richness in the face of high regional diversity (Viljanen, 2009a; Viljanen *et al.*, 2010a), of species replacing each other along elevational gradients (Viljanen, 2009a; 2009b; Viljanen *et al.*, 2010b) and of range expansion following competitive release after the switch to a new resource (cattle dung; Hanski *et al.*, 2008; *cf.* below). Strong competition indeed seems typical of dung beetle communities across the tropics (Hanski & Cambefort, 1991 and references therein).

For northern temperate areas like Finland, the evidence for competition is less clear (summarized by Finn & Gittings, 2003). That one dung beetle species will lower the local abundance of another is possible (Hanski, 1991; Hanski & Cambefort, 1991), but such competitive effects will hardly extend to one species excluding another from a local community (Finn & Gittings, 2003). Indeed, we find a very different pattern in the range of species within Finland and Madagascar (compare Figure 11.2a *versus* 11.2b). In Finland, species richness is highest in the south, with distinctly fewer species towards the north (Figure 11.2c; Roslin, 2001b). In the species-rich south, the fraction of species coexisting locally is exceptionally high (75 per cent of the national pool; Figure 11.1), whereas, in northern Finland, even areas of 2,500 km^2 will typically sustain only one or a few species (*cf.* Figure 11.2c).

The differences in species diversity and range size between tropical Madagascar and temperate Finland are consistent with broader patterns across taxa. The species diversity of almost all animal groups is highest in the tropics (Rosenzweig, 1995; Willig *et al.*, 2003; Lewinsohn & Roslin, 2008), with aphids (Dixon *et al.*, 1987) and ichneumonids (Owen & Owen, 1974; Gauld *et al.*, 1992) being notable insect exceptions. Likewise, the average size of species' geographical range typically decreases towards the equator, a pattern established for vertebrate taxa and known as Rapoport's effect (Stevens, 1989; but see Novotny *et al.*, 2007).

At least six different hypotheses have been proposed to account for latitudinal gradients in species richness (Pianka, 1966; Brown & Gibson, 1983; Begon *et al.*,

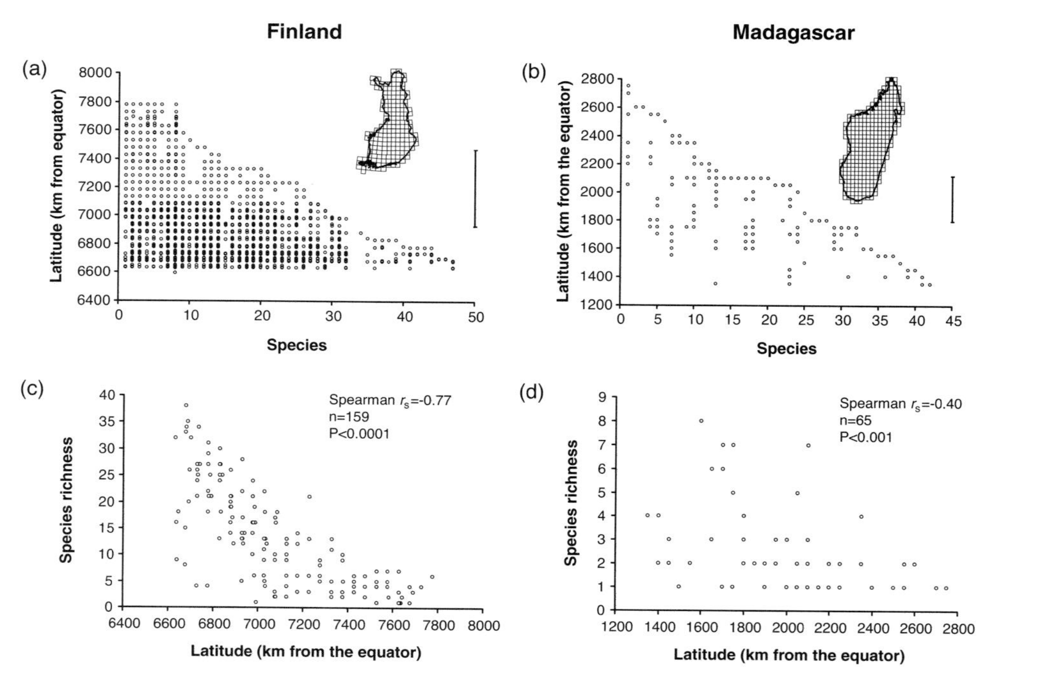

Fig. 11.2 Latitudinal extent of dung beetles in Finland and Madagascar. For clarity, we present data on species pools of comparable size: for Finland, we show data on the full fauna of 47 species; for Madagascar, we present data on 42 species in the genus *Nanos* (twelve of which are still undescribed). In both cases, the figures are based on the presence of species (a and b) or species counts (c and d), for grid squares of approximately 50 × 50 km (see inset maps). In panels (a) and (b), open circles shows the latitude for each map square where a species was encountered, with species sorted along the *x*-axis in decreasing order of maximum latitude. The interval on the right shows average range size in the north-south dimension (ANOVA of log-transformed range size Finland versus Madagascar: $F_{1,87} = 13.4$, $P < 0.001$)[i]. In (c) and (d), we show the total number of species encountered in each grid square vs. latitude[ii].

[i] For species recorded from a single grid cell, we put range size equal to the height of the cell i.e. 50 km. [ii] In both cases, we have excluded the map squares with zero counts ('no species encountered'), since those sites have likely never been sampled for dung beetles, or lack the appropriate habitat.

2006; and see Chapter 10). One of the most frequently cited posits that habitat heterogeneity increases with decreasing latitude, and that the greater the diversity of habitats, the more species can coexist (Pianka, 1966). Yet this explanation cannot account for a clear-cut latitudinal trend within a single uniform habitat, such as cattle dung within the open pastures of Finland (*cf.* Davidowitz & Rosenzweig, 1998).

A more likely answer lies in the history of Finnish dung beetle populations. All Finnish insects have spread there after the glacial retreat and the gradual uplift of major parts of the mainland some 10,000–7,000 years ago (Donner, 1995), and there has been little time for evolution or speciation within Finland. As a consequence, the range of a species in Finland reflects the species' evolutionary history elsewhere, and species with a southern distribution in Finland are generally those with a preference for warm regions and habitats in southern and central Europe. On the other hand, species with a relatively northern distribution in Finland are typically restricted to cool mountainous regions south of the country (Horion, 1958; Chapter 10 provides a discussion of thermal tolerances and their implications for dung beetle distributions). Regional patterns of species richness within Finland are thus linked to biogeographical and evolutionary processes acting over spatio-temporal scales vastly beyond the history and extent of the current land mass (Roslin, 2001b), a relationship likely shared with many other taxa (Ricklefs & Schluter, 1993; Gaston & Williams, 1996).

In sharp contrast, patterns in dung beetle range sizes within Madagascar are reflective of processes confined to the island itself. Here, the species diversity of today has arisen through adaptive radiation from a few ancestral species arriving in the area some tens of millions of years ago (Wirta *et al.*, 2008). That the dominant processes have been similar across taxa is suggested by a coherence in pattern; within a major dung beetle phylum, there is no detectable difference in range size among clades (Wirta *et al.*, 2008). We can thus envisage local forces, including competition, as the drivers behind Malagasy patterns.

To conclude, the range size of dung beetle species within Madagascar and Finland appears to have been moulded by different factors. As a result, the species attracted locally to a given dropping are drawn from quite different patterns of variation in the regional species pool (Figure 11.2). Our examination of these patterns sets an important baseline for the sections to come, since what we can find locally is always conditional on what species are available in the area (Zobel, 1992; 1997; Pärtel *et al.*, 1996). We will turn next to examining how the beetles present in a given area select their resources and habitats at a finer spatial scale, as this will determine what parts of the landscape they actually use and how local abundances vary in space.

11.4 Habitat and resource selection

Of the 47 dung beetle species encountered in Finland, almost any can turn up in our focal cow pat – provided that it was dropped in southern Finland (see Figure 11.2). Indeed, given the short history of dung beetle species in Finland, and the long association between large ungulates and these beetle species elsewhere (*cf.* Section 11.2.1), most dung beetle species can be found in the dung of large herbivores. This makes cattle dung a key resource for most species. A subset of species will turn up on the dung of other herbivores, whereas the dung of omnivores like bear will attract only few species (Muona & Viramo, 1995; Roslin *et al.*, unpublished data), as will carrion, decaying mushrooms, compost, etc. (Roslin & Heliövaara, 2007). Detailed work on the life cycles of individual species has resolved finer differences in resource use (e.g. White, 1960; Holter, 1982; Gittings & Giller, 1997), but differences in terms of exact resource preferences still appear relatively minor (see Chapter 5 for a discussion on olfactory preferences).

More important than the specific dung type is the habitat that surrounds it (Rainio, 1966; Roslin, 1999; Roslin & Heliövaara, 2007). While cattle dung occurs in both open and forested habitats, the species complement of these two environments is rather different (Landin, 1961; Rainio, 1966; Koskela & Hanski, 1977). Such differences can be related to the temperature requirements of developing larvae, where early work by Landin (1961) suggested diverging optima among species. In practice, the differences observed will imply that some species can only develop in shady forest habitats, whereas others will thrive in the very warmest parts of sun-drenched pastures (*cf.* Roslin *et al.*, 2009; for a classification of Finnish species with respect to habitat preference, see Roslin, 2001b).

In Madagascar, patterns of resource use are substantially different. Given the long history of dung beetle species within the island, and the previous scarcity of dung-producing ungulates, cattle dung is still a largely neglected resource. Although overall cattle densities are actually higher than those of Finland, only a few beetle species have switched to using this dung supply. In fact, not a single dung beetle species of the wet forests uses cattle dung as its primary food resource, and only few species are ever encountered in cattle dung within closed-canopy habitats (Rahagalala *et al.*, 2009). In open habitats, there are approximately 20 cattle dung-using species, including introduced species and several small Aphodiidae (Rahagalala *et al.*, 2009). The local communities have 1–12 species, which is almost an order of magnitude less than in similar communities in mainland Africa (Rahagalala *et al.*, 2009). In terms of absolute numbers, species richness on cattle dung is then similar to that of Finland – but it is much lower in relation to the overall species pool.

Among resources other than cattle dung, the majority of Malagasy species display a broad preference (Viljanen, 2004; 2009a; Viljanen *et al.*, 2010b). In experiments comparing different types of bait, roughly four-fifths of the species were shown to feed on both dung and carrion (Table 11.1). While tropical beetles are frequently presumed to show rather generalist habits (Hanski & Cambefort, 1991), the observed figure is relatively high compared to other tropical regions, where approximately one-third of the dung beetle species have been classified as generalists (Hanski, 1983; Hanski & Krikken, 1991; Cambefort & Walter, 1991; Feer & Pincebourde, 2005; Gill, 1991).

Nonetheless, among Malagasy species – as among other tropical (Halffter & Matthews, 1966; Hanski & Cambefort, 1991) and Finnish (above) dung beetles – the level of specialization does form a continuum (Viljanen, 2004; 2009a Viljanen *et al.* 2010b), spanning the full spectrum from species specialized on the droppings of a given lemur species to species essentially attacking most organic waste (*cf.* Table 11.1).

Table 11.1 Diet choice of dung beetle species in the wet forest area of Talatakely in the Ranomafana National Park (RNP), south-eastern Madagascar. N gives the total number of individuals caught in the experiment, whereas the other figures represent the percentage of all individuals encountered on a given bait type[i]

Species	*N*	*Carrion*	*Faeces of:*								*Diet*
			Cattle	*Pig*	*Brown Mouse Lemurs*	*Red-bellied Lemur*	*Golden Bamboo Lemur*	*Eastern Lesser Bamboo Lemur*	*Greater Bamboo Lemur*	*Milne-Edward's Sifaka*	
Helictopleurus corruscus	21									100	Dung specialist
Helictopleurus dorbignyi	13									100	Dung specialist
Helictopleurus fasciolatus	12	37								63	Generalist
Helictopleurus rudicollis	111	11	30					42	18		Generalist
Helictopleurus semivirens	90								9	91	Dung specialist
Helictopleurus steineri	1	100									Generalist*
Apotolamprus quadrinotatus	117	62	11	24						3	Generalist
Epactoides frontalis	6		12					79		9	Generalist*
Epactoides major	4	6								94	Generalist*
Epilissus apotolamproides	104	25	10	23				35		7	Generalist
Epilissus delphinensis	122	79							14	7	Generalist
Epilissus genieri	21		1	11			83		3	2	Dung specialist
Epilissus mantasoae	33	75								25	Generalist
Nanos rubromaculatus	4	100									Generalist
Nanos viettei	1911	45	5	3		14		2	23	9	Generalist
Pseudoarachnodes hanskii	190		4	31	19	19		2	1	25	Dung specialist
Aphodius humerosanquineum	2243	2	7		7	17	6	9	10	42	Dung specialist/ generalist*
Aphodius new species	491		4		1	12		13	4	66	Dung specialist
Aphodius ranomadryensis	63		1			1		2	11	85	Dung specialist

[i] Beetles were trapped with baited pitfall-traps (plastic cups, 1.5 dl) during the rainy season in November and December in years 2003–04 for a total of 1,587 trap-days. A bait of approximately 3 cm^3 was wrapped in gauze and hung from a stick above the trap. Note that the column 'Diet' draws on some additional information beyond that presented in the table: species caught in low numbers on a single bait type (faeces) in the current study, but in carrion-baited traps elsewhere (or some other time), have been classified as generalists. These species are identified by asterisks (*). Seventeen additional species encountered in other, carrion-baited trap designs within the RNP are excluded from the table. These species are most likely generalists, since no carrion-specialists are known from Madagascar, and dung specialists rarely come to carrion-baited traps.

In terms of macrohabitat preferences, Malagasy species are more specialized than their Finnish counterparts. Most of the endemic forest-dwelling species are strictly confined to this habitat (Koivulehto, 2004) and only few species will appear in other habitats (Rahagalala *et al.*, 2009). Similar habitat specialization is observed in many tropical regions, where species composition changes strongly across habitat types (e.g. Spector & Ayzama, 2003), and traps put in open habitats within forested regions will typically catch only very few species (Peck & Forsyth, 1982; Klein, 1989; Estrada & Coates-Estrada, 2002; Quintero & Roslin, 2005). This presumed sensitivity to habitat type has made tropical dung beetles a popular target for biodiversity assessments in changing environments (Favila & Halffter, 1997; Estrada *et al.*, 1998; Davis *et al.*, 2001; Halffter & Arellano, 2002; McGeoch *et al.*, 2002; Escobar, 2004; Andresen, 2005; Arellano *et al.*, 2005; Avendaño-Mendoza et al., 2005; Pineda *et al.*, 2005; and see Chapter 13).

How these differences in dung beetle specialization between Madagascar and Finland compare to broader patterns of insect specialization between the temperate and the tropical zone is hard to establish, since there is no consensus regarding what those broader patterns really are (see Lewinsohn & Roslin, 2008, with references therein). Yet, from a dung beetle perspective, the patterns described above have clear implications for how we might expect populations to be structured at the landscape level. The high habitat specialization typical of Malagasy dung beetles implies that open areas offer unsuitable habitat, thus separating different populations from each other. Within these forests, variation in the resources used causes variation in the exact fraction of habitat available to a species.

In Finland, variation in habitat preference adds to variation caused by differences in resource use (Roslin, 1999; Roslin & Koivunen, 2001; Roslin, 2001b). Of the beetles found co-occurring in a single dung pat, some will represent generalist species, forming part of large populations stretching through multiple habitats and occurring on several resources in the surrounding landscapes. Others are part of more localized populations confined to open pastures, the most extreme of which are actually restricted to the very warmest fraction of the overall landscape (Roslin *et al.*, 2009). A third group consists largely of cool-preferring species (*cf.* Landin, 1961) 'spilling over' from populations in the surrounding forest matrix (Roslin, 1999; 2000; 2001a).

Taken together, this implies that the individuals turning up on our cow pat or lemur pellet will have been drawn from populations of different spatial structures. Depending on the surrounding landscape, we may then expect to find very different assemblages on the same resource and habitat – an issue to which we will return in Section 11.7, where we discuss dung beetle responses to landscape change.

11.5 Dung beetle movement

In Section 11.4, we examined how differences in resource use *sensu lato* creates differences in species distribution. Many insect species occupy only the tiniest fraction of modern landscapes (Cowley *et al.*, 1999), and increasing specialization typically restricts the available parts even further (MacArthur & Levins, 1964; van Nouhuys, 2005; Roslin *et al.*, 2009). Hence, with increasing ecological specialization, we expect dung beetle species to form local populations more isolated from each other. However, the degree to which local populations exchange individuals depends not only on their position in the landscape but also on the dispersal capacity of the species relative to this landscape configuration (Moilanen & Hanski, 2001; Moilanen & Nieminen, 2002). Indeed, understanding how animal movements are affected by landscape structure has been identified as a critical gap in current ecological knowledge, and has been advanced as a research priority for the management and conservation of species (Wiens, 1989; 1990; Ims, 1995).

What then do we know about the dispersal capacity of dung beetles as a group, and about variation in dispersal processes among species and landscapes? The short answer is 'very little'. While the distribution and ephemeral nature of dung resources *a priori* suggests that these beetles are likely good dispersers, quantitative descriptions of dung beetle dispersal are few. Results from Finland still offer some insight into the movements of the dung beetles of this region.

For a Finnish dung beetle species occurring on cattle dung, the landscape is distinctly structured at several different scales. The fundamental resource, dung, appears in distinct entities (dung pats), and these pats are confined to well-defined habitat patches (i.e. cattle pastures). At the smallest of these scales (i.e. pats within pastures), movements appear very frequent. Of the more than 3,200 marked beetles of twelve species released by Roslin (2000) in dung pats, virtually all (98 per cent of individuals) left the original dropping to enter a new one within one to three weeks (typically 7–12 days) from release. This occurred even though the released individuals had been caught as adults and thus had moved at least once from their site of hatching to the point of capture. We can infer that within Finnish pastures, individuals of most dung beetle species are likely to mix freely among dung pats.

At a larger spatial scale (i.e. pastures within landscapes), larger beetle species conform to the *a priori* expectation of high movement rates. For *Aphodius fossor* (average dry weight 26.1 mg $\pm$ SD 9.8 mg – a large species by Finnish standards), a detailed set of calculations reveal that for each generation, as many as one-fifth (18 per cent) of dung beetle individuals encountered in a given pasture had actually hatched at another pasture (Roslin, 2001a). Nonetheless, rates of movement differ substantially between species. The mark-release recapture studies of Roslin (2000) revealed that movements between pastures are more frequent the larger the species, the more specific its occurrence in relation to pat age, and the more specialized it is on cow dung and open pasture habitats.

Of these effects, the one related to body size deserves particular attention. What it implies is that even among species confined to the very same resource and habitat, local populations are connected by dissimilar levels of dispersal. This difference can be illustrated by a comparison between the big species *A. fossor* (above), and a small species such as *Aphodius pusillus* (1.0 $\pm$ 0.3 mg). In the mark-release-recapture study referred to above, about one-third of the recaptured *A. fossor* individuals were observed to move between experimental pastures separated by 300 m, but the figure fell well below one-tenth for *Aphodius pusillus*, even though both species are specialized on open pastures. Hence, variation in movement patterns adds to variation in resource selection in creating differences in spatial population structures. In the Finnish countryside, neighbouring populations of *A. fossor* will have a considerable impact on each other's dynamics, whereas local populations of *A. pusillus* are dynamically less coupled (Roslin, 2000; 2001b).

For Madagascar, direct studies of dung beetle movement are scarce, so we will have to turn to other tropical systems for added support. Scattered evidence at first seems to support the general expectation of dung beetles as strong dispersers. Working in Ecuador, Peck & Forsyth (1982) observed substantial movement of marked beetles within a continuous rainforest; as a record, an individual of *Oxysternon conspicillatus* was recaptured 1 km from the release site within two days of release. However, the maximal (and also the mean) distances observed in any movement study are conditional on the trapping design, on what distances one *could* have observed, and with what probability (*cf.* Roslin *et al.*, 2009). Another complication relates to the attraction range of the traps used. Passive movement to the vicinity of the trap is hard to disentangle from active attraction to the trap.

An example of why such a distinction might be needed is offered by the study of Larsen & Forsyth (2005). Working in Venezuela, these authors aimed to quantify the range from which *Canthon acutus* (7.0 ± 3.1 mg) is attracted to traps. For this purpose, they released marked individuals at different distances from baited pitfall traps, then examined the rate of decline in the fraction of individuals recovered as a function of distance. Importantly, if the beetles move far but are attracted from only the close vicinity of the trap, the patterns recovered will be more reflective of the movement than the attraction process. Assuming the latter, the study design is conceptually similar to that of trap-lines outside pastures (*cf.* Roslin, 2000), where, instead of measuring dispersal from a source (the pasture) to multiple traps at variable distances, we measure dispersal from multiple sources (release points) at variable distances to a single target trap.

Fitting an exponential decay function $e^{-\alpha d}$ to both patterns, where d is the distance between source and trap, we may use the parameter α as a measure of the characteristic scale of dispersal. Here, a high value of α implies poor dispersal. For *C. acutus* in forest, Larsen & Forsyth (2005, p. 323) report $\alpha = 0.11$, whereas Roslin (2001b) found $\alpha = 0.0016$ for temperate *Aphodius fossor* in open terrain. For *C. acutus*, most of the decline in trap-catches will then occur within some tens of metres, and for *A. fossor* within hundreds of metres. Assuming the attraction range of traps to be similar across systems, this points to a striking difference in dispersal distances between the tropical *C. acutus* and the slightly larger, temperate *A. fossor*. It suggests that *C. acutus* is actually a rather weak disperser. Nonetheless, it should be noted that the trap design differed between the two studies, and consequently the direct comparison offered above may represent an oversimplification of the results.

Given the problems with interpreting distributions of dispersal distances from any given sampling design, recent methodological work has been focused on teasing apart the observation process from the movement process observed (Ovaskainen, 2004; Ovaskainen *et al.*, 2008a). In fact, the only study on dung beetle dispersal conducted in Madagascar to date was based on Ovaskainen's (2004) technique. Focusing on the small ($\approx$6 mm) canthonine species *Nanos viettei*, Viljanen (2009b) inferred limited rates of mobility: the data seemed most compatible with a daily recapture probability of 0.1 and a diffusion rate of $50\,m^2\,d^{-1}$. Even individuals recaptured after a year had only moved some tens of metres.

Again, these figures appear surprisingly low, as the resource basis within most tropical rain forests would appear rather scarce when examined at the scale of less than a hectare. Nevertheless, *N. viettei* is apparently a long-lived species, and low rates of translocation may be combined with relatively low resource requirements per time unit and with occasional 'gorge feeding' when suitable opportunity arises (Viljanen, 2009b).

The movement studies summarized above were conducted in uniform habitats. That habitat discontinuities form strong barriers to dispersal is shown by the fact that few dung beetles can typically be trapped outside of their preferred habitat (*cf.* Section 11.4). This implies that the forest-dwelling dung beetles show strong edge-mediated behaviour, essentially turning back at any forest edge. Responses to habitat discontinuities have also been reported from Los Tuxtlas, Mexico, where forest dung beetle species like *Ateuchus illaesum*, *Ontherus mexicanus* and *Copris laeviceps* show attraction to edges and other wooded habitats (Montes de Oca, 2001).

How responses to habitat structure affect realized movement patterns was demonstrated by another study from southern Mexico. Here, Arellano *et al.* (2008) again used the technique of Ovaskainen (Ovaskainen, 2004; Ovaskainen *et al.*, 2008a), this time to quantify dispersal through different parts of the landscape. Focusing on the dung beetle *Canthon cyanellus cyanellus* (26.0 ± 2.3 mg), they showed that males generally moved faster than females, but also moved faster within forests and hedgerows than within pastures.

Interestingly, the diffusion coefficients estimated for tropical dung beetles by Arellano *et al.* (2008) appear similar to those estimated for other insects (i.e. fritillaries and apollo butterflies) in the temperate zone. The order of magnitude is consistently 10^4–$10^5\,m^2\,d^{-1}$ (Ovaskainen, 2004; Ovaskainen *et al.*, 2008a; 2008b). Nonetheless, the dung beetle species they examined live for much longer and their mortality rates are thus lower (Arellano *et al.*, 2008; Viljanen, 2009b). Hence, the expected lifetime displacement is much higher for the dung beetles than for the butterflies, supporting the *a priori* expectation of dung beetles as relatively mobile insects from a *per generation* perspective.

Differences in movement patterns among habitats are further elaborated by patterns observed within single habitats. Returning to Finland, observations suggest that dung beetle movement may be influenced by microclimatic gradients within open landscapes. Here, at the northern edge of its range, the species *Onthophagus gibbulus* shows a stronger tendency to move towards warmer than colder parts of the landscape, even when these habitats are located at the same distance and within the same macrohabitat. These effects are markedly strong: relatively minor differences in microhabitat (as characterized by incident solar radiation) will easily result in a two-fold difference in the probability of moving to one of two sites located at the same distance (Roslin *et al.*, 2009). Thus, aggregates of individuals in different parts of the landscape may exchange more or fewer individuals, depending on what microhabitats fall between them.

Patterns like these are only now being explored, and may affect our impression of how populations are structured in space. A given distance in one direction may offer a quite different barrier to dispersal than the same distance in another direction, even within what seems to be the same overall habitat.

The studies described so far were conducted at a local scale and concern direct observations of marked individuals. Interestingly, the resulting estimates of dispersal rates seem rather modest when compared to rates of spread for species introduced in new environments. Here, Hanski & Cambefort (1991, with references therein) provide figures for two medium-sized tunnellers, *Digitonthophagus (Onthophagus) gazella* and *Onthophagus taurus*, both open-habitat species in the tribe Onthophagini.

For *D. gazella*, rates of spread after an introduction to Australia and another to North America appear similar (50–80 vs. 58 km/year, respectively) and are comparable to those for *O. taurus*, as introduced in North America (129 km/year; see Chapter 12). These expansion rates are not entirely incompatible with, but are certainly at the higher end of, movement estimates derived from the mark-release-recapture studies quoted above. However, similar discrepancies have been observed for patterns of, for example, postglacial range expansion in North American trees, as compared to direct observations of dispersal distances of seed (e.g. Clark *et al*, 1998).

One explanation advanced in the latter context is that dispersal rates are strongly influenced by the rare case of long-distance dispersal (Kot *et al*., 1996; Lewis, 1997; Clark *et al*, 1998; Turchin, 1998), which is hard to quantify in local studies. Whether or not patterns observed for dung beetles might be attributed to particularly high dispersal capacities of onthophagines (as suggested by Hanski & Cambefort, 1991), or to differences between species typical of forests and open habitats, or whether it suggests a more general mismatch between studies conducted at local and larger scales, is a matter worthy of further investigation.

Taken together, available observations of dung beetle movement point to differences in dispersal capacity among species and to substantial impacts of landscape structure on dung beetle movement patterns. The few species examined so far seem compatible with our *a priori* expectation that dung beetles, as a group, would generally be relatively strong dispersers – although two tropical studies (Larsen & Forsyth, 2005 and Viljanen, 2009b) suggest surprisingly short dispersal distances.

Based on the few studies available to date, there appears to be no systematic differences in dispersal between dung beetles at temperate sites (such as Finland), and tropical sites (such as Madagascar), nor between dung beetles and Lepidoptera – but this may so far be more reflective of the lack of evidence than the evidence for any true lack of difference. Under all circumstances, variation among dung beetle species within a single area (Finland) seems pronounced, and it certainly contributes to differences in spatial population structures across species. That more dispersal studies are needed is evident, and these should be focused on general rates of movement, on dispersal in different habitats and on how landscape structure modifies the flow of individuals.

11.6 The genetic structure of dung beetle populations

The imprint of movements – both of ongoing gene flow and of past population expansions and retractions – is reflected in current patterns of genetic diversity (Slatkin, 1985; 1987; Templeton, 1998). Here, differences between Finland and Madagascar become of particular interest, given differences in the history of the respective systems as reviewed in Section 11.2. What processes can we infer from contemporary patterns of genetic variation? What can we learn regarding the evolution of the system and about the forces connecting current populations within species?

In Finland, the direct estimates of dispersal discussed in Section 11.5 suggest that local populations of large, dispersive dung beetle species will be well connected at a large spatial scale. Such gene flow should efficiently homogenize the genetic make-up of populations (e.g. Slatkin, 1985; 1987). Indeed, for the relatively large Finnish species *Aphodius fossor*, we find low levels of genetic differentiation across the country and no detectable signs of isolation by distance (Figure 11.3a; Roslin, 2001a; *cf.* Slatkin, 1993). Assuming an island-model of dispersal, the observed level of genetic differentiation is compatible with populations exchanging *at least* 14 effective migrants per generation for nuclear genes and eight effective female migrants for mitochondrial markers – a figure well compatible with estimates of more than one hundred effective migrants per generation, as extrapolated from a smaller-scale mark-release-recapture-study. Only between Åland (a group of islands separated from the Finnish mainland by tens of kilometres of open sea (see Figure 11.1)) and the mainland do we find detectable differentiation ($F_{ST} = 0.09$ for mtDNA markers).

Large local population size, extreme haplotype diversity (Figure 11.3a) and a high regional incidence of *A. fossor* all testify against recurrent population turnover (Roslin, 2001a). Patterns of genetic differentiation are then attributable to spatial processes occurring in the *A. fossor* population of today: both direct and indirect measures of dispersal suggest that the entire mainland population of the species functions as one large, relatively well-connected albeit 'patchy' population (*sensu*- Harrison, 1994; Harrison & Taylor, 1997; but see Ovaskainen & Hanski, 2004; Gripenberg *et al.*, 2008) across the tens of thousands of pastures that it comprises (Roslin, 2001a).

Interestingly, given the history of the Finnish *A. fossor* population, most of the genetic material reshuffled by current migration derives from a time period prior to its postglacial expansion into the area. The mtDNA haplotypes currently found within single pastures often display major sequence divergence, and haplotypes encountered within a given pasture will typically differ as much from each other as will haplotypes sampled in Finland and Spain (Figure 11.3a). Indeed, Roslin (2001b) estimates that the joint ancestor of current mtDNA haplotypes occurred some 300,000 years ago, and that much genetic variation will have survived over multiple phases of population expansion and retraction. Hence, the scenario best compatible with present-day patterns of genetic variation is that of the gradual expansion and retraction of a large ancestral population, as followed by strong ongoing gene flow.

For Madagascar, the patterns are quite different. Here, DNA-based phylogenies suggest that most of the genetic variation that we find today has evolved within the island itself. Hence, Wirta *et al.* (2008) infer that all species of Helictopleurina derive from a single colonization of the island some 37 to 23 million years ago, followed by evolutionary radiation coinciding with that of the lemurs. This evolution has also included a shift from the open habitats used

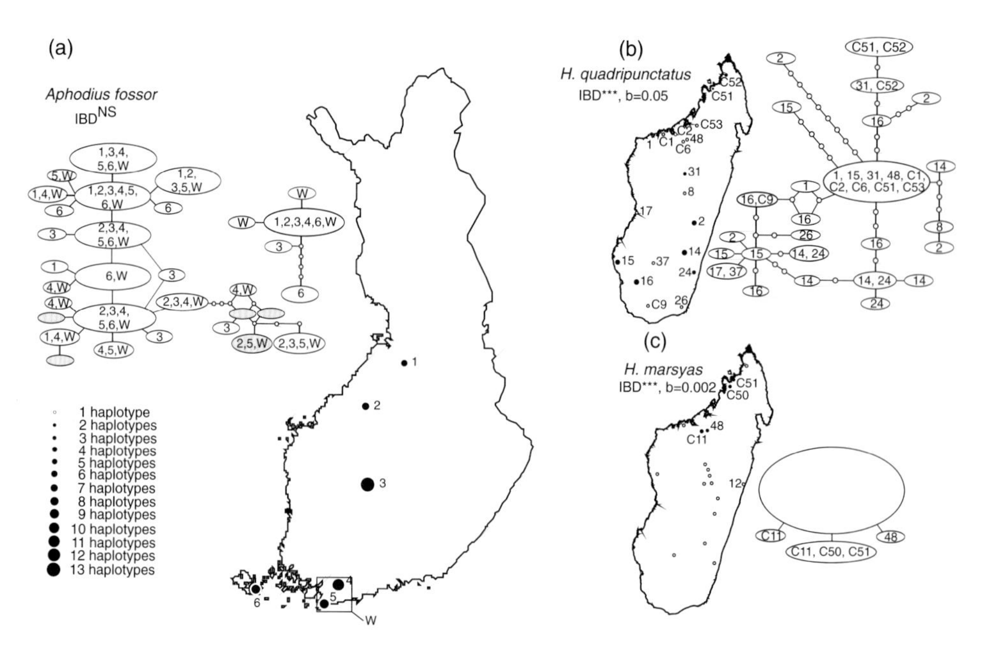

Fig. 11.3 Distribution of haplotypes (maps) and haplotype networks in the Finnish dung beetle species *Aphodius fossor* (a)[i] and in six Malagasy species (b-g)[ii], three of which are cattle dung-using *Helictopleurus* species (b–d), two of which are forest-dwelling *Helictopleurus* species (e and f) and one of which is a forest-dwelling Canthonini in genus *Nanos* (g). Within the maps, the size of the symbol indicates the number of different haplotypes recorded per sampling locality (see the legend in the figure). For each species, 'IBD' identifies the relationship between geographical distance and genetic differentiation, with associated significance levels[iii]. For significant IBD, the value of b defines the slope of the regression line of percentage sequence divergence on distance in degrees (ca 112 km at the equator). Within haplotypes, letter and number codes refer to the occurrence of the respective haplotype at the sites identified in the juxtaposed map. In the haplotype network of (a), haplotypes shaded in grey have been encountered in both Finland and Spain, whereas site 'W' refers to a compound sample from multiple farms within an intensively sampled region identified by a square. In the Malagasy networks, common haplotypes with no specific locations codes have been encountered at all sampling sites examined. Redrawn, with permission, from Roslin (2001b) and Hanski et al (2008). © Wiley-Blackwell.

[i] Redrawn, with permission, from Roslin (2001b), adhering to the exact methodology of Hanski *et al.*, 2008. [ii] Redrawn, with permission, from Hanski *et al.* (2008). [iii] NS $P > 0.05$; * $P<0.05$; ** $P<0.01$; ***$P<0.001$.

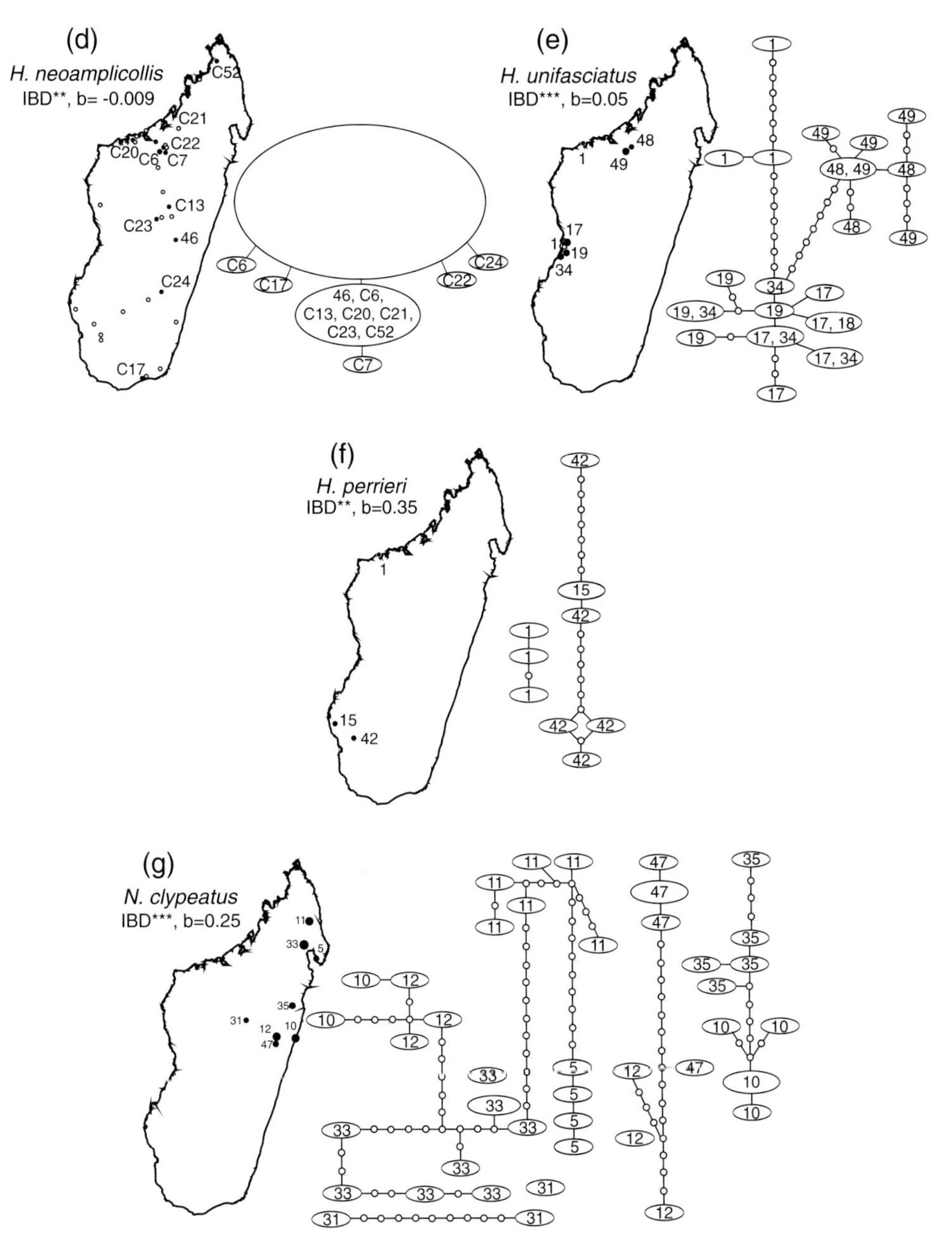

Fig. 11.3 (*Continued*)

by ancestral species to the forested habitats preferred by most Malagasy species of today.

Among the species that evolved in Madagascar, patterns of genetic diversity suggest striking differences in recent population history and important consequences of changes in resource selection (*cf.* Section 11.4). Hanski *et al.* (2008)

offer a revealing comparison between six species endemic to Madagascar, three of which are forest-dwelling generalists (*Helictopleurus unifasciatus*, *H. perrieri* and *Nanos clypeatus*), and three of which can currently be found on cattle dung in open localities (*H. neoamplicollis*, *H. marsyas* and *H. quadripunctatus*). Of these, the forest-dwelling species show substantial haplotype diversity, with geographical differences in the occurrence of different haplotypes (Figure 11.3e–g). Genetic differentiation among the sampled populations increases quickly with distance – a sign of successively decreasing gene flow between population pairs separated by increasing intervals (Figure 11.3e–g; Hanski *et al.*, 2008).

In the three species using cattle dung, we find very different patterns: In *H. neoamplicollis* and *H. marsyas*, most individuals represent just a single haplotype, and differentiation increases only slowly (Figure 11.3c), or not at all (Figure 11.3d), with distance. In the third species, *H. quadripunctatus*, there is, in fact, substantial haplotype diversity – in particular in southern Madagascar (Figure 11.3b). Nonetheless, one of the haplotypes is quantitatively dominant across the range, and this is also the only haplotype encountered in several samples from north-west Madagascar (Figure 11.3b; Hanski *et al.*, 2008).

The different patterns observed in different species point to different causal processes. For the forest dwellers (Figure 11.3e–g), genetic patterns seem most indicative of relatively large and/or stable populations (sustaining local genetic diversity) combined with ongoing but restricted gene flow within the range of the species (causing isolation-by-distance; *cf.* Slatkin, 1993). For two species now occurring on cattle dung (*H. neoamplicollis* and *H. marsyas*), the disproportionate dominance of given haplotypes and low sequence divergence among extant haplotypes (Figure 11.3c,d) suggest a recent and quick expansion of the population (see e.g. Merilä *et al.*, 1996; 1997). For the third species (*H. quadripunctatus*; Figure 11.3b), the pattern seems more indicative of a gradual shift to cattle dung across multiple parts of the range and also of range expansion towards the north (as evidenced by the loss of haplotype diversity in this part of the range).

Importantly, the three *Helictopleurus* species moving to cattle dung have a much greater range size today than do the forest-dwelling species – a pattern shared by other, distantly related dung beetle species performing the same resource switch (Wirta *et al.*, 2008; Rahagalala *et al.*, 2009). This difference illustrates how the change to a new resource and habitat may allow a species to escape competition and spread widely into new parts of the landscape (Hanski *et al.*, 2008; Rahagalala *et al.*, 2009). Again, we see a close interaction between evolution, large-scale population structure and population dynamics – a theme brought up repeatedly by our comparison between Finland and Madagascar.

11.7 Consequences: spatial population structures and responses to habitat loss

Our comparison between Madagascar and Finland highlights variation between dung beetle species. Both within and between regions, species differ in terms of range size (Section 11.3), habitat and resource selection (Section 11.4), dispersal capacity (Section 11.5) and population genetic structure (Section 11.6). An added dimension is local population size, where some species are locally rare, whereas others form denser local populations (e.g. Hanski, 1991; Roslin & Koivunen, 2001). Hence, while we have mostly focused on beetles attracted to a single patchy and ephemeral resource within each region – the cow pat in Finland or the lemur dropping in Madagascar – species seem to perceive and use the surrounding landscape quite differently. As a consequence, the individuals arriving at the key resource have been drawn from populations of different spatial structure.

For the Finnish system, Roslin (2001a) and Roslin & Koivunen (2001) envisage species as falling into a range of different population structures (Figure 11.4). Whereas specialist species with a strong preference for pasture habitats are structured into assemblages of more or less discrete local populations, generalist species are more evenly distributed within and between pastures. However, even among species with the same habitat and resource preferences, the level of dispersal among local populations will vary with the size of the species. We therefore infer that generalist *Aphodius* species, and specialist species with high dispersal powers, occur as large 'patchy' populations in the landscape (*sensu* Harrison & Taylor, 1997). As noted above, local populations then largely consist of temporary aggregates of individuals, a substantial fraction of which are reshuffled among habitat patches each generation. In contrast, small pasture specialist species with limited dispersal powers form classical metapopulations; here, local populations are less coupled by dispersal and some species may persist in a dynamic equilibrium between local colonization and extinction events (Roslin, 1999; 2000; 2001a; Roslin & Koivunen, 2001). Collectively, these differences should translate into differing sensitivity to landscape composition, with poorly dispersive specialist species being particularly sensitive to changes which increase distances between patches of suitable habitat (Figure 11.4).

For the Malagasy system, we see variation along the same ecological axes. While most endemic species are confined to closed forest habitat, some occur in multiple habitats and a minority even prefer open habitats (Rahagalala *et al.*, 2009). That expansion into open habitats allows populations to expand into larger parts of the landscape has already been described above. Thus, the fact that variation in resource and habitat selection results in variation in landscape-level distribution offers a significant parallel across the Finnish and Malagasy system. Nonetheless, in Madagascar, disparities are also to be found in the larger-scale distribution of species. Even among the forest-dwelling beetles, we find significant variation in the general size of the distribution range (Figure 11.2b). While most species in this group have small ranges (Hanski *et al.*, 2007; Viljanen, 2009a; 2009b; Viljanen *et al.*, 2010a), others are broadly distributed across the island (Hanski *et al.*, 2007; 2008).

Given this variation, we can generate some clear-cut predictions:

- First, mere species-area considerations suggest that massive loss of prime habitat (pastures in Finland, forests in Madagascar) should result in general changes in the species composition on remaining resources: the loss of habitat should shrink

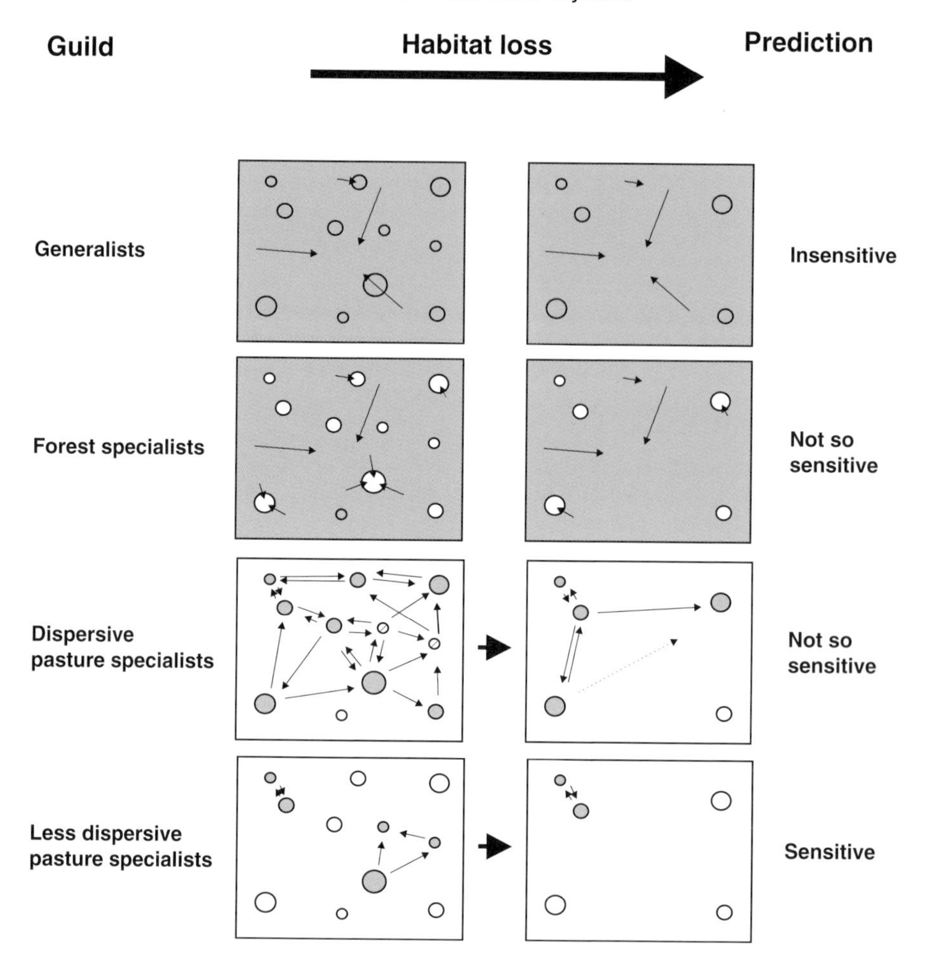

Fig. 11.4 Spatial population structure of Finnish dung beetles. Here, circles show cattle farms within the landscape. For each guild defined on the left, we show the hypothetical distribution of species within the landscape by grey shading. Dung beetle movements are shown by arrows. Habitat loss and fragmentation progresses from the left to the right and the response of the respective group is predicted in the right-hand margin.

the regional species pool in rough proportion to the square root of area (MacArthur & Wilson, 1967; Ovaskainen & Hanski, 2003).

- Second, variation in the general patterns of species loss should be reflective of variation in patterns of habitat loss: the loss of species should be concentrated in areas where changes in landscape composition have been most pronounced.
- Third, some species should be more affected than others, with interspecific differences in response reflecting the differences in spatial population structures postulated above.

Unfortunately, these predictions have been tested only all too literally in both the Malagasy and the Finnish systems. In Finland, 93 per cent of cattle farms

(Figure 11.5a) and 53 per cent of cattle have been lost over the past 50 years. Of the remaining cattle stock, current legislation requires only one-third to be pastured, whereas dung produced by the rest is largely lost from a dung beetle perspective. Hence, the overall reduction in resources amounts to ≈90 per cent as measured by both loss of habitat and loss of dung. In Madagascar, the changes have been equally dramatic. Since the arrival of humans, most of the Malagasy landscape has been deforested (Harper *et al.*, 2007; Dufils, 2003; Hanski *et al.*, 2007), and current

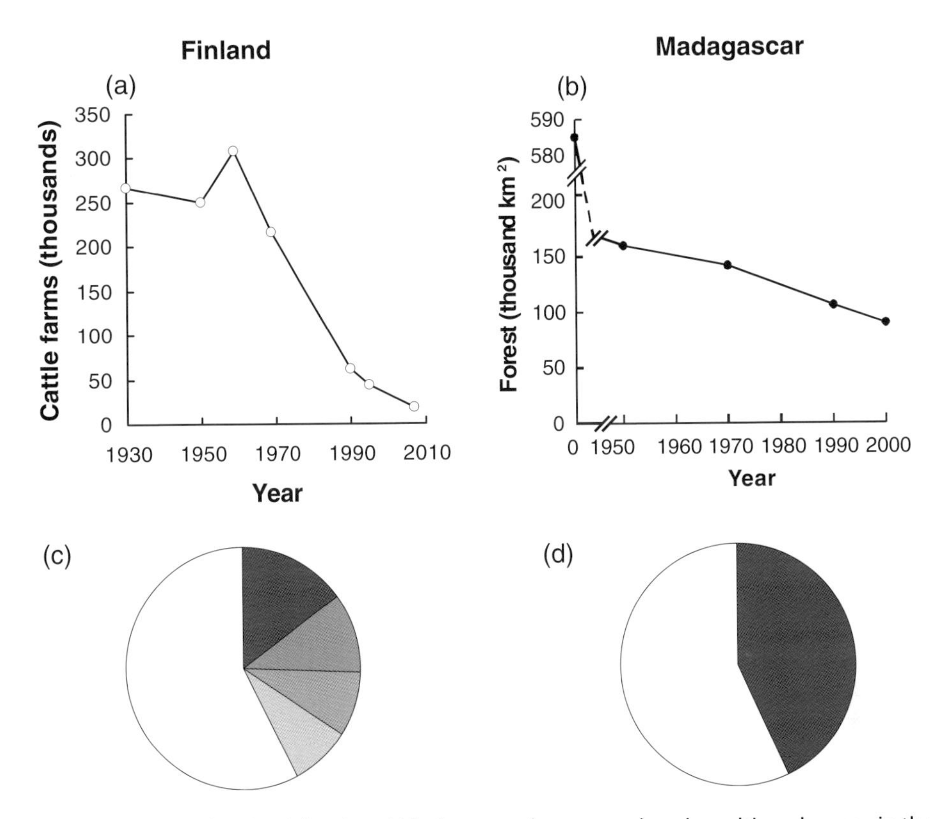

Fig. 11.5 Habitat loss in Finland and Madagascar (upper row) and resulting changes in the dung beetle fauna (lower row). For Finland (a), we show changes in the number of cattle farms between 1930 and 1995[i]. For Madagascar (b), we show changes in the extent of forest cover[ii]. Here, the data point on the y-axis ('Year 0') is based on the assumption that before the arrival of humans, the entire island was covered by forest. With respect to changes in the dung beetle fauna, panel (c) shows the fraction of species classified as threatened in Finland by Rassi *et al.* (2001). Progressively darker shades of grey indicate increasing threat status, from vulnerable (lightest shading) through endangered and critically endangered to regionally extinct (darkest shading). (d) For Madagascar, the figure represents the fraction of apparent extinctions among 51 endemic species examined by Hanski *et al.* (2007).

[i] Data source: National Board of Agriculture (1920–1985); Information Centre of the Ministry of Agriculture and Forestry (1986–2008); Anonymous (2008). [ii] Data source: Humbert & Cours Darne (1965); Harper *et al.* (2007).

efforts focus on protecting ≈10 per cent of what is assumed to have been the original forest cover (Figure 11.5b).

For Finland, the response of the dung beetle fauna has been examined by Roslin (2001a; and in preparation) and Roslin & Koivunen (2001), and for Madagascar by Hanski *et al.* (2007). The patterns emerging support the predictions. First, across both systems, we see a general change in species composition and the loss of many species. For Finland, collection of beetles by entomologists over the last few centuries offers some indication of population trends (Biström *et al.*, 1991; Roslin, 2001b, in preparation), supporting the current classification of seven dung beetle species as nationally extinct, five as critically endangered, four as endangered, four as vulnerable and five as near threatened (Rassi *et al.*, 2001). Hence, 42 percent of the species are considered threatened (Figure 11.5c) and an additional 11 per cent near-threatened. Interestingly, the changes are clearly concentrated on the majority of species dependent on cattle dung, whereas the relatively few forest specialist species have – if anything – increased with an increase in the national elk population (Roslin, 2001b; Roslin & Heliövaara, 2007). For Madagascar, there has been no long-term monitoring by local entomologists, but a gradual build-up of records from more scattered expeditions. A similar comparison can then be made between historical species records aggregated through time and modern records obtained through a single, intensive bout of sampling aimed at revealing the *status quo*. Based on this rationale, Hanski *et al.* (2007) compared the set of Helictopleurini species found during intensive sampling in 2002–2006 to that observed in samples aggregated between 1875 and 1990. Of 51 endemic species for which past locality information was available, modern sampling failed to detect 22 (43 per cent; Figure 11.5d). Comparing the situation in Finland and Madagascar, the proportion of disappearing species is then notably similar across areas (compare Figure 11.5c vs. 11.5d). Indeed, species-area considerations suggest that with a loss of 90 per cent of the habitat, roughly half of the species should disappear. Assuming that species currently classified as threatened in Finland are on their way to extinction, and that species not detected in Madagascar are effectively extinct, these figures are remarkably consistent with the prediction (*cf.* Hanski *et al.*, 2007).

Second, regional variation in the pattern observed at the national level seems reflective of regional variation in habitat loss. Within Finland, we find a disproportional loss of habitat-specialist species from areas where the network of cattle pastures has thinned out most strongly (Roslin, 2001b; Roslin & Koivunen, 2001). For Madagascar, we see that species are more likely to disappear, the higher the fraction of forest lost within their former range (Hanski *et al.*, 2007).

Third, how a specific species responds appears to be dictated by its spatial population structure. For Finland, we observe major changes in the distribution of different dung beetle guilds over time, with (small) pasture-specialist species declining the most (Roslin, 2001b; see also Roslin & Koivunen, 2001). These temporal patterns are consistent with signals in terms of contemporary distribution, where the occurrence of the small specialist species *Aphodius pusillus* can clearly be linked to the density of the surrounding network of farms (Roslin, 2001b; Roslin & Koivunen, 2001). For Madagascar, we find an imprint of other aspects of spatial population structure. Here, species which were previously more common and widespread are less likely to disappear than species which were originally rare and local (Hanski *et al.*, 2007).

Some changes in landscape composition affect the relative abundance and distribution of species without causing species loss (Roslin & Koivunen, 2001). An interesting question, then, is whether landscape impacts on population processes and whether community composition will ultimately affect ecosystem services such as dung decomposition, seed dispersal and burial. That habitat fragmentation affects such functional processes has been shown in other tropical systems (Klein, 1989; Andresen, 2003; see also Quintero & Roslin, 2005) but not yet for Madagascar.

For Finland, a study conducted by Rosenlew & Roslin (2008) found mixed results. In a microcosm experiment, we used our previous observations of dung beetle assemblage structure in fragmented and intact landscapes (Roslin & Koivunen, 2001) to create realistic differences in *Aphodius* assemblages, asking whether such differences will affect ecosystem functioning. As measured by changes in dung fresh weight, we found no detectable effects of the relatively minor changes in *Aphodius* assemblages induced by current levels of fragmentation. Nonetheless, manipulating the presence of *Geotrupes stercorarius* – a large dung beetle species declining in the landscapes of today – had an overriding impact, with the amount of dung remaining at any one time doubling when this species was excluded.

Importantly, this finding suggests that not only will species react differently to changes in landscape composition but that the functional consequences will vary substantially, depending on which species are lost. In particular, large-bodied species are likely to be functionally important and their loss will result in disproportionate loss of function (Larsen *et al.*, 2005; 2008; Slade *et al.*, 2007). Such idiosyncrasies among species underscores the importance of our analysis of how taxa differ in their response to landscape modification. For a more in-depth review of the ecosystem services offered by dung beetles in other parts of the world, see Chapter 12.

11.8 Perspectives

Our comparison of dung beetle populations in two different biomes identifies a diversity of factors affecting local community structure. Most fundamentally, it illustrates how the general species pool available in a region will be determined by long-term and large-scale processes, and that these processes may vary between regions. Here, the comparison between the old endemic fauna of Madagascar, and the younger, more recently immigrated fauna of Finland shows how evolutionary history may constrain patterns of range size, habitat and resource use emerging in different parts of the world.

At the landscape level, we have argued that ecological traits such as these – compounded by dispersal capacity and local population size – affect how populations are structured, and that interspecific variation in such traits determines how local assemblage structure co-varies with the structure of the surrounding landscape. Most importantly, the variation uncovered among the dung beetle species colonizing our focal resources runs in stark contrast with the idea that populations of all species would function similarly over the same set of sites (*cf.* Pandit *et al.*, 2009). The general realization that different population processes will dominate the landscape-level dynamics of different species, and that such variation should be built into conceptual models, is now penetrating the field of metacommunity ecology (Driscoll, 2008; Driscoll & Lindenmayer, 2009; Pandit *et al.*, 2009), with dung beetles adding significantly to the evidence.

While the studies accumulated to date allow us to draft a first picture of how dung beetle populations are structured in space, much work remains to be done. Whether there are more general trends in how dung beetle populations are structured in different biomes is a question that we can only address when additional species have been examined in further detail. Yet, an answer is needed if we want to understand how dung beetle populations and assemblages react to global habitat change. In this context, we particularly need more studies of dung beetle dispersal in different parts of the world. By understanding how dung beetles traverse the landscape, we will be better placed to understand how they respond to changes in landscape composition.

We also need more studies linking dung beetle population processes to ecosystem services – studies examining how changes in landscape composition will eventually affect what dung beetles do best, i.e. decompose dung. Ultimately, the emerging understanding of how dung beetle populations are structured in space will help us to understand how we should manage them (see Chapter 13) and what patterns to expect in terms of population dynamics and microevolution.

Acknowledgements

We are indebted to Ilkka Hanski, to the editors of this volume and to Saul Cunningham for perceptive comments on previous versions of this manuscript, and also to IH for making unpublished material from Madagascar available to us. Alison Cameron, Evgeniy Meyke, Sami Ojanen and Helena Wirta helped us with information and technical know-how for the figures, and Otso Ovaskainen commented on some estimates of dung beetle dispersal. Financial support from the Academy of Finland (grant number 129636 to the Centre of Excellence in Metapopulation Research 2009–2011, and grant number 129142 to TR) is gratefully acknowledged.

12

Biological Control: Ecosystem Functions Provided by Dung Beetles

T. James Ridsdill-Smith[1] and Penny B. Edwards[2]

[1]*School of Animal Biology, University of Western Australia, Crawley, Western Australia*
[2]*Maleny, Queensland, Australia*

12.1 Introduction

When exotic species become pests in new regions, classical biological control may be attempted through the introduction of natural enemies to control the pest populations. Usually the targets are pests of disturbed or managed perennial agricultural systems, but they may also be species that have become pests in natural habitats. Neoclassical biological control is used where endemic species become pests as a result of changing farming or management practices, and imported natural enemies provide a novel association to control the pests (Ehler, 2000). In a review of 87 biological control programmes of arthropod pests in Australia, in nearly half the cases, introductions of biological control agents have resulted in substantial reductions in pest abundance (Waterhouse & Sands, 2001).

Dung beetle introduction programmes aim to improve nutrient recycling as well as the control of dung-breeding flies (Bornemissza 1976; Fincher, 1981). Scarabaeine dung beetles have been deliberately introduced to Australia (Bornemissza, 1976, Doube *et al.*, 1991; Legner, 1995; Waterhouse & Sands, 2001), to Hawaii (Legner, 1995) and to North America (Fincher, 1986). In Australia, 23 species are known to be established, of which the first was recovered in 1969 and the last in 1991 (Edwards, 2007).

Ecologists are concerned that any benefits from introduced biological control agents should not be offset by adverse impacts on populations of non-target species through pest replacement, disruption of existing biological control or extinction at a local or global scale of rare and endangered species (Ehler, 2000). Non-target effects of introducing dung beetles could include adverse impacts on native dung beetles and on the activity of introduced and native predator species (Legner, 1995).

Ecology and Evolution of Dung Beetles, First Edition. Edited by Leigh W. Simmons and T. James Ridsdill-Smith. Published 2011 by Blackwell Publishing Ltd.

Australia is a large continental island of some 7.7 million km^2, where 27.3 million cattle and 76.9 million sheep (Australian Bureau of Statistics, 2008; www.abs.gov.au) graze on native and improved pastures. These herbivorous animals drop considerable quantities of dung on pastures all year round. A cow is estimated to drop 10–12 pads a day (Fincher, 1981) of 2.9 l each (Ridsdill-Smith, 1993). In addition, about 50 million kangaroos of the four most common species (Department of Foreign Affairs and Trade, 1997; www.dfat.gov.au) feed and drop pellets in areas of natural vegetation and to a lesser extent in pastures.

In this chapter, we assess the changes to the dung beetle community in Australian pastures over the 40 years since exotic species were first released. The impact of introduced dung beetles on dung burial and control of pest flies in pastures is considered, as well as possible non-target impacts on native dung beetles. The Australian biological control programme is used as an example to address the question of how the benefits of a created dung beetle community could be maximized to utilize the dung resource effectively.

12.2 Functions of dung beetles in ecosystems

Scarabaeine dung beetles belong to an abundant and species-rich worldwide family, members of which are specialized to feed as adults and as larvae on herbivore dung (Chapter 2). Ecosystem functions provided by scarabaeine dung beetles include nutrient cycling, mixing of dung into the soil, plant growth enhancement, secondary seed dispersal (Chapter 5), parasite control and fly control (Nichols *et al.*, 2008). The fast dispersal and burial of fresh dung from the soil surface by adult dung beetles has been shown to increase nutrient cycling and to reduce numbers of dung-breeding flies and diseases associated with dung in India (Oppenheimer, 1977), Australia (Bornemissza, 1976) and North America (Fincher, 1981). Seasonal growth of plants, which affects the quality of herbivore dung and impacts upon the seasonal patterns of reproduction both of dung beetles and dung-breeding flies, is influenced by temperature and rainfall (Ridsdill-Smith & Hall, 1984a; Matthiessen *et al.*, 1986; Galante *et al.*, 1991; Anduaga, 2004).

12.2.1 Dung burial and nutrient cycling

In summer rainfall regions of northern Queensland, Australia, native dung beetles buried little cattle dung in pastures in 1967 and 1968, and much of the dung lay on the pasture virtually undisturbed for three to twelve months of the year (Ferrar, 1975). In south-eastern Australia between 1970 and 1973, dung burial by native dung beetles was greatest just after winter, but was generally low and varied between years (Hughes, 1975). By 1990–1993, two native dung beetle species were active, together with a variable number of introduced species at different sites. Average dung dispersal per week was six per cent at a site where one introduced species was established, 12 per cent at a site with two species, 16 per cent with four species and 30 per cent where five species were present (Tyndale-Biscoe, 1994). The increased rate of dung burial and shredding occurred mostly during the summer.

In winter rainfall regions of south-western Western Australia, the native species *Onthophagus ferox* has one generation a year. Beetles emerge in autumn (May) and bury dung masses for feeding. As temperatures rise in spring (from October to December), they bury dung for brood masses. *Onthophagus ferox* buried 20–30 per cent of cattle dung for a period in autumn and 20–30 per cent in spring (Ridsdill-Smith & Hall, 1984a; Ridsdill-Smith, 1993a; T.J. Ridsdill-Smith, unpublished). The introduced species *O. binodis* has two generations a year. The main breeding period occurs in spring, when temperatures rise, and the overwintering adults become active and bury dung to make brood masses (Ridsdill-Smith & Hall, 1984a; Ridsdill-Smith, 1993a). In summer, newly emerged beetles feed in the dung pads, causing shredding, and a second, smaller breeding period occurs in late summer (Ridsdill-Smith, 1993a; T.J. Ridsdill-Smith, unpublished). At a site when *O. binodis* was the only introduced beetle present, in spring they buried 45 per cent of available cattle dung, and in late summer they buried 7 per cent of available dung to construct brood masses (Ridsdill-Smith, 1993a; T.J. Ridsdill-Smith, unpublished data).

Studies in grasslands demonstrate a strongly positive role of consumers in returning limiting labile nutrients to plants, including nitrogen, potassium and phosphorus (Loreau, 1995). At Tsavo National Park in Kenya, during peak seasonal dung beetle activity, a 2 kg pad of elephant dung was removed completely by 17,000 beetles in two hours (Coe, 1977). Dung beetles are not active in the dry season in tropical regions, where termites are important in breaking down dung (Ferrar, 1975; Coe, 1977).

In many ecosystems, nutrients associated with mammalian dung are made available to plants predominantly through the actions of dung beetles (Nichols *et al.*, 2008). Under laboratory conditions, the addition of scarabaeine dung beetles to dung placed on the soil surface of pots where plants are growing improves the nutrient levels in the soil and increases plant yield (Bornemissza & Williams, 1970; Fincher *et al.*, 1981; Yokoyama *et al.*, 1991; Bertone *et al.*, 2006).

In Queensland, only a small proportion (2.5 per cent) of pasture is actually covered by cattle dung over a year (Jones & Ratcliff, 1983), and the impact of patchy dung burial on the whole pasture productivity is not easy to determine. Consumer-nutrient cycling dynamics is also difficult to determine in tropical forests (Feeley & Terborgh, 2005; John *et al.*, 2007), yet the generally low nutrient levels and highly localized differences in soil fertility indicate that dung beetles may play an important role in nutrient cycling. Further study is required to quantify these impacts (Nichols *et al.*, 2008).

12.2.2 Control of dung-breeding flies

Dung-breeding flies lay eggs in fresh dung, where the larvae complete their development. Flies targeted for control by dung beetles include species that are nuisance pests of grazing animals and people, such as bush fly (*Musca vetustissima*) in Australia, face fly (*M. autumnalis*) in North America and biting flies that feed on blood of grazing animals and horses, such as buffalo fly (*Haematobia irritans exigua*) in northern Australia, horn fly (*H. irritans*) in North America and the African buffalo fly (*H. thirouxi potans*) in Southern Africa (Bornemissza, 1976;

Fincher, 1981; Doube *et al.*, 1988b). Any reduction in the abundance of these flies brought about by dung beetles would be considered an ecosystem service, especially for bush fly and face fly, which are nuisance pests of people.

Under controlled conditions in the laboratory and field, when sufficient numbers of dung beetles are added to experimental pads, the numbers of flies emerging are reduced for horn fly (Macqueen & Beirne, 1975), buffalo fly (Roth *et al.*, 1988), African buffalo fly (Doube & Moola, 1988), face fly (Fincher, 1986), bush fly (Hughes *et al.*, 1978; Wallace & Tyndale-Biscoe, 1983; Ridsdill-Smith *et al.*, 1986; Tyndale-Biscoe & Vogt, 1991; Ridsdill-Smith, 1993b) and *Musca hervei* (Hasegawa & Yamashita, 1985).

However, species vary in their responses to dung quality. Scarabaeine dung beetles produce more brood masses, and flies are larger and more fecund and their larvae survive better when feeding on good quality dung from cattle feeding on pasture growing under favourable seasonal conditions (Ridsdill-Smith, 1986; 1991; Macqueen *et al.*, 1986). Bush flies are more responsive to dung quality than buffalo fly (Macqueen *et al.*, 1986), and the introduced scarabaeine dung beetle *O. binodis* is more responsive than *Onitis alexis* (*Onitis* are abbreviated to *On.* in this chapter to distinguish them from *Onthophagus* abbreviated as *O.*) (Ridsdill-Smith, 1986). Disturbance of poor quality dung by dung beetle feeding is also effective in controlling flies (Ridsdill-Smith *et al.*, 1986). The quantity of dung buried by beetles is an important predictor of fly mortality, but not the only one.

In semi-natural field experiments in Australia, dung beetles were excluded from control dung pads, either by using mesh covers or by placing pads on two-metre poles, which allows oviposition by flies in the absence of beetles. Numbers of flies emerging from these control dung pads were then contrasted with numbers from dung pads to which beetles had normal access. Mortality of both buffalo fly and bush fly was measured in the presence of predatory fauna plus dung beetles (Doube *et al.*, 1988b; Tyndale-Biscoe & Vogt, 1991). These experiments demonstrated that bush fly and buffalo fly survival is reduced when sufficient dung beetles (native or introduced) are naturally attracted to cattle dung (Wallace & Tyndale-Biscoe, 1983; Ridsdill-Smith & Matthiessen, 1984; Roth *et al.*, 1988; Doube *et al.*, 1988b; Tyndale-Biscoe & Vogt, 1991). For example:

- At Busselton (south-western Western Australia), an average of 679 bush flies emerged from pads when only the native species *O. ferox* was present, against 1,019 when beetles were excluded (Ridsdill-Smith & Matthiessen, 1984). A mean of 32 *O. ferox* was caught in 24-hour pitfall traps at this time (Ridsdill-Smith & Hall, 1984a).
- At Geraldton (652 km north of Busselton), an average of 29 bush flies emerged from pads in the presence of *On. alexis*, against 70 flies when beetles were excluded (Ridsdill-Smith & Matthiessen, 1984). A mean of 47 *On. alexis* was caught per pitfall trap over 24 hours (Ridsdill-Smith & Hall, 1984a).
- At Dardanup (south-western Western Australia), an average of two bush flies emerged when *O. binodis* was the dominant species present, but 128 when beetles were excluded. A mean of 147 beetles was caught per pitfall trap over 24 h (Ridsdill-Smith & Matthiessen, 1988).

There was no evidence that the effectiveness of the predatory and parasitic fauna in dung pads decreased when introduced dung beetles were added to experimental pads (Wallace & Tyndale-Biscoe, 1983).

The impact of dung beetles on natural fly populations and on fly nuisance is less easy to measure. The exotic dung beetle *Digitonthophagus gazella* (in earlier literature, this species is called *O. gazella*) is well established in Californian pastures, but horn fly numbers remain unacceptably high (Legner & Warkentin, 1991). At low beetle densities in Australia, it is often difficult to determine an impact of beetles on bush fly abundance, although at high beetle densities it is easier (Hughes *et al.*, 1978; Wallace & Tyndale-Biscoe, 1983; Tyndale-Biscoe & Vogt, 1996).

Bush fly is found throughout Australia in the summer, but it repopulates the southern third of the continent during late winter by wind-borne long distance movements (Waterhouse & Sands, 2001). In the inland tropics of northern Australia, populations build up to a peak in late summer and autumn and decline in winter. In the south, they build up to a peak in late spring and early summer but also decline in the wetter, colder winter (Waterhouse & Sands, 2001).

Bush fly populations have been assessed by netting flies that approach the collector (a measure of their nuisance value). In summer, when bush fly populations exceed an abundance index of 10, they are considered significant pests (Ridsdill-Smith & Matthiessen, 1988; Tyndale-Biscoe & Vogt, 1996). At three sites in temperate south-eastern Australia, in 1981–1982, bush fly populations during summer comprised wind-borne migrants (determined by dissection), and dung beetle activity (largely the native species *O. granulatus*) had little effect on their numbers. However, in 1980–1981, the populations comprised locally bred flies, and dung beetle activity had a much bigger impact (Feehan *et al.*, 1985). Following the establishment of introduced beetles in the same region, it was shown that the nuisance value of bush flies in early summer (November to December) was reduced when the dry weight of dung beetles present exceeded 3 g per dung pad (caught in dung baited pitfall traps over 24 hours; Tyndale-Biscoe & Vogt, 1996).

In the winter rainfall region of south-western Western Australia, seasonal patterns of bush fly and dung beetle abundance (Ridsdill-Smith & Matthiessen, 1988) are more predictable. In 1980–1981, in the presence of only the native beetle, *O. ferox,* the bush fly Abundance Index (AI) was similar in December (AI = 46) and January (AI = 58). In 1983–1984, after two exotic species became established, the bush fly abundance was unchanged in December (AI = 50) but in January was 88 per cent lower (AI = 7), while the dry weight of dung beetles was 63 fold greater (Ridsdill-Smith & Matthiessen, 1988). A mean of 1,673 *O. binodis* plus *Euoniticellus pallipes* was caught per pitfall trap over 24 hours in January of 1983 and 1984 (equivalent to 31 g dry weight of beetles). January is a major holiday period in the Busselton region, so dung beetles provide a major ecosystem service at this time.

In summary, control of the dung-breeding pest, bush fly *M. vetustissima,* has occurred in pastures over significant regions of Australia as a result of the introduction of exotic scarabaeine dung beetles. While another dung-breeding fly, the buffalo fly *H. irritans exigua*, remains a pest, some local control of this has also occurred due to introduced dung beetles (Waterhouse & Sands, 2001). A number of insect and mite predators and parasites have also been introduced to control the fly pests (Wallace & Tyndale-Biscoe, 1983; Doube *et al.*, 1988b; Tyndale-Biscoe &

Vogt, 1991; Legner, 1995). In addition, insecticides are used to kill adult flies, and some chemical drenches or their breakdown products are excreted in the dung. These may kill both flies and dung beetles or impact negatively on their breeding success (Ridsdill-Smith, 1988).

12.2.3 Control of parasites

Under controlled experimental conditions, the addition of *D. gazella* to dung pads caused a reduction in the numbers of infective larvae of gastrointestinal nematodes of cattle on grass surrounding dung pads (Bryan, 1976). Numbers moving from dung to grass were also influenced by rainfall events. The interactions between dung beetles and the midge *Culicoides brevitarsis*, an arbovirus vector, were investigated in laboratory and field conditions in coastal New South Wales (Bishop *et al.*, 2005). While the potential for beetle activity to reduce *C. brevitarsis* survival in cattle dung was established, a decline in midge numbers was only significant in one out of nine field trials (Bishop *et al.*, 2005). These studies did not establish clear evidence of broad-scale substantial control of nematodes or midges in cattle dung under natural conditions.

12.3 Dung beetles in pasture habitats

Dung beetle diversity varies among different ecosystems. In most tropical forests around the world, there is a rich dung beetle community of around 50 species (Howden & Nealis, 1975; Halffter, 1991; Hanski & Cambefort, 1991), while in tropical savannas there can be up to 100 species (Halffter, 1991). In contrast, the faunas of cattle pastures are characterized by low beetle diversity, with few of the forest-dwelling species from adjacent habitats (Halffter, 1991; Nichols *et al.*, 2007; Horgan, 2007; see Chapters 11 and 13). In Mexico, Hawaii and South Africa, pastures grazed by cattle support around eight dung beetle species (Harris *et al.*, 1982; Davis, 1987; Anduaga, 2004; Horgan, 2007). The species found in pastures are often open habitat specialists favoured by human activity (Horgan, 2007); many have smaller overall body size and some are very abundant (Nichols *et al.*, 2007).

One notable characteristic of many of the numerically abundant introduced species in pastures is their capacity for long-distance dispersal. It is only with newly established species that it is possible to determine the rate of spread to new areas:

- Following their initial introductions, *D. gazella* and *Euoniticellus intermedius* dispersed naturally over most of northern Australia in just a few years (Bornemissza, 1976), and *D. gazella* and *O. taurus* spread over much of the southern USA in an equivalent period (Fincher *et al.*, 1983).
- Rates of dispersal for *D. gazella* are estimated to be over 50 km a year, with a maximum of 800 km, and flights of 30 km over ocean to islands have been recorded (Bornemissza, 1976; Fincher *et al.*, 1983; Barbero & Lopez-Guerrero, 1992; Montes de Oca & Halffter, 1998).
- *Euoniticellus intermedius* can disperse 50 km a year, with a maximum of 480 km (Montes de Oca & Halffter, 1998).

- *Onthophagus taurus* can disperse 300 km a year (Fincher *et al.*, 1983), and has been found in the guts of fish caught at sea (Berry, 1993).

Other dung beetle species found in pastures spread more slowly. For instance, in Queensland, *On. caffer* is estimated to have spread approximately 70 km in 20 years (3.5 km a year), while *Copris elphenor* dispersed 50 km in 20 years (2.5 km a year) (Edwards, 2007). In Tasmania, observations indicate that *Geotrupes spiniger* (an introduced dung-breeding Geotrupidae) spread at 1 km a year, while *E. fulvus* spreads 6 km a year (D. Kershaw & G. Stevenson (personal communication); estimates of dung beetle movement are also discussed in Chapter 11).

Over 100 species of scarabaeine dung beetles have been deliberately introduced into new countries in order to disperse the apparently under-utilized cattle dung resource in pastures (Hanski & Cambefort, 1991). By enhancing the breakdown of dung, introduced beetles are expected to improve the ecosystem functions in pastures and grasslands grazed by cattle. The African dung beetle *D. gazella* now has a wide global distribution, having been deliberately introduced to Hawaii in 1958, to Australia in 1968, to North America in 1972, to South America in 1990, and also to Easter Island, New Caledonia, and Vanuatu (Barbero & Lopez-Guerrero, 1992).

In the continental United States, seven species are established that were introduced through Texas, and three species from Texas have become established in Puerto Rico (Fincher, 1986). The introduced dung beetle species *D. gazella* and *E. intermedius* have spread naturally from the USA into Mexico, where they are now abundant in cattle dung in pastures (Barbero & Lopez-Guerrero, 1992; Montes de Oca & Halffter, 1998; Anduaga, 2004). *Copris incertus* has been introduced from Mexico to New Zealand (Blank *et al.*, 1983). Introductions of further exotic species are being considered for Australia, Chile, and New Zealand (J. Wright, S. Forgie, personal communication).

A rich community of native scarabaeine beetles is found in Australia in woodland and heath, but few of them have become abundant in the new pasture habitats where cattle dung accumulates on the pasture surface. In Queensland in 2001–2002, at 72 sites in pastures, an average of nine native species per site was collected at cattle dung baited traps (Edwards, 2009), while in southern Australia, at 10 sites in pastures, 1–2 native species were trapped per site over a year (Hughes, 1975; Tyndale-Biscoe, 1994; Ridsdill-Smith & Hall, 1984a).

In 1964, George Bornemissza proposed the introduction of exotic scarabaeine dung beetles to Australia to enhance recycling of cattle dung in pastures. He suggested that about 160 species of scarabaeine beetles would be required, occupying complementary niches (Bornemissza, 1976). We will now describe the community of native scarabaeine dung beetles found in Australia, then look at the impact that the introduced beetles have had on these native communities.

12.4 Seasonal occurrence and abundance of native dung beetles in Australia

Australia has a relatively rich fauna of 437 known species of native scarabaeine dung beetles, mostly found in natural forest and heath habitats (T.A. Weir, personal communication), of which 355 species are described (Cassis *et al.*, 2002). Most are endemic and feed on the dung of kangaroos and other native animals.

The distribution of the tribe Onthophagini in Australia falls into 11 patterns which can be defined by temperature and rainfall (Matthews, 1972). The tribe Scarabaeini follow the same patterns, but they are more prevalent in forest habitats and tend to be confined to the wetter and warmer areas of Australia, near to the coast (Matthews, 1974). The third and smallest tribe are the Coprini (Matthews, 1976). Half of the native species are in the tribe Onthophagini (234 species), a few of which play a significant role dispersing cattle dung in pastures. *Onthophagus* species disperse readily over water and often are the only scarabaeine species on islands (Matthews, 1972). In 1836, during his visit to Tasmania on the *Beagle*, Charles Darwin found native *Onthophagus* feeding in cattle dung in areas where cattle had been introduced 33 years previously (Darwin, 1845).

In the subtropics and tropics of Queensland there are 325 species of native dung beetles (G. Monteith, personal communication), and 73 of these (22 per cent) have been caught in traps baited with cattle dung in pastures (Edwards, 2009). Fifty-nine species (81 per cent of 73) belong to the genus *Onthophagus*. Of the native species in cattle dung in pastures in Queensland, the highest diversity occurs in regions where the annual rainfall is between 500 and 1200 mm. In regions where annual rainfall exceeds 1,800 mm, species diversity is low, perhaps because mortality of dung beetle larvae is greater in soils that are wet for prolonged periods.

A 16-month study of seasonal abundance and geographical range of dung beetle species in pastures was undertaken in 2001–2002 at 119 sites throughout Queensland by monthly pitfall trapping using cattle dung baits (Edwards, 2009). The 22 sites in Queensland's arid west receive less than 500 mm annual rainfall. The three other regions all receive more than 500 mm annual rainfall; 32 sites were in the tropical north (north of 20° S), 34 sites were in central Queensland (between 20° S and 25° S) and 31 sites were in the south-east (south of 25° S). In all four Queensland regions, most activity of native dung beetles occurred in the warmer, wetter summer months between October and May (Figure 12.1a). There were two exceptions to this general pattern: *O. pentacanthus* was more active from late autumn to spring than it was in summer; and *O. tamworthi* was also frequently collected in the cooler months (Edwards, 2009).

In the arid west (<500 mm rainfall), average abundance of native beetles was low. But surprisingly, abundance in central Queensland (500 mm to 1,600 mm rainfall) was even lower than in the arid west. Abundance in the tropical north and the south-east was higher, but the variation within these regions was considerable.

Of the 15 sites throughout Queensland where native dung beetles were most abundant, nine sites were in the tropical north and six were in the south-east. The nine tropical north sites were in the Einasleigh Upland region, where 7–23 species were recorded at each site, and the six southern sites were in the Darling Downs area of the southern Brigalow Belt region, where 6–17 species were recorded at each site. It is likely that the higher diversity and abundance of native dung beetles in these two regions is linked to vegetation, soil and climate parameters, plus dung supply. The Einasleigh Uplands and the Brigalow Belt were both recognized as biodiversity hotspots by the Australian Government in 2003 (www.environment.gov.au/biodiversity/hotspots).

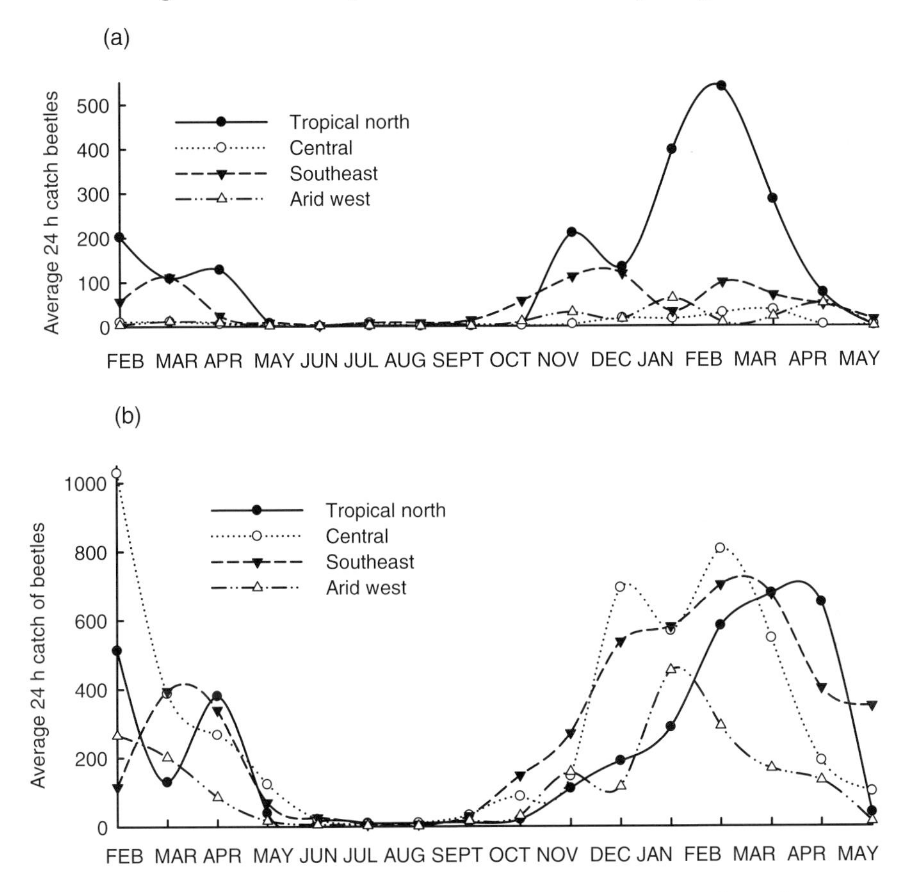

Fig. 12.1 Seasonal activity of scarabaeine dung beetles at 119 sites throughout Queensland in 2001–2002 by pitfall trapping using cattle dung baits for 24 hours (Edwards, 2009). Monthly abundance of beetles per trap is given for each of four regions. (**a**) The native species (**b**) The introduced species

Onthophagus consentaneus was recorded at 69 of the 119 sites in Queensland (Edwards, 2009). This is one of the most widely distributed species across northern Australia, found commonly in open grassland and savanna habitats (Matthews, 1972). *Onthophagus consentaneus* also coexists with introduced species at Alice Springs in central Australia (Matthiessen *et al.*, 1986). It is not found in marsupial dung, but under decaying mushrooms and vertebrate carcasses, and it is attracted to human and cow dung. Matthews suggested its great abundance: '*appears not to be supported by the indigenous marsupial fauna, but, directly or indirectly, by man through his own excrement and that of his cattle and through the roadside slaughter that is a consequence of motor traffic.*' Of the native species in Australia, those adapted to open grassland and savanna are more likely to become abundant in pastures.

While many native species are small, the *atrox* group of native *Onthophagus* species are quite large, usually 15–20 mm long, comparable in size to the introduced

Onitis species, and are found generally in open or semi-woodland habitats in drier regions (Matthews, 1972). There are 15 species of the *atrox* group found in different regions of Australia (G. Monteith, personal communication), and they include several species found commonly in cattle dung in pastures. This is one group of native species for which pastures provide suitable habitat, and cattle dung may provide a more suitable food source for them than the smaller-pelleted dung of most kangaroos and other marsupials.

In Western Australia there are 75 species of native dung beetles, of which 48 species (64 per cent) belong to the genus *Onthophagus,* and a further 24 species (32 per cent) are Scarabaeini (T. Weir, personal communication). Seven species of the tribe Onthophagini (Matthews, 1972) and nine species of the tribe Scarabaeini are endemic to the winter rainfall agricultural regions of south-western Western Australia (Matthews, 1974). *Onthophagus* are more abundant in heath than in open forest and are commonly trapped with dung baits, whereas Scarabaeini abundance does not vary between these undisturbed habitats and four out of five species were more abundant in traps baited with carrion than dung (Ridsdill-Smith *et al.*, 1983). However, in annual pastures across south-western Western Australia, only four endemic *Onthophagus* species and no Scarabaeini were trapped at cattle dung baits (Ridsdill-Smith & Hall, 1984a). One of these species, *O. ferox,* a member of the *atrox* group, is common in pastures.

In the winter rainfall areas of south-western Western Australia, the native dung beetles are active in autumn and spring, in contrast to the situation in Queensland, where the native beetles are active from spring to early autumn. For O. *ferox*, there is a peak of feeding activity in May to June (at the start of the winter rainfall period) and a peak of breeding activity in October and November (spring) (Figure 12.2a). This is similar to that for other endemic species in Western Australia (Ridsdill-Smith & Hall, 1984a). When temperatures are low, in mid-winter, few beetles fly to traps.

In the northern warmer parts of south-western Western Australia, the native species emerge in May to June but breed in August and September, earlier than in the south, in response to higher temperatures (Ridsdill-Smith & Hall, 1984b).

12.5 Distribution and seasonal occurrence of introduced dung beetles in Australia

The first beetles deliberately released in Australia were five scarabaeine species introduced via Hawaii in 1968–69, of which three became established (Edwards, 2007). Further species were subsequently introduced from Africa and Europe, and 43 species had been released by 1984. Twenty-three species, 53 per cent of those released, were known to be established in 2007 (Edwards, 2007). The number of beetles released appears to be a major factor in successful establishment and, for most species that did establish, at least 500 beetles were released at five or more sites, with the total released being more than 8,000 beetles (Edwards, 2007).

The current distributions of introduced dung beetles in Australia are given by plotting the number of species known to occur for each square degree (an area of one degree of latitude by one degree of longitude) (Figure 12.3). At least

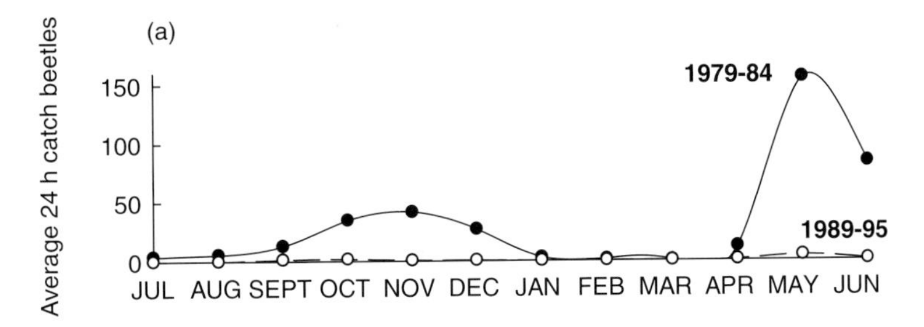

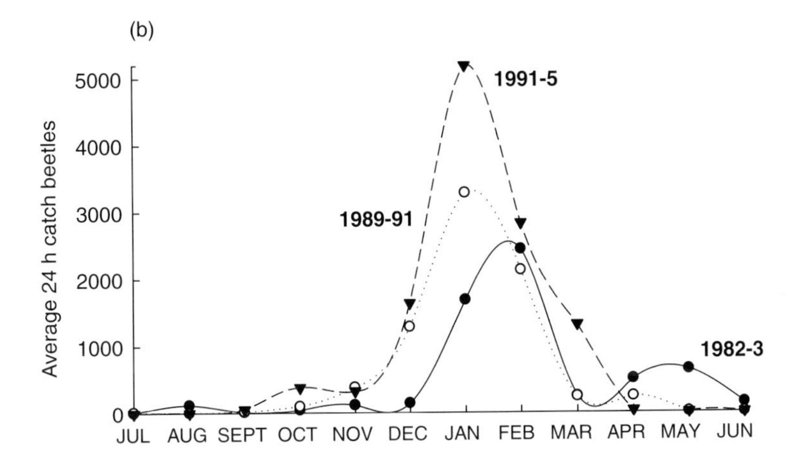

Fig. 12.2 Seasonal activity of scarabaeine dung beetles at one site (Busselton) in SW Western Australia by pitfall trapping using cattle dung baits for 24 hours. Monthly abundance of beetles per trap averaged over the years. (**a**) 1979–84 for a single native species, *O. ferox,* before introduced beetles became abundant (Ridsdill-Smith & Hall 1984a) and 1989–95, after introduced species became abundant (I. Dadour, personal communication). (**b**) The first individual *O. binodis* was sampled in November 1980, and data shows the changes over time in total introduced beetle numbers in 1982–83 (Ridsdill-Smith & Hall, 1984a), 1988–91 and 1991–95 (I. Dadour, personal communication).

one introduced dung beetle species occurs virtually everywhere there are cattle in Australia.

The highest species diversity, of between 9–13 species per square degree, occurs in south-eastern Queensland. In this region, several temperate species are close to the northern limits of their distribution, including *E. africanus*, *O. binodis,* and *On. pecuarius*, while some tropical species are near the southern limits of their distribution, e.g. *Liatongus militaris*, *On. viridulus*, *Sisyphus rubrus*, and *S. spinipes*. None of these temperate or tropical species is abundant in this south-east region, presumably because the climate is suboptimal for both groups. Abundance and complementary seasonal activity may be more important than number of species in determining the success of a created dung beetle community.

Sampling for introduced beetles was conducted throughout Queensland in 2001–2002, in the same traps as the native beetles (Edwards, 2009). Fifteen introduced species were caught, whose activity was mainly restricted to the wet

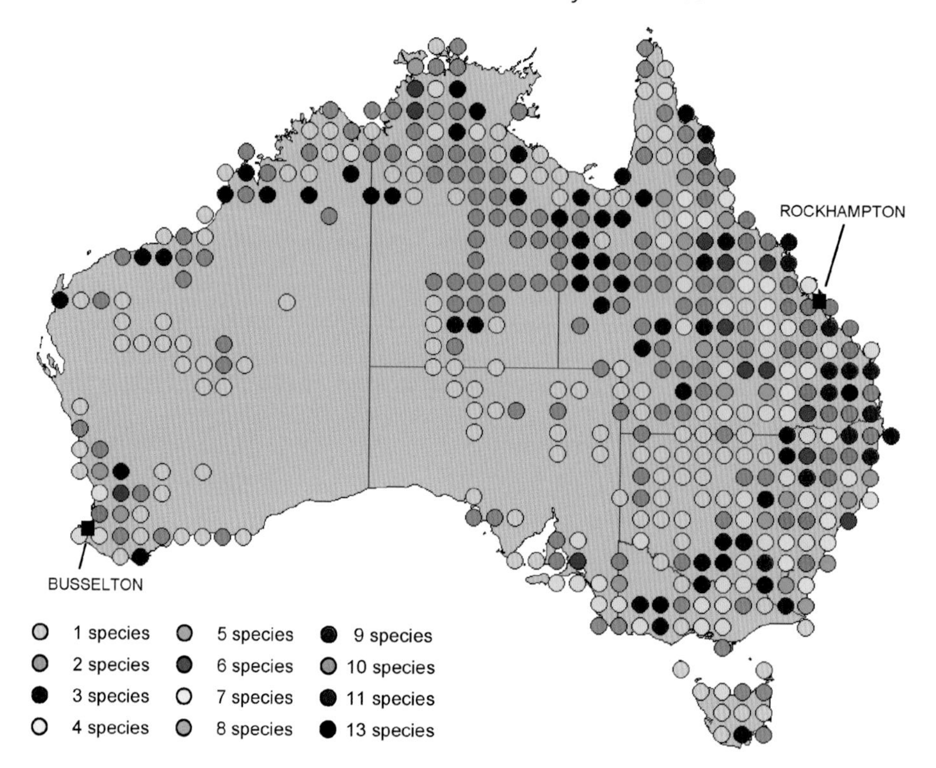

Fig. 12.3 Distribution of introduced dung beetles in Australia in 2007, expressed as number of species for each square degree (Edwards, 2007).

warm months of October to April (spring to early autumn) (Figure 12.1b). The differences in beetle abundance between the four geographical zones were not as marked as for the native beetles. Numbers were lowest in the arid west zone. The main peak of adult activity was later and for a shorter period in the tropical north than in other regions, which is consistent with the shorter, later and more intense wet season that occurs there.

Edwards (2009) reported *D. gazella* from all 119 sites in Queensland and *E. intermedius* at 118 sites. The third most widespread species was *On. alexis*, which occurred at 92 sites, being absent from the wettest sites and from the northern Mitchell grasslands. These three species exhibit adaptability to a broad range of climatic and soil conditions, and they are also the three most widespread introduced species in Australia (Edwards, 2007). All three species have a wide distribution throughout Africa south of the Sahara, with *On. alexis* extending into southern Europe (Edwards, 2007).

Some species with less widespread distributions in Queensland have clear climatic limits to their distributions, while others have yet to spread to all suitable regions. In the former group are species whose distributions align with rainfall limits, such as *L. militaris*, which is restricted to areas with more than 600 mm annual rainfall, and *O. sagittarius*, which occurs in regions with more than 800 mm. Most species are absent from or less abundant in the extremely high rainfall areas; however, two species, *O. nigriventris* and *On. vanderkelleni*, are restricted to these areas, where rainfall may be over 2,000 mm a year. Both species originated from the moist tropical highlands of Africa and were introduced from Kenya. Another high rainfall species from tropical Africa, *O. obliquus*, has established in the Cooktown region of north Queensland, where it occurs at very low density (P. Edwards, unpublished data).

Twelve species of introduced dung beetles are established in Western Australia. Eight of these are restricted to the south-west of the state, two (*D. gazella* and *On. viridulus*) have only been found in the summer-rainfall northern part of the state, and two (*E. intermedius* and *On. alexis*) have a broad distribution across much of the state (Edwards, 2007). *Onthophagus taurus, O. binodis, E. fulvus* and *E. pallipes* were the most abundant species in pastures trapped with cattle dung baits at eight sites in dairying areas of south-western Western Australia from 1990–1994 (Dadour & Cook, 1997). *Onthophagus taurus* made up 51–94 per cent of the community at seven sites, but *O. binodis* and *E. fulvus* were equally abundant at the remaining site, where *O. taurus* was absent. *Euoniticellus pallipes* was relatively more abundant in the north of the region than *E. fulvus*. Three other species established in the south-west of WA, *On. caffer, On. aygulus* and *Bubas bison*, have become sufficiently locally abundant to be harvested and redistributed. A fourth species, *C. hispanus*, is not yet well enough established to be redistributed.

Introduced dung beetles disperse naturally, but deliberate cropping and redistribution of beetles can speed up this process. Cattle farmers have shown great interest in obtaining exotic dung beetles to improve the recycling of dung on their pastures. Between 1989 and 1995, more than 3.5 million dung beetles were redistributed across southern Australia, supported by funding from the dairy industry (Dadour & Cook, 1997; Edwards, 2007), and between 1993 and 2009, a private company redistributed over six million dung beetles of 18 species (J. Feehan, personal communication). Species were identified that would benefit from further redistribution by comparing their current range in Australia with their predicted distribution based on the climate in their native range (Edwards, 2007).

12.6 Long-term studies of establishment and abundance

Long-term studies on opposite sides of the Australian continent provide insights into the establishment and abundance of introduced dung beetles. Here we discuss and contrast data collected from sites in Queensland (summer rainfall climate) and in Western Australia (winter rainfall climate). Standard methods were used to assess occurrence and abundance of introduced dung beetles, using pitfall traps baited for 24 hours with one litre of fresh cattle dung wrapped in muslin. In the Queensland studies, baits were replaced after 12 hours, but in Western Australia the dung bait was left unchanged for 24 hours.

Flight activity is influenced by ambient temperature, wind speed and the incidence of rain during the trapping period, and the number of beetles caught in a pitfall trap provides an index of the number of beetles flying to fresh dung. Data are summarized from several studies between the 1970s and the 2000s (Ridsdill-Smith & Hall, 1984a; Ridsdill-Smith & Matthiessen, 1988; Doube *et al.*, 1991; Dadour & Cook, 1997; I. Dadour (personal communication); Edwards, 2009;

A. Macqueen (personal communication)). Information is available for the greatest number of years for the summer period of beetle activity, but data is also given for a smaller number of years when trapping was continued for the full 12 months of the year. Since some species are active in different seasons, the 12-month trapping data provides a better comparison of species abundance.

12.6.1 Summer rainfall climate area of Queensland

Beetles were sampled at Belmont Research Station, near Rockhampton in Central Queensland (23°8′ S, 150°13′ E) (Macqueen *et al.*, 1986). The property contains both native and introduced pasture species. It has an annual rainfall of 795 mm, with a summer peak, and mean annual minimum and maximum temperatures of 16.6 and 28.3 °C. Average summer 24-hour catches are presented for each species for the 16 years with four summer months of data (December to March) from 1972 to 1986 (A. Macqueen, unpublished), and 2001 to 2002 (Edwards, 2009), together with the average annual 24-hour catch for the 11 years with 12 months of data. The four-month summer data are used to discuss the establishment and abundance of species at this site.

The introduced beetles at Rockhampton are summer-active. When trapping commenced in 1972, two introduced species were established, and by year six, in 1977, six species of introduced beetles were present, of which *D. gazella* was dominant. From year seven onwards, further species became established, increasing to eight species by 2001 (not shown in Table 12.1), of which seven were caught in 2002. In this second period, 1978–2002, total numbers of beetles were higher than in the first period, and in different years *D. gazella, E. intermedius,* and *S. spinipes* were the most abundant species (Table 12.1). There were considerable fluctuations of total beetle numbers between years; average 24-hour catches over summer exceeded 1,000 beetles per trap in 1979, 1984 and 1986. Years when beetle numbers were high in spring and summer were characterized by good rain in the preceding autumn, which promoted survival of adults and larvae over winter, followed by good spring rain (A. Macqueen, personal communication). These effects can be seen in the relative differences between average beetle counts in summer versus the full 12 months.

The abundance of two of the dominant species at Rockhampton, *D. gazella* and *E. intermedius*, was affected similarly in different years. However, the third dominant species, *S. spinipes,* followed a different pattern. During the peaks in dung beetle numbers in 1979 and 1984, *S. spinipes* was the most abundant species, but during the peak in 1986 it ranked third and, by 2002, *S. spinipes* ranked fifth (Table 12.1).

Sisyphus spinipes leaves its brood balls above ground, attached to vegetation. This makes them vulnerable to desiccation, depending on pasture type, grazing intensity and rainfall. The decline in abundance may have been due partly to the fact that the trapping site was moved from improved pasture in 1972–1986 to a new adjacent site in native pasture in 2002. The improved pasture included a tussock species that provided good habitat for *S. spinipes* broods (A. Macqueen, personal communication). *Sisyphus spinipes* was less abundant than *S. rubrus* in 2001–2002 at nearly all sites in Queensland where both species occurred. *Sisyphus rubrus* was released in Queensland after *S. spinipes*, so it may have contributed to the decline of

Table 12.1 Average summer 24-hour trap catch (Dec to Mar) for eight species of introduced dung beetles from traps baited with fresh cattle dung at Rockhampton, Queensland between 1972 and 2002. Average annual 24-hour catch for Jan to Dec given for years where trapping occurred in all 12 months. (1972–1986: A. Macqueen, personal communication. 2002: Edwards, 2009).

Species	*1972*	*1973*	*1974*	*1975*	*1976*	*1977*	*1978*	*1979*	*1980*	*1981*	*1982*	*1983*	*1984*	*1985*	*1986*	*2002*
D. gazella	63	56	63	147	159	103	119	230	68	132	230	65	253	87	511	107
O. sagittarius	5	3	33	34	29	14	3	15	5	5	4	1	3	1	7	3
L. militaris			1	8	19	42	21	78	21	40	169	9	139	35	112	12
E. intermedius					32	37	51	324	66	180	192	59	1344	233	902	486
S. spinipes					11	36	35	483	48	34	70	1	1807	618	303	7
Onitis viridulus								<1		1	1		7	6	7	
S. rubrus															8	73
Onitis alexis																<1
Average summer 24-hr trap catch (Dec–Mar)	67	58	97	189	251	231	229	1129	206	393	666	135	3552	979	1847	688
Average annual 24-hr trap catch (Jan–Dec)	47	36	85	130	123	114	541	247			230	807	2235			

S. spinipes. However, the Rockhampton data indicate that the decline in *S. spinipes* preceded the arrival of *S. rubrus* at that site. The reason for the apparent state-wide decline of *S. spinipes* remains unknown.

Onthophagus sagittarius, although never very abundant, formed a significant part of the fauna at Rockhampton for the first seven years, until 1979. Thereafter it declined to become a very minor component from 1980 onwards. During the 2001–2002 survey, it was not abundant anywhere in Queensland and was restricted to a narrow coastal belt of less than 100 km in width (Edwards, 2007).

12.6.2 Mediterranean climate area of south Western Australia

Beetles were sampled at a farm, near Busselton in south-west Western Australia (33° 45′S, 115°17′E), in a sown annual pasture grazed by cattle, on sandy soils and surrounded by woodland (Ridsdill-Smith & Hall, 1984a). The annual rainfall of 823 mm falls mainly in winter, and the annual mean minimum and maximum temperatures are 11.1°C and 22.1°C. Initially, only the native species *O. ferox* was present, constituting 99 per cent of the dung beetle community from May 1979–November 1983, and 24-hour catches over 12 months averaged 31, 35 and 43 beetles (Table 12.2). *Onthophagus ferox* is active from autumn to spring. After the introduced beetles became abundant at this site, the average 24-hour catch of *O. ferox* over 12 months fell to 0.6, 0.5, 0.5 and 1.8 for the four years 1990–1993 (I. Dadour, personal communication; Table 12.2). In 2005, *O. ferox* was still present, but at very low levels.

The introduced beetles in the west are active from spring to early autumn, as in Queensland. In 1983, four years after sampling commenced, two introduced species were established, of which *O. binodis* was the dominant species. Years six to ten were not sampled, but from year 11 onwards five species were present, of which four were introduced species. In these later years from 1991–1995, the summer activity was greater again, and *O. taurus* was the dominant species making

Table 12.2 Average summer 24-hour trap catch for Jan and Feb for four species of introduced dung beetles and one native species from dung-baited traps at Busselton, Western Australia between 1980 and 1995. Average 24 catch for 12 months given for years where trapping occurred in all 12 months. (1980–1984: Ridsdill-Smith & Hall, 1984a. 1990–1995: I. Dadour, personal communication).

Species	*1980*	*1981*	*1983*	*1984*	*1990*	*1991*	*1992*	*1993*	*1994*	*1995*
O. ferox (native)	1	2	3	1	<1	<1	<1	<1	<1	<1
O. binodis		1	873	2867	659	1013	178	274	97	41
E. pallipes			63	302	335	140	157	189	105	87
E. fulvus					319	195	621	516	317	283
O. taurus					1168	1568	2397	4353	4191	2200
Average 24-hr catch (Jan–Feb)	1	3	939	3170	2481	2916	3353	5332	4710	2611
Average 24-hr catch (12 months)	31	35	312		664	495	1011	1070		

up 81 per cent of the total beetle catches (range 71–89 per cent) (Dadour & Cook, 1997; I. Dadour (personal communication); Table 12.2). Total beetle numbers were lower in 1995 than 1994.

In 1983 and 1984, the time of the major peak of *O. binodis* activity was in February, with a smaller peak in April to May (Figure 12.2b). In 1989–1991, when *O. taurus* and *O. binodis* were both abundant, the main peak of activity for all introduced dung beetles was in January, with a second smaller peak of *O. binodis* in April in 1990 and March in 1991 (Figure 12.2b). In 1991–1995 *O. taurus* was the dominant species and the peaks were in January and February.

There was substantial activity of all introduced species in March 1992 and a smaller peak in October (Figure 12.2b). *Euonticellus pallipes* fell from 11 per cent of the total catch in 1983 to 3 per cent in 1993, while *E. fulvus* remained at around 10 per cent of the catch between 1990 and 1993. In 2005, ten years later, there were six species present, including *On. alexis*. *Onthophagus taurus* was even more dominant (91 per cent of sample), and *E. fulvus* remained more abundant than *E. pallipes* (J. Matthiessen, personal communication). The year-to-year fluctuation in total beetle numbers was high.

12.6.3 Long-term population trends

The long-term data for Rockhampton (13 years) and Busselton (14 years) have been combined with data from a site near Townsville, in the tropical north of Queensland (6 years) (A. Macqueen, personal communication), to show the change in number of beetles trapped per 24-hour catch averaged for 12 months as a function of the number of years since the start of sampling at each site. For the first six years, the numbers remained around 100 beetles, but after this there was a significant increase to over 1,000 beetles per trap by year 14 (Figure 12.4).

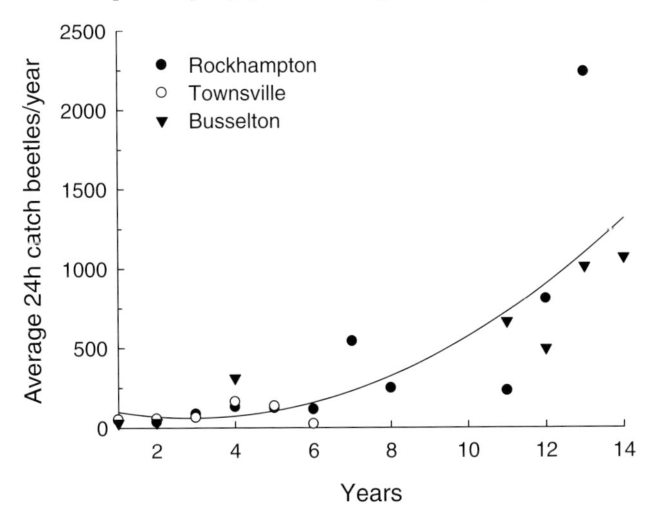

Fig. 12.4 Average annual dung beetle abundance at three sites as a function of number of years since establishment of introduced beetles. Mean monthly abundance for 12-month periods from 24-hour trap catches fitted by a quadratic curve ($Y = 148.0 - 59.4X + 10.2X^2$, $R^2 = 0.666$).

During the early 1980s, when only single species were present at different pasture sites in south-western Australia (Ridsdill-Smith & Hall, 1984a), seasonal peak numbers of *O. ferox* buried about 20 per cent of the dung, and for *O. binodis* this was about 45 per cent (see Section 12.2.1). Presumably, intraspecific competition prevented them from fully utilizing the available dung (Ridsdill-Smith, 1991). As new species became established at the long-term study sites, total numbers increased; however, the total population was not equal to the sum of the numbers each species would have achieved if it had been the sole species at that site, probably due to interspecific competition.

There appears to be a lag of several years after a species becomes established until it becomes more abundant (Tables 12.1 and 12.2). Since some of the dung beetle species – especially the large species – spread slowly, and their populations build up slowly, it will be many more years before a regional equilibrium will develop for all of the introduced species.

12.7 Competitive exclusion

The richness of scarabaeine dung beetle communities in forests and savannas in Africa and South America, where large herbivores are common, was an incentive to develop biological control programmes to introduce dung beetles to pastures grazed by cattle where there were few native beetle species (Section 12.3). Dung beetles co-existing in fresh dung occupy many different niches, and interspecific competition is likely to have played a significant role in the evolution of niche divergence. Species that evolve to exploit alternative niches can escape the interspecific competition that reduces individual fitness (see Chapter 1). Alternative niches include habitat preferences, seasonal activity patterns, dung burial behaviour and time of flight.

While the importance of interspecific competition for species fitness is clear from laboratory studies, there are few quantitative field studies of interspecific competition in action. If a new introduced species utilizes a niche that is already occupied, interspecific competition is likely to occur. The introduction programme has thus provided a unique opportunity to look for evidence of competitive exclusion in natural populations.

An important niche for dung beetles is habitat. In June and July 1980, only native beetle species were trapped in forest and heath habitats in the southern areas of south-western Western Australia (Ridsdill-Smith *et al.*, 1983). Most Onthophagini were in heath habitats. In the northern areas of south-western Western Australia, scarabaeine beetles were sampled at monthly intervals from November 1982 to October 1984, both in National Parks and in an adjacent pasture (G.P. Hall, personal communication). Six dung beetle species trapped at omnivore dung in the National Parks were active from autumn to spring, and 99.9 per cent were endemic species. In the adjacent pasture site, three species were trapped at cattle dung, and the endemic *O. ferox* made up 16 per cent of the population and was active from autumn to spring. Two introduced species, *E. intermedius* and *On. alexis,* made up 84 per cent of the population and were active mainly from summer to early autumn. In this situation, native and introduced beetles occupy different habitats and are active at different times of the year, so competition is thereby reduced.

At the Busselton site in south-west Western Australia, abundance of the native species *O. ferox* was measured before and after the arrival of introduced beetles. As discussed in Section 12.6.2, *O. ferox* numbers averaged 36 beetles per 24-hour catch over 12 months from 1979–1983, with a peak of 166 beetles in May 1980 (autumn) (Table 12.2; Ridsdill-Smith & Hall, 1984a). In the years 1990–1993, an average of over 500 introduced beetles were trapped per 24 hours over 12 months, and there was an average of less than one *O. ferox* per trap over 12 months during the same period, with the greatest 24-hour catch of 14 *O. ferox* in May 1993 (I. Dadour, personal communication).

This reduction in abundance of over 97 per cent seems to provide clear evidence of competitive exclusion of *O. ferox* following establishment of introduced beetles. This occurred even though the two species have predominantly different seasonal patterns of activity, apart from an overlapping period in spring. Furthermore, *O. binodis* flies during the day and *O. ferox* flies at night, with peak numbers of *O. binodis* (84 per cent) flying from 1200 (noon) to 1600 (Figure 12.5a), and peak numbers of *O. ferox* (78 per cent) flying from 1800 (dusk) to 2100 (Figure 12.5b) (T.J. Ridsdill-Smith, unpublished data).

With this considerable niche separation, it is interesting to speculate on the mechanism whereby the introduced species have impacted negatively on the native *O. ferox*. Cattle feed and produce dung mostly during the day, with main peaks in early morning and mid-afternoon, and a minor peak shortly after midnight (D. Thomas, unpublished data). Little cattle dung therefore is produced when *O. ferox* is flying. It can be hypothesized that when both species are breeding in

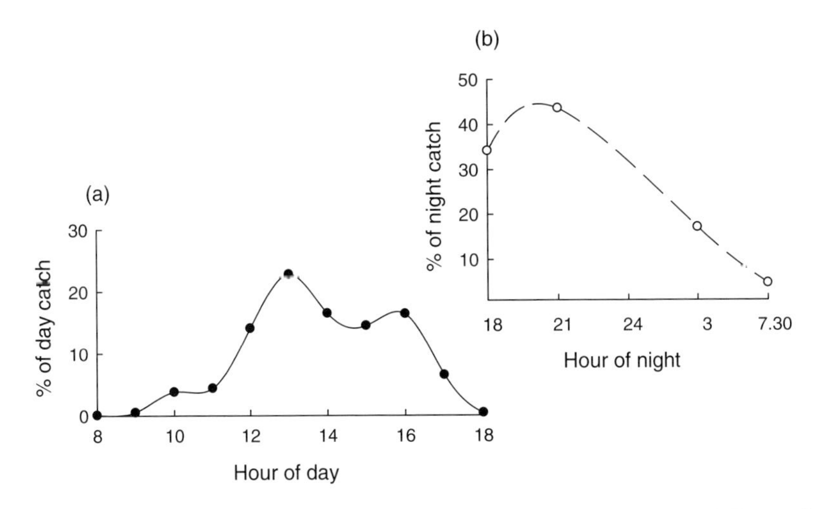

Fig. 12.5 Diel flight activity of dung beetles in SW Western Australia in pitfall traps baited with cattle dung. (**a**) Average percentage of the daily catch of *O. binodis* sampled hourly once in February and twice in October (total of 2901 beetles). (**b**) Average percentage of the nightly catch of *O. ferox* sampled at intervals from 3 to 6 hours in May and October (total of 158 beetles). Trapping was from one hour before dusk, dusk to 9 pm, 9 pm to 3 am and 3 am to 7.30 am. Neither species flew around dawn.

spring (see Figure 12.2 a & b), the pads of cattle dung will be largely occupied by *O. binodis* (and in later years by *O. taurus*) during the day, excluding the night-flying *O. ferox* through overcompensating density-dependent competition.

Of the introduced beetles at Busselton, *O. binodis* was dominant in 1983 and 1984. After *O. taurus* became established at this site, *O. binodis* and *O. taurus* had similar abundance in 1990 and 1991. Subsequently, however, *O. taurus* has become the dominant species and *O. binodis* abundance has remained low from 1992 onwards (Table 12.2). It is reasonable to assume that *O. taurus* contributed to the competitive exclusion of *O. binodis*, although the effect was not as strong as that of *O. binodis* on *O. ferox*. Abundance of *O. binodis* has decreased elsewhere in Australia, most notably in Tasmania (Kershaw & Stevenson, personal communication), possibly also in response to the abundance of *O. taurus*.

However, competition is not the only reason that species decrease in abundance. Although the reason for the decline in *S. spinipes* abundance throughout Queensland is not known, it is thought not to be due to competitive exclusion by *S. rubrus*, since its decline occurred at Rockhampton before *S. rubrus* became established (see Section 12.6.1).

12.8 Optimizing the benefits of biological control

The establishment of introduced species in Australia has clearly resulted in substantially increased dung beetle numbers in pastures, particularly in summer. It has thereby increased dung dispersal (an ecosystem function) and, at times, pest fly control (an ecosystem service) (see Chapter 13). More native and introduced dung beetle species are found in pastures in the warmer northern regions of Queensland than in the more temperate south-west Western Australia. In 2001–2002, an average of 6.3 introduced species were recorded per site in Queensland ($n = 72$) (Edwards, 2009), while in 1990–1994, an average of 3.5 introduced species was recorded per site ($n = 8$) in south-west Western Australia (Dadour & Cook, 1997). Is this a sufficient number of introduced dung beetle species in Australian pastures, or could the presence of more species improve the ecosystem functions?

Communities containing many species are predicted to be more resilient and stable, with better maintained ecosystem functions than communities with fewer species (Walker, 1992; Naeem, 1998; Cardinale *et al.*, 2006; Dunne & Williams, 2009). To understand the functioning of species rich communities, species have been grouped by their activities. For example, dung beetle communities have been described by grouping species according to their dung burial behaviour (Hanski & Cambefort, 1991).

An analysis of beetle biomass data from a transect across a South African savanna (Doube, 1983) showed a strong interaction between functional groups of beetles and soil type. Large rollers were more abundant on sand, while slow-burying tunnellers were more numerous on clay-loam soil (Hanski & Cambefort, 1991).

In most cases, the selection of meaningful groups has proved difficult (Walker, 1992; Naeem, 1998; Cardinale *et al.*, 2006; Wright *et al*, 2006; McGill *et al*, 2006). One characteristic that seems to be important for achieving high rates of dung burial is body size (McGill *et al*, 2006; Dunne & Williams, 2009). In Malaysian Borneo forests, large nocturnal tunnellers were the most important functional group in terms of dung and seed removal (Slade *et al.*, 2007).

In the laboratory, when the larger *O. ferox* (16 mm long) and the smaller *O. binodis* (9 mm long) were added to cattle dung at the same time, the competition was asymmetric, with the *O. ferox* reducing brood mass production of *O. binodis* more than vice versa, as well as controlling bush fly better (Ridsdill-Smith, 1993b). This was achieved because of pre-emptive dung burial by *O. ferox,* which, in the first day, buried 58 per cent of the total dung buried over 6 days, compared with 15 per cent by *O. binodis* (Ridsdill-Smith, 1993b). As discussed above, the opposite outcome occurred in the field, where *O. binodis* (and later *O. taurus*) displaced the native *O. ferox,* probably as a result of differences in temporal activity.

Although the largest species in a community tend to be functionally more important, they are also more prone to extinction (Larsen *et al.*, 2005; Slade *et al.*, 2007). This has been illustrated in the field by *O. ferox* and *O. binodis.* It may be that the abundance of the small, fast-dispersing, day-flying species in pastures is one factor preventing the build-up of numbers of large, slower-dispersing, night-flying species in cattle dung.

Of the 23 exotic species established in Australia, 13 are considered small species (under 13 mm in length) and 10 are considered large species (over 15 mm in length). Most of the abundant introduced dung beetles in Australian pastures are the small species (*O. taurus, O. binodis, D. gazella, E. intermedius*). The larger species established in Australia are *B. bison, Geotrupes spiniger, Copris elphenor, C. hispanus* and six species of *Onitis*. The larger species were not abundant at the long-term study sites.

In Chapter 9, Byrne and Dacke note that many large dung beetle species are night-flying, while the smaller species are day-flying, and it is suggested that perhaps the large night-flying species have trouble locating cattle dung not already occupied by small species. However, high biomass (and numbers) of the large species have been reported at a few sites in Australia, apparently successfully co-existing with the smaller species. In May 2009, over 100 *B. bison* were caught in traps baited with cattle dung for 24 hours at several sites in Victoria, and at the same sites over 1,000 beetles of the smaller summer-active species, *O. taurus* and *E. fulvus*, were trapped in December and January 2010 (B. Pearce, personal communication – see http://northeast.landcarevic.net.au/dungbeetle). In May 2001, up to 40 *On. caffer* were trapped in 24 hours at Highfields in Queensland, followed by summer catches of several hundred beetles from the smaller summer-active species *D. gazella* and *E. intermedius* (Edwards, 2009). Other large species, such as *C. elphenor* at Jambin, Queensland, and *B. bison* at Kojonup, Western Australia, are present at a few sites at numbers high enough for these beetles to be cropped and redistributed. We still do not have sufficient information to predict under what conditions large beetles will become abundant in the presence of small species.

Dung beetle species richness and abundance generally decreases with increasing modification and fragmentation from forest to pastures (Nichols *et al.*, 2007; Horgan, 2007). On islands in a lake flooded for a hydroelectric dam in Venezuela, the number of dung beetle species fell with decreasing island size and, thus, patch size of tropical forest (Larsen *et al.*, 2005). In Spain, beetles were trapped weekly for 12 months on pasture grazed year round by cattle under holm oak woodland cleared of understorey, and on open pasture 150 m distant (Galante *et al.*, 1991). Eighteen scarabaeine species were found, with the abundance and biomass of beetles of larger size being greater in the oak woodland, and abundance and biomass of beetles of smaller size being greater in the open pasture.

Some elements or fragments of the woodland vegetation may assist large species to maintain higher populations. Horgan (2007) has suggested that, to preserve dung beetle species diversity in pasture landscapes, forest fragments should be maintained (see Chapters 11 and 13). Nichols *et al.* (2007) have also identified vegetation structure as likely to play a role in determining dung beetle community structure in fragmented landscapes, but they point to a lack of studies that measure these variables. For conservation studies and for biological control, it is important that we undertake these studies.

Acknowledgements

We are grateful to Angus Macqueen for supplying the dung beetle data from Rockhampton and Townsville between 1972 and 1986, to Ian Dadour for supplying dung beetle data from Busselton for 1990–1994 and to Geoff Monteith and Tom Weir for advice on the current species lists of beetles for Australia. We would also like to acknowledge the many former members from the CSIRO dung beetle programme and farmers who have helped to make the programme such a success over many years.

13

Dung Beetles as a Candidate Study Taxon in Applied Biodiversity Conservation Research

Elizabeth S. Nichols[1] and Toby A. Gardner[2]

[1]*Center for Biodiversity and Conservation, Invertebrate Conservation Program, American Museum of Natural History, NY, USA*
[2]*Department of Zoology, University of Cambridge, Cambridge, UK*

13.1 Introduction

To safeguard a significant proportion of the world's biodiversity, it is necessary to integrate conservation efforts with other human activities. On the one hand, protected areas represent an important, yet grossly inadequate component of a wider conservation strategy, with many endangered species (Rodrigues *et al.*, 2004) and ecoregions (Schmitt *et al.*, 2008) falling outside the existing reserve networks. On the other hand, it is also clear that conservation is not an 'all or nothing' game, and many human land-uses (e.g. regenerating secondary forests, and responsibly managed agro-forestry systems) are compatible with conserving at least part of the native biota of a given region (see Chazdon *et al.*, 2009 and references therein).

Given this reality, a central problem faced by conservation science is to understand how different types of human-modified land use are able to support the maintenance of native biota and associated ecological and evolutionary processes (Chazdon *et al.*, 2009; Gardner *et al.*, 2009). To improve our understanding of conservation opportunities in the face of rapid land-use intensification and severe funding limitations, our research methods must be effective, efficient and practical. Increasingly, applied biodiversity research challenges are met through the use of ecological indicator assemblages – suites of species whose presence and abundance in a given area provide a useful gauge for measuring and interpreting changing environmental conditions.

Ecology and Evolution of Dung Beetles, First Edition. Edited by Leigh W. Simmons and T. James Ridsdill-Smith.

In this chapter, we review the case for the use of Scarabaeine dung beetle assemblages as an ecological indicator taxon. We outline a general framework for selecting indicator taxa that can deliver robust and cost-effective information, and we build upon the insights of previous researchers to demonstrate why dung beetles represent such an appealing candidate study group (Davis *et al.*, 2001; Spector, 2006b; Spector & Forsyth, 1998; Halffter & Favila, 1993; Davis *et al.*, 2004). We draw upon existing research from a diverse array of ecosystems, but particularly tropical forests, where the majority of our own field experience lies. Finally, we consider the conservation status of dung beetles themselves and discuss some of the practical opportunities and challenges that lie ahead for the conservation of approximately 6,000 Scarabaeine species described to date.

13.2 Satisfying data needs to inform conservation practice

Two basic allied concepts form the conceptual foundation of modern conservation science – biodiversity and ecological integrity (Noss, 1990; Noss, 2004). Biodiversity can be defined as the variety of life-forms at all levels of biological systems (i.e. from genes through species, populations and ecosystems) (Wilcox, 1984). Given the intractable nature of such a broad concept, many conservation projects target only a subset of biodiversity, e.g. specific species or vegetation types that are associated with a particular conservation value.

In contrast to the notion of biodiversity, ecological integrity is defined broadly as an ecosystem's capacity to maintain biotic communities that have a structural, compositional and functional organization comparable to that of relatively undisturbed ecosystems in the same region (Karr, 1991; Karr, 1993). The maintenance of ecological integrity invokes a much broader challenge for conservation than a more narrow focus on the preservation of particular biodiversity elements (Noss, 2004; Folke *et al.*, 2004).

For the purposes of conserving biodiversity in human-dominated landscapes, distinguishing between these two concepts is important, since they underpin distinct approaches for management and monitoring (Gardner, 2010; Lindenmayer & Hobbs, 2007; Lindenmayer *et al.*, 2007). While it is often easier to draw attention to the needs of individual species (which frequently represent the cornerstone of efforts to mobilize interest and investment in conservation action), it is impossible to develop species-based approaches to satisfy the conservation requirements of all taxa. Indeed, for most of the world, we have little idea of the *identity* of species in need of conservation attention, let alone their specific resource requirements, or how they interact in important ways with other taxa. Alternative conservation approaches, that instead focus on the integrity or condition of human-modified ecosystems themselves, can provide an arguably much richer source of information for which to guide ecological management (Gardner, 2010; Angermeier & Karr, 1994)

Nevertheless, measuring and interpreting changing patterns of ecological integrity remains a major scientific challenge. We generally have a very poor understanding of the processes and functions that are necessary to maintain resilient ecological systems (Angermeier & Karr, 1994; Naeem, 2008). It is often the case that carefully selected biological species can themselves provide the most effective indicators of ecological integrity. This is because we have some understanding of the factors that drive changes in species distribution and abundance patterns, and because individual organisms (compared to process rates) are often less expensive to monitor, they may provide a more effective gauge of the impact of human activities than ecosystem processes or functions themselves (Angermeier & Karr, 1994).

Taxa that serve this function are termed 'ecological indicators' (McGeoch, 1998; McGeoch, 2007) or, more specifically, 'ecological disturbance indicators' (Caro & Girling, 2010; Gardner, 2010). The purpose of ecological disturbance indicators is to provide reliable and interpretable information on the ecological consequences of human activities (compared against some acceptable reference condition) for a measured component of biodiversity. In a practical sense, the concept of such taxa is to help us to move beyond a simple classification of areas based on physical habitat measurements and to translate the meaning of land use changes into measures that capture changes in the ecological integrity of the system. This is only possible for taxa that can be sampled cost-effectively, and for which we have some *a priori* ecological understanding of disturbance response patterns.

The concept of ecological disturbance indicators is quite distinct from the notion of 'biodiversity indicators' (McGeoch, 2007), 'biodiversity surrogates' (Moreno *et al.*, 2007) or 'cross-taxon response indicators' (Caro & Girling, 2010), all of which depend upon the (largely unfounded) belief that observations of one species group can provide reliable inferences about changes in other (unstudied) species (Cushman *et al.*, 2010).

Despite the existence of various selection criteria for ecological disturbance indicators (Pearson & Cassola, 1992; Pearson, 1994; Noss, 1990; McGeoch, 1998; Kremen, 1992; Halffter & Favila, 1993; Greenslade & Greenslade, 1987; Davis *et al.*, 2004), very few studies have adopted a systematic approach that accounts for both the practical and theoretical factors influencing the value of such data in addressing conservation problems (Spector, 2006b; McGeoch, 1998; Gardner, 2010). Figure 13.1 outlines three basic criteria that can be used systematically to assess the potential of any candidate ecological disturbance indicator group, namely *viability*, *reliability* and *interpretability* (Gardner, 2010):

- First and foremost, it is necessary for any candidate indicator group to be *viable* for study. Do the necessary field, laboratory and taxonomic expertise exist to ensure that a project has a viable chance of success?
- Second, a potential candidate indicator group must provide a *reliable* and practically relevant measure of the ecological consequences of human activity. By reliable, we mean both responsive to human-induced environmental heterogeneity at spatial and temporal scales commensurate with human management practices, and also measurable within a standardized sampling protocol and limited budget. Ideally, such a species group will be comprised of individual species that vary significantly in their sensitivity to human activities (i.e. response diversity, *sensu* Elmqvist *et al.* 2003), thereby allowing the evaluation of a wide range of land management practices.
- A third general criterion for selecting ecological disturbance indicators is that biodiversity sample data must be interpretable, insofar as we have sufficient

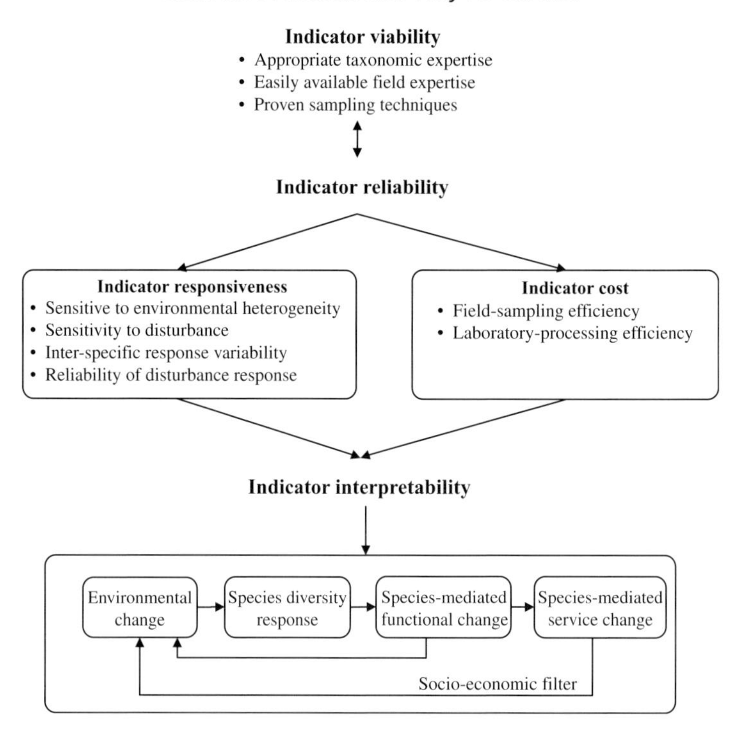

Fig. 13.1 A general framework for selecting high-value ecological disturbance indicator taxa for monitoring changes in ecological condition.

prior ecological knowledge to understand something about observed patterns of abundance and occupancy and to be able to link changes in these patterns to measured environmental variables. An understanding of such cause-effect relationships is ultimately necessary to develop a predictive capacity for linking human activities to changing patterns of ecological integrity (Landres *et al.*, 1988). In situations where species-mediated ecological functions have demonstrable effects on human wellbeing, we can further use biodiversity sample data to evaluate the consequences of human activities for the provision of ecosystem services – an area that remains a major knowledge gap in conservation science (Nicholson *et al.*, 2009).

13.3 The role of dung beetles in applied biodiversity research in human-modified landscapes

While no single group of species can fully satisfy the three criteria of viability, reliability and interpretability, Scarabaeine dung beetles present a very strong candidate. Accordingly, they have received substantial interest in applied biodiversity research (Davis *et al.*, 2001; Spector, 2006b; Spector & Forsyth, 1998; Halffter & Favila, 1993). In this section, we employ this same three-tiered selection framework (Figure 13.1) to discuss why.

13.3.1 Dung beetles as a viable candidate for biodiversity research

Comparable and standardized estimates of local species distribution and abundance form the cornerstone of applied biodiversity research. The employment of standardized, cost-effective sampling approaches can go a long way towards maximizing the information return gained from the field surveys that are typically time-limited and budget constrained (Gardner *et al.*, 2008a). In this regard, dung beetles are appealing because representative samples of a given locality can be collected within days rather than weeks. This contrasts starkly with the challenges associated with sampling many other species groups. In the case of terrestrial vertebrates, for example, rapid assessment of the distribution of species or patterns of abundance is often hampered by high detection biases and a lack of adequately trained experts (Landres *et al.*, 1988; Feinsinger, 2001).

Standardized sampling methods can also help to ensure a minimum level of methodological consistency to support the kinds of meta-analyses that are needed to draw generalizations at regional or global scales. While, to date, no single dung beetle collection protocol has been uniformly adopted, there is considerable coherence in the methodologies employed across different studies. The vast majority of comparative dung beetle studies are conducted with human dung-baited pitfall traps (comprising a simple collecting vessel sunk flush with the ground, with the bait suspended above). Bovine or pig dung is often used in African and Asian biogeographical contexts (Davis & Philips, 2005; Boonrotpong *et al.*, 2004), and trap arrays of faeces and carrion (and occasionally fruit and fungus) are increasingly popular in the Neotropics (Larsen *et al.*, 2008; Horgan, 2008; Escobar *et al.*, 2007).

Efforts to investigate the quantitative efficiencies of different collecting protocols (Larsen & Forsyth, 2005) in order to create an actual standardized sampling protocol are under way (ScarabNet, 2007). It is important to note that not all Scarabaeine beetles are attracted to baited pitfall traps. Where the objective is to conduct a species inventory, it is necessary to employ a suite of complementary methods, including hand collection and passive flight intercept traps (FITs) (Davis *et al.*, 2000). That said, the global use of human dung-baited pitfall traps has created a truly tremendous potential data source from which to explore patterns in beta-diversity (Viljanen *et al.*, 2010), as well as community (Nichols *et al.*, 2007) and trait-based responses (Nichols *et al.*, unpublished manuscript) to land use change.

Once material has been collected, the ability to identify specimen material, reliably and consistently, to the species level is a critical requirement for any study group used in biodiversity research. Taxonomic challenges represent a major barrier to the development of time-efficient and cost-effective field research projects on the ecology of many hyper-diverse invertebrate groups (Samways, 2002). While Scarabaeinae dung beetles are diverse, levels of both local and global diversity are tractable. A global taxon database managed by the Scarabaeine Research Network currently has just over 5,700 valid species names from 225 genera (ScarabNet, 2009), and an ever-increasing variety of identification tools are becoming available to facilitate the accurate processing of new material by specialist and non-specialist workers (Vaz-de Mello & Edmonds, 2007; Mann, 2008; Larsen & Génier, 2008).

13.3.2 Dung beetles as reliable indicators of environmental change

Given our current understanding of the sensitivity in response of dung beetles to: (i) natural environmental gradients as a basis for disentangling the importance of human disturbances in driving observed patterns; (ii) land-use change and intensification gradients in tropical forests (an ecosystem that has received particular research attention within the dung beetle research community in recent years); and (iii) declines in resource diversity and availability as a consequence of mammal hunting, dung beetles can provide a valuable gauge of changes in ecological integrity in human-modified systems. Our ability to collect cost-effective and representative samples of underlying patterns of Scarabaeinae distribution and abundance further enhances their utility as an ecological disturbance indicator taxon.

Dung beetle response patterns across natural environmental gradients

The impacts of human activities on biodiversity – whether local, regional or global, negative or positive – are invariably conditioned by the natural environmental and biophysical characteristics of the spatial scale of interest. Dung beetles are highly responsive to environmental heterogeneities across multiple scales and levels of ecological organization (Hanski & Cambefort, 1991). These associations provide the basis for understanding the structure of local species assemblages in both natural and human-modified ecosystems.

At the regional scale, dung beetle associations with specific climate and edaphic conditions often demonstrate strong patterns of fidelity or biogeographical distinctness (Davis, 1993; Davis, 1994; Davis, 1997; Davis *et al.*, 2000; Hanksi & Krikken, 1991; Davis & Dewhurst, 1993). Indeed, biogeographical circumstances play an important role in filtering those species that are capable of surviving in human-modified lands (see Chapter 12). This was very clearly demonstrated by Davis & Phillips (2009), who found high densities of dung beetles inhabiting the matrix of plantation and farmland in fragmented forest landscapes of the Ivory Coast that are dominated largely by a subset of species originating from neighbouring savannah areas (Figure 13.2). Similarly, Scarabaeine communities in areas of introduced pastures adjacent to native Brazilian savannah (Cerrado) are more likely to overlap in their community structure and composition (Vidaurrre and Louzada, unpublished manuscript) than communities found in introduced pastures in Central America, due to the historical lack of such native open areas in the latter region (Horgan, 2007).

Understanding how dung beetle assemblage structure changes, in response to natural vegetation gradients at finer spatial scales, can also provide vital clues as to the appropriate spatial scales at which to assess species-abundance relationships in disturbed habitats. Remarkably high rates of species turnover have been found across neighbouring habitat types, often over very short distances – hundreds of metres.

For example, from directly adjacent savannah and forest sites in Bolivia, Spector & Ayzama (2003) reported that 24 of the 50 most common species were restricted to a single vegetation type. Only two species were present in both vegetation types, with patterns being remarkably consistent over time. Working in Sabah Borneo, Davis *et al.* (2001) reported not only a clear distinction in species composition for

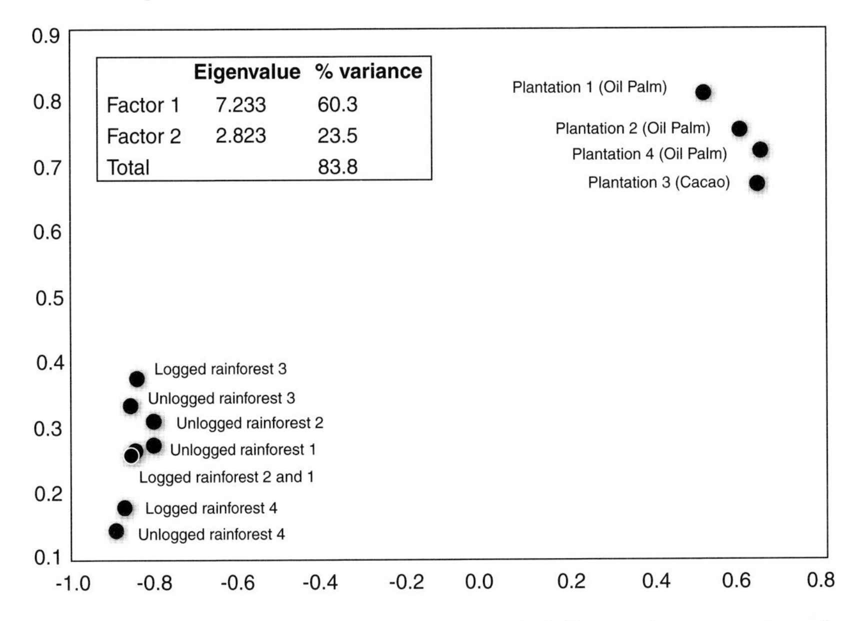

Fig. 13.2 Ordination plot demonstrating the statistical distance between plantation and rainforest dung beetle faunas along the southern edge of Ankasa Conservation Area, south-west Ghana. Figure redrawn from Davis & Philips (2005).

dung beetle assemblages sampled in riverine and interior rainforest, but also a finer-scale subdivision within riverine forest, with species being clustered into river-edge, river-bank, and riverine non-edge/bank components, each occupying a slightly different microclimate.

Edaphic differences likely underlie many of the patterns associated with vegetation changes as described above (Davis, 1997). As soil structure and consistency can directly affect reproductive site selection (Vessby & Wiktelius, 2003) and reproductive success of individual dung beetle species (Lumaret & Kirk, 1987; Sowig, 1995), soil plays a direct role in structuring local dung beetle assemblages and determining local population viabilities, even in the absence of any changes in vegetation (Nealis, 1977; Doube, 1991; Lumaret & Kirk, 1987). As with vegetation, the fidelity of beetle-soil associations varies across species. Some species demonstrate obligate associations to certain soil types, while with others there are marked differences in species density, and for still others there is little change in abundance (see Doube, 1991 and references therein for a range of South African examples).These are associations that can be manifest over a range of spatial scales.

Dung beetle response patterns across land use intensification gradients in tropical forests

Understanding differences in species-response patterns to a given type of human-associated disturbance or land use change is essential if we are to generate reliable information on the causes of biodiversity loss. In some areas (e.g. tropical land use change), our understanding has benefited from both a large number of individual studies and also from meta-analyses of key similarities and differences in community

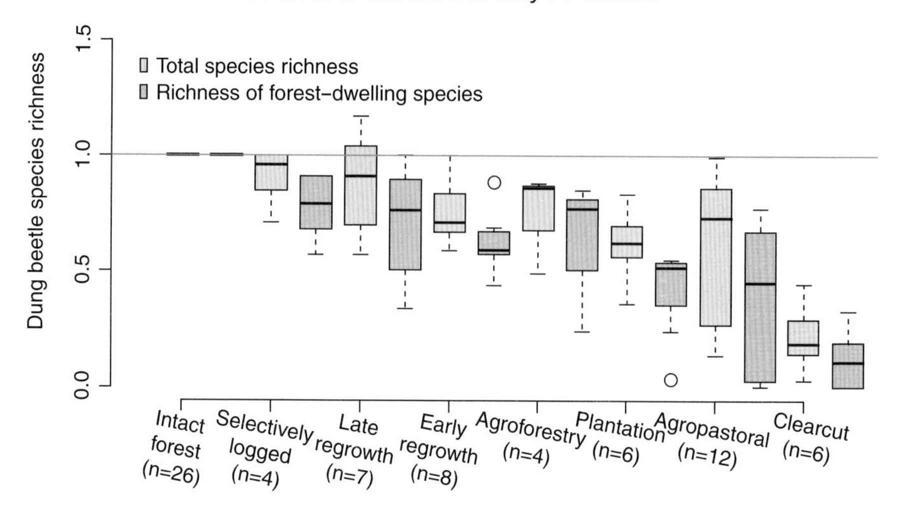

Fig. 13.3 Generalized declines in dung beetle species richness across common anthropogenic land uses of moist tropical forest. Redrawn from a global meta-analysis by Nichols *et al.* (2007) that combined 26 independent studies of species response patterns to forest modification. Here the total number of species recorded in any given land use (light gray) is distinguished from the proportion of species recorded in that land use that were also captured in that study's intact forest (dark gray). Within a given land use, the difference between this 'complete' and 'intact forest' beetle assemblage is an indicator of turnover from forest-restricted to open-habitat adapted species.

response patterns. The relative lack of such synthetic assessments of dung beetle responses to other disturbance types (i.e. ivermectin impacts on agropastoral dung beetles) more reflects a lack of research attention than a lack of importance *per se*.

In a recent meta-analysis, Nichols *et al.* (2007) brought together 33 individual studies to synthesize current knowledge concerning the responses of dung beetles to land use change in human-modified tropical forests (Figure 13.3). Land uses with a high degree of forest cover, such as selectively logged forest, secondary and agroforests, were shown to support dung beetle communities with similar community attributes to those found in intact tropical forest (Pineda *et al.*, 2005). This indicates that these more structurally complex habitats could make an important contribution to mitigating biodiversity loss from deforestation (Wright & Muller-Landau, 2006; Dunn, 2004).

In contrast, heavily modified habitats with little or no tree cover were shown to support species-poor dung beetle communities with high rates of species turnover, dramatically altered abundance distributions, and smaller overall body size compared with species found in intact forest (see Chapter 11). Dung beetle communities in highly modified habitats are commonly characterized by the hyper-abundance of a few small-bodied species, including species in the genera *Trichillium* in central Amazonia (Scheffler, 2005) and *Tiniocellus* in West Africa (Davis & Philips, 2005; Davis & Philips, 2009).

Nichols *et al.* (2007) found that the extent of species turnover is generally greatest in open managed fields (annually cropped fields and cattle pastures) and is likely dependent upon differences in the landscape or regional context (Howden &

Nealis, 1975; Davis *et al.*, 2000). The spatial extent of disturbance may also play a critical role in determining estimates of species loss following disturbance (Shahabuddin *et al.*, 2005; Avendano-Mendoza *et al.*, 2005).

Nichols *et al.* (2007) also found that forest fragments displayed similarly consistent dung beetle community response patterns to those found in human-modified areas, with fragments tending to be characterized by reduced levels of richness, abundance, community similarity and species evenness relative to intact forest. Most often, these parameters vary positively with fragment size, although the make-up of the wider landscape-matrix can also play a very important role (Gardner *et al.*, 2009).

Dung beetle response patterns to shifting dung resource availability

Because of their broad dependency on mammalian faeces as a larval and adult food resource, dung beetles are expected to be highly sensitive to shifts in mammal community composition and structure (Nichols *et al.*, 2009). Despite substantially less documentation on their response to resource shifts than is available for changes to physical vegetation structure, there is strong evidence for global patterns of dung beetle co-decline and co-extinction following changes in native mammal assemblages through persistent human hunting pressure (Andresen & Laurance, 2007) and altered grazing regimes (Carpaneto *et al.*, 2005).

In tropical forests, persistent mammal hunting often affects even the most remote of protected areas (Peres & Lake, 2003) and uninhabited areas hundreds of kilometres from remote urban centres (Parry *et al.*, 2010). As, typically, large frugivorous primates and ungulates are preferentially hunted first, persistent hunting can result in massive local reductions in overall mammal biomass (Jerozolimski & Peres, 2003). The resultant declines in local abundance of ungulates and large primates have the effect of reducing the availability of large amounts of moist faeces and of possibly increasing the dry, pelleted dung produced by non-hunted rodents, small armadillos and small primates (Peres & Dolman, 2000). Currently, only a single study has undertaken a preliminary investigation of such hunting-mediated resource shifts. It reported significant declines in dung beetle species richness, combined with a sharp reduction in individual abundances for over two-thirds of beetle species (Andresen & Laurance, 2007).

The cost-effectiveness and reliability of dung beetles as ecological disturbance indicator taxa

Many species of dung beetle are highly susceptible to baited pitfall traps, leading to a relatively low level of false-negative recordings such as plague so many biodiversity studies, especially those focused on vertebrates (Tyre *et al.*, 2003). In a similar sense, dung beetle sampling using baited pitfall traps is relatively insensitive to variability in study design with respect to the number and distribution of individual trapping events (Figure 13.4).

Together with the intrinsic sensitivity to environmental change of the species themselves, these characteristics render dung beetle sample data a particularly valuable source of information for applied ecological research. Moreover, field sampling techniques and specimen processing are comparatively low-cost, underpinning the fact that field research on dung beetles is far more cost- (and time-)

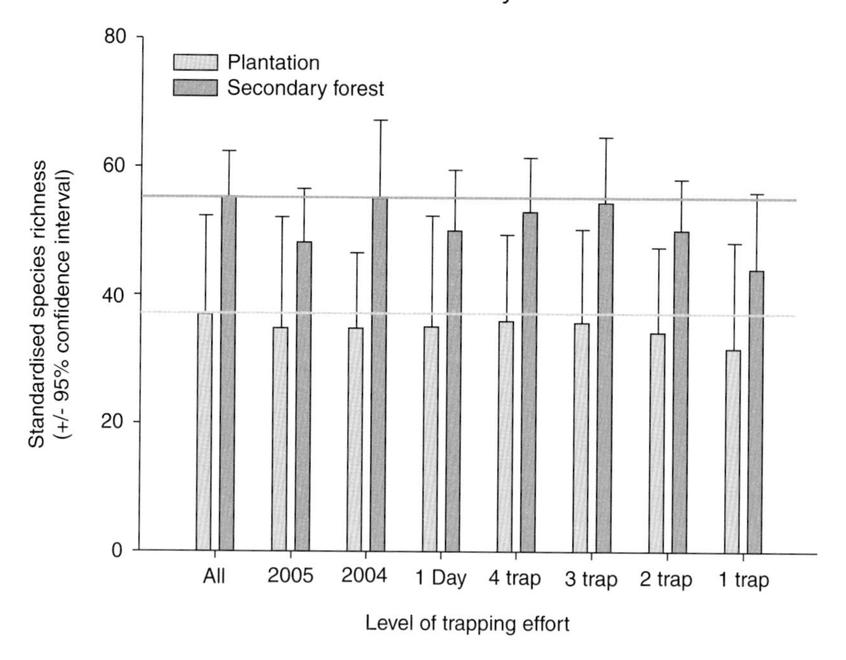

Fig. 13.4 Robustness of standardized effect sizes to differences in sampling effort. Plot shows the reduction in number of dung beetle species from primary forest to plantations (light shading) and primary to secondary forest (dark shading) in Jari, Brazilian Amazonia. Different bars represent the same measurements, using subsets of the data that draw on different levels of sampling effort, including only using data from one year (2004 or 2005), only using one temporal sub-sample from each site visit (rather than two) and only using data from a decreasing number of spatial sub-samples (1, 2, 3 or 4 traps rather than 5). Redrawn from Gardner (2010).

effective than many other commonly studied species groups (Gardner *et al.*, 2008a) (Figure 13.5).

13.3.3 Interpreting disturbance response patterns: application of a trait-based framework for ecological research

The extensive (and growing) knowledge of Scarabaeine ecology provides an invaluable basis from which to draw meaning from sample data and to isolate the drivers that are responsible for observed changes. Understanding the long-term drivers and consequences of biodiversity change ultimately requires a research framework within which we can translate changes in abundance or species composition into an enhanced understanding of ecological processes and, ultimately, changes in ecological integrity of the study system (Chapin *et al.*, 2000). Species traits directly mediate fluxes of energy and material as well as interact with the abiotic variables that indirectly regulate ecological process rates. Consequently, within a given community, the species present, their relative abundances, the interactions among species, and the temporal and spatial variation in all of the

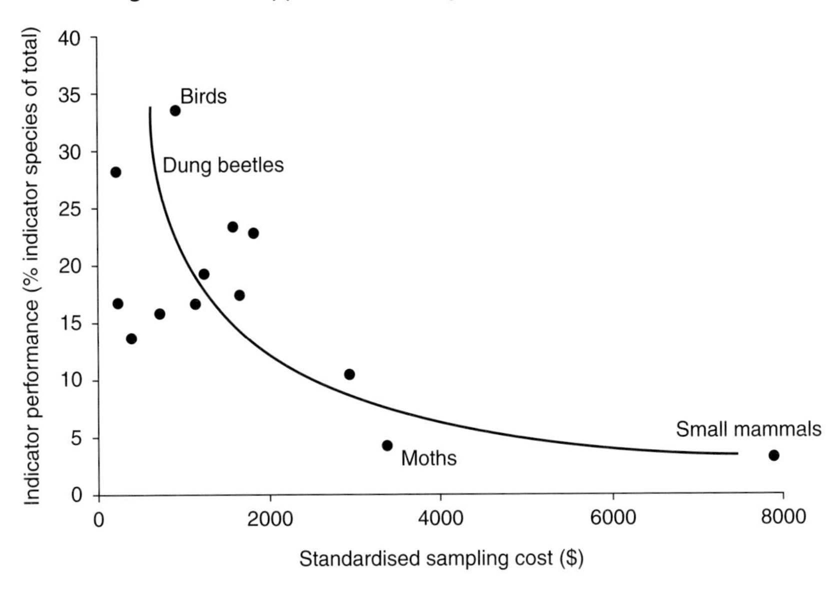

Fig. 13.5 Comparing patterns of indicator value and standardized survey cost across 14 taxa sampled in an area of primary rainforest in Jari, Brazilian Amazonia. From Gardner *et al.*, 2008b.

above, influence ecological functioning by determining the overall expression of organismal traits within a local community. Only by developing an integrated understanding of the ways in which species respond to a changing environment, and how such changes feed back into altered ecological processes, is it possible to scale up inferences drawn from individual studies into a broader understanding of how human activities can influence the structure and function of ecosystems (Didham *et al.*, 1996).

While animal ecologists have traditionally used species traits to predict extinction risk (Cole, 1954; Davidson *et al.*, 2009), the finer-grained natural history information typically available for plants has driven the use of a more sophisticated framework that distinguishes between response and effect based traits (Suding *et al.*, 2008; Lavorel & Garnier, 2002). Response traits are those associated with a given species' Grinnelian niche; they relate species' resource or environmental needs with species performance. Impact traits are expressions of an Eltonian niche concept, associated with the impacts of a species on its environment, and they are often measured in terms of biotic interactions or abiotic consequences (Devictor *et al.*, 2010).

Taking the view that local species communities are the product of a hierarchy of biotic (predation, competition, resource availability, habitat configuration) and abiotic (climate, geology) filters that determine the make-up of a regional species pool, species traits can be used to develop a mechanistic and predictive understanding of how species will respond to the filters represented by human impacts, and of what these responses mean in ecological terms. The degree to which these response and impact traits overlap has enormous implications for the long-term persistence of biodiversity and ecological integrity in human-modified landscapes (e.g. Larsen *et al.* 2005).

Dung beetle ecologists have long documented a series of easily measured morphological and behavioural differences among species and individuals. These more easily discernible 'soft' traits (Hodgson *et al.*, 1999) often represent groups of underlying, yet more elusive, 'hard' traits that are challenging to measure, yet represent actual functional mechanisms associated with fitness and ecological function (Table 13.1; see also Chapter 10).

Dung beetle ecologists have also tended to focus on two related filters as drivers of changes in dung beetle communities: changes in vegetation structure and associated microclimates; and changes in the availability of food and breeding resources (typically dung, but also carrion and other materials) (see Chapter 11). Disentangling how the soft and hard species traits that are manifest in a given regional species pool interact with novel filters represented by local human activities, and are further conditioned by the wider biogeographical context, is at the cutting edge of ecological function research. It provides an increasingly valuable basis from which to draw meaning from biodiversity field data.

Trait-based responses to changes in vegetation structure and associated microclimate

Understanding how inherent trade-offs between different life history traits (e.g. body size and reproductive rate or heat-dissipation capacity) relate to differences in the structure and composition of dung beetle assemblages found across vegetation and resource gradients can greatly facilitate the interpretation of observed species patterns. Human-induced changes to the structure and complexity of forest canopies and understorey vegetation can impose dramatic changes on local microclimatic conditions, increasing levels of radiant heat, light intensity and air and soil temperature while decreasing humidity (Halffter & Edmonds, 1982; Duncan & Byrne, 2000). The combined effects of a relatively narrow physiological tolerance to temperature changes (a feature shared by many species; Chown, 2001, and see Chapter 10), together with the influence of changes to solar radiation on adult activity patterns (Lobo *et al.*, 1998) and soil moisture content on larval survival (Sowig, 1996b), likely represent strong environmental filters which restructure local dung beetle assemblages following disturbance (Table 13.1).

Dung beetles span four orders of magnitude in size (Larsen *et al.*, 2008). Large body size is increasingly cited as a response trait that confers greater risk of local extinction in the face of forest fragmentation (Klein, 1989; Larsen *et al.*, 2005), conversion to agriculture (Shahabuddin *et al.*, 2010; Gardner *et al.*, 2008b; Nichols *et al.*, unpublished data) and deforestation (Scheffler, 2005). These effects may occur as a consequence of physiological intolerance to thermal stress, size-dependent response to declining diversity or abundance of dung resources, or their combined effects. As body size can have a strong effect on community assembly order (Horgan & Fuentes, 2005) and interspecific competition (Horgan, 2005), the early loss of large-bodied beetles may have significant secondary consequences for community structure and subsequent patterns of ecological function (Larsen *et al.*, 2005; Slade *et al.*, 2007).

Table 13.1 Dung beetle traits relevant to species responses to environmental change and impacts on ecosystems. The leftmost column indicates the easily measured 'soft' traits commonly surveyed as part of natural history and ecological field studies. All other columns briefly describe some of the example mechanisms or 'hard' traits for which soft traits are a presumed proxy. Response traits are those associated with a given species' resource or environmental needs with species performance, while impact traits are associated with the impacts of a species on its environment. Nidification behaviour can be broken down into nest location (superficial or subterranean), position relative to food source (adjacent, distant or within) and complexity (simple or compound). Though not immediately intuitive, nidification behaviour may be functionally linked to larval nutritional needs, as some species are restricted to nesting with dung of a given morphology or type. These dung types may also be deposited only at certain times of day, linking diel activity with larval nutritional needs

	Mechanism: hard trait				
	Response		*Impact*		
Soft trait	Habitat disturbance	Food resource shift	Waste removal	Seed dispersal	Secondary productivity
Body size	Thermoregulation: adult metabolism Conservation: adult nutritional needs Avoidance: physical escape	Reproduction: larval nutritional needs Reproduction: reproductive rate Reproduction: resource detection strategy	Consumption: brood ball size Consumption: adult nutritional needs Consumption: burial depth	Incidental: brood ball size Incidental: burial depth Incidental: cumulative seed dispersal	Production: biomass
	Dispersal: colonization capacity	Conservation: resource detection strategy	Consumption: cumulative waste removal		
Diel activity	Thermoregulation: adult metabolism Avoidance: palatability Avoidance: phenology	Reproduction: larval nutritional needs Reproduction: phenology	Probability of presence: phenology		Production: predation
Wing loading	Dispersal: colonization capacity	Reproduction: resource detection strategy	Probability of presence: colonization capacity	Production: predation	Production: predation

(continued)

Table 13.1 *(Continued)*

	Mechanism: hard trait				
	Response		*Impact*		
	Avoidance: physical escape	Conservation: resource detection strategy			
Nidification behaviour	Reproduction: larval thermoregulation Avoidance: physical escape	Reproduction: larval nutritional needs Reproduction: reproductive rate	Consumption: burial depth Incidental: burial depth	Incidental: probability of seed removal	Production: predation
Brood care		Reproduction: reproductive rate	Consumption: cumulative waste removal	Incidental: cumulative seed dispersal	
Habitat breadth	Thermoregulation: adult metabolism Reproduction: larval thermoregulation	Reproduction: larval nutritional needs Reproduction: resource detection strategy			
	Reproduction: larval nutritional needs	Conservation: resource detection strategy			
	Conservation: adult nutritional needs				
Dietary breadth	Conservation: adult nutritional needs Reproduction: larval nutritional needs Reproduction: adult nutritional needs	Reproduction: larval nutritional needs	Probability of presence and waste removal		

** Brood care is binary, fully absent or present for most species and is related to reproductive rate.
*** A mobility trait is one that influences the probability of a mobile species being present in a given site. As this is intimately connected with its potential ecosystem impact, these traits can be considered an impact trait, an important difference between mobile and sessile (plant-based) trait frameworks.

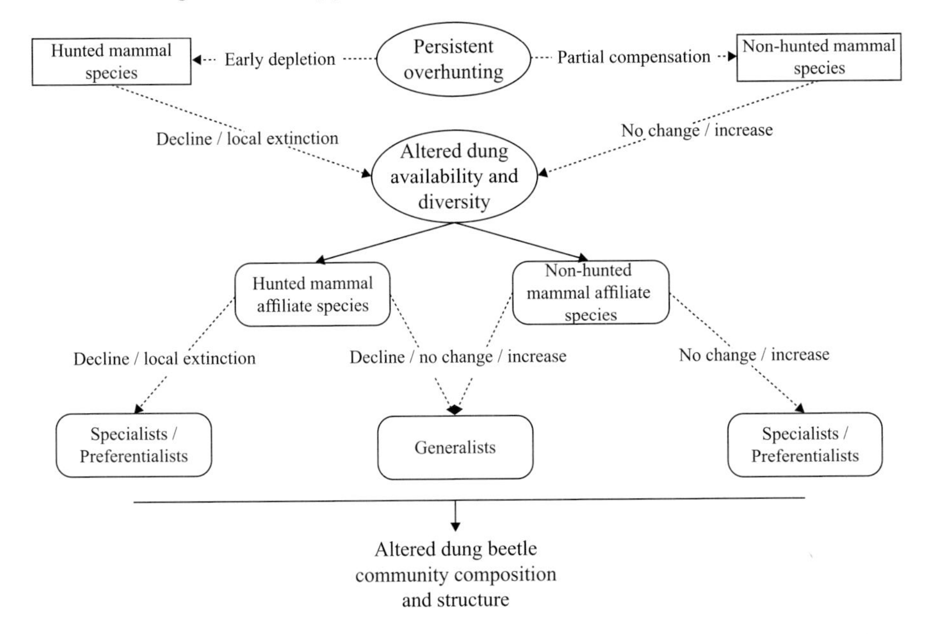

Fig. 13.6 Conceptual diagram of the mechanisms likely to drive changes in coprophagus dung beetle community structure as a consequence of hunting-mediated resource shifts. Modified from Nichols *et al* (2009).

Trait-based responses to changing resource availability

Variability in both the amount and diversity of dung (and other resources such as carrion) are also important factors in explaining the distribution of local dung beetle assemblages (Nichols *et al.*, 2009; Figure 13.6). There is a growing body of evidence linking the nutritional quality of different dung types with variable parental investment in brood ball size (Kanda *et al.*, 2005; Hunt & Simmons, 2004; and see Chapter 8) and several aspects of dung beetles development, from adult body size (Emlen, 1994) to diet-induced allometric plasticity in certain polyphenic species (Rowland & Emlen, 2009).

The strength of any cascading effects on resource dependent dung beetle assemblages will depend upon species-specific differences in dietary specialization, dietary plasticity over ecological timescales and the relationship between diet and fitness (see Chapters 7 and 8). Yet further linking these physiological 'hard trait' responses to soft, easily measurable traits, as well as to short and long-term population dynamics, remains a major research challenge as we generally lack a species-specific understanding of dung beetle diet breadth and plasticity (Holter & Scholtz, 2007; Nichols *et al.*, 2009).

This challenge is particularly great in tropical forests, where adult dung beetle diets appear to be more diverse (Gill, 1991; Hanksi & Krikken, 1991; Cambefort & Walter, 1991) and mammalian and food resources are very hard to study. Most coprophagous dung beetles may be attracted to several dung types, though adult attraction does not necessarily equate to optimal larval nutrition (Barbero *et al.*, 1999; Kanda *et al.*, 2005; Hunt & Simmons, 2004). This somewhat compromises the reliability of the information that can be gained from 'buffet' style dung-preference studies.

Even generalist species have been shown to discriminate differences in water and/or fibre content (Verdú & Galante, 2004; Lopez-Guerrero & Zunino, 2007), nutritional value (Verdú & Galante, 2004), dung shape (Gordon & Cartwright, 1974) and dung size (Peck & Howden, 1984). Still others are highly specialized or obligate to a single host species (Larsen *et al.*, 2006; Cambefort, 1991a).

Finally, a large variety of non-mammalian dung food resources are utilized by many dung beetle species, including bird, insect and reptile faeces, carrion, fungi and rotting fruits (Gill, 1991; Falqueto *et al.*, 2005; Young, 1981; Halffter & Halffter, 2009; and Chapter 2). Understanding the importance of these different resources to adult and larval feeding of dung beetles is the key to understanding the long-term resilience of dung beetle assemblages in the face of complex patterns of environmental change.

Large body size is also likely to be linked to a high level of sensitivity to declining dung availability. Persistently hunted systems have lower large mammal biomass (Peres & Palacios, 2007) and are likely to support disproportionately fewer large-bodied, active-foraging dung beetle species. The fitness of these larger-bodied species may be compromised by reduced encounter rates of the large individual resource patches needed to construct viable brood balls (Holter & Scholtz, 2007; Nichols *et al.*, 2009), or by an overall reduction in the density of larger dung pats (generated by larger vertebrates) because of elevated levels of interspecific competition within each pat (Horgan & Fuentes, 2005).

By contrast, small-bodied 'sit and wait' style foragers that dominate tropical forest beetle assemblages (Gill, 1991) are likely to be less affected by hunting, because they are capable of exploiting smaller and more ephemeral resource pulses from small-bodied vertebrates that are of little value to humans as food. Preliminary work along an extensive gradient of hunting pressure in the western Brazilian Amazon suggests that persistently hunted areas have fewer large-bodied species, and individuals at the smaller range of body size within a species, relative to areas with reduced current-day hunting pressures (Nichols, unpublished data).

To move towards a more predictive framework that links biodiversity responses to ecological processes in human-modified landscapes, a clear understanding of how response traits interact with different environmental filters is key. Yet, in the case of dung beetles (and doubtless other groups), habitat modification seldom occurs without accompanying changes to the availability and diversity of (dung) resources, resulting in multiple, potentially interacting filters that will interact with response traits and influence the structure and composition of local dung beetle assemblages (Table 13.1, also see Lavorel & Garnier, 2002).

The effects of vegetation structure and subsequent microclimate appear to be consistently stronger determinants of beetle presence or absence than availability of food resources for the majority of species. However, studies that simultaneously track species occupancy and abundance across habitat and resource gradients (i.e. Macagno & Palestrini, 2009; Jay-Roberts et al., 2008; Barbero *et al.*, 1999) are critically important to disentangling these two sets of explanatory factors.

Widespread changes in dung availability and diversity associated with changing rural economies in European Alpine and Mediterranean regions have driven altered composition and abundance of native (Carpaneto *et al.*, 2005; Lumaret *et al.*, 1992; Carpaneto *et al.*, 2007) and introduced herbivores (Carpaneto *et al.*, 2005; Hanski *et al.*, 2008; Jay-Robert *et al.*, 2008). These resource shifts are often accompanied by extensive reforestation or succession in areas of abandoned pastures. In turn, these changes in landscape composition have led to severe declines in native open-habitat-associated dung beetle faunas (Macagno & Palestrini, 2009; Lobo, 2001).

Research aimed at partitioning these effects has found that, beyond local declines in overall dung availability with the shift from livestock to native grazers, the habitat preference of native pigs and deer for closed habitats has the effect of further reducing dung resources, limiting their distribution to woody successional areas that are suboptimal for the region's open-habitat restricted dung beetle fauna (Jay-Robert *et al.*, 2008; Barbero *et al.*, 1999; Macagno & Palestrini, 2009). Within biogeographical regions where open habitats represent predominantly novel land cover types, often the few native, open-habitat-tolerant dung beetle species demonstrate sufficient dietary plasticity to thrive on introduced livestock faeces (Louzada & Carvalho e Silva, 2009; Hanski *et al.*, 2008).

Trait-based correlates of dung beetle mediated ecological functions

Overall patterns of ecological function are determined by the combination of species-level impact traits present in a local community, species abundance distributions and interactions among species. Ecological functions can be classified into three categories, based on the mechanism of energy flow across trophic levels, namely consumption, production and incidence functions (Figure 13.7a). Consumption functions are those that result from an organism's consumption of resources (typically from one trophic level below, i.e. herbivory), while production functions result from the secondary production of organismal biomass (Clark, 1946). However, many important ecological functions come neither from direct consumption nor secondary production; these are best considered to be incidental functions or by-products of these two primary function classes. For dung beetles, they include functions such as secondary seed dispersal (a by-product of faeces consumption) and alteration of parasite transmission rates (often a joint product of faeces consumption and biomass production through dung beetle predation; Figure 13.7a).

As the selection pressures for these three function classes varies, their corresponding impacts may be quite different. Large body size is likely the principal impact trait related to the consumption function of waste removal for Scarabaeine dung beetles, and multiple lines of evidence suggest that larger-bodied beetles can remove disproportionately more dung than smaller-bodied beetles (Slade *et al.*, 2007; Nichols *et al.*, 2008). While nesting strategy itself may confer less variation in brood ball size, many of the largest-sized beetles in a community tend to have tunnelling morphologies (e.g. Slade *et al.*, 2007).

Impact traits relative to the production of dung beetle biomass are most likely to be those that motivate adult beetles to select the appropriate dung type or to create sufficiently large brood balls (Table 13.1). Impact traits most relevant to incidental functions are likely to be highly context and function-specific. For example, nesting strategy plays a stronger role in secondary seed dispersal than in disruption of parasite transmission.

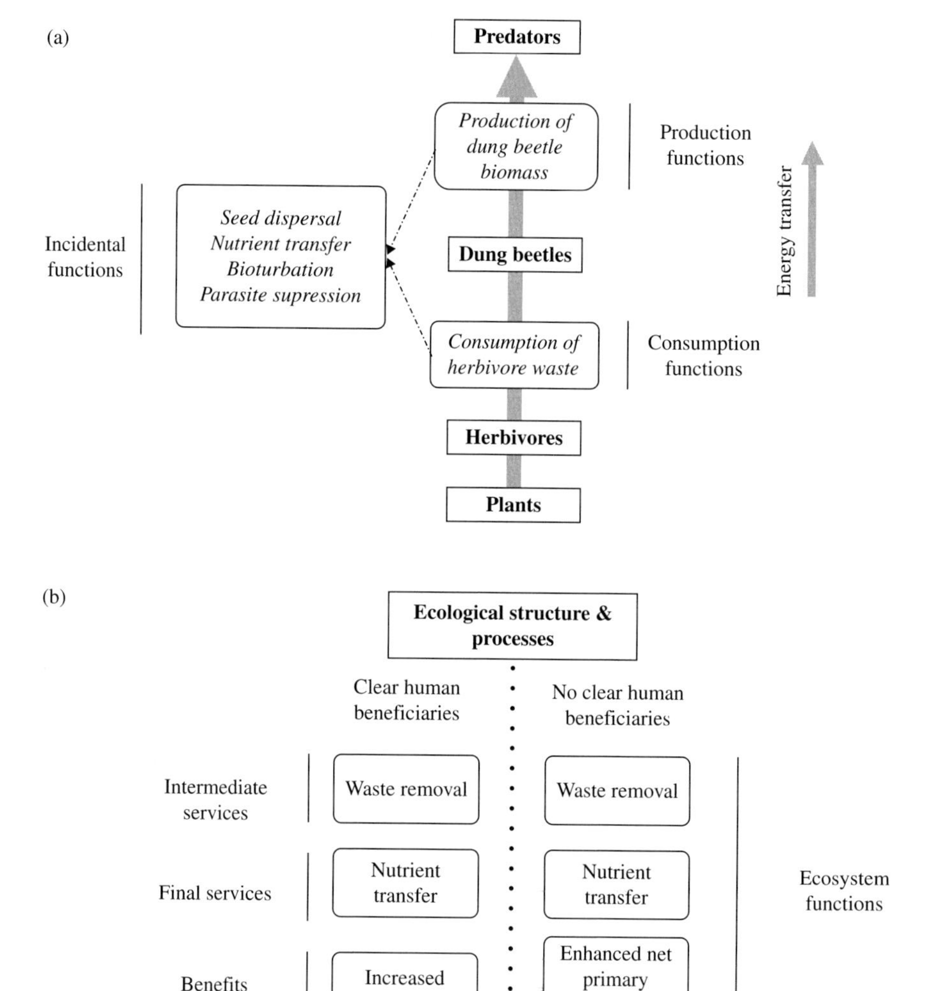

Fig. 13.7 Example of how the ecological functions mediated by Scarabaeine dung beetles can be categorized as consumption, production and incidental functions (a), and an illustration of the differences between ecological functions and ecosystem services, and how one single set of dung beetle-mediated ecological functions may be categorized as services, given the appropriate socio-economic context (b).

Challenges in a trait-based understanding of dung beetle species-function relationships

Scarabaeinae beetles provide an effective mobile animal study system with which to decipher biodiversity-ecosystem function (BEF) relationships (Nichols *et al.*, 2008). Experimental work by Slade *et al.* (2007) to isolate the functional contributions of different elements of a dung beetle assemblage represented a novel step in this direction. They reported that waste and seed removal rates increased with the number of different functional groups represented, although individual functional groups had largely idiosyncratic relationships with function rates, suggesting that:

(i) dung beetle functional groups can be highly complementary;
(ii) maximum function is seldom achieved without the representation of all functional groups;
(iii) any loss of any aspect of dung beetle biodiversity appears to precipitate some loss of functional capacity (Slade *et al.*, 2007).

Moving from community-level patterns, such as these, towards a mechanistic understanding of the functional consequences of individual species responses, will require an understanding of the impact of functional linkages and trade-offs between various traits that govern resource use (Goldberg *et al.*, 2008). One drawback to the current emphasis on the collection and analysis of more easily measurable soft trait information to explain species response and function (Larsen *et al.*, 2005; 2008; Slade *et al.*, 2007) is the one-to-many relationship of functional linkages between many soft and hard traits (Table 13.1; and see Lavorel & Garnier, 2002). This challenges our ability to tease out the actual mechanism behind species response or impact. For example, populations of larger-bodied beetles may be at higher extinction risk for multiple reasons (Table 13.1), yet determining analytically which hard trait interacts with a given environmental filter is challenging.

Across species within a given community, interactions among impact traits may exhibit facilitation, complementarity or even antagonism across different ecological functions. For example, large body size in dung beetles confers greater rates of waste removal and is therefore likely to be positively associated with incidental functions like fly suppression (see Chapter 12). However, large beetles also may carry small seeds down to such a depth that they are often unable to germinate successfully (Nichols *et al.*, 2008; Andresen & Feer, 2005), generating a fly-suppression/small seed dispersal trade-off for communities with large beetles present.

The distribution of impact traits across the entire dung beetle community is also important (Slade *et al*, 2007). For example, species of ball-roller typically dig much shallower nests (Halffter & Edmonds, 1982), potentially improving their strengths as providers of seed dispersal functions over tunnelling species, which may in turn have stronger impacts on nutrient cycling to lower soil horizons.

Another consideration in species-function relationships of dung beetles (and other mobile animal communities), is that food and nesting resources appear as discrete units in both space and time (Finn, 2001; O'Hea *et al.*, 2010). Consequently, the factors that increase or decrease the likelihood of colonization of a given resource play a major role in determining both species composition and abundance structure (Table 13.1).The growing need to include such 'mobility-based' traits in biodiversity-ecological function work is borne out of efforts to increase realism in manipulative and observational studies (Naeem, 2008), by expanding from artificially controlled arable plant and microbial community work to an understanding of the functional importance of mobile animals and real-world field conditions (Naeem, 2008; Tylianakis *et al.*, 2008; Kremen *et al.*, 2007).

As a final point, we emphasize the distinction between ecological functions and the increasingly popular 'ecosystem services' paradigm (Armsworth *et al.*, 2007). Functions and services are nested, but not synonymous concepts, with ecosystem services most appropriately defined as the subset of ecological functions that directly or indirectly yield an improvement in human well-being (Fisher *et al.*, 2009; Figure 13.7b). While every act of waste removal by a dung beetle is itself an important ecological function that helps maintain the integrity of the ecological system, not all such functions are immediately and directly pertinent to human well-being at all times (Nichols *et al.*, 2008).

13.3.4 Dung beetles as ecological disturbance indicator taxa: applied examples

Because of the many favourable attributes outlined in this chapter, dung beetles present an attractive and cost-effective option for conservation assessments and monitoring (Barbero *et al.*, 1999; McGeoch *et al.*, 2002; Celi & Dávalos, 2001; Gardner *et al.*, 2008b). However, only a few projects have actually moved from rhetoric to on-the-ground monitoring programmes (Table 13.2). Most of the studies listed in Table 13.2 looked at the viability of implementing field-based monitoring programmes that link dung beetle community change to various types of management practice. The goals of these studies range from validating remote-sensing data on changes to forest structure, following selective logging under different management scenarios (Aguilar-Amuchastegui & Henebry, 2007), to (more commonly) providing additional data on the specificity of dung beetle species-environment relationships under various disturbance or management regimes (McGeoch *et al.*, 2002; Davis *et al.*, 2001; Davis *et al.*, 2004; Celi & Dávalos, 2001).

At least one large-scale monitoring programme (The Amazon Land-Use Change and Biodiversity Project – Jari Brazilian Amazonia) is currently collecting longitudinal data on Scarabaeine dung beetles across multiple native and plantation forest management regimes (Gardner & Barlow, unpublished). There is an urgent need to expand efforts to include dung beetles in a greater number of biodiversity monitoring programmes, and publish case studies of the successes and challenges involved in their use.

13.4 Dung beetle conservation

Basic distribution and natural history information is frequently missing for poorly studied and diverse invertebrate groups (Samways, 2002; Pawar, 2003), which are often plagued by a set of challenges, ranging from taxonomic chauvinism, lack of expertise in the description of hyper-diverse taxa, and associated bioinformatics (Samways, 1994; 2004; 2006; Pawar, 2003), to a general 'public relations' crisis (Kellert, 1993). Despite their charisma and well-resolved taxonomy (Chapter 2), dung beetles are far from immune to these roadblocks.

Table 13.2 Existing applications of Scarabaeinae dung beetles as ecological disturbance indicators. IndVal refers to the species indicator value method developed by Dufrêne & Legendre (1997).

Study	*Location*	*Study purpose*	*Management type*	*Method*	*Conclusions*
Davis *et al.*, 2004	South Africa	Review species-environment relationships for two disturbance types.	Agricultural conversion, endectocide application.	None, review of published studies.	When sampled at the appropriate spatial scale, Scarabaeine dung beetles can effectively discriminate between types of disturbance, and hence assist in guiding management improvements.
Celi & Davalos, 2001	Ecuador	Generate data on species-environment relationship in zones of varying management, develop a tool for local inhabitants to assess the environmental effects of management practices.	Two selective logging practices (3/4 trees/ha and all large trees/ha), agricultural conversion.	Coupled human dung and carrion baited pitfall traps along linear transects.	Dung beetle community structure and composition was differentially impacted by different management practices. Some, but not all, species were robust indicators of ecological change.
Davis *et al.*, 2001	Borneo	Generate data on species-environment relationships in disturbed and undisturbed areas.	Logging, conversion to plantation forest.	Human dung baited pitfall traps, flight intercept traps.	Species associations demonstrate high biotope fidelity, even over extremely short physical distances.

(continued)

Table 13.2 *(Continued)*

Study	*Location*	*Study purpose*	*Management type*	*Method*	*Conclusions*
McGeoch *et al.*, 2002	South Africa	Generate data on species-environment relationships in disturbed and undisturbed areas and understand the reliability of these relationships over time, and utility of the IndVal metric to select candidate indicator species.	Human presence.	Elephant dung-baited pitfall traps.	Some, not all species were robust indicators of ecological change within a given habitat and across time. IndVal is a useful tool to select ecological disturbance indicator species.
Aguilar-Amuchastegui & Henebry, 2007	Costa Rica	Establish proof of concept for use of dung beetles as ecological disturbance indicators of selective logging regimes and minimum cutting schedules.	Selective logging.	Human dung-baited pitfall traps.	Dung beetle community structure and composition can be clearly linked to changes in forest management and used to validate remote-sensing data and propose a minimum logging intensity ($\leq$5 trees/ha).

These problems are compounded by the fact that survey efforts for insect taxa are often insufficiently replicated in space, time and diversity of methodologies to complete dependable species lists for a given area (Samways & Grant, 2007; Cotterill *et al.*, 2008). As a consequence, most insect groups are likely to reflect a tremendous mismatch between global threats to their conservation and the extent of their inclusion in the IUCN Red List (New, 2002; Maudsley & Stork, 1995). For example, in the entire IUCN Red List (IUCN, 2009), there are only 2,619 species of insect listed in any of the threat categories (of which 1,989 are Odonata and only 72 are Coleoptera). Despite the well-documented decline and loss of several species, particularly across the Mediterranean region (Hanski & Cambefort, 1991), there is currently not a single Scarabaeine dung beetle species on the global IUCN Red List.

In the case of dung beetles, one ambitious step towards rectifying these gaps has recently begun in the form of their inclusion in the IUCN Sampled Red List Index (Baillie *et al.*, 2008). The Sampled Red List Index (SRLI) programme seeks to conduct global assessments for 1,500 randomly selected species of several taxonomic groups that are currently underrepresented on the IUCN Red List. By re-conducting these global assessments at regular intervals, the SRLI programme can measure the rate at which species move through extinction risk categories over time. As this volume goes to press, expert assessors from The Scarabaeine Research Network (a National Science Foundation funded, international consortium of Scarabaeine systematists, ecologists and conservationists) have just completed the first round of assessments. Consequently, we now have some preliminary numbers on the global extinction risk faced by the nearly 6,000 species found around the globe.

Extrapolating from the 1,500 species randomly selected for assessment, over 12 per cent of all dung beetle species on Earth are currently threatened with extinction, in IUCN categories EN, NE or NT. The majority of these species are known to be extremely range-restricted, are often limited to areas of high forest biomass subject to logging or charcoal production, or have an extremely narrow diet breadth, and also extensive documentation exists of the complete loss, near loss, or extensive fragmentation of their range (S. Spector, personal communication).

In Africa, elephant, rhino and lemur specialists frequently appear in one of these three threat categories. A further nine per cent of assessed dung beetle species were classified as vulnerable to extinction (VU). Often, these species are reported as dietary specialists, restricted to areas of high vegetative biomass, or occurring in significantly reduced population densities.

A preliminary look at many of the VU species suggests that, once reviewed, many will move towards one of the three threatened categories. This could mean that as many as one in five dung beetles is currently in a state of extreme conservation concern (S. Spector, personal communication). The remaining species are near evenly split between those of least concern (LC, 37 per cent) and those for which we have insufficient data to assess their status (42 per cent) – very often represented only by a single (type) specimen.

The modern-day biodiversity crisis is most often portrayed in vertebrate terms, yet it is overwhelmingly a loss of invertebrate life. Those species with tight resource dependencies on other organisms face a particularly heightened risk of extinction

(Dunn, 2005). At the global level, this preliminary stage of the IUCN listing process suggests that extinction risk for dung beetles is higher for those species historically and currently restricted to forest areas where human needs for timber and biomass for cooking are high, and for those species with diets limited to large-bodied mammal species in areas historically and currently subject to over-hunting. The huge number of data-deficient species also highlights that our understanding of historical and current dung beetle distributions is extremely poor, and that sample and distribution data is often difficult to access where it exists.

How can we develop conservation strategies for Scarabaeinae dung beetles (Koch *et al.*, 2000)? As is often the case, the key to success may be in plurality of action. The co-dependency of dung beetles on (predominantly) intact mammal communities *and* tight habitat requirements necessitates system-level conservation approaches. Ecosystem or habitat-level strategies both include a variety of interventions, from large-scale conservation planning to more responsible approaches to natural resource management. For dung beetles, their incorporation, alongside data on other species groups and environmental surrogates, will lead naturally to their increased representation in conservation efforts.

Ecosystem-level efforts should be complemented by indirect approaches that inspire the attention and imagination of public, funding bodies and decision-makers to inspire investment in the protection of entire ecosystems. Such approaches often depend upon the existence of flagship species that operate as 'social hooks' to motivate political and community engagement in difficult conservation problems (Lindenmayer *et al.*, 2007; Caro and O'Doherty, 1999). While flagship species are most commonly charismatic megafauna that are easy to 'sell', ample opportunities exist for educating about the importance of conserving dung beetles. These range from economic arguments (Losey and Vaughan, 2006), potential links to human and domestic animal health (Nichols *et al.*, 2008; du Toit *et al.*, 2008) and cultural factors (Hanski, 1988; Scholtz, 2008).

13.5 Some ways forward

As we have hopefully demonstrated in this chapter, our understanding of the ways in which dung beetles respond to human-induced disturbances and are linked to ecological processes can make a significant contribution to our understanding of opportunities for biodiversity conservation in human-modified landscapes. Many of the challenges facing applied dung beetle research could be significantly ameliorated through an increased effort to collect standardized datasets on species and functional disturbance response patterns. Over the past four years, work by individuals associated with The Scarabaeinae Research Network has made considerable progress towards closing some of these database gaps. This has included the creation of a wide range of publicly accessible resources, including an online global catalogue of the Scarabaeinae, new keys to genera and species, standard survey protocols and trait information (www.scarabnet.org).

Since Halffter & Matthews' (1966) seminal treatise on dung beetle natural history, scarabaeine workers and enthusiasts have benefited from a series of exceptional advances in our understanding of dung beetle natural history (Halffter

& Matthews, 1966), nesting behaviours (Halffter & Edmonds, 1982), basic and applied ecology (Hanski & Cambefort, 1991) and more recently, evolutionary biology and conservation (Scholtz *et al.*, 2009). It is to these earlier works that the expanding collection of studies on dung beetle evolution and ecology, highlighted throughout this volume, owes both their inspiration and their foundation.

Since *Dung Beetle Ecology* was published in 1991, the number of studies documenting dung beetle community responses to land use change has literally exploded. In conjunction with our continuously improved understanding of the actual ecological mechanisms that underlie individual species responses to habitat change (see Chapters 10, 11 and 12), this information base now provides a valuable opportunity to develop an integrated model study system to enhance our understanding of the drivers of biodiversity persistence in human-modified systems (Didham *et al.*, 1996).

In this International Year of Biodiversity, when the relative successes and failures of conservation are in sharp focus, the need to improve our capacity to understand the consequences of global environmental change cannot be overemphasized.

Acknowledgements

The authors would like to acknowledge useful comments by Sacha Spector and Júlio Louzada.

References

Abrams, P.A., Rueffler, C. & Kim, G. (2008) Determinants of the strength of disruptive and/or divergent selection arising from resource competition. *Evolution* **62**, 1571–1586.

Addo-Bediako, A., Chown, S.L. & Gaston, K.J. (2000) Thermal tolerance, climatic variability and latitude. *Proceedings of the Royal Society B* **267**, 739–745.

Addo-Bediako, A., Chown, S.L. & Gaston, K.J. (2001) Revisiting water loss in insects: a large scale view. *Journal of Insect Physiology* **47**, 1377–1388.

Addo-Bediako, A., Chown, S.L. & Gaston, K.J. (2002) Metabolic cold adaptation in insects: a large-scale perspective. *Functional Ecology* **16**, 332–338.

Agashe, D. & Bolnick, D.I. (2010) Intraspecific genetic variation and competition interact to influence niche expansion. *Proceedings of the Royal Society B* doi: 1098/rspb.2010.0232.

Aguilar-Amuchastegui, N. & Henebry, G.M. (2007) Assessing sustainability indicators for tropical forests: Spatio-temporal heterogeneity, logging intensity, and dung beetles communities. *Forest Ecology and Management* **253**, 56–67.

Alarcón, D.L., Halffter, G. & Vaz-de-Mello, F.Z. (2009) Nesting Behavior in *Trichillum* Harold, 1868 and related genera (Coleoptera: Scarabaeidae: Scarabaeinae: Ateuchini: Scatimina): A primitive process or a loss of nidification? *Coleopterists Bulletin* **63**, 289–297.

Albone, E.S. (1984) *Mammalian semiochemistry; the investigation of chemical signals between mammals.* Wiley, Toronto.

Alcock, J. (1979) The evolution of intraspecific diversity in male reproductive strategies in some bees and wasps. In: Blum, M.S.& Blum, N.A. (eds.) *Sexual selection and reproductive competition in insects.* Academic Press, New York.

Alcock, J. (1994) Post insemination associations between males and females in insects: the mate-guarding hypothesis. *Annual Review of Entomology* **39**, 1–21.

Alcock, J. (1996) The relation between male body size, fighting, and mating success in Dawson's burrowing bee, *Amegilla dawsoni* (Apidae, Apinae, Anthophorini). *Journal of Zoology* **239**, 663–675.

Alcock, J. (2009) *Animal Behaviour: An evolutionary approach*, Sinauer Associates, USA.

Anderson, J.C. & Laughlin, S.B. (2000). Photoreceptor performance and the coordination of achromatic and chromatic inputs in the fly visual system. *Vision Research* **40**, 13–31.

Andersson, M. (1994) *Sexual Selection*, Princeton University Press, Princeton, NJ.

Andersson, M. & Simmons, L.W. (2006) Sexual selection and mate choice. *Trends in Ecology and Evolution* **21**, 296–302.

Andresen, E. (2003) Effect of forest fragmentation on dung beetle communities and functional consequences for plant regeneration. *Ecography* **26**, 87–97.

Andresen, E. (2005) Effects of season and vegetation type on community organization of dung beetles in a tropical dry forest. *Biotropica* **37**, 291–300.

Andresen, E. & Feer, F. (2005) The role of dung beetles as secondary seed dispersers and their effect on plant regeneration in tropical rainforests. In: Forget, P.M., Lambert, J.E., Hulme, P.E.& Vander Wall, S.B. (eds.) *Seed fate: Predation, dispersal and seedling establishment.* CABI International, Wallingford, Oxfordshire, UK.

Ecology and Evolution of Dung Beetles, First Edition. Edited by Leigh W. Simmons and T. James Ridsdill-Smith.

Andresen, E. & Laurance, S.G.W. (2007) Possible indirect effects of mammal hunting on dung beetle assemblages in Panama. *Biotropica* **39**, 141–146.

Anduaga, S. (2004) Impact of the activity of dung beetles (Coleoptera: Scarabaeidae: Scarabaeinae) inhabiting pasture land in Durango, Mexico. *Environmental Entomology* **33**, 1306–1312.

Anduaga, S., Halffter, G. & Huerta, C. (1987) Adaptaciones ecológicas de la reproducción en *Copris* (Coleoptera: Scarabaeidae: Scarabaeinae). *Bollettino del Museo Regionale di Scienze Naturali Torino* **5**, 45–65.

Angelini, D.R. & Kaufman, T.C. (2004) Functional analyses in the hemipteran *Oncopeltus fasciatus* reveal conserved and derived aspects of appendage patterning in insects. *Developmental Biology* **271**, 306–321.

Angermeier, P.L. & Karr, J.R. (1994) Biological integrity versus biological diversity as policy directives – protecting biotic resources. *Bioscience* **44**, 690–697.

Anonymous (2008) *Rural Business Register.* Data base administered by the Information Centre of the Finnish Ministry of Forestry and Agriculture, Helsinki.

Apol, M.E.F., Etienne, R.S. & Olff, H. (2008) Revisiting the evolutionary origin of allometric scaling in biology. *Functional Ecology* **22**, 1070–1080.

Arellano, L., Favila, M.E. & Huerta, C. (2005) Diversity of dung and carrion beetles in Mexican fragmented tropical montane cloud forests and shade coffee plantations. *Biodiversity and Conservation* **14**, 601–615.

Arellano, L., León-Cortés, J. L. & Ovaskainen O. (2008) Patterns of abundance and movement in relation to landscape structure – a study of a common scarab (*Canthon cyanellus cyanellus*) in Southern Mexico. *Landscape Ecology* **23**, 69–78.

Arillo, A. & Ortuño, V.M. (2008) Did dinosaurs have any relation with dung-beetles? (The origin of coprophagy), *Journal of Natural History* **42**, 1405–1408.

Armsworth, P.R., Chan, K.M.A., Daily, G.C., Ehrlich, P.R., Kremen, C., Ricketts, T.H. & Sanjayan, M.A. (2007) Ecosystem-service science and the way forward for conservation. *Conservation Biology* **21**, 1383–1384.

Arnqvist, G. & Nilsson, T. (2000) The evolution of polyandry: multiple mating and female fitness in insects. *Animal Behaviour* **60**, 145–164.

Arnqvist, G. & Rowe, L. (2005) *Sexual Conflict*, Princeton University Press, NJ.

Arrow, G.J. (1951) *Horned Beetles*, Dr. W. Junk, The Hague.

Aubin-Horth, N. & Dodson, J.J. (2004) Influence of individual body size and variable thresholds on the incidence of a sneaker male reproductive tactic in Atlantic salmon. *Evolution* **58**, 136–144.

Australian Bureau of Statistics (2008) www.abs.gov.au.

Avendano-Mendoza, C., Moron-Rios, A., Cano, E.B. & Leon-Cortes, J.L. (2005) Dung beetle community (Coleoptera: Scarabaeidae: Scarabaeinae) in a tropical landscape at the Lachua Region, Guatemala. *Biodiversity and Conservation* **14**, 801–822.

Axelrod, R. & Hamilton, W.D. (1981) The evolution of cooperation. *Science* **211**, 1390–1396.

Baillie, J.E.M., Collen, B., Amin, R., Akcakaya, H.R., Butchart, S.H.M., Brummitt, N., Meagher, T.R., Ram, M., Hilton-Taylor, C. & Mace, G.M. (2008) Toward monitoring global biodiversity. *Conservation Letters* **1**, 18–26.

Baird, E., Srinivasan, M.V., Zhang, S. & Cowling, A. (2005) Visual control of flight speed in honeybees. *Journal of Experimental Biology* **208**, 3895–3905.

Baird, E., Warrant, E.J., Scholtz, C.H., Byrne, M.J. & Dacke, M. (2008) *Ball-room Dancing: The Dung Beetle Dance*. The Second International Conference on Invertebrate Vision. August 1–8, 2008. Bäckaskog Castle, Sweden.

Baird, E., Byrne, M.J., Scholtz, C.H., Warrant, E.J. & Dacke, M. (2010). The choice of rolling direction in dung beetles. *Journal of Comparative Physiology A.* **196**, 801–806.

Bakker, T.C.M. (1993) Positive genetic correlation between female preference and preferred male ornament in sticklebacks. *Nature* **363**, 255–257.

Ballerio, A. (1999) Revision of the genus *Pterorthochroaetes* first contribution (Coleoptera: Scarabaeoidea: Ceratocanthidae). *Folia Heyrovskyana* **7**, 221–228.

Balthasar, V. (1963a) *Monographie der Scarabaeidae und Aphodiidae der Palaearktischen und Orientalischen Region (Coleoptera; Lamellicornia).* Volume 1, Scarabaeidae. Verlag der Tschechoslovakischen Akademie der Wissenschaft, Prague.

Balthasar, V. (1963b) *Monographie der Scarabaeidae und Aphodiidae der Palaearktischen und Orientalischen Region (Coleoptera; Lamellicornia).* Volume 2, Coprinae. Verlag der Tschechoslovakischen Akademie der Wissenschaft, Prague.

Barbero, E. & Lopez-Guerrero, Y. (1992) Some considerations on the dispersal power of *Digitonthophagus gazella* (Fabricius 1787) in the New World (Coleoptera Scarabaeidae Scarabaeinae). *Tropical Zoology* 5, 115–120.

Barbero, E., Palestrini, C. & Rolando, A. (1999) Dung beetle conservation: effects of habitat and resource selection (Coleoptera: Scarabaeoidea). *Journal of Insect Conservation* 3, 75–84.

Barker, M.S., Demuth, J.P. & Wade, M.J. (2005) Maternal expression relaxes constraint on innovation of the anterior determinant, bicoid. *PloS Genetics* 1, 527–530.

Bartholomew, G.A. & Casey, T.M. (1977) Endothermy during terrestrial activity in large beetles. *Science* 195, 882–883.

Bartholomew, G.A. & Heinrich, B. (1978) Endothermy in African dung beetles during flight, ball making, and ball rolling. *Journal of Experimental Biology* 73, 65–83.

Bateman, A.J. (1948) Intrasexual selection in *Drosophila*. *Heredity* 2, 349–368.

Beebe, W. (1947) Notes on the hercules beetle, *Dynastes hercules* (Linn.) at Rancho Grande, Venezuela, with special references to combat behavior. *Zoologica* 32, 109–116.

Begon, M., Townsend, C.R. & Harper, J.L. (2006) *Ecology: from Individuals to Ecosystems*. 4th edition. Blackwell, Oxford.

Beketov, M.A. (2009) The Rapoport effect is detected in a river system and is based on nested organization. *Global Ecology and Biogeography* 18, 498–506.

Bell, G. (2001) Ecology-neutral macroecology. *Science* 293, 2413–2418.

Bernardo, J. (1996) The particular maternal effect of propagule size, especially egg size: patterns, models, quality of evidence, and interpretations. *American Zoologist* 36, 216–236.

Bernon, G. (1981) Species abundance and diversity of the Coleoptera component of a South African cow dung community, and associated insect predators. PhD thesis, Bowling Green State University, USA.

Berry, P. (1993) From cow pat to frying pan: Australian herring (*Arripes gerogianus*) feed on an introduced dung beetle (Scarabaeidae). *Western Australian Naturalist* 19, 241–242.

Bertone, M.A., Green, J.T., Washburn, S.P., Poore, M.H. & Watson, D.W. (2006) The contribution of tunnelling dung beetles to pasture soil nutrition. *Forage and Grazinglands*, DOI: 10.1094/FG-2006–0711–02–RS.

Bilde, T., Foged, A., Schilling, N. & Arnqvist, G. (2009) Postmating sexual selection favors males that sire offspring with low fitness. *Science* 324, 1705–1706.

Birkhead, T.R. (1998) Cryptic female choice: Criteria for establishing female sperm choice. *Evolution* 52, 1212–1218.

Birkhead, T.R. & Møller, A.P. (1998) *Sperm Competition and Sexual Selection*. Academic Press, London.

Bishop, A.L., McKenzie, H.J., Spohr, L.J. & Barchia, I.M. (2005) Interactions between dung beetles (Coleoptera: Scarabaeidae) and the arbovirus vector *Culicoides brevitarsis* Kieffer (Diptera: Ceratopogonidae). *Australian Journal of Entomology* 44, 89–96.

Biström, O., Silfverberg, H. & Rutanen, I. (1991) Abundance and distribution of coprophilous Histerini (Histeridae) and *Onthophagus* and *Aphodius* (Scarabaeidae) in Finland. *Entomologica Fennica* 27, 53–66.

Bitsch, C. & Bitsch, J. (2005) Evolution of eye structure and arthropod phylogeny. In: Koenemann, S.& Jenner, R.A. (eds) *Crustacea and Arthropod Relationships*. Taylor and Francis, London.

Blank, R.H., Black, H. & Olson, M.H. (1983) Preliminary investigations of dung removal and flight biology of the Mexican dung beetle *Copris incertus* in Northland (Coleoptera: Scarabaeidae). *New Zealand Entomologist* 7, 360–364.

Blomquist, G.J. & Bagneres, A.-G. (2010) *Insect Hydrocarbons: Biology, Biochemistry, and Chemical Ecology*. Cambridge University Press, Cambridge, UK.

Blows, M.W. (2002) Interaction between natural and sexual selection during the evolution of mate recognition. *Proceedings of the Royal Society of London B* 269, 1113–1118.

Blum, M.S. (1981) *Chemical Defences of Arthropods*. Academic Press, New York.

Bonduriansky, R. (2007) Sexual selection and allometry: A critical reappraisal of the evidence and ideas. *Evolution* 61, 838–849.

Bonduriansky, R. & Day, T. (2003) The evolution of static allometry in sexually selected traits. *Evolution* 57, 2450–2458.

Boonrotpong, S., Sotthibandhu, S. & Pholpunthin, C. (2004) Species composition of dung beetles in the primary and secondary forestes at Ton Nga Chan Wildlife Sanctuary. *ScienceAsia* 30, 59–65.

Bordat, P., Paulian, R. & Pittino, R. (1990) *Faune de Madagascar, 74: Insectes, Coléoptères Aphodiidae*. Muséum National d'Histoire Naturelle, Paris.

Bornemissza, G.F. (1971) Mycetophagous breeding in the Australian dung beetle, *Onthophagus dunningi*, *Pedobiologia* **11**, 133–142.

Bornemissza, G.E. (1976) The Australian dung beetle project – 1965–1975. *Australian Meat Research Committee Review* **30**, 1–30.

Bornemissza, G.F. & Williams, C.H. (1970) An effect of dung beetle activity on plant yield. *Pedobiologia* **10**, 1–7.

Botes, A., McGeoch, M.A. & van Rensburg, B.J. (2006) Elephant- and human-induced changes to dung beetle (Coleoptera: Scarabaeidae) assemblages in the Maputaland Centre of Endemism. *Biological Conservation* **130**, 573–583.

Boyd, R. (1989) Mistakes allow evolutionary stability in the repeated Prisoner's Dilemma game. *Journal of Theoretical Biology* **136**, 47–56.

Bradley, T.J. (2006) Discontinuous ventilation in insects: protecting tissues from O_2. *Respiratory Physiology and Neurobiology* **154**, 30–36.

Bradshaw, A.D. (1965) Evolutionary significance of phenotypic plasticity in plants. *Advances in Genetics* **13**, 115–155.

Bretman, A. & Tregenza, T. (2005) Measuring polyandry in wild populations: a case study using promiscuous crickets. *Molecular Ecology* **14**, 2169–2179.

Brett, M.T. (2004) When is a correlation between non-independent variables 'spurious'? *Oikos* **105**, 647–656.

Brigandt, I. (2002) Homology and the origin of correspondence. *Biology and Philosophy* **17**, 389–407.

Briggs, J.C. (2003) The biogeographic and tectonic history of India. *Journal of Biogeography* **30**, 381–388.

Brines, M.L. & Gould, J.L. (1982) Skylight polarization patterns and orientation. *Journal of Experimental Biology* **96**, 69–91.

Brody, M.S. & Lawlor, L.R. (1984) Adaptive variation in offspring size in the terrestrial isopod *Armadillium vulgare*. *Oecologia* **61**, 55–59.

Brø-Jørgensen, J. (2007) The intensity of sexual selection predicts weapon size in male bovids. *Evolution* **61**, 1316–1326.

Brooks, R. & Couldridge, V. (1999) Multiple sexual ornaments coevolve with multiple mating preferences. *American Naturalist* **154**, 37–45.

Brown, J.H. & Gibson, A.C. (1983) *Biogeography*. The C.V. Mosby Company, St Louis, MO.

Brown, J.H., Gillooly, J.F., Allen, A.P., Savage, V. & West, G.B. (2004) Toward a metabolic theory of ecology. *Ecology* **85**, 1771–1789.

Brown, L. & Bartalon, J. (1986) Behavioral correlates of male morphology in a horned beetle. *American Naturalist* **127**, 565–570.

Browne, J. & Scholtz, C.H. (1998) Evolution of the scarab hindwing articulation and wing base: A contribution toward the phylogeny of the Scarabaeidae (Scarabaeoidea: Coleoptera). *Systematic Entomology* **23**, 307–326.

Brühl, C. & Krell, F.-T. (2003) Finding a rare resource: Bornean Scarabaeoidea (Coleoptera) attracted by defensive secretions of Diplopoda. *Coleopterists Bulletin* **57**, 51–55.

Brussaard, L. (1983) Reproductive behaviour and development of the dung beetle *Typhaeus typhoeus* (Coleoptera, Geotrupidae). *Tijdschrift voor Entomologie* **126**(10), 203–231.

Bryan, R.P. (1976) Effect of dung beetle, *Onthophagus gazella,* on ecology of infective larvae of gastrointestinal nematodes of cattle. *Australian Journal of Agricultural Research* **27**, 567–574.

Buck J., Keister, M. & Specht, H. (1953) Discontinuous respiration in diapausing *Agapema* pupae. *Anatomical Record* **117**, 541.

Bull, J.J. (1983) *Evolution of sex determining mechanisms*. Benjamin Cummings Pub. Co., Advanced Book Program, Menlo Park, CA.

Burger, B.V. (2005) Mammalian Semiochemicals, in S. Schultz (ed.) *The Chemistry of Pheromones and Other Semiochemicals*. pp. 231–278. Topics in Current Chemistry, Volume 240. Springer, Berlin.

Burger B.V. & Petersen, W.G.B. (1991) Semiochemicals of the Scarabaeinae, III. Identification of an attractant for the dung beetle *Pachylomerus femoralis* in the fruit of the spineless monkey orange tree, Strychnos madagascariensis. *Zeitschrift für Naturforschung* **46c**, 1073–1079.

Burger, B.V. & Petersen, W.G.B. (2002) Semiochemicals of the Scarabaeinae VI. Identification of EAD-active constituents of abdominal secretion of male dung beetle *Kheper nigroaeneus*. *Journal of Chemical Ecology* **28**, 501–513.

Burger, B.V., Munro, Z., Röth, M., Spies, H.S.C., Truter, V., Tribe, G.D. & Crewe, R.M. (1983) Composition of the heterogeneous sex attracting secretion of the dung beetle, *Kheper lamarcki. Zeitschrift für Naturforschung* **38c**, 848–855.

Burger, B.V., Munro, Z. & Brandt, W.F. (1990) Pheromones of the Scarabaeinae II. Composition of the pheromone disseminating carrier material secreted by male dung beetles of the genus *Kheper. Zeitschrift für Naturforschung* **45c**, 863–872.

Burger, B.V., Petersen, W.G.B. & Tribe, G.D. (1995a) Semiochemicals of the Scarabaeinae, IV: Identification of an attractant for the dung beetle *Pachylomerus femoralis* in the abdominal secretion of the dung beetle *Kheper lamarcki. Zeitschrift für Naturforschung* **50c**, 675–680.

Burger, B.V., Petersen, W.G.B. & Tribe, G.D. (1995b) Semiochemicals of the Scarabaeinae, V: Characterization of the defensive secretion of the dung beetle *Oniticellus egregius. Zeitschrift für Naturforschung* **50c**, 681–684.

Burger, B.V., Petersen, W.G.B., Weber, W.G. & Munro, Z.M. (2002) Semiochemicals of the Scarabaeinae VII. Identification and synthesis of EAD-active constituents of abdominal sex attracting secretion of the male dung beetle *Kheper subaeneus. Journal of Chemical Ecology* **28**, 2527–2539.

Burger, B.V., Petersen, W.G.B., Ewig, B.T., Neuhaus, J., Tribe, G.D., Spies, H.S.C. & Burger, W.J.G. (2008) Semiochemicals of the Scarabaeinae VIII. Identification of active constituents of the abdominal sex-attracting secretion of the male dung beetle, *Kheper bonellii*, using gas chromatography with flame ionization and electroantennographic detection in parallel. *Journal of Chromatography A* **1186**, 245–253.

Burley, N. (1988) The differential allocation hypothesis: an experimental test. *American Naturalist* **132**, 611–628.

Burney, D.A., Burney, L.P., Godfrey, L.R., Jungers, W.L., Goodman, S.M., Wright, H.T. & Jull, A.J.T. (2004) A chronology for late prehistoric Madagascar. *Journal of Human Evolution* **47**, 25–63.

Byrne, M.J., Dacke, M., Nordström, P. Scholtz, C.H. & Warrant, E.J. (2003) Visual cues used by ball-rolling dung beetles for orientation. *Journal of Comparative Physiology A* **189**, 411–418.

Byrne, M.J., Nadel, R.L., Dacke, M., Warrant, E.J., Baird, E. & Frederiksen, R. (2008) Allometry of dung beetle eyes in relation to their ecology and behaviour. *The Second International Conference on Invertebrate Vision*. August 1–8, 2008. Bäckaskog Castle, Sweden.

Byrne, M.J., Dacke, M., Warrant, E.J. & Baird, E. (2009) Flying in the dark: dung beetle allometry, in time and space. *The 16th Congress of the Entomological Society of Southern Africa. 5–8 July 2009*. Stellenbosch, South Africa.

Cabrera-Walsh, G. & Gandolfo, C. (1996) Nidification of thirteen common Argentine dung beetles (Scarabaeidae: Scarabaeinae). *Annals of the Entomological Society of America* **89**, 4581–588.

Cabrero-Sañudo, F.J. (2007) The phylogeny of Iberian Aphodiini species (Coleoptera, Scarabaeoidea, Scarabaeidae, Aphodiinae) based on morphology. *Systematic Entomology* **32**, 156–175.

Cabrero-Sañudo, F.J. & Zardoya, R. (2004) Phylogenetic relationships of Iberian Aphodiini (Coleoptera: Scarabaeidae) based on morphological and molecular data, *Molecular Phylogenetics and Evolution* **31**, 1084–1100.

Cain, A.J. & Sheppard, P.M. (1954) Natural selection in *Cepea. Genetics* **39**, 89–116.

Calosi, P., Bilton, D.T., Spicer, J.I. & Atfield, A. (2008) Thermal tolerance and geographical range size in the *Agabus brunneus* group of European diving beetles (Coleoptera: Dytiscidae). *Journal of Biogeography* **35**, 295–305.

Calosi, P., Bilton, D.T., Spicer, J.I., Votier, S.C. & Atfield, A. (2010) What determines a species' geographical range? Thermal biology and latitudinal range size relationships in European diving beetles (Coleoptera: Dytiscidae). *Journal of Animal Ecology* **79**, 194–204.

Cambefort, Y. (1982) Nidification behavior of Old World Oniticellini (Coleoptera: Scarabaeidae) In: Halffter, G.& Edmonds, W.D. (eds.) *The nesting behavior of dung beetles (Scarabaeinae): An ecological and evolutionary approach*, pp. 141–45. Institudo Ecología, Mexico.

Cambefort, Y. (1987) *Faune de Madagascar. 69. Insectes Coléoptères Aulonocnemidae*. Muséum national d'Histoire naturelle, Paris.

Cambefort, Y. (1991a) From Saprophagy to Coprophagy. In: Hanski I.& Cambefort, Y. (eds.) *Dung Beetle Ecology*, pp. 22–35. Princeton University Press, Princeton, NJ.

Cambefort, Y. (1991b) Biogeography and evolution. In: Hanski I.& Cambefort, Y. (eds.) *Dung Beetle Ecology*, pp. 51–67, Princeton University Press, Princeton, NJ.

Cambefort, Y. & Lumaret, J.-P. (1984) Nidification et larves des Oniticellini afrotropicaux (Col. Scarabaeidae), *Bulletin Société Entomologique de France* **88**, 542–69.

Cambefort, Y. & Walter, P. (1985) Description du nid et de la larve de *Paraphytus aphodiodes* Boucomont et notes sur l'origine de la coprophagie et l'évolution des Coléoptères Scarabaeidae s.str. *Annales Société Entomologique de France (N.S.)* **21**, 351–56.

Cambefort, Y. & Walter, P. (1991) Dung beetles in tropical forest in Africa In: Hanski, I.& Cambefort, Y. (eds.) *Dung Beetle Ecology*, pp 198–210, Princeton University Press, Princeton, NJ.

Candolin, U. & Heuschele, J. (2008) Is sexual selection beneficial during adaptation to environmental change? *Trends in Ecology and Evolution* **23**, 446–452.

Cantoni, D. & Brown, R.E. (1997) Paternal investment and reproductive success in the California mouse, *Peromyscus californicus*. *Animal Behaviour* **54**, 377–386.

Cardinale, B.J., Srivastava, D.S., Duffy, J.E., Wright, J.P., Downing, A.L., Snakaran, M. & Jouseau, C. (2006) Effects of biodiversity on the functioning of trophic groups and ecosystems. *Nature* **442**, 989–992.

Caro, T. & Girling, S. (2010). *Conservation by Proxy: Indicator, Umbrella, Keystone, Flagship, and Other Surrogate Species*. Island Press, Washington DC.

Caro, T.M. & O'Doherty, G. (1999) On the use of surrogate species in conservation biology. *Conservation Biology* **13**, 805–814.

Caro, T.M., Graham, C.M., Stoner, C.J. & Flores, M.M. (2003) Correlates of horn and antler shape in bovids and cervids. *Behavioral Ecology and Sociobiology* **55**, 32–41.

Carpaneto, G.M., Mazziotta, A. & Piattella, E. (2005) Changes in food resources and conservation of scarab beetles: from sheep to dog dung in a green urban area of Rome (Coleoptera, Scarabaeoidea). *Biological Conservation* **123**, 547–556.

Carpaneto, G.M., Mazziotta, A. & Valerio, L. (2007) Inferring species decline from collection records: roller dung beetles in Italy (Coleoptera: Scarabaeidae). *Diversity and Distributions* **13**, 903–919.

Carroll, S.B., Grenier, J.K. & Weatherbee, S.D. (2005) *From DNA to diversity. Molecular genetics and the evolution of animal design*. 2nd edn. Blackwell. Malden, MA, USA.

Cassis, G., Weir, T.A. & Calder, A.A. (2002) Subfamily Scarabaeinae. Australian Faunal Directory. www.environment.gov.au/biodiversity/abrs

Cattermole, P.J. (2000) *Building Planet Earth: Five Billion Years of Earth History*. Cambridge University Press, Cambridge.

Caveney, S. (1985) The phylogenetic significance of ommatidium structure in the compound eyes of polyphagan beetles. *Canadian Journal of Zoology* **64**, 1787–1819.

Caveney, S. & McIntyre, P.D. (1981) Design of graded-index lenses in the superposition eyes of scarab beetles. *Philosophical Transactions of the Royal Society of London B* **294**, 589–632.

Caveney, S., Scholtz, C.H. & McIntyre, P.D. (1995) Patterns of daily flight activity in onitine dung beetles. *Oecologia* **103**, 444–452.

Celi, J. & Dávalos, A. (2001). *Manuel de monitero. Los escarabajos peloteros como indicatores de la calidad ambiental*. EcoCiencia, Quito, Ecuador.

Chapin, F.S.I., Zavaleta, E.S., Eviners, V.T., Naylor, R.L., Vitousek, P.M., Reynolds, H.L., Hooper, D.U., Lavorel, S., Sala, O.E., Hobbie, S.E., Mack, M.C. & Díaz, S. (2000) Consequences of changing biodiversity. *Nature* **405**, 234–242.

Chapman, T. & Davies, S.J. (2004) Functions and analysis of the seminal fluid proteins of male *Drosophila melanogaster* fruit flies. *Peptides* **25**, 1477–1490.

Chapman, T., Liddle, L.F., Kalb, J.M., Wolfner, M.F. & Partridge, L. (1995) Cost of mating in *Drosophila melanogaster* females is mediated by male accessory gland products. *Nature* **373**, 241–244.

Chappell, M.A. (1984) Thermoregulation and energetics of the green fig beetle (*Cotinis texana*) during flight and foraging behavior. *Physiological Zoology* **57**, 581–589.

Charlesworth, B. (1994) Evolution in age-structured populations. *Cambridge Studies in Mathematical Biology* **13**, 1–306.

Chase, I.D. (1980) Cooperative and noncooperative behaviour in animals. *American Naturalist* **115**, 827–857.

Chave, J., Muller-Landau, H.C. & Levin, S. (2002) Comparing classical community models: theoretical consequences for patterns of diversity. *American Naturalist* **159**, 1–23.

Chazdon, R.L., Harvey, C.A., Komar, O., Griffith, D.M., Ferguson, B.G., Martinez-Ramos, M., Morales, H., Nigh, R., Soto-Pinto, L., Van Breugel, M. & Philpott, S.M. (2009) Beyond reserves: A research agenda for conserving biodiversity in human-modified tropical landscapes. *Biotropica* **41**, 142–153.

Chefaoui, R.M., Hortal, J. & Lobo, J.M. (2005) Potential distribution modelling, niche characterization and conservation status using GIS tools: a case study of Iberian *Copris* species. *Biological Conservation* **122**, 327–338.

Chin, K. & Gill, B.D. (1996) Dinosaurs, dung beetles, and conifers: participants in a Cretaceous food web, *Palaios* **11**, 280–285.

Chittka, L., Skorupski, P. & Raine, N.E. (2009) Speed-accuracy tradeoffs in animal decision making. *Trends in Ecology and Evolution* **24**, 400–407.

Choi, J.-H., Kijimoto, T. Snell-Rood, E.C., Tae, H., Yang, Y.-I. Moczek, A.P. & Andrews, J. (2010). Gene discovery in the horned beetle *Onthophagus taurus*. BMC Genomics. **11**, 703.

Chown, S.L. (2001) Physiological variation in insects: hierarchical levels and implications. *Journal of Insect Physiology* **47**, 649–660.

Chown, S.L. (2002) Respiratory water loss in insects. *Comparative Biochemistry and Physiology A* **133**, 791–804.

Chown, S.L. & Davis, A.L.V. (2003) Discontinuous gas exchange and the significance of respiratory water loss in scarabaeine beetles. *Journal of Experimental Biology* **206**, 3547–3556.

Chown, S.L. & Gaston, K.J. (1999) Exploring links between physiology and ecology at macro-scales: the role of respiratory metabolism in insects. *Biological Reviews* **74**, 87–120.

Chown, S.L. & Gaston, K.J. (2008) Macrophysiology for a changing world. *Proceedings of the Royal Society B* **275**, 1469–1478.

Chown, S.L. & Gaston, K.J. (2010) Body size variation in insects: a macroecological perspective. *Biological Reviews* **85**, 139–169.

Chown, S.L. & Klok, C.J. (2003) Water balance characteristics respond to changes in body size in sub-Antarctic weevils. *Physiological and Biochemical Zoology* **76**, 634–643.

Chown, S.L. & Nicolson, S.W. (2004) *Insect Physiological Ecology. Mechanisms and Patterns*. Oxford University Press, Oxford.

Chown, S.L. & Scholtz, C.H. (1993) Temperature regulation in the nocturnal melolonthine *Sparrmannia flava*. *Journal of Thermal Biology* **18**, 25–33.

Chown, S.L. & Terblanche, J.S. (2007) Physiological diversity in insects: ecological and evolutionary contexts. *Advances in Insect Physiology* **33**, 50–152.

Chown, S.L., Scholtz, C.H., Klok, C.J., Joubert, F.J. & Coles, K.S. (1995) Ecophysiology, range contraction and survival of a geographically restricted African dung beetle (Coleoptera: Scarabaeidae). *Functional Ecology* **9**, 30–39.

Chown, S.L., Pistorius, P. & Scholtz, C.H. (1998) Morphological correlates of flightlessness in southern African Scarabaeinae (Coleoptera: Scarabaeidae): testing a condition of the water-conservation hypothesis. *Canadian Journal of Zoology* **76**, 1123–1133.

Chown, S.L., Gibbs, A.G., Hetz, S.K., Klok, C.J., Lighton, J.R.B. & Marais, E. (2006) Discontinuous gas exchange in insects: a clarification of hypotheses and approaches. *Physiological and Biochemical Zoology* **79**, 333–343.

Chown, S.L., Marais, E., Terblanche, J.S., Klok, C.J., Lighton, J.R.B. & Blackburn, T.M. (2007) Scaling of insect metabolic rate is inconsistent with the nutrient supply network model. *Functional Ecology* **21**, 282–290.

Chown, S.L., Jumbam, K.R., Sørensen, J.G. & Terblanche, J.S. (2009) Phenotypic variance, plasticity and heritability estimates of critical thermal limits depend on methodological context. *Functional Ecology* **23**, 133–140.

Clark, C.A. & Sheppard, P.M. (1960) The evolution of mimicry in the butterfly *Papilio dardanus*. *Heredity* **14**, 163–173.

Clark, G.L. (1946) Dynamics of production in a marine area. *Ecological Monographs* **16**, 321–335.

Clark, J.S., Fastie, C., Hurtt, G., Jackson, S.T., Johnson, C., King, G.A., Lewis, M., Lynch, J., Pacala, S., Prentice, C., Schupp, E.W., Webb, T. III & Wyckoff, P. (1998) Reid's paradox of rapid plant migration. *BioScience* **48**, 13–24.

Clarke, A. (1993) Seasonal acclimatization and latitudinal compensation in metabolism: do they exist? *Functional Ecology* **7**, 139–149.

Cloudsley-Thomson, J.L. (1964) On the function of the sub-elytral cavity in desert Tenebrionidae. *Entomologists Monthly Magazine* **100**, 148–157.

Clusella-Trullas, S. & Chown, S.L. (2008) Investigating onychophoran gas exchange and water balance as a means to inform current controversies in arthropod physiology. *Journal of Experimental Biology* **211**, 3139–3146.

Clutton-Brock, T.H. (1991) *The Evolution of Parental Care*. Princeton University Press, Princeton.

Cockburn, A. (2006) Prevalence of different modes of parental care in birds. *Proceedings of the Royal Society of London B* **273**, 1375–1383.

Coe, M. (1977) The role of termites in the removal of elephant dung in the Tsavo (East) National Park Kenya. *East African Journal of Wildlife* **49**, 49–55.

Cole, L. (1954) The population consequences of life history phenomena. *The Quarterly Review of Biology* **29**, 103–137.

Collett, T.S. & Land, M.F. (1974) Visual control of flight behaviour in the hoverfly *Syritta pipiens* L. *Journal of Comparative Physiology* **99**, 1–66.

Connor, J. (1989) Density-dependent sexual selection in the fungus beetle, *Bolitotherus cornutus*. *Evolution* **43**, 1378–1386.

Cook, D.F. (1987) Sexual selection in dung beetles. 1. A multivariate study of the morphological variation in two species of dung beetle *Onthophagus* (Scarabaeidae: Onthophagini). *Australian Journal of Zoology* **35**, 123–132.

Cook, D.F. (1988) Sexual selection in dung beetles. 2. Female fecundity as an estimate of male reproductive success in relation to horn size, and alternative behavioural strategies in *Onthophagus binodis* Thunberg (Scarabaeidae: Onthophagini). *Australian Journal of Zoology* **36**, 521–532.

Cook, D.F. (1990) Differences in courtship, mating and postcopulatory behavior between male morphs of the dung beetle *Onthophagus binodis* Thunberg (Coleoptera: Scarabaeidae). *Animal Behaviour* **40**, 428–436.

Costa, M. (1969) The association between mesostigmatic mites and coprid beetles. *Acarologia* **11**, 411–428.

Cottenie, K. (2005) Integrating environmental and spatial processes in ecological community dynamics. *Ecology Letters* **8**, 1175–1182.

Cotter, S.C., Beveridge, M. & Simmons, L.W. (2007) Male morph predicts investment in larval immune function in the dung beetle, *Onthophagus taurus*. *Behavioral Ecology* **19**, 331–337.

Cotterill, F.P.D., Al-Rasheid, K.A.S. & Foissner, W. (2008) Conservation of protists: is it needed at all? *Biodiversity and Conservation* **17**, 427–443.

Cowley, M.J.R., Thomas, C.D., Thomas, J.A. & Warren, M.S. (1999) Flight areas of British butterflies: assessing species status and decline. *Proceedings of the Royal Society of London B* **266**, 1587–1592.

Cruickshank, T. & Wade, M.J. (2008) Microevolutionary support for a developmental hourglass: gene expression patterns shape sequence variation and divergence in *Drosophila*. *Evolution and Development* **10**, 583–590.

Cruz, M. & Martinez, I. (1998) Effect of male mesadene secretions on females of *Canthon cyanellus cyanellus* (Coleoptera, Scarabaeidae). *Florida Entomologist* **81**, 23–30.

Cunningham, E.J.A. & Russell, A.F. (2000) Egg investment is influenced by male attractiveness in the mallard. *Nature* **404**, 74–76.

Curtsinger, J.W. (1991) Sperm competition and the evolution of multiple mating. *American Naturalist* **138**, 93–102.

Cushman, S., McKelvey, K., Noon, B.R., McGarigal, K. (2010) Use of abundance of one species as a surrogate for abundance of others. *Conservation Biology* **24**, 830–840.

Dacke, M. (2003) *Celestial orientation in dim light*. PhD thesis. Lund University, Sweden.

Dacke, M., Nilsson, D-E., Warrant, E.J., Blest, A.D., Land, M.F. & O'Carroll, D.C. (1999) Built-in polarizers form part of a compass organ in spiders. *Nature* **401**, 470–472.

Dacke, M., Thuy, A.D. & O'Carroll, D.C. (2001) Polarized light detection in spiders. *Journal of Experimental Biology* **204**, 2481–2490.

Dacke, M., Nordström, P., Scholtz, C.H. & Warrant, E.J. (2002) A specialized dorsal rim area for polarized light detection in the compound eye of the scarab beetle *Pachysoma striatum*. *Journal of Comparative Physiology* **188**, 211–216.

Dacke, M., Nordström, P. & Scholtz, C.H. (2003a) Twilight orientation to polarised light in the crepuscular dung beetle *Scarabaeus zambesianus*. *Journal of Experimental Biology* **206**, 1535–154.

Dacke, M., Nilsson, D-E., Scholtz, C.H., Byrne, M. & Warrant, E.J. (2003b) Insect orientation to polarized moonlight. *Nature* **424**, 33.

Dacke, M., Byrne, M., Scholtz, C.H. & Warrant, E.J. (2003c) Lunar orientation in a beetle. *Philosophical Transactions of the Royal Society of London B* **271**, 361–365.

Dacke, M., Byrne, M.J., Baird, E., Scholtz, C.H. & Warrant, E.J. (2011) *How dim is dim? The dung beetle polarisation compass in moonlight compared to sunlight. Philosophical Transactions of the Royal Society of London B.*

Dadour, I. & Cook, D. (1997) Cropping and distribution of introduced dung beetles in south-western Australia. Dairy Research and Development Corporation DAW22. Department of Agriculture and Food, Perth, Australia.

Danielsson, I. (2000) Antagonistic pre- and post-copulatory sexual selection on male body size in a water strider (*Gerris lacustris*). *Proceedings of the Royal Society of London B* **268**, 77–81.

Darwin, C.D. (1845) *Journal of researches into the Natural History and Geology of the countries visited during the voyage of H.M.S. Beagle round the world.* John Murray, London.

Darwin, C. (1871) *The Descent of Man and Selection in Relation to Sex*, John Murray, London.

David, I., Bodin, L., Gianola, D., Legarra, A., Manfredi, E. & Robert-Granie, C. (2009) Product versus additive threshold models for analysis of reproduction outcomes in animal genetics. *Journal of Animal Science* **87**, 2510–2518.

Davidowitz, G. & Rosenzweig, M.L. (1998) The latitudinal gradient of species diversity among North American grasshoppers (Acrididae) within a single habitat: a test of the spatial heterogeneity hypothesis. *Journal of Biogeography* **25**, 553–560.

Davidson, A.D., Hamilton, M.J., Boyer, A.G., Brown, J.H. & Ceballos, G. (2009) Multiple ecological pathways to extinction in mammals. *Proceedings of the National Academy of Sciences USA* **106**, 10702–10705.

Davidson, E.H. & Erwin, D.H. (2006) Gene regulatory networks and the evolution of animal body plans. *Science* **311**, 796–800.

Davis, A.L.V. (1987) Geographical distribution of dung beetles (Coleoptera: Scarabaeidae) and their seasonal activity in south-western Cape Province. *Journal of the Entomological Society of South Africa* **50**, 275–285.

Davis, A.L.V. (1989) Nesting of Afrotropical *Oniticellus* (Coleoptera: Scarabaeidae) and its evolutionary trend from soil to dung. *Ecological Entomology* **14**, 11–21.

Davis, A.L.V. (1993) Biogeographical groups in a southern African, winter rainfall, dung beetle assemblage (Coleoptera: Scarabaeidae) – consequences of climatic history and habitat fragmentation. *African Journal of Ecology* **31**, 306–327.

Davis, A.L.V. (1994) Compositional differences between dung beetle (Coleoptera: Scarabaeidae *s. str.*) assemblages in winter and summer rainfall climates. *African Entomology* **2**, 45–51.

Davis, A.L.V. (1996) Diel and seasonal community dynamics in an assemblage of coprophagous, Afrotropical, dung beetles (Coleoptera: Scarabaeidae *s. str.*, Aphodiidae, and Staphylinidae: Oxytelinae). *Journal of African Zoology* **110**, 291–308.

Davis, A.L.V. (1997) Climatic and biogeographical associations of southern African dung beetles (Coleoptera: Scarabaeidae *s. str.*). *African Journal of Ecology* **35**, 10–38.

Davis, A.L.V. & Dewhurst, C.F. (1993) Climatic and biogeographical associations of Kenyan and northern Tanzanian dung beetles (Coleoptera: Scarabaeidae). *African Journal of Ecology* **31**, 290–305.

Davis, A.L.V. & Philips, T.K. (2005) Effect of deforestation on a southwest Ghana dung beetle assemblage (Coleoptera: Scarabaeidae) at the periphery of Ankasa conservation area. *Environmental Entomology* **34**, 1081–1088.

Davis, A.L.V. & Philips, T.K. (2009) Regional fragmentation of rain forest in West Africa and its effect on local dung beetle assemblage structure. *Biotropica* **41**, 215–220.

Davis, A.L.V. & Scholtz, C.H. (2001) Historical vs. ecological factors influencing global patterns of scarabaeine dung beetle diversity. *Diversity and Distributions* **7**, 161–174.

Davis, A.L.V., Chown, S.L. & Scholtz, C.H. (1999) Discontinuous gas-exchange cycles in *Scarabaeus* dung beetles (Coleoptera: Scarabaeidae): mass-scaling and temperature dependence. *Physiological and Biochemical Zoology* **72**, 555–565.

Davis, A.L.V., Chown, S.L., McGeoch, M.A. & Scholtz, C.H. (2000) A comparative analysis of metabolic rate in six *Scarabaeus* species (Coleoptera: Scarabaeidae) from southern Africa: further caveats when inferring adaptation. *Journal of Insect Physiology* **46**, 553–562.

Davis, A.J., Holloway, J.D., Huijbregts, H., Krikken, J., Kirk-Spriggs, A.H. & Sutton, S.L. (2001) Dung beetles as indicators of change in forests of Northern Borneo. *Journal of Applied Ecology* **38**, 593–616.

Davis, A.J., Huijbregts, H. & Krikken, J. (2000) The role of local and regional processes in shaping dung beetle communities in tropical forest plantations in Borneo. *Global Ecology and Biogeography* **9**, 281–292.

Davis, A.L.V. (1977) *The endocoprid dung beetles of southern Africa (Coleoptera: Scarabaeidae)*. M.Sc. thesis, Rhodes University, Grahamstown, South Africa.

Davis, A.L.V., Scholtz, C.H. & Philips, T.K. (2002) Historical biogeography of scarabaeine dung beetles. *Journal of Biogeography* **29**, 1217–1256.

Davis, A.L.V., van Aarde, R.J., Scholtz, C.H. & Delport, J.H. (2002a) Increasing representation of localized dung beetles across a chronosequence of regenerating vegetation and natural dune forest in South Africa. *Global Ecology and Biogeography* **11**, 191–209.

Davis, A.L.V., van Aarde, R.J., Scholtz, C.H. & Delport, J.H. (2002b) Convergence between dung beetle assemblages of a post-mining vegetational chronosequence and unmined dune forest. *Restoration Ecology* **11**, 29–42.

Davis, A.L.V., Scholtz, C.H., Dooley, P.W., Bharm, N. & Kryger, U. (2004) Scarabaeine dung beetles as indicators of biodiversity, habitat transformation and pest control chemicals in agro-ecosystems. *South African Journal of Science* **100**, 415–424.

Davis, A.L.V., Scholtz, C.H. & Deschodt, C. (2008a) Multi-scale determinants of dung beetle assemblage structure across abiotic gradients of the Kalahari-Nama Karoo ecotone, South Africa. *Journal of Biogeography* **35**, 1465–1480.

Davis, A.L.V., Frolov, A.V. & Scholtz, C.H. (2008b) *The African Dung Beetle Genera*. Protea Book House, Pretoria.

Davis, A.L.V., Brink, D.J., Scholtz, C.H., Prinsloo, L.C. & Deschodt, C.M. (2008c). Functional implications of temperature-correlated colour polymorphism in an iridescent, scarabaeine dung beetle. *Ecological Entomology* **33**, 771–779.

Day, T. & Young, K.A. (2004) Competitive and facultative evolutionary diversification. *BioScience* **54**, 101–109.

de Wit, M.J. (2003) Madagascar: heads it's a continent, tails it's an island. *Annual Review of Earth Planetary Sciences* **31**, 213–248.

Dellacasa, M. (1987) Contribution to a world-wide Catalogue of Aegialiidae, Aphodiidae, Aulonocnemidae, Termitotrogidae. (Part I). *Memorie della Società Entomologica italiana, Genova* **66**, 1–455.

Demary, K.C. & Lewis, S.M. (2007) Male courtship attractiveness and paternity success in *Photinus greeni* fireflies. *Evolution* **61**, 431–439.

Dempster, E.R. & Lerner, I.M. (1950) Heritability of threshold characters. *Genetics* **35**, 212–236.

Demuth, J.P. & Wade, M.J. (2007) Maternal expression increases the rate of bicoid evolution by relaxing selective constraint. *Genetica* **129**, 37–43.

Denholm-Young, P.A. (1978) *Studies of decomposing cattle dung and its associated fauna*. Ph.D. thesis, Oxford University, UK.

Department of Foreign Affairs and Trade (Australia) (1997) www.dfat.gov.au.

Deschodt, C.M., Kryger, U., Scholtz, C.H. (2007) New taxa of relictual Canthonini dung beetles (Scarabaeidae: Scarabaeinae) utilizing rock hyrax middens as refuges in South-western Africa. *Insect Systematics and Evolution* **38**, 4361–367.

Devictor, V., Clavel, J., Julliard, R., Lavergne, S., Mouillot, D., Thuiller, W., Venail, P., Villéger, S. & Mouquet, N. (2010) Defining and measuring ecological specialization. *Journal of Applied Ecology* **47**, 15–25.

DeWitt, T.J. & Scheiner, A.M. (2003) *Phenotypic plasticity. Functional and conceptual approaches*. Oxford University Press, Oxford.

Didham, R.K., Ghazoul, J., Stork, N.E. & Davis, A.J. (1996) Insects in fragmented forests: A functional approach. *Trends in Ecology and Evolution* **11**, 255–260.

Dixon, A.F.G., Kindlmann, P., Leps, J. & Holman, J. (1987) Why there are so few species of aphids, especially in the tropics. *American Naturalist* **129**, 580–592.

Donnell, D.M. & Strand, M.R. (2006) Caste-based differences in gene expression in the polyembryonic wasp *Copidosoma floridanum*. *Insect Biochemistry and Molecular Biology* **36**, 141–153.

Donner, J. (1995) *The Quaternary history of Scandinavia*. Cambridge University Press, Cambridge.

Dormont, L., Epinat, G. & Lumaret, J.P. (2004) Trophic preferences mediated by olfactory cues in dung beetles colonizing cattle and horse dung. *Environmental Entomology* **33**, 370–377.

Dormont, L., Rapior, S., McKey, D.B. & Lumaret, J.P. (2007) Influence of dung volatiles on the process of resource selection by coprophagous beetles. *Chemoecology* **17**, 23–30.

Doube, B.M. (1983) The habitat preference of some bovine dung beetles (Coleoptera, Scarabaeidae) in Hluhluwe game reserve, South-Africa. *Bulletin of Entomological Research* **73**, 357–371.

Doube, B.M. (1987) Spatial and temporal organization in communities associated with dung pads and carcasses. In: Gee, J.H.R.& Giller, P.S. (eds.) *Organization of Communities Past and Present*, pp. 255–280, Blackwell Scientific Publications, Oxford.

Doube, B.M. (1990) A functional classification for analysis of the structure of dung beetle assemblages. *Ecological Entomology* **15**, 371–383.

Doube, B.M. (1991) Dung Beetles of Southern Africa. In: Hanski, I.& Cambefort, Y. (eds.) *Dung Beetle Ecology*. Princeton University Press, Princeton, NJ.

Doube, B.M. & Moola, F. (1988) The effect of the activity of the African dung beetle *Catharsius tricornutus* De Geer (Coleoptera, Scarabaeidae) on the survival and size of the African buffalo fly, *Haematobia thirouxi*

potans (Bezzi) (Diptera, Muscidae), in bovine dung in the laboratory. *Bulletin of Entomological Research* 78, 63–73.
Doube, B.M., Giller, P.S. & Moola, F. (1988a) Dung burial strategies in some South-African coprine and onitine dung beetles (Scarabaeidae, Scarabaeinae). *Ecological Entomology* 13, 251–261.
Doube, B.M., Macqueen, A. & Fay, H.A.C. (1988b) Effects of dung fauna on survival and size of buffalo flies (*Haematobia* spp) breeding in the field in South-Africa and Australia. *Journal of Applied Ecology* 25, 523–536.
Doube, B.M., Macqueen, A., Ridsdill-Smith, T.J. & Weir, T.A. (1991) Native and Introduced Dung Beetles in Australia. In: Hanski, I.& Cambefort, Y. (eds.) *Dung Beetle Ecology*. pp. 255–278. Princeton University Press, Princeton, NJ.
Drent, D.H. & Daan, S. (1980) The prudent parent: energetic adjustments in avian breeding. *Ardea* 68, 225–252.
Driscoll, D.A. (2008) The frequency of metapopulations, metacommunities and nestedness in a fragmented landscape. *Oikos* 117, 297–309.
Driscoll, D.A. & Lindenmayer, D.B. (2009) Empirical tests of metacommunity theory using an isolation gradient. *Ecological Monographs* 79, 485–501.
du Toit, C.A., Scholtz, C.H. & Hyman, W.B. (2008) Prevalence of the dog nematode *Spirocerca lupi* in populations of its intermediate dung beetle host in the Tshwane (Pretoria) Metropole, South Africa. *Onderstepoort Journal of Veterinary Research* 75, 315–321.
Dufils, J.-M. (2003) Remaining forest cover. In: Goodman, S.M.& Benstead, J.P. (eds.) *The Natural History of Madagascar*, pp. 88–96. The University of Chicago Press, Chicago.
Dufrêne, M. & Legendre, P. (1997) Species assemblages and indicator species: the need for a flexible asymetrical approach. *Ecological Monographs* 67, 345–366.
Dugatkin, L.A. (2001) *Model Systems in Behavioural Ecology*. Princeton University Press, Princeton, NJ.
Dunbar, R.I.M., Korstjens, A.H. & Lehmann, J. (2009) Time as an ecological constraint. *Biological Reviews* 84, 413–429.
Duncan, F.D. (2002) The role of the subelytral cavity in the flightless dung beetle, *Circellium bacchus* (F.). *European Journal of Entomology* 99, 253–258.
Duncan, F.D. (2003) The role of the subelytral cavity in a tenebrionid beetle, *Onymacris multistriata* (Tenebrionidae: Adesmiini). *Journal of Insect Physiology* 49, 339–346.
Duncan, F.D. & Byrne, M.J. (2000) Discontinuous gas exchange in dung beetles: patterns and ecological implications. *Oecologia* 122, 452–458.
Duncan, F.D. & Byrne, M.J. (2002) Respiratory airflow in a wingless dung beetle. *Journal of Experimental Biology* 205, 2489–2497.
Duncan, F.D. & Byrne, M.J. (2005) The role of the mesothoracic spiracles in respiration in flighted and flightless dung beetles. *Journal of Experimental Biology* 208, 907–914.
Duncan, F.D. & Dickman, C.R. (2009) Respiratory strategies of tenebrionid beetles in arid Australia: does physiology beget nocturnality? *Physiological Entomology* 34, 52–60.
Duncan, R.P., Cassey, P. & Blackburn, T.M. (2009) Do climate envelope models transfer? A manipulative test using dung beetle introductions. *Proceedings of the Royal Society B* 267, 1449–1457.
Duncan, F.D., Förster, T.D. & Hetz, S.K. (2010) Pump out the volume –The effect of tracheal and subelytral pressure pulses on convective gas exchange in a dung beetle, *Circellium bacchus* (Fabricius). *Journal of Insect Physiology* in press.
Dunn, R.R. (2004) Recovery of faunal communities during tropical forest regeneration. *Conservation Biology* 18, 302–309.
Dunn, R.R. (2005) Modern insect extinctions, the neglected majority. *Conservation Biology* 19, 1030–1036.
Dunne, J.A. & Williams, R.J. (2009) Cascading extinctions and community collapse in model food webs. *Philosophical Transactions of the Royal Society B* 364, 1711–1723.
Eberhard, W.G. (1977) Fighting behavior of male *Golofa porteri* beetles (Scarabaeidae: Dynastinae). *Psyche* 84, 292–298.
Eberhard, W.G. (1979) The function of horns in *Podischnus agenor* (Dynastinae) and other beetles. In: Blum, M.S.& Blum, N.A. (eds.) *Sexual selection and reproductive competition in insects*. pp. 231–258. Academic press, London.
Eberhard, W.G. (1985) *Sexual Selection and Animal Genitalia*, Harvard University Press, Cambridge, Massachusetts.
Eberhard, W.G. (1990) Animal genitalia and female choice. *American Scientist* 78, 134–141.

Eberhard, W.G. (1996) *Female Control: Sexual Selection by Cryptic Female Choice.* Princeton University Press, Princeton.

Eberhard, W.G. (2009) Postcopulatory sexual selection: Darwin's omission and its consequences. *Proceedings of the National Academy of Science USA* **106**, 10025–10032.

Eberhard, W.G. & Gutierrez, E.E. (1991) Male dimorphisms in beetles and earwigs and the question of developmental constraints. *Evolution* **45**, 18–28.

Eberhard, W.G., Garcia-C., J.M. & Lobo, J. (2000) Size-specific defensive structures in a horned weevil confirm a classic battle plan: avoid fights with larger opponents. *Proceedings of the Royal Society of London B* **267**, 1129–1134.

Edmonds, W.D. (1972) Comparative skeletal morphology and evolution of the phanaeine dung beetles (Coleoptera: Scarabaeidae). *University of Kansas Science Bulletin* **49**, 731–874.

Edwards, P.B. (1986) Development and larval diapause in the southern African dung beetle *Onitis caffer* Boheman (Coleoptera, Scarabaeidae). *Bulletin of Entomological Research* **76**, 109–117.

Edwards, P.B. (1988) Field ecology of a brood-caring dung beetle *Kheper nigroaeneus* – habitat predictability and life history strategy. *Oecologia* **75**, 527–534.

Edwards, P.B. (2007) *Introduced dung beetles in Australia 1967–2007 – current status and future directions.* Dung Beetles for Landcare Farming Committee.

Edwards, P.B. (2009) *Improving sustainable land management systems in Queensland using dung beetles. Final Report of the 2001–2002 Queensland Dung Beetle Project.* Dung Beetles for Landcare Farming Committee (Original report 2003).

Edwards, P.B. & Aschenborn, H.H. (1988) Male reproductive behaviour of the African ball-rolling dung beetle, *Kheper nigroaeneus* (Coleoptera: Scarabaeidae). *The Coleopterist's Bulletin* **42**, 17–27.

Edwards, P.B. & Aschenborn, H.H. (1989) Maternal care of a single offpsring in the dung beetle *Kheper nigroaeneus*: the consequences of extreme parental care. *Journal of Natural History* **23**, 17–27.

Ehler, L.E. (2000) Critical issues related to nontarget effects in classical biological control of insects. In: Follett, P.A.& Duan, J.J. (eds.) *Nontarget effects of biological control.* pp. 3–13. Kluwer Academic Publishers, Boston, MA.

Ehrlich, P.R. & Hanski, I. (2004) *On the Wings of Checkerspots: A Model System for Population Biology.* Oxford University Press, Oxford.

Ekman, J. & Ericson, Per G.P. (2006) Out of Gondwanaland: the evolutionary history of cooperative breeding and social behaviour among crows, magpies, jays and allies. *Proceedings of the Royal Society of London B* **273**, 1117–1125.

Elmqvist, T., Folke, C., Nystrom, M., Peterson, G., Bengtsson, J., Walker, B. & Norberg, J. (2003) Response diversity, ecosystem change, and resilience. *Frontiers In Ecology And The Environment* **1**, 488–494.

Emlen, D.J. (1994) Environmental control of horn length dimorphism in the beetle *Onthophagus acuminatus* (Coleoptera: Scarabaeidae). *Proceedings of the Royal Society of London B* **256**, 131–136.

Emlen, D.J. (1996) Artificial selection on horn length body size allometry in the horned beetle *Onthophagus acuminatus* (Coleoptera: Scarabaeidae). *Evolution* **50**, 1219–1230.

Emlen, D.J. (1997a) Alternative reproductive tactics and male-dimorphism in the horned beetle *Onthophagus acuminatus* (Coleoptera: Scarabaeidae). *Behavioural Ecology and Sociobiology* **41**, 335–342.

Emlen, D.J. (1997b) Diet alters male horn allometry in the beetle *Onthophagus acuminatus* (Coleoptera: Scarabeidae). *Proceedings of the Royal Society of London B* **264**, 567–574.

Emlen, D.J. (2000) Integrating development with evolution: a case study with beetle horns. *Bioscience* **50**, 403–418.

Emlen, D.J. (2001) Costs and the diversification of exaggerated animal structures. *Science* **291**, 1534–1536.

Emlen, D.J. (2008) The evolution of animal weapons. *Annual Review of Ecology and Systematics* **39**, 387–413.

Emlen, D.J. & Allen, C.E. (2004) Genotype to phenotype: physiological control of trait size and scaling in insects. *Integrated Comparative Biology* **43**, 617–634.

Emlen, D.J. & Nijhout, H.F. (1999) Hormonal control of male horn length dimorphism in the dung beetle *Onthophagus taurus* (Coleoptera: Scarabaeidae). *Journal of Insect Physiology* **45**, 45–53.

Emlen, D.J. & Nijhout, H.F. (2001) Hormonal Control of male horn length dimorphism in the dung beetle *Onthophagus taurus* (Coleoptera: Scarabaeidae): A second critical period of sensitivity to juvenile hormone. *Journal of Insect Physiology* **47**, 1045–1054.

Emlen, S.T. & Oring, L.W. (1977) Ecology, sexual selection, and the evolution of mating systems. *Science* **197**, 215–223.
Emlen, D.J. & Philips, T.K. (2006) Phylogenetic evidence for an association between tunnelling behavior and the evolution of horns in dung beetles (Coleoptera: Scarabaeidae: Scarabaeinae). *Coleopterist Society Monograph* **5**, 47–56.
Emlen, D.J., Hunt, J. & Simmons, L.W. (2005a) Evolution of sexual dimorphism in the expression of beetle horns: phylogenetic evidence for modularity, evolutionary lability, and constraint. *American Naturalist* **166**, S42–S68.
Emlen, D.J., Marangelo, J., Ball, B. & Cunningham, C.W. (2005b) Diversity in the weapons of sexual selection: horn evolution in the beetle genus *Onthophagus* (Coleoptera: Scarabaeidae). *Evolution* **59**, 1060–1084.
Emlen, D.J., Szafran, Q., Corley, L.S. & Dworkin, I. (2006) Insulin signaling and limb-patterning: candidate pathways for the origin and evolutionary diversification of beetle 'horns'. *Heredity* **97**, 179–191.
Emlen, D.J., Corley Lavine, L. & Ewen-Campen, B. (2007) On the origin and evolutionary diversification of beetle horns. *Proceedings of the National Academy of Sciences USA* **104**, 8661–8668.
Erwin, D.H. (2000) Macroevolution is more than repeated rounds of microevolution. *Evolution and Development* **2**, 78–84.
Esch, H.E. & Burns, J.E. (1995) Honeybees use optic flow to measure the distance of a food source. *Naturwissenschaften* **82**, 38–40.
Esch, H.E. & Burns, J.E. (1996) Distance estimation by foraging honeybees. *The Journal of Experimental Biology* **199**, 155–162.
Escobar, F. (2004) Structure and composition of dung beetle (Scarabaeinae) assemblages in an Andean landscape, Colombia. *Tropical Zoology* **17**, 123–136.
Escobar, F., Halffter, G. & Arellano, L. (2007) From forest to pasture: an evaluation of the influence of environment and biogeography on the structure of dung beetle (Scarabaeinae) assemblages along three altitudinal grafients in the Neotropical region. *Ecography* **30**, 193–208.
Estrada, A. & Coates-Estrada, R. (2002) Dung beetles in continuous forest, forest fragments and in an agricultural mosaic habitat island at Los Tuxtlas, Mexico. *Biodiversity and Conservation* **11**, 1903–1918.
Estrada, A., Coates-Estrada, R., Anzures, A. & Cammarano, P. (1998) Dung and carrion beetles in tropical rain forest fragments and agricultural habitats at Los Tuxtlas, Mexico. *Journal of Tropical Ecology* **14**, 577–593.
Evans, J.D. & Wheeler, D.E. (1999) Differential gene expression between developing queens and workers in the honey bee, *Apis mellifera*. *Proceedings of the National Academy of Science USA* **96**, 5575–5580.
Evans, J.D. & Wheeler, D.E. (2001a) Expression profiles during honeybee caste determination. *Genome Biology* **2**, Research 0001.
Evans, J.D. & Wheeler, D.E. (2001b) Gene expression and the evolution of insect polyphenisms. *Bioessays* **23**, 62–68.
Evans, J.P. & Simmons, L.W. (2008) The genetic basis of traits regulating sperm competition and polyandry: can selection favour the evolution of good- and sexy-sperm? *Genetica* **134**, 5–19.
Evans, J.P., Zane, L., Francescato, S. & Pilastro, A. (2003) Directional postcopulatory sexual selection revealed by artificial insemination. *Nature* **421**, 360–363.
Exner, S. (1891) *The Physiology of the Compound Eyes of Insects and Crustaceans*. Translated by R.C. Hardic (ed.) (1989) Springer-Verlag, Berlin.
Fabre, J.H. (1918) *The Sacred Beetle and Others*, Hodder and Stoughton, London.
Falconer, D.S. (1965) The inheritance of liability to certain diseases, estimated from the incidence among relatives. *Annals of Human Genetics* **29**, 51–76.
Falconer, D.S. & Mackay, T.F.C. (1996) *Introduction to Quantitative Genetics*. Longman, New York.
Falqueto, S.A., Vaz-De-Mello, F.Z. & Schoereder, J.H. (2005) Are fungivorous Scarabaeidae less specialist? *Ecologia Austral* **15**, 17–22.
Favila, M.E. (1988) Chemical labeling of the food ball during rolling by males of the subsocial coleopteran *Canthon cyanellus cyanellus* Leconte (Scarabaeidae). *Insectes Sociaux* **35**, 125–129.
Favila, M.E. (1993) Some ecological factors affecting the life- style of *Canthon cyanellus cyanellus* (Coleoptera Scarabaeidae): an experimental approach. *Ethology, Ecology and Evolution* **5**, 319–328.
Favila, M. E & Díaz, A. (1996) *Canthon cyanellus cyanellus* LeConte (Coleoptera: Scarabaeidae) makes a nest in the field with several brood balls. *The Coleopterists Bulletin* **50**, 52–60.
Favila, M.E. & Halffter, G. (1997) The use of indicator groups for measuring biodiversity as related to community structure and function. *Acta Zoológica Mexicana* **72**, 1–25.

Feehan, J., Hughes, R.D., Bryce, M.A. & Runko, S. (1985) Bush fly abundance and population events in relation to dung beetle catches on the south coast of New-South-Wales. *Journal of the Australian Entomological Society* **24**, 37–43.

Feeley, K.J. & Terborgh, J.W. (2005) The effects of herbivore density on soil nutrients and tree growth in tropical forest fragments. *Ecology* **86**, 116–124.

Feer, F. & Pincebourde, S. (2005) Diel flight activity and ecological segregation within an assemblage of tropical forest dung and carrion beetles. *Journal of Tropical Ecology* **21**, 21–30.

Feinsinger, P. (2001) *Designing field studies for biodiversity conservation*. The Nature Conservancy and Island Press, Washington, DC.

Ferrar, P. (1975) Disintegration of dung pads in north Queensland before the introduction of exotic dung beetles. *Australian Journal of Experimental Agriculture and Animal Husbandry* **15**, 325–329.

Fincher, G.T. (1981) The potential value of dung beetles in pasture ecosystems. *Journal of the Georgia Entomological Society* **16**, 316–333.

Fincher, G.T. (1986) Importation, colonization, and release of dung-burying scarabs. *Miscellaneous Publications of the Entomological Society of America* **61**, 69–76.

Fincher, G.T., Monson, W.G. & Burton, G.W. (1981) Effects of cattle feces rapidly buried by dung beetles on yield and quality of coastal Bermuda grass. *Agronomy Journal* **73**, 775–779.

Fincher, G.T., Stewart, T.B. & Hunter, J.S. (1983) The 1981 distribution of *Onthophagus gazella* Fabricius from releases in Texas and *Onthophagus taurus* Schreber from an unknown release in Florida (Coleoptera: Scarabaeidae). *The Coleopterists Bulletin* **37**, 159–163.

Finn, J.A. (2001) Ephemeral resource patches as model systems for diversity-function experiments. *Oikos* **92**, 363–366.

Finn, J.A. & Giller, P.S. (2000) Patch size and colonisation patterns: an experimental analysis using north temperate coprophagous dung beetles. *Ecography* **23**, 315–327.

Finn, J.A. & Gittings, T. (2003) A review of competition in north temperate dung beetle communities. *Ecological Entomology* **28**, 1–13.

Finnish Forest Research Institute (2005) *Finnish Statistical Yearbook of Forestry*. Finnish Forest Research Institute, Vantaa.

Fisher, B., Turner, R.K. & Morling, P. (2009) Defining and classifying ecosystem services for decision making. *Ecological Economics* **68**, 643–653.

Fisher, D.O., Double, M.C., Blomberg, S.P., Jennions, M.D. & Cockburn, A. (2006) Post-mating sexual selection increases lifetime fitness of polyandrous females in the wild. *Nature* **444**, 89–92.

Fisher, R.A. (1915) The evolution of sexual preference. *Eugenics Review* **7**, 184–192.

Fisher, R.A. (1930) *The Genetical Theory of Natural selection*, Clarendon Press, Oxford.

Fitzpatrick, J.L., Montgomerie, R., Desjardins, J.K., Stiver, K.A., Kolm, N. & Balshine, S. (2009) Female promiscuity promotes the evolution of faster sperm in cichlid fishes. *Proceedings of the National Academy of Science USA* **106**, 1128–1132.

Folke, C., Carpenter, S., Walker, B., Scheffer, M., Elmqvist, T., Gunderson, L. & Holling, C.S. (2004) Regime shifts, resilience, and biodiversity in ecosystem management. *Annual Review of Ecology Evolution and Systematics* **35**, 557–581.

Forgie, S.A., Grebennikov, V.V. & Scholtz, C.H. (2002) Revision of *Sceliages* Westwood, a millipede-eating genus of southern African dung beetles (Coleoptera: Scarabaeidae). *Invertebrate Systematics* **16**, 931–955.

Forgie, S.A., Philips, T.K. & Scholtz, C.H. (2005) Evolution of the Scarabaeini (Scarabaeidae: Scarabaeinae). *Systematic Entomology* **30**, 60–96.

Forgie, S.A., Kryger, U., Bloomer, P. & Scholtz, C.H. (2006) Evolutionary relationships among the Scarabaeini (Coleoptera: Scarabaeidae) based on combined molecular and morphological data. *Molecular Phylogenetics and Evolution* **40**, 662–678.

Forsyth, A. & Alcock, J. (1990) Female mimicry and resource defence polygyny by males of a tropical rove beetle, *Leistotrophus versicolor* (Coleoptera: Staphylinidae). *Behavioral Ecolology and Sociobiology* **26**, 325–330.

Fox, C.W., Nilsson, J.A. & Mousseau, T.A. (1997) The ecology of diet expansion in a seed-feeding beetle: pre-existing variation, rapid adaptation and maternal effects? *Evolutionary Ecology* **11**, 183–194.

Francke, W. & Dettner, K. (2005) Chemical signalling in beetles. In: S. Schulz (ed.) *The chemistry of pheromones and other semiochemicals*. pp. 85–166. Topics in Current Chemistry, Volume 240. Springer, Berlin.

Frankino, W.A., Stern D.S. & Brakefield P.M. (2005) Developmental constraints and natural selection in the evolution of allometries. *Science* **307**, 718–720.

Frankino, W.A., Emlen, D.J. & Shingleton, A.W. (2008) Experimental approaches to studying the evolution of animal form: The shape of things to come. In: Garland, Jr., T.& Rose, M.R. (eds.) *Experimental Evolution: Concepts, Methods, and Applications of Selection Experiments.* pp. 419–478. University of California Press, California.

Frantsevich, L., Govardovski, V., Gribakin, F., Nikolajcv, G., Pichka, V., Polanovsky, A., Shevchenk, V. & Zolotov, V. (1977) Astroorientation in *Lethrus* (Coleoptera, Scarabaeidae). *Journal of Comparative Physiology* **121**, 253–271.

Freckleton, R.P. & Jetz, W. (2009) Space versus phylogeny: disentangling phylogenetic and spatial signals in comparative data. *Proceedings of the Royal Society of London B* **276**, 21–30.

Freestone, A.L. & Inouye, B.D. (2006) Dispersal limitation and environmental heterogeneity shape scale-dependent diversity patterns in plant communities. *Ecology* **87**, 2425–2432.

Fricke, C. & Arnqvist, G. (2007) Rapid adaptation to a novel host in a seed beetle (*Callosobruchus maculatus*): the role of sexual selection. *Evolution* **61**, 440–454.

Fricke, C., Wigby, S., Hobbs, R. & Chapman, T. (2009) The benefits of male ejaculate sex peptide transfer in *Drosophila melanogaster. Journal of Evolutionary Biology* **22**, 275–286.

Galante, E., Garciaroman, M., Barrera, I. & Galindo, P. (1991) Comparison of spatial distribution patterns of dung feeding scarabs (Coleoptera, Scarabaeidae, Geotrupidae) in wooded and open pastureland in the Mediterranean 'Dehesa' area of the Iberian peninsula. *Environmental Entomology* **20**, 90–97.

García-González, F. & Simmons, L.W. (2007) Shorter sperm confer higher competitive fertilization success. *Evolution* **61**, 816–824.

García-González, F., Núñez, Y., Ponz, F., Roldán, E.R.S. & Gomendio, M. (2003) Sperm competition mechanisms, confidence of paternity, and the evolution of paternal care in the golden egg bug (*Phyllomorpha laciniata*). *Evolution* **57**, 1078–1088.

Gardner, T.A. (2010) *Monitoring Forest Biodiversity: Improving Conservation Through Ecologically-Responsible Management.* Earthscan Publications Ltd., London.

Gardner, T.A., Barlow, J., Araujo, I.S., Avila-Pires, T.C.S., Bonaldo, A.B., Costa, J.E., Esposito, M.C., Ferreira, L.V., Hawes, J., Hernandez, M.I.M., Hoogmoed, M., Leite, R.N., Lo-Man-Hung, N.F., Malcolm, J.R., Martins, M.B., Mestre, L.A.M., Miranda-Santos, R., Nunes-Gutjahr, A.L., Overal, W.L., Parry, L.T.W., Peters, S.L., Ribeiro-Junior, M.A., Da Silva, M.N.F., Da Silva Motta, C. & Peres, C. (2008a) The cost-effectiveness of biodiversity surveys in tropical forests. *Ecology Letters* **11**, 139–150.

Gardner, T.A., Hernandez, M.I.M., Barlow, J. & Peres, C.A. (2008b) Understanding the biodiversity consequences of habitat change: the value of secondary and plantation forests for neotropical dung beetles. *Journal of Applied Ecology* **45**, 883–893.

Gardner, T.A., Barlow, J., Chazdon, R.L., Ewers, R., Harvey, C.A., Peres, C.A. & Sodhi, N.S. (2009) Prospects for tropical forest biodiversity in a human-modified world. *Ecology Letters* **12**, 561–582.

Garland, T., Jr., Dickerman, A.W. Janis, C.M. & Jones, J.A. (1993) Phylogenetic analysis of covariance by computer simulation. *Systematic Biology* **42**, 265–292.

Gaston, K.J. (2003) *The Structure and Dynamics of Geographic Ranges.* Oxford University Press, Oxford.

Gaston, K.J. (2009) Geographic range limits: achieving synthesis. *Proceedings of the Royal Society of London B* **276**, 1395–1406.

Gaston, K.J. & Chown, S.L. (1999a) Elevation and climatic tolerance: a test using dung beetles. *Oikos* **86**, 584–590.

Gaston, K.J. & Chown, S.L. (1999b) Why Rapoport's rule does not generalise. *Oikos* **84**, 309–312.

Gaston, K.J. & Williams, P.H. (1996) Spatial patterns in taxonomic diversity. In: Gaston, K.J. (ed.) *Biodiversity.* pp. 202–229. Blackwell, Oxford.

Gaston, K.J., Blackburn, T.M. & Spicer, J.I. (1998) Rapoport's rule: time for an epitaph? *Trends in Ecology and Evolution* **13**, 70–74.

Gaston, K.J., Chown, S.L. & Evans, K.L. (2008) Ecogeographic rules: elements of a synthesis. *Journal of Biogeography* **35**, 483–500.

Gaston, K.J., Chown, S.L., Calosi, P., Bernardo, J., Bilton, D.T., Clarke, A., Clusella-Trullas, S., Ghalambor, C. K., Konarzewski, M., Peck, L.S., Porter, W.P., Pörtner, H.-O., Rezende, E.L., Schulte, P.M., Spicer, J.I., Stillman, J.H., Terblanche, J.S. & van Kleunen, M. (2009) Macrophysiology: a conceptual re-unification. *American Naturalist* **174**, 595–612.

Gauld, I.D., Gaston, K.J. & Janzen, D.H. (1992) Plant allelochemicals, tritrophic interactions and the anomalous diversity of tropical parasitoids: the 'nasty' host hypothesis. *Oikos* **65**, 353–357.

Gavrilets, S. (2000) Rapid evolution of reproductive barriers driven by sexual conflict. *Nature* **403**, 886–889.
Gavrilets, S. & Waxman, D. (2002) Sympatric speciation by sexual conflict. *Proceedings of the National Academy of Sciences USA* **99**, 10533–10538.
Genise, J.F., Mangano, M.G., Buatois, L.A., Laza, J.H., Verde, M. (2000) Insect trace fossil association in paleosols; the *Coprinisphaera* ichnofacies. *Palaios* **15**, 49–64.
Gibbs, A.G. (2002) Water balance in desert *Drosophila*: lessons from non-charismatic microfauna. *Comparative Biochemistry and Physiology A* **133**, 781–789.
Gibbs, A.G., Fukuzato, F. & Matzkin, L.M. (2003) Evolution of water conservation mechanisms in *Drosophila*. *Journal of Experimental Biology* **206**, 1183–1192.
Gibson, G. & Muse, S.V. (2009) *A primer of genome science. 3rd edition*. Sinauer, MA.
Gilbert, S.F. (2006) *Developmental Biology*. Sinauer, MA.
Gilbert, S.F. & Epel, D. (2009) *Ecological Developmental Biology and Epigenesis: An Integrated Approach to Embryology, Evolution, and Medicine*. Sinauer, MA.
Gilburn, A.S., Foster, S.P. & Day, T.H. (1993) Genetic correlation between a female mating preference and the preferred male character in seaweed flies (*Coelopa frigida*). *Evolution* **47**, 1788–1795.
Gill, B.D. (1991) Dung beetles in tropical american forests. In: Hanski I.& Cambefort, Y. (eds.) *Dung Beetle Ecology*. pp. 211–229. Princeton University Press, Princeton, NJ.
Giller, P.S. & Doube, B.M. (1989) Experimental-analysis of interspecific and intraspecific competition in dung beetle communities. *Journal of Animal Ecology* **58**, 129–142.
Gillett, C.P.D.T., Edmonds, W.D. & Villamarin, S. (2009) Distribution and biology of the rare scarab beetle *Megatharsis buckleyi* Waterhouse, 1891 (Coleoptera: Scarabaeinae: Phanaeini). *Insecta Mundi* **0080**, 1–8.
Gittings, T. & Giller, P.S. (1997) Life history traits and resource utilisation in an assemblage of north temperate *Aphodius* dung beetles (Coleoptera: Scarabaeidae). *Ecography* **20**, 55–66.
Glazier, D.S. (2005) Beyond the '3/4-power law': variation in the intra- and interspecific scaling of metabolic rate in animals. *Biological Reviews* **80**, 611–662.
Glazier, D.S. (2008) Effects of metabolic level on the body size scaling of metabolic rate in birds and mammals. *Proceedings of the Royal Society of London B* **275**, 1405–1410.
Glazier, D.S. (2010) A unifying explanation for diverse metabolic scaling in animals and plants. *Biological Reviews* **85**, 111–138.
Gokan, N. (1989a) The compound eye of the dung beetle *Geotrupes auratus* (Coleoptera: Scarabaeidae). *Applied Entomology and Zoology* **24**, 133–146.
Gokan, N. (1989b) Fine structure of the compound eye of the dung beetle *Aphodius haroldianus* (Coleoptera: Scarabaeidae). *Applied Entomology and Zoology* **24**, 483–486.
Gokan, N. (1989c). Fine structure of the compound eye of the dung beetle *Ochodaeus maculatus* (Coleoptera: Scarabaeidae). *Japanese Journal of Entomology* **57**, 823–830.
Gokan, N. (1990) Fine structure of the compound eye of the dung beetle *Onthophagus lenzii* (Coleoptera, Scarabaeidae). *Japanese Journal of Entomology* **4**, 823–830.
Gokan, N. & Meyer-Rochow, V.B. (1990) The compound eye of the dung beetle, *Onthophagus posticus* (Coleoptera: Scarabaeidae). *New Zealand Entomologist* **13**, 7–15.
Gokan, N. & Meyer-Rochow, V.B. (2000) Morphological comparisons of compound eyes in Scarabaeoidea (Coleoptera) related to the beetles' daily activity maxima and phylogenetic positions. *Journal of Agricultural Science* **45**, 15–61.
Gokan, N., Meyer-Rochow, B., Nakazawa, A. & Iida, K. (1998) Compound eye ultrastructures in six species of ecologically diverse stag-beetles (Coleoptera, Scarabaeoidea, Lucanidae). *Applied Entomology and Zoology* **33**, 157–169.
Goldberg, D., Wildová, R. & Herben, T. (2008) Consistency vs. contingency of trait–performance linkages across taxa. *Evolutionary Ecology* **22**, 477–481.
Gomendio, M. & Roldan, E.R.S. (1991) Sperm competition influences sperm size in mammals. *Proceedings of the Royal Society of London B* **243**, 181–185.
González-Megías, A. & Sánchez-Piñero, F. (2003) Effects of brood parasitism on host reproductive success: evidence from larval interactions among dung beetles. *Oecologia* **134**, 195–202.
González-Megías, A. & Sánchez-Piñero, F. (2004) Response of host species to brood parasitism in dung beetles: importance of nest location by parasitic species. *Functional Ecology* **18**, 914–924.
Gordon, R.D. & Cartwright, O.L. (1974) Survey of food preferences of some North American Canthonini (Coleoptera: Scarabaeidae). *Entomological News* **85**, 181–185.
Gould, S.J. (1990) *Wonderful Life; the Burgess Shale and the Nature of History*. Hutchinson Radius. UK.

Gould, S.J. & Vrba, E.S. (1982) Exaptation: a missing term in the science of form. *Paleobiology* 8, 4–15.

Grafen, A. (1980) Opportunity cost, benefit and the degree of relatedness. *Animal Behaviour* 28, 967–968.

Greenslade, P. & Greenslade, P.J.M. (1987) Ecological strategies in Collembola: a new approach ot the use of terrestrial invertebrates in environmental assessment. In: Striganova, B.R. (ed.) *Soil Fauna and Soil Fertility.* Nuaka, Moscow.

Grimaldi, D. & Engel, M.S. (2005) *Evolution of the Insects*. Cambridge University Press, Cambridge, UK.

Gripenberg, S., Ovaskainen, O., Morriën, E. & Roslin, T. (2008) Spatial population structure of a specialist leaf-mining moth. *Journal of Animal Ecology* 77, 757–767.

Gross, M.R. (1996) Alternative reproductive tactics: diversity within sexes. *Trends in Ecology and Evolution* 11, 92–98.

Gustafsson, L. (1986) Lifetime reproductive success and heritability: empirical support for Fisher's fundamental theorem. *American Naturalist* 128, 761–764.

Hadley, N.F. (1994) *Water Relations of Terrestrial Arthropods*. Academic Press, San Diego, CA.

Halffter, G. (1991) Historical and ecological factors determining the geographical distribution of beetles (Coleoptera: Scarabaeidae: Scarabaeinae). *Folia Entomologica Mexicana* 82, 195–238.

Halffter, G. & Arellano, L. (2002) Response of dung beetle diversity to human-induced changes in a tropical landscape. *Biotropica* 34, 144–154.

Halffter, G. & Edmonds, W.D. (1982) *The nesting behavior of dung beetles (Scarabaeinae): an ecological and evolutive approach*. Instituto de Ecología, Mexico.

Halffter, G. & Favila, M.E. (1993) The Scarabaeinae (Insecta: Coleoptera) an animal group for analyzing, inventorying and monitoring biodiversity in tropical rainforest and modified landscapes. *Biology International* 27, 15–21.

Halffter, G. & Halffter, V. (1989) Behavioral evolution of the non-rolling roller beetles (Coleoptera: Scarabaeidae: Scarabaeinae). *Acta Zoologica Mexicana (Nueva Serie)* 32, 1–53.

Halffter, G. & Halffter, V. (2009) Why and where coprophagous beetles (Coleoptera: Scarabaeinae) eat seeds, fruits or vegetable detritus. *Boletín Sociedad Entomológica Aragonesa* 45, 1–22.

Halffter, G. & Matthews, E.G. (1966) The natural history of dung beetles of the subfamily Scarabaeinae (Coleoptera: Scarabaeidae). *Folia Entomologica Mexicana* 12–14, 1–312.

Halffter, G., Halffter, V. & Huerta, C. (1980) Mating and nesting behavior of *Eurysternus* (Coleoptera: Scarabaeinae). *Quaestiones Entomologicae* 16, 599–620.

Halffter, G., Huerta, C. & Lopez-Portillo, J. (1996) Parental care and offspring survival in *Copris incertus* Say, a sub-social beetle. *Animal Behaviour* 52, 133–139.

Hamilton, W.D. (1964) The genetical evolution of social behaviour. *Journal of Theoretical Biology* 7, 1–16.

Hamilton, W.D. (1979). Wingless and fighting males in fig wasps and other species. In: Blum, M.S.& Blum, N.A. (eds.) *Reproductive competition, mate choice and sexual selection in insects*. pp. 167–220. Academic Press, New York.

Hanski, I. (1983) Distributional ecology and abundance of dung and carrion–feeding beetles (Scarabaeidae) in tropical rain forest in Sarawak, Borneo. *Acta Zoologica Fennica* 167, 1–45.

Hanski, I. (1986) Individual behaviour, population dynamics and community structure in *Aphodius* (Scarabaeidae) in Europe. *Acta Oecologica, Oecologia Generalis* 7, 171–187.

Hanski, I. (1988) Are the pyramids deified dung pats? *Trends in Ecology and Evolution* 3, 34–36.

Hanski, I. (1991) North Temperate Dung Beetles. In: Hanski I.& Cambefort, Y. (eds.) *Dung Beetle Ecology*, pp. 75–96. Princeton University Press, Princeton, NJ.

Hanski, I. (1999) *Metapopulation Ecology.* Oxford University Press, Oxford.

Hanski, I. & Cambefort, Y. (1991) *Dung Beetle Ecology.* Princeton University Press, Princeton, NJ.

Hanski, I. & Gaggiotti, O. (2004) *Ecology, Genetics, and Evolution of Metapopulations*. Elsevier Academic Press, Amsterdam.

Hanski, I. & Krikken, J. (1991) Dung beetles in tropical forest in South–East Asia. In: Hanski I.& Cambefort, Y. (eds.) *Dung Beetle Ecology.* pp. 179–197. Princeton University Press, Princeton.

Hanski, I. & Simberloff, D. (1997) The metapopulation approach, its history, conceptual domain, and application to conservation. In: Hanski, I.& Gilpin, M.E. (eds.) *Metapopulation Biology: Ecology, Genetics and Evolution*. pp. 5–26. Academic Press, San Diego, CA.

Hanski, I., Kouki, J. & Halkka, A. (1993) Three explanations of the positive relationship between distribution and abundance of species. In: Ricklefs, R.E.& Schluter, D. (eds.) *Species Diversity in Ecological Communities. Historical and Geographical Perspectives*. pp. 108–116. University of Chicago Press, Chicago.

Hanski, I., Koivulehto, H., Cameron, A. & Rahagalala, P. (2007) Deforestation and apparent extinctions of endemic forest beetles in Madagascar. *Biology Letters* **3**, 344–347.

Hanski, I., Wirta, H., Nyman, T. & Rahagalala, P. (2008) Resource shifts in Malagasy dung beetles: contrasting processes revealed by dissimilar spatial genetic patterns. *Ecology Letters* **11**, 1208–1215.

Hardie, J. & Lees, A.D. (1981) Endocrine control of polymorphism and polyphenism. In: Kerkut, G.A.& Gilbert, L.I. (eds.) *Comprehensive insect physiology, biochemistry and pharmacology.* Volume 8, pp. 441–490. Pergamon, New York.

Hardy, I.C.W. & Field, S.A. (1998) Logistic analysis of animal contests. *Animal Behaviour* **56**, 787–792.

Harper, G.J., Steininger, M.K., Tucker, C.J., Juhn, D. & Hawkins, F. (2007) Fifty years of deforestation and forest fragmentation in Madagascar. *Environmental Conservation* **34**, 325–333.

Harris, R.L., Onaga, K., Blume, R.R., Roth, J.P. & Summerlin, J.W. (1982) Survey of beneficial insects in undisturbed cattle droppings on Oahu, Hawaii. *Proceedings Hawaiian Entomological Society* **24**, 91–95.

Harrison, F., Barta, Z., Cuthill, I. & Székely, T. (2009) How is sexual conflict over parental care resolved? A meta-analysis. *Journal of Evolutionary Biology* **22**, 1800–1812.

Harrison, J. du G. & Philips, T.K. (2003) Phylogeny of *Scarabaeus* (*Pachysoma* MacLeay) *stat.n.*, and related flightless Scarabaeini (Scarabaeidae: Scarabaeinae). *Annals of the Transvaal Museum* **40**, 47–71.

Harrison, S. (1994) Metapopulations and conservation. In: Edwards, P.J., Webb, N.R.& May, R.M. (eds.) *Large-Scale Ecology and Conservation Biology.* pp 111–128. Blackwell, Oxford.

Harrison, S. & Taylor, A.D. (1997) Empirical evidence for metapopulation dynamics. In: Hanski, I.& Gilpin, M.E. (eds.) *Metapopulation Biology: Ecology, Genetics, and Evolution.* pp 27–42. Academic Press, San Diego, CA.

Hartfelder, K. & Emlen, D.J. (2005) Endocrine control of insect polyphenism. In: Gilbert L.I., Iatrou, K.& Gill, S.S. (eds.) *Comprehensive Molecular Insect Science.* pp. 651–703. Elsevier, Oxford.

Hasegawa, T. & Yamashita, N. (1985) Semifield experiments on the control of the dung breeding fly, *Musca hervei* (Diptera: Muscidae) by the activity of native dung beetle, *Onthophagus lenzii* (Coleoptera: Scarabaeidae). *Annual Report Plant Protection, North Japan.* pp. 110–113. North Japan.

Hazel, W.N. (1977) The genetic basis of pupal colour dimorphism and its maintenance by natural selection in *Papillio polyxenes* (Papilionidae: Lepidoptera). *Heredity* **59**, 227–236.

Hazel, W.N. (2002) The environmental and genetic control of seasonal polyphenism in larval color and its adaptive significance in a swallowtail butterfly. *Evolution* **56**, 342–348.

Hazel, W.N. & Smock, R. (1993) Modeling selection on conditional strategies in stochastic environments. In: Yoshimura, J.& Clark, C. (eds.) *Adaptation in Stochastic Environments.* Springer-Verlag, Berlin.

Hazel, W.N. & West, D.A. (1982) Pupal colour dimorphism in swallowtail butterflies as a threshold trait – selection in *Eurytides-marcellus* (Cramer). *Heredity* **49**, 295–301.

Hazel, W.N., Smock, R. & Johnson, M.D. (1990) A polygenic model for the evolution and maintenance of conditional strategies. *Proceedings of the Royal Society of London B* **242**, 181–187.

Hazel, W.N., Smock, R. & Lively, C.M. (2004) The ecological genetics of conditional strategies. *American Naturalist* **163**, 888–900.

Heinrich, B. (1993) *The Hot-blooded Insects: Strategies and Mechanisms of Thermoregulation.* Harvard University Press, Cambridge, MA.

Heinrich, B. & Bartholomew, G.A. (1979) Roles of endothermy and size in inter- and intraspecific competition for elephant dung in an African dung beetle, *Scarabaeus laevistriatus. Physiological Zoology* **52**, 484–496.

Heming, B.S. (2003) *Insect Development and Evolution.* Cornell University Press, New York.

Hernández, M.I.M. (2002) The night and day of dung beetles (Coleoptera, Scarabaeidae) in the Sierra do Japi, Brazil: elytral colour related to daily activity. *Revista Brasileira de Entomologia* **46**, 597–600.

Herrera, E.R.T., Vulinec, K., Knogge, C., Heymann, E.W. (2002) Sit and wait at the source of dung—an unusual strategy of dung beetles. *Ecotropica* **8**, 87–88.

Herrling, P.L. (1976) Regional distribution of three ultrastructural retinula types in the retina of *Cataglyphis bicolor* Fabr. (Formicidae, Hymenoptera). *Cell Tissue Research* **169**, 247–266.

Herzmann, D. & Labhart, T. (1989) Spectral sensitivity and absolute threshold of polarization vision in crickets: a behavioral study. *Journal of Comparative Physiology A* **165**, 315–319.

Hetz, S.K. & Bradley, T.J. (2005) Insects breathe discontinuously to avoid oxygen toxicity. *Nature* **433**, 516–519.

Heywood, V.H. (1995) *Global Biodiversity Assessment. United Nations Environment Programme*. Cambridge University Press, Cambridge, UK.

Higgie, M., Chenoweth, S. & Blows, M.W. (2000) Natural selection and the reinforcement of mate recognition. *Science* **290**, 519–521.

Hodgson, J.G., Wilson, P.J., Hunt, R., Grime, J.P. & Thompson, K. (1999) Allocating C-S-R plant functional types: a soft approach to a hard problem. *Oikos* **85**, 282–294.

Hoffmann, A.A., Sørensen, J.G. & Loeschcke, V. (2003) Adaptation of *Drosophila* to temperature extremes: bringing together quantitative and molecular approaches. *Journal of Thermal Biology* **28**, 175–216.

Hoffman, E.A. & Goodisman, M.A.D. (2007) Gene expression and the evolution of phenotypic diversity in social wasps. *BMC Biology* **5**, 23.

Holland, B. (2002) Sexual selection fails to promote adaptation to a new environment. *Evolution* **56**, 721–730.

Holland, B. & Rice, W.R. (1999) Experimental removal of sexual selection reverses intersexual antagonistic coevolution and removes a reproductive load. *Proceedings of the National Acadamy of Sciences USA* **96**, 5083–5088.

Hölldobler, B. & Wilson, E.O. (1990) *The Ants*. Belknap Press. Harvard University Press. Cambridge, MA.

Holloway, B.A. (1969) Further studies on generic relationships in Lucanidae (Insecta: Coleoptera) with special reference to the ocular canthus. *New Zealand Journal of Science* **12**, 958–977.

Holm, E. & Kirsten, J.F. (1979) Pre-adaptation and speed mimicry among Namib Desert scarabaeids with orange elytra. *Journal of Arid Environments* **2**, 263–271.

Holt, R.D. (1993) Ecology at the mesoscale: the influence of regional processes on local communities. In: Ricklefs, R.E.and Schluter, D. (eds.) *Species diversity in ecological communities*, pp. 77–88. University of Chicago Press, Chicago.

Holt, R.D. (1996) Demographic constraints in evolution: towards unifying the evolutionary theories of senescence and niche conservatism. *Evolutionary Ecology* **6**, 433–447.

Holt, R.D. (1997) From metapopulation dynamics to community structure: some consequences of spatial heterogeneity. In: Hanski, I.& Gilpin, M.E. (eds.) *Metapopulation Biology: Ecology, Genetics, and Evolution*. pp. 149–164. Academic Press, San Diego, CA.

Holter, P. (1982) Resource utilization and local coexistence in a guild of scarabaeid dung beetles (*Aphodius* spp.) *Oikos* **39**, 213–227.

Holter, P. & Scholtz, C.H. (2005) Are ball-rolling (Scarabaeini, Gymnopleurini, Sisyphini) and tunnelling scarabaeine dung beetles equally choosy about the size of ingested dung particles? *Ecological Entomology* **30**, 700–705.

Holter, P. & Scholtz, C.H. (2007) What do dung beetles eat? *Ecological Entomology* **32**, 690–697.

Holter, P., Scholtz, C.H. & Stenseng, L. (2009) Desert detritivory: Nutritional ecology of a dung beetle (*Pachysoma glentoni*) subsisting on plant litter in arid South African sand dunes. *Journal of Arid Environments* **73**, 1090–1094.

Holyoak, M., Leibold, M.A. & Holt, R.D. (2005) *Metacommunities: Spatial Dynamics and Ecological Communities*. The University of Chicago Press, Chicago.

Horgan, F.G. (2005) Aggregated distribution of resources creates competition refuges for rainforest dung beetles. *Ecography* **28**, 603–618.

Horgan, F.G. (2007) Dung beetles in pasture landscapes of Central America: proliferation of synanthropogenic species and decline of forest specialists. *Biodiversity and Conservation* **16**, 2149–2165.

Horgan, F.G. (2008) Dung beetle assemblages in forests and pastures of El Salvador: a functional comparison. *Biodiversity and Conservation* **17**, 2961–2978.

Horgan, F.G. & Fuentes, R.C. (2005) Asymmetrical competition between Neotropical dung beetles and its consequences for assemblage structure. *Ecological Entomology* **30**, 182–193.

Horion, A. (1958) *Faunistik der mitteleuropäischen Käfer. Band VI: Lamellicornia*. Kommissionsverlag Buchdruckerei Aug. Feyel, Uberlingen-Bodensee.

Hosken, D.J., Garner, T.W.J., Tregenza, T., Wedell, N. & Ward, P.I. (2003) Superior sperm competitors sire higher-quality young. *Proceedings of the Royal Society of London B* **270**, 1933–1938.

Hosken, D.J., Taylor, M.L., Hoyle, K., Higgins, S. & Wedell, N. (2008) Attractive males have greater success in sperm competition. *Current Biology* **18**, R553–R554.

Houde, A.E. (1994) Effect of artificial selection on male colour patterns on mating preference of female guppies. *Proceedings of the Royal Society of London B* **256**, 125–130.

Houde, A.E. & Endler, J.A. (1990) Correlated evolution of female mating preferences and male color patterns in the guppy *Poecilia reticulata*. *Science* **248**, 1405–1408.

Houle, D. (1992) Comparing evolvability and variability of quantitative traits. *Genetics* **130**, 195–204.

House, C.M. & Simmons, L.W. (2003) Genital morphology and fertilization success in the dung beetle *Onthophagus taurus*: an example of sexually selected male genitalia. *Proceedings of the Royal Society of London B* **270**, 447–455.

House, C.M. & Simmons, L.W. (2005) Relative influence of male and female genital morphology on paternity in the dung beetle *Onthophagus taurus*. *Behavioral Ecology* **16**, 889–897.

House, C.M. & Simmons, L.W. (2007) No evidence for condition-dependent expression of male genitalia in the dung beetle *Onthophagus taurus*. *Journal of Evolutionary Biology* **20**, 1322–1332.

Houston, A.I. & Davies, N.B. (1985) The evolution of cooperation and life history in the dunnock *Prunella modularis*. In: Sibley, R.M.& Smith, R.H. (eds.) *Behavioral Ecology: Ecological Consequences of Adaptive Behaviour*. pp 471–487. Blackwell Scientific, Oxford.

Houston, A.I. & McNamara, J.M. (2002) A self-consistent approach to paternity and parental effort. *Proceedings of the National Academy of Sciences USA* **357**, 351–362.

Houston, W.W.K. (1986) Exocrine glands in the forelegs of dung beetles in the genus *Onitis* F. *(Coleoptera: Scarabaeidae)*. *Journal of the Australian Entomology Society* **25**, 161–169.

Howden, H.F. & Nealis, V.G. (1975) Effects of clearing in a tropical rain forest on the composition of the coprophagous scarab beetle fauna (Coleoptera). *Biotropica* **7**, 77–83.

Howden, H.F. & Scholtz, C.H. (1987) A revision of the African genus *Odontoloma* Boheman (Coleoptera: Scarabaeidae: Scarabaeinae). *Journal of the South African Entomological Society* **50**, 155–192.

Howden, H.F. & Storey, R.I. (1992) Phylogeny of the Rhyparini and the new tribe Stereomerini, with descriptions of new genera and species (Coleoptera; Scarabaeidae; Aphodiinae). *Canadian Journal of Zoology* **70**, 1810–1823.

Hubbell, S.P. (2001) *The Unified Neutral Theory of Biodiversity and Biogeography*. Princeton University Press, Princeton.

Huerta, C., Halffter, G. & Halffter, V. (2005) Nidification in *Eurysternus foedus* Guérin-Meneville: its relationship to other dung beetle nesting patterns (Coleoptera: Scarabaeidae, Scarabaeinae). *Folia Entomológica Mexicana* **44**, 75–84.

Hughes, R.D. (1975) Assessment of the burial of cattle dung by Australian dung beetles. *Journal of the Australian Entomological* **14**, 129–134.

Hughes, R.D., Tyndale-Biscoe, M. & Walker, J. (1978) Effects of introduced dung beetles (Coleoptera-Scarabaeidae) on breeding and abundance of Australian bushfly, *Musca vetustissima* Walker (Diptera: Muscidae). *Bulletin of Entomological Research* **68**, 361–372.

Humbert, H. & Cours Darne, G. (1965) *Notice de la Carte Madagascar*. Section Scientifique et Technique de L'Institut Francais de Pondichéry, Puducherry, India.

Humphries, S., Evans, J.P. & Simmons, L.W. (2008) Sperm competition: linking form to function. *BMC Evolutionary Biology* **8**, 319.

Hunt, J. & Simmons, L.W. (1997) Patterns of fluctuating asymmetry in beetle horns: an experimental examination of the honest signalling hypothesis. *Behavioral Ecology and Sociobiology* **41**, 109–114.

Hunt, J. & Simmons, L.W. (1998a) Patterns of parental provisioning covary with male morphology in a horned beetle (*Onthophagus taurus*) (Coleoptera: Scarabaeidae). *Behavioral Ecology and Sociobiology* **42**, 447–451.

Hunt, J. & Simmons, L.W. (1998b) Patterns of fluctuating asymmetry in beetle horns: no evidence for reliable signalling. *Behavioral Ecology* **9**, 465–470.

Hunt, J. & Simmons, L.W. (2000) Maternal and paternal effects on offspring phenotype in the dung beetle *Onthophagus taurus*. *Evolution* **54**, 936–941.

Hunt, J. & Simmons, L.W. (2001) Status-dependent selection in the dimorphic beetle *Onthophagus taurus*. *Proceedings of the Royal Society of London B* **268**, 2409–2414.

Hunt, J. & Simmons, L.W. (2002a) The genetics of maternal care: direct and indirect genetic effects in the dung beetle *Onthophagus taurus*. *Proceedings of the National Academy of Sciences USA* **99**, 6828–6832.

Hunt, J. & Simmons, L.W. (2002b) Behavioural dynamics of biparental care in the dung beetle *Onthophagus taurus*. *Animal Behaviour* **64**, 65–75.

Hunt, J. & Simmons, L.W. (2002c) Confidence of paternity and paternal care: covariation revealed through the experimental manipulation of the mating system in the beetle *Onthophagus taurus*. *Journal of Evolutionary Biology* **15**, 784–795.

Hunt, J. & Simmons, L.W. (2004) Optimal maternal investment in the dung beetle *Onthophagus taurus*. *Behavioural Ecology and Sociobiology* **42**, 447–451.

Hunt, J., Kotiaho, J.S. & Tomkins, J.L. (1999) Dung pad residence time covaries with male morphology in the dung beetle *Onthophagus taurus*. *Ecological Entomology* **24**, 174–180.

Hunt, J., Simmons, L.W. & Kotiaho, J.S. (2002) A cost of maternal care in the dung beetle *Onthophagus taurus*? *Journal of Evolutionary Biology* **15**, 57–64.

Hunt, J., Breuker, C.J., Sadowski, J.A. & Moore, A.J. (2009) Male-male competition, female mate choice and their interaction: determining total sexual selection. *Journal of Evolutionary Biology* **22**, 13–26.

Hunt, T., Bergsten, J., Levkanicova, Z., Papadopoulou, A., St. John, O., Wild, R., Hammond, P.M., Ahrens, D., Balke, M., Caterino, M.S., Gómez-Zurita, J., Ribera, I., Barraclough, T.G., Bocakova, M., Bocak, L. & Vogler, A.P. (2007) A comprehensive phylogeny of beetles reveals the evolutionary origins of a super-radiation. *Science* **318**, 1913–1916.

Huxley, J.S. (1932) *Problems of Relative Growth*. Methuen and Co., London.

Ims, R.A. (1995) Movement patterns related to spatial structures. In: Hansson, L., Fahrig, L.& Merriam, G. (eds.) *Mosaic landscapes and ecological processes*, pp 85–109. Chapman and Hall, London.

Information Centre of the Ministry of Agriculture and Forestry (1986–2008). *Yearbooks of Farm Statistics 1986–1995*. Helsinki.

Inouchi, J., Shibuya, T. & Hatanaka, T. (1988) Food odor responses of single antennal olfactory cells in the Japanese dung beetle, *Geotrupes auratus* (Coleoptera: Geotrupidae). *Applied Entomology and Zoology* **23**, 167–174.

Irlich, U., Terblanche, J.S., Blackburn, T.M. & Chown, S.L. (2009) Insect rate-temperature relationships: environmental variation and the metabolic theory of ecology. *American Naturalist* **174**, 819–835.

Itzkowitz, M., Santangelo, N. & Richter, M. (2001) Parental division of labour and the shift from minimal to maximal role specializations: an examination using a biparental fish. *Animal Behaviour* **61**, 1237–1245.

IUCN (2009) *IUCN Red List of Threatened Species*. ⟨www.iucnredlist.org⟩.

Jacobs J., Nole, I., Palminteri, S. & Ratcliffe, B. (2008) First come, first serve: 'sit and wait' behavior in dung beetles at the source of primate dung. *Neotropical Entomology* **37**, 1–7.

Jacobson, B. & Peres-Neto, P.R. (2010) Quantifying and disentangling dispersal in metacommunities: how close have we come? How far is there to go? *Landscape Ecology*, in press, DOI: 10.1007/s10980–009–9442–9.

Jay-Robert, P., Niogret, J., Errouissi, F., Labarussias, M., Paoletti, E., Vazques Luis, M. & Lumaret, J.-P. (2008) Relative efficiency of extensive grazing vs. wild ungulates management for dung beetle conservation in a heterogeneous landscape from Southern Europe (Scarabaeinae, Aphodiinae, Geotrupinae). *Biological Conservation* **141**, 2879–2887.

Jerozolimski, A. & Peres, C. (2003) Bringing home the biggest bacon: a cross-site analysis of the structure of hunter-kill profiles in Neotropical forests. *Biological Conservation* **111**, 415–425.

Jeschke, J.M. & Strayer, D.L. (2008) Usefulness of bioclimatic models for studying climate change and invasive species. *Annals of the New York Academy of Sciences* **1134**, 1–24.

Jiggins, F.M., Hurst, G.D.D. & Majerus, M.E.N. (2000) Sex-ratio distorting Wolbachia causes sex-role reversal in its butterfly host. *Proceedings of the Royal Society B* **267**, 69–73.

Johansson, B.G. & Jones, T.M. (2007) The role of chemicals communication in mate choice. *Biological Reviews* **82**, 265–289.

John, R., Dalling, J.W., Harms, K.E., Yavitt, J.B., Stallard, R.F., Mirabello, M., Hubbell, S.P., Valencia, R., Navarrete, H., Vallejo, M. & Foster, R.B. (2007) Soil nutrients influence spatial distributions of tropical tree species. *Proceedings of the National Academy of Sciences USA* **104**, 864–869.

Johnstone, R.A. & Hinde, C.A. (2006) Negotiation over offspring care – how should parents respond to each other's efforts? *Behavioral Ecology* **17**, 818–827.

Jones, A.G. & Ratterman, N.L. (2009) Mate choice and sexual selection: what have we learned since Darwin? *Proceedings of the National Academy of Science USA* **106**, 10001–10008.

Jones, A.G., Moore, G.I., Kvarnemo, C., Walker, D. & Avise, J.C. (2003) Sympatric speciation as a consequence of male pregnancy in seahorses. *Proceedings of the National Academy of Sciences USA* **100**, 6598–6603.

Jones, K.M., Ruxton, G.D. & Monaghan, P. (2002) Model parents: is full compensation for reduced partner nest attendance compatible with stable biparental care? *Behavioral Ecology* **13**, 838–843.

Jones, R.M. & Ratcliff, D. (1983) Patchy grazing and its relation to deposition of cattle dung pats in pastures in coastal subtropical Queensland. *Journal of the Australian Institute of Agricultural Science* **49**, 109–111.

Kahlke, R.-D. (1994) Die Entstehungs-, Entwicklungs- und Verbreitungsgeschichte des oberpleistozänen *Mammuthus*-Coelodonta-Faunenkomplexes in Eurasien (Großsäuger). *Abhandlungen der senckenbergischen naturforschenden Gesellschaft* **546**, 1–164.

Kanda, N., Yokota, T., Shibata, E. & Sato, H. (2005) Diversity of dung-beetle community in declining Japanese subalpine forest caused by an increasing sika deer population. *Ecological Research* **20**, 135–141.

Kaplan, R.H. & Cooper, W.S. (1984) The evolution of developmental plasticity in reproductive characteristics: an application of the 'adaptive coin-flipping' principle. *American Naturalist* **123**, 393–410.

Karino, K., Niiyama, H. & Chiba, M. (2005) Horn length is the determining factor in the outcomes of escalated fights among male Japanese horned beetles, *Allomyrina dichotoma* L. (Coleoptera: Scarabaeidae). *Journal of Insect Behavior* **18**, 805–815.

Karr, J.R. (1991) Biological integrity – a long-neglected aspect of water-resource management. *Ecological Applications* **1**, 66–84.

Karr, J.R. (1993) Measuring biological integrity: lessons from streams. In Woodley, S., Kay, J.& Francis, G. (eds.) *Ecological integrity and the management of ecosystems*. St. Lucie Press, Ottawa.

Kearney, M. & Porter, W. (2009) Mechanistic niche modelling: combining physiological and spatial data to predict species' ranges. *Ecology Letters* **12**, 334–350.

Keller, L. & Reeve, H.K. (1995) Why do females mate with multiple males? The sexually selected sperm hypothesis. *Advances in the Study of Behavior* **24**, 291–315.

Kellert, S.R. (1993) Values and Perceptions of Invertebrates. *Conservation Biology* **7**, 845–855.

Kestler, P. (1985) Respiration and respiratory water loss. In: Hoffmann, K.H. (ed.) *Environmental Physiology and Biochemistry of Insects*. pp. 137–186. Springer, Berlin.

Kielan-Jaworowska, Z., Cifelli, R.L. & Luo, Z.-X. (2004) *Mammals from the Age of Dinosaurs – origins, evolution, and structure*. Columbia University Press, New York.

Kijimoto, T., Costello, J., Tang, Z., Moczek, A.P. & Andrews, J. (2009) Candidate genes for the development and evolution of beetle horns. *BMC Genomics* **10**, 504.

Kijimoto, T., Andrews, J. & Moczek, A.P. (2010) Programmed cell death shapes the expression of horns within and between species of horned beetles. *Evolution & Development*, **12**, 449–458.

Kingston, T.J. & Coe, M. (1977) The biology of the giant dung-beetle (*Heliocopris dilloni*) (Coleoptera, Scarabaeidae). *Journal of the Zoological Society of London* **181**, 243–63.

Kirkpatrick, M. (1982) Sexual selection and the evolution of female choice. *Evolution* **36**, 1–12.

Kirkpatrick, M. & Lande, R. (1989) The evolution of maternal characters. *Evolution* **43**, 485–503.

Kirkpatrick, M. & Ryan, M.J. (1991) The evolution of mating preferences and the paradox of the lek. *Nature* **350**, 33–38.

Kishi, S. & Nishida, T. (2006) Adjustment of parental investment in the dung beetle *Onthophagus atripennis* (Coleoptera: Scarabaeidae). *Ethology* **112**, 1239–1245.

Kishi, S. & Nishida, T. (2008) Optimal investment in sons and daughters when parents do not know the sex of their offspring. *Behavioral Ecology Sociobiology* **62**, 607–616.

Kishi, S. & Nishida, T. (2009) Adjustment of parental expenditure in sympatric *Onthophagus* beetles (Coleoptera: Scarabaeidae). *Journal of Ethology* **27**, 59–65.

Klein, B. (1989) Effects of forest fragmentation on dung and carrion beetle communities in Central Amazonia. *Ecology* **70**, 1715–1725.

Klemperer, H.G. (1982) Parental behaviour in *Copris lunaris* (Coleoptera, Scarabaeidae): care and defense of brood balls and nest. *Ecological Entomology* **7**, 155–167.

Klemperer, H.G. (1983) Subsocial behaviour in *Oniticellus cinctus* (Coleoptera: Scarabaeidae): effect of the brood on parental care and oviposition. *Physiological Entomology* **8**, 393–402.

Klemperer, H.G. (1983) The evolution of parental behavior in Scarabaeinae (Coleoptera, Scarabaeidae): an experimental approach. *Ecological Entomology* **8**, 49–59.

Klingenberg, C.P. (1998) Heterochrony and allometry: the analysis of evolutionary change in ontogeny. *Biological Reviews* **73**, 79–123.

Klingenberg, C.P. & Nijhout, H.F. (1998) Competition among growing organs and developmental control of morphological asymmetry. *Proceedings of the National Academy of Sciences USA* **265**, 1135–1139.

Klok, C.J. (1994) *Desiccation Resistance in Dung-feeding Scarabaeinae*. M.Sc. Thesis, University of Pretoria, South Africa.

Klok, C.J., Mercer, R.D. & Chown, S.L. (2002) Discontinuous gas exchange in centipedes and its convergent evolution in tracheated arthropods. *Journal of Experimental Biology* **205**, 1031–1036.

Klok, C.J., Sinclair, B.J. & Chown, S.L. (2004) Upper thermal tolerance and oxygen limitation in terrestrial arthropods. *Journal of Experimental Biology* **207**, 2361–2370.

Knell, R.J. (2009a) On the analysis of non-linear allometries. *Ecological Entomology* **34**, 1–11.

Knell, R.J. (2009b) Population density and the evolution of male aggression. *Journal of Zoology* **278**, 83–90.

Knell, R.J. & Simmons, L.W. (2010). Mating tactics determine patterns of condition-dependence in a dimorphic horned beetle. *Proceedings of the Royal Society B-Biological Sciences B* **277**, 2347–2353.

Koch, S.O., Chown, S.L., Davis, A.L.V., Endrody-Younga, S. & Van Jaarsveld, A.S. (2000) Conservation strategies for poorly surveyed taxa: a dung beetle (Coleoptera, Scarabaeidae) case study from southern Africa. *Journal of Insect Conservation* **4**, 45–56.

Kodric-Brown, A., Sibly, R.M. & Brown, J.H. (2006) The allometry of ornaments and weapons. *Proceedings of the National Academy of Sciences USA* **103**, 8733–8738.

Kohlmann, B. & Solís, A. (2001) El género *Onthophagus* (Coleoptera: Scarabaeidae) en Costa Rica. *Giornale Italiano de Entomologia* **9**, 159–261.

Koivulehto, H. (2004) *Madagascar's dung beetles – rain forest species avoid open areas*. MSc Thesis, University of Helsinki, Helsinki.

Kojima, T. (2004) The mechanism of *Drosophila* leg development along the proximodistal axis. *Development, Growth and Differentiation* **46**, 115–129.

Kokko, H. & Jennions, M. (2003a) It takes two to tango. *Trends in Ecology and Evolution* **18**, 103–104.

Kokko, H. & Jennions, M. (2003b) Response to McDowall: In defense of the caring male. *Trends in Ecology and Evolution* **18**, 611–612.

Kokko, H. & Rankin, D.J. (2006) Lonely hearts or sex in the city? Density-dependent effects in mating systems. *Philosophical Transactions of the Royal Society of London Series B* **361**, 319–334.

Kokko, H., Jennions, M.D. & Brooks, R. (2006) Unifying and testing models of sexual selection. *Annual Review of Ecology Evololution and Systematics* **37**, 43–66.

Koskela, H. & Hanski, I. (1977) Structure and succession in a beetle community inhabiting cow dung. *Annales Zoologici Fennici* **14**, 204–223.

Kot, M., Lewis, M.A. & van der Driessche, P. (1996) Dispersal data and the spread of invading organisms. *Ecology* 77, 2027–2042.

Kotiaho, J.S. (1999) Estimating fitness: comparison of body condition indices revisited. *Oikos* **87**, 399–400.

Kotiaho, J.S. (2002) Sexual selection and condition dependence of courtship display in three species of horned dung beetle. *Behavioral Ecology* **13**, 791–799.

Kotiaho, J.S. & Simmons, L.W. (2001) Effects of *Macrocheles* mites on longevity of males of the dimorphic dung beetle *Onthophagus binodis*. *Journal of Zoology* **254**, 441–445.

Kotiaho, J.S. & Tomkins, J.L. (2001) The discrimination of alternative male morphologies. *Behavioral Ecology* **12**, 553–557.

Kotiaho, J.S., Simmons, L.W. & Tomkins, J.L. (2001) Towards a resolution of the lek paradox. *Nature* **410**, 684–686.

Kotiaho, J.S., Simmons, L.W., Hunt, J. & Tomkins, J.L. (2003) Males influence maternal effects that promote sexual selection: A quantitative genetic experiment with dung beetles *Onthophagus taurus*. *American Naturalist* **161**, 852–859.

Kotiaho, J.S., LeBas, N.R., Puurtinen, M. & Tomkins, J.L. (2008) On the resolution of the lek paradox. *Trends in Ecology and Evolution* **23**, 1–3.

Kozielska, M., Krzeminska, A. & Radwan, J. (2003) Good genes and the maternal effects of polyandry on offspring reproductive success in the bulb mite. *Proceedings of the Royal Society of London B* **271**, 165–170.

Kozłowski, J., Konarzewski, M. & Gawelczyk, A.T. (2003) Cell size as a link between noncoding DNA and metabolic rate scaling. *Proceedings of the National Academy of Sciences of the USA* **100**, 14080–14085.

Krapp, H.G. (2009) Ocelli. *Current Biology* **19**, R435–R437.

Krell, F.-T. (1999) Southern African dung beetles (Coleoptera: Scarabaeidae) attracted by defensive secretions of Diplopoda. *African Entomology* 7, 287–288.

Krell, F.-T. (2006) Fossil Record and Evolution of Scarabaeoidea (Coleoptera: Polyphaga). In: Jameson, M.L., Ratcliffe, B.C. (eds.) *Scarabaeoidea in the 21st Century: A Festchrift Honoring Henry F. Howden. Coleopterists Society Monograph* 5, 120–143.

Krell, F.-T., Schmitt, T., Linsenmair, K.E. (1997) Diplopod defensive secretions as attractants for necrophagous scarab beetles (Diplopoda; Insecta, Coleoptera: Scarabaeidae). In: Enghoff, H. (ed.) *Entomologica Scandinavica Supplementum 51, Many-legged animals – A collection of papers on Myriapoda and Onychophora. Proceedings of the Tenth International Congress of Myriapodology, Copenhagen, 29 July – 2 August, 1996*. pp. 281–285. Lund.

Krell, F-T., Schmitt, T., Dembele, A. & Linsenmair, K.E. (1997) Repellents as attractants – extreme specialization in Afrotropical dung beetles (Coleoptera: Scarabaeidae) as competition avoiding strategy. *Zoology, Analysis of Complex Systems* **101**, Supplement 1: 12.

Krell, F.-T., Krell-Westerwalbesloh, S., Weiß, I, Eggleton, P. & Linsenmair, K.E. (2003) Spatial separation of Afrotropical dung beetle guilds: a trade-off between competitive superiority and energetic constraints (Coleoptera: Scarabaeidae). *Ecography* **26**, 210–222.

Kremen, C. (1992) Assessing the indicator properties of species assemblages for natural areas monitoring. *Ecological Applications* **2**, 203–217.

Kremen, C., Williams, N.M., Aizen, M.A., Gemmill-Herren, B., Lebuhn, G., Minckley, R., Packer, L., Potts, S. G., Roulston, T., Steffan-Dewenter, I., Vazquez, D.P., Winfree, R., Adams, L., Crone, E.E., Greenleaf, S.S., Keitt, T.H., Klein, A.M., Regetz, J. & Ricketts, T.H. (2007) Pollination and other ecosystem services produced by mobile organisms: a conceptual framework for the effects of land-use change. *Ecology Letters* **10**, 299–314.

Krikken, J. & Huijbregts, J. (2006) Miniature scarabs of the genus *Haroldius* on Sulawesi, with notes on their relatives (Coleoptera: Scarabaeidae). *Tijdschrift voor Entomologie* **149**, 167–187.

Krogh, A. & Zeuthen, E. (1941) The mechanism of flight preparation in some insects. *Journal of Experimental Biology* **18**, 1–10.

Kryger, U., Coles, K.S., Tukker, R. & Scholtz, C.H. (2006) Biology and ecology of *Circellium bacchus* (Fabricius 1781) (Coleoptera: Scarabaeidae), a South African dung beetle of conservation concern. *Tropical Zoology* **19**, 185–207.

Kvarnemo, C. & Ahnesjö, I. (1996) The dynamics of operational sex ratios and competition for mates. *Trends in Ecology and Evolution* **11**, 404–408.

Kvarnemo, C. & Ahnesjö, I. (2002). Operational sex ratios and mating competition. In: Hardy, I.C.W. (ed.) *Sex ratios: concepts and research methods*. pp. 366–382. Cambridge University Press, Cambridge.

Labhart, T. & Meyer, E.P. (1999) Detectors for polarized skylight in insects: a survey of ommatidial specializations in the dorsal rim area of the compound eye. *Microscopy Research and Techniques* **47**, 368–379.

Labhart, T. & Nilsson, D.-E. (1995) The dorsal eye of the dragonfly *Sympetrum*: specializations for prey detection against the blue sky. *Journal of Comparative Physiology A* **176**, 437–453.

Lailvaux, S.P., Hathway, J., Pomfret, J.C. & Knell, R.J. (2005) Horn size predicts physical performance in the beetle *Euoniticellus intermedius* (Coleoptera: Scarabaeidae). *Functional Ecology* **19**, 632–639.

Land, M.F. (1988) The optics of animal eyes. *Contemporary Physics* **29**, 435–455.

Land, M.F. (1997) Visual acuity in insects. *Annual Review of Entomology* **42**, 147–77.

Land, M.F. & Nilsson D.-E. (2002) *Animal Eyes*. Oxford University Press. New York. USA.

Lande, R. (1981) Models of speciation by sexual selection on polygenic traits. *Proceedings of the National Academy of Sciences USA* **78**, 3721–3725.

Landin, B.-O. (1961) Ecological studies on dung beetles. *Opuscula Entomologica, Supplement* **19**, 1–228.

Landres, P.B., Verner, J. & Thomas, J.W. (1988) Ecological uses of vertebrate indicator species – a critique. *Conservation Biology* **2**, 316–328.

Larsen, T. & Génier, F. (2008) *Dung Beetles of Cocha Cashu Biological Station*. Madre de Dios, Peru.

Larsen, T.H. & Forsyth, A. (2005) Trap spacing and transect design for dung beetle biodiversity studies. *Biotropica* **37**, 322–325.

Larsen, T.H., Williams, N.M. & Kremer, C. (2005) Extinction order and altered community structure rapidly disrupt ecosystem functioning. *Ecology Letters* **8**, 538–547.

Larsen, T.H., Lopera, A. & Forsyth, A. (2006) Extreme trophic and habitat specialization by Peruvian dung beetles (Coleoptera: Scarabaeidae: Scarabaeinae). *Coleopterists Bulletin* **60**, 315–324.

Larsen, T.H., Lopera, A. & Forsyth, A. (2008) Understanding trait-dependent community disassembly: dung beetles, density functions, and forest fragmentation. *Conservation Biology* **22**, 1288–1298.

Larsen, T.H., Lopera, A., Forsyth, A. & Génier, F. (2009) From coprophagy to predation: a dung beetle that kills millipedes. *Biology Letters* **5**, 152–155.

Laughlin, S.B. (2001) The metabolic cost of information – a fundamental factor in visual ecology. In: Barth, F.G.& Schmid, A. (eds.) *Ecology of Sensing*. Springer, Berlin.
Lavorel, S. & Garnier, E. (2002) Predicting changes in community composition and ecosystem functioning from plant traits: revisiting the Holy Grail. *Functional Ecology* **16**, 545–556.
Lazarus, J. (1989) The logic of animal desertion. *Animal Behaviour* **39**, 672–684.
Leal, W.S. (1998) Chemical ecology of phytophagous scarab beetles. *Annual Review of Entomology* **43**, 39–61.
Leal, W.S., Mochizuki, F., Wakamura, S. & Yasuda, T. (1992). Electroantennographic detection of *Anomala cuprea* Hope (Coleoptera: Scarabaeidae) sex pheromone. *Applied Entomology and Zoology* **27**, 289–291.
Leal, W.S., Kuwahara, X., Shi, H., Higuchi, H., Marino, C.E.B., Ono, M. & Meinwald, J. (1998). Male-released sex pheromone of the stink bug *Piezodorus hybneri*. *Journal of Chemical Ecology* **24**, 1817–1829.
Lebis, E. (1953) *Revision des Canthoninae de Madagascar [Col. Scarabaeidae]* Mémoires de l'institut scientifique de Madagascar, Série E – Tome III, pp. 107–252.
Lee, J.M. & Peng, Y.-S. (1981) Influence of adult size of *Onthophagus gazella* on manure pat degradation, nest construction, and progeny size. *Environmental Entomology* **10**, 626–630.
Leimar, O. (1997) Reciprocity and communication of partner quality. *Proceedings of the Royal Society of London B* **264**, 1209–1215.
Legner, E.F. (1995) Biological control of Diptera of medical and veterinary importance. *Journal for Vector Ecology* **20**, 59–120.
Legner, E.F. & Warkentin, R.W. (1991) Influence of *Onthophagus gazella* on hornfly, *Haematobia irritans* density in irrigated pastures. *Entomophaga* **36**, 547–553.
Leibold, M.A., Holyoak, M., Mouquet, N., Amarasekare, P., Chase, J.M., Hoopes, M.F., Holt, R.D., Shurin, J. B., Law, R., Tilman, D., Loreau, M. & Gonzalez, A. (2004) The metacommunity concept: a framework for multi-scale community ecology. *Ecology Letters* **7**, 601–613.
Lewinsohn, T. & Roslin, T. (2008) Four ways toward tropical herbivore megadiversity. *Ecology Letters* **11**, 398–416.
Lewis, M.A. (1997) Variability, patchiness, and jump dispersal in the spread of an invading population. In: Tilman, D.& Kareiva, P. (eds.) *Spatial Ecology: the Role of Space in Population Dynamics and Interspecific Interactions*, pp. 46–74. Princeton University Press, Princeton.
Lewis, S.M. & Austad, S.N. (1994) Sexual selection in flour beetles: the relationship between sperm precedence and male olfactory attractiveness. *Behavioral Ecology* **5**, 219–224.
Lichtenstein, G. & Sealy, S.G. (1998) Nestling competition, rather than supernormal stimulus, explains the success of parasitic brown-headed cowbird chicks in yellow warbler nests. *Proceedings of the Royal Society of London B* **265**, 249–254.
Lichti, D. (1937) *Canthon dives* Harold (Col. Copridae), predator das femeas de *Atta laevigata* Smith (Hym. Formicidae). *Revista Entomologia* **7**, 117–118.
Lighton, J.R.B. (1985) Minimum cost of transport and ventilatory patterns in three African beetles. *Physiological Zoology* **58**, 390–399.
Lighton, J.R.B. (1994) Discontinuous ventilation in terrestrial insects. *Physiological Zoology* **67**, 142–162.
Lighton, J.R.B. (1996) Discontinuous gas exchange in insects. *Annual Review of Entomology* **41**, 309–324.
Lighton, J.R.B. (1998) Notes from the underground: towards ultimate hypotheses of cyclic, discontinuous gas-exchange in tracheate arthropods. *American Zoologist* **38**, 483–491.
Lighton, J.R.B. (2002) Lack of discontinuous gas exchange in a tracheate arthropod, *Leiobunum townsendi* (Arachnida, Opiliones). *Physiological Entomology* **27**, 170–174.
Lighton, J.R.B. & Fielden, L.J. (1996) Gas exchange in wind spiders (Arachnida, Solphugidae): independent evolution of convergent control strategies in solphugids and insects. *Journal of Insect Physiology* **42**, 347–357.
Lighton, J.R.B. & Joos, B. (2002) Discontinuous gas exchange in the pseudoscoprion *Garypus californicus* is regulated by hypoxia, not hypercapnia. *Physiological and Biochemical Zoology* **75**, 345–349.
Lighton, J.R.B. & Ottesen, E.A. (2005) To DGC or not to DGC: oxygen guarding in the termite *Zootermopsis nevadensis* (Isoptera: Termopsidae). *Journal of Experimental Biology* **208**, 4671–4678.
Lighton, J.R.B. & Turner, R.J. (2008) The hygric hypothesis does not hold water: abolition of discontinuous gas exchange cycles does not affect water loss in the ant *Camponotus vicinus*. *Journal of Experimental Biology* **211**, 563–567.
Lindén, M. & Møller, A.P. (1989) Costs of reproduction and covariation of life-history traits in birds. *Trends in Ecology and Evolution* **4**, 367–371.

Lindenmayer, B.D. & Hobbs, R. (2007) *Managing and Designing Landscapes for Conservation*. Blackwell, London.
Lindenmayer, D.B., Fischer, J. & Hobbs, R. (2007) The need for pluralism in landscape models: a reply to Dunn and Majer. *Oikos* **116**, 1419–1421.
Lloyd, M. (1967) Mean crowding. *Journal of Animal Ecology* **36**, 1–30.
Lobo, J.M. (2001) Decline of roller dung beetle (Scarabaeinae) populations in the Iberian peninsula during the 20th century. *Biological Conservation* **97**, 43–50.
Lobo, J.M. & Montes De Oca, E. (1997) Spatial microdistribution of two introduced dung beetle species *Digonthophagus gazella (F.)* and *Euoniticellus intermedius* (Reiche) Coleoptera Scarabaeidae) in an arid region of northern Mexico (Durango, Mexico). *Acta Zoologica Mexico* **71**, 17–32.
Lobo, J.M., Lumaret, J.P. & Jay-Robert, P. (1998) Sampling dung beetles in the French Mediterranean area: Effects of abiotic factors and farm practices. *Pedobiologia* **42**, 252–266.
Lobo, J.M., Lumaret, J.-P. & Jay-Robert, P. (2002) Modelling the species richness distribution of French dung beetles (Coleoptera, Scarabaeidae) and delimiting the predictive capacity of different groups of explanatory variables. *Global Ecology and Biogeography* **11**, 265–277.
Lobo, J.M., Verdú, J.R. & Numa, C. (2006) Environmental and geographical factors affecting the Iberian distribution of flightless *Jekelius* species (Coleoptera: Geotrupidae). *Diversity and Distributions* **12**, 179–188.
Locatello, L., Rasotto, M.B., Evans, J.P. & Pilastro, A. (2006) Colourful male guppies produce faster and more viable sperm. *Journal of Evolutionary Biology* **19**, 1595–1602.
Lohmann, I., McGinnis, N., Bodme,r M. & McGinnis, W. (2002) The *Drosophila* Hox gene *Deformed* sculpts head morphology via direct regulation of the apoptosis activator reaper. *Cell* **110**, 457–466.
Lopez-Guerrero, I. & Zunino, M. (2007) Consideraciones acerca de la evolución de las piezas bucales en los Onthophagini (Coleoptera: Scarabaeidae) en relacion con diferentes regimenes alimenticios. *Interciencia* **32**, 482–489.
Lorch, P.D., Proulx, S.R., Rowe, L. & Day, T. (2003) Condition-dependent sexual selection can accelerate adaptation. *Evolutionary Ecology Research* **5**, 869–881.
Loreau, M. (1995) Consumers as maximizers of matter and energy flow in ecosystems. *American Naturalist* **145**, 22–42.
Losey, J.E. & Vaughan, M. (2006) The economic value of ecological services provided by insects. *BioScience* **56**, 311–323.
Louzada, J.N.C. & Carvalho e Silva, P.R. (2009) Utilisation of introduced Brazilian pastures ecosystems by native dung beetles: diversity patterns and resource use. *Insect Conservation and Diversity* **2**, 45–52.
Lozano, G.A. & Lemon, R.E. (1996) Male plumage, parental care and reproductive success in yellow warblers, *Dendroica petechia*. *Animal Behaviour* **51**, 265–272.
Lumaret, J.P. (1995) Desiccation rate of excrements: a selective pressure on dung beetles (Coleoptera, Scarabeoidea) In: Roy J., Aronson J., Castri, F. Di. (eds.) *Time scales of biological responses to water constraints*, pp. 105–118. SPB Academic Publishing, Amsterdam.
Lumaret, J.P. & Errouissi, F. (2002). Use of anthelmintics in herbivores and risk evaluation for the non-target fauna of pastures. *Veterinary Research* **33**, 547–562.
Lumaret, J.P. & Kirk, A.A. (1987) Ecology of dung beetles in the French Meditteranean Region (Coleoptera: Scarabaeidae). *Acta Zoologica Mexicana* **24**, 1–55.
Lumaret, J.P., Kadiri, N. & Bertrand, M. (1992) Changes in resources – consequences for the dynamics of dung beetle communities. *Journal of Applied Ecology* **29**, 349–356.
Lumaret, J.P., Galante, E., Lumbreras, C., Mena, J., Bertrand, M., Bernal, J.L., Cooper, J.F., Kadiri, N. & Crowe, D. (1993) Field effects of Ivermectin residues on dung beetles. *Journal of Applied Ecology* **30**, 428–436.
Luzzatto, M. (1994) *Tattiche comportamentali nella riproduzione degli Scarabaeidi degradatori (Coleoptera: Scarabaeoidea). Un approccio evolutivo e filogenetico*. Ph.D. thesis, University of Pavia and University of Torino, Italy.
Lynch, M. & Walsh, B. (1998) *Genetics and Analysis of Quantitative Traits*, Sinauer, Sunderland, MA.
Macagno, A.L.M. & Palestrini, C. (2009) The maintenance of extensively exploited pastures within the Alpine mountain belt: implications for dung beetle conservation (Coleoptera: Scarabaeoidea). *Biodiversity and Conservation* **18**, 3309–3323.
MacArthur, R.H. & Levins, R. (1964) Competition, habitat selection, and character displacement in a patchy environment. *Proceedings of the National Academy of Sciences USA* **51**, 1207–120.

MacArthur, R.H. & Wilson, E.O. (1967) *The Theory of Island Biogeography.* Princeton University Press, Princeton.

Macqueen, A. & Beirne, B.P. (1975) Dung burial activity and fly control potential of *Onthophagus nuchicornis* (Coleoptera Scarabaeinae) in British-Columbia. *Canadian Entomologist* **107**, 1215–1220.

Macqueen, A., Wallace, M.M.H. & Doube, B.M. (1986) Seasonal changes in favourability of cattle dung in central Queensland for three species of dung-breeding insects. *Journal of the Australian Entomological Society* **25**, 23–29.

Madewell, R. & Moczek, A.P. (2006) Horn possession reduces maneuverability in the horn-polymorphic beetle, *Onthophagus nigriventris. Journal of Insect Science* **6**, 1–10.

Makarieva, A.M., Gorshkov, V.G., Li, B.-L., Chown, S.L., Reich, P.B. & Gavrilov, V.M. (2008) Mean mass-specific metabolic rates are strikingly similar across life's major domains: evidence for life's metabolic optimum. *Proceedings of the National Academy of Sciences of the USA* **105**, 16994–16999.

Mallock, A. 1894. Insect sight and the defining power of compound eyes. *Proceedings of the Royal Society of London B* **55**, 85–90.

Mann, D. (2008) *An annotated bibliography for Scarabaeinae identification.* Hope Entomological Collections, Oxford.

Marais, E., Klok, C.J., Terblanche, J.S. & Chown, S.L. (2005) Insect gas exchange patterns: a phylogenetic perspective. *Journal of Experimental Biology* **208**, 4495–4507.

Markman, S., Yom-Tov, Y. & Wright, J. (1995) Male parental care in the orange-tufted sunbird: behavioural adjustments in provisioning and nest guarding effort. *Animal Behaviour* **50**, 655–669.

Markman, S., Yom-Tov, Y. & Wright, J. (1996) The effect of male removal on female parental care in the orange-tufted sunbird. *Animal Behaviour* **52**, 437–444.

Markow, T.A. & O'Grady, P.M. (2006) *Drosophila. A Guide to Species Identification and Use.* Academic Press, London.

Martin, R.D. (1981) Relative brain size and metabolic rate in terrestrial vertebrates. *Nature* **293**, 57–60.

Martínez, A. (1952) Scarabaeidae nuevos o poco conocidos III. *Mision de Estudios de Patologya Regional Argentina* **23**, 53–118.

Martínez. A. (1959) Catálogo de los Scarabaeidae argentines (Coleoptera). *Revista del Museo Argentino de Ciencias Naturales Bernardino Rivadavia* **5**, 1–126.

Martinez, M.M. & Magdalena, C.R. (1999) The effects of male glandular secretions on female endocrine centers in *Canthon cyanellus cyanellus* LeConte (Coleoptera: Scarabaeidae, Scarabaeinae). *The Coleopterists Bulletin* **53**, 208–216.

Mate, J.F. (2007) New species of *Aphodius* (*Neomadiellus*) described from Madagascar (Coleoptera: Scarabaeidae). *Nouvelle revue d'entomologie* **23**, 267–276.

Matthews, E.G. (1963) Observations on the ball-rolling behaviour of *Canthon pilularius* (L.) (Coleoptera: Scarabaeidae). *Psyche* **70**, 75–93.

Matthews, E.G. (1965) The taxonomy, geographical distribution, and feeding habits of the canthonines of Puerto Rico (Coleoptera: Scarabaeidae). *Transactions of the American Entomological Society* **91**, 431–465.

Matthews, E.G. (1972) A revision of the Scarabaeine dung beetles of Australia. 1. Tribe Onthophagini. *Australian Journal of Zoology. Supplementary Series* **9**, 1–330.

Matthews, E.G. (1974) A revision of the Scarabaeine dung beetles of Australia. 11. Tribe Scarabaeini. *Australian Journal of Zoology. Supplementary Series* **24**, 1–211.

Matthews, E.G. (1976) A revision of the Scarabaeine dung beetles of Australia. 11. Tribe Coprini. *Australian Journal of Zoology. Supplementary Series* **38**, 1–52.

Matthews, E.G. & Stebnicka, Z.T. (1986) A review of *Demarziella* Balthasar, with a transfer from Aphodiinae to Scarabaeinae (Coleoptera: Scarabaeidae). *Australian Journal of Zoology* **34**, 449–61.

Matthiessen, J.N., Hall, G.P. & Chewings, V.H. (1986) Seasonal abundance of *Musca vetustissima* Walker and other cattle dung fauna in Central Australia. *Journal of the Australian Entomological Society* **25**, 141–147.

Matzkin, L.M., Watts, T.D. & Markow, T.A. (2009) Evolution of stress resistance in *Drosophila*: interspecific variation in tolerance to desiccation and starvation. *Functional Ecology* **23**, 521–527.

Maudsley, N.A. & Stork, N.E. (1995) Species extinctions in insects: ecological and biogeographical considerations. In Harrington, R.& Stork, N.E. (eds.) *Insects in a changing environment.* Academic Press, London.

Mavrogenis, A.P. & Papachristoforou, C. (2001) Genetic and phenotypic relationships between milk production and body weight in Chios sheep and Damascus goats. *Livestock Production Science* 67, 81–87.

Maynard Smith, J. (1977) Parental Investment: a prospective analysis. *Animal Behaviour* **25**, 1–9.

Maynard Smith, J. (1978) *The Evolution of Sex*. Cambridge University Press, Cambridge, UK.

Maynard Smith, J. (1982) *Evolution and the Theory of Games*. Cambridge University Press, Cambridge, UK.

McGeoch, M.A. (1998) The selection, testing and application of terrestrial insects as bioindicators. *Biological Reviews* **73**, 181–201.

McGeoch, M.A. (2007) Insects and bioindication: theory and progress. In: Stewart, A.J.A., New, T.R. & Lewis, O.T. (eds.) *Insect Conservation Biology*. pp. 144–174 CABI Publishing, Oxfordshire, UK.

McGeoch, M.A., van Rensburg, B.J. & Botes, A. (2002) The verification and application of bioindicators: A case study of dung beetles in a savanna ecosystem. *Journal of Applied Ecology* **39**, 661–672.

McGill, B.J., Enquist, B.J., Weihner, E. & Westoby, M. (2006) Rebuilding community ecology from functional traits. *Trends in Ecology and Evolution* **21**, 177–185.

McGinley, M.A. & Charnov, E.L. (1988) Multiple resources and the optimal balance between size and number of offspring. *Evolutionary Ecology* **3**, 150–156.

McGinley, M.A., Temme, D.H. & Geber, M.A. (1987) Parental investment in offspring in variable environments: theoretical and empirical considerations. *American Naturalist* **130**, 370–398.

McIntyre, P.D. & Caveney, S. (1985) Graded-index optics are matched to optical geometry in the superposition eyes of scarab beetles. *Philosophical Transactions of the Royal Society of London B* **311**, 237–269.

McIntyre, P.D. & Caveney, S. (1998) Superposition optics and the time of flight in Onitine dung beetles. *Journal of Comparative Physiology A* **183**, 45–60.

McNamara, J.M., Gasson, C.E. & Houston, A.I. (1999) Incorporating rules for responding into evolutionary games. *Nature* **401**, 368–371.

Medina, C. & Scholtz, C.H. (2005) Systematics of the southern African genus *Epirinus* Reiche (Coleoptera: Scarabaeinae: Canthonini): descriptions of new species and phylogeny. *Insect Systematics and Evolution* 36, 2145–60.

MEFT, USAID & CI (2009) *Evolution de la couverture de forêts naturelles à Madagascar, 1990–2000–2005*. Ministere de L'Environment, des Forêts et du Tourisme (MEFT), The United States Agency for International Development (USAID), Conservation International (CI).

Meinwald, J., Meinwald, Y.C. & Mazzocchi, P.H. (1969) Sex pheromone of the Queen butterfly; Chemistry. *Science* **153**, 1174–1175.

Merilä, J., Björklund, M. & Baker, A.J. (1996) Genetic population structure and gradual northward decline of genetic variability in the greenfinch (*Carduelis chloris*). *Evolution* **50**, 2548–2557.

Merilä, J., Björklund, M. & Baker, A.J. (1997) Historical demography and present day population structure of the greenfinch, *Carduelis chloris* — an analysis of mtDNA control-region sequences. *Evolution* **51**, 946–956.

Meyer-Rochow, V.B. (1978) Retina and dioptic apparatus of the dung beetle *Euonticellus africanus*. *Journal of Insect Physiology* **24**, 165–179.

Meyer-Rochow, V.B. & Gal, J. (2004) Dimensional limits for arthropod eyes with superposition optics. *Vision research* **44**, 2213–2223.

Moczek, A.P. (1999) Facultative paternal investment in the polyphenic beetle *Onthophagus taurus*: the role of male morphology and social context. *Behavioral Ecology* **10**, 641–647.

Moczek, A.P. (2002) Allometric plasticity in a polyphenic beetle. *Ecological Entomology* **27**, 58–67.

Moczek, A.P. (2003) The behavioral ecology of threshold evolution in a polyphenic beetle. *Behavioral Ecology* **14**, 841–854.

Moczek, A.P. (2005) The evolution and development of novel traits, or how beetles got their horns. *BioScience* **55**, 937–951.

Moczek, A.P. (2006a) Integrating micro- and macroevolution of development through the study of horned beetles. *Heredity* **97**, 168–178.

Moczek, A.P. (2006b) Pupal remodeling and the development and evolution of sexual dimorphism in horned beetles. *American Naturalist* **168**, 711–729.

Moczek, A.P. (2008) On the origin of novelty in development and evolution. *Bioessays* **5**, 432–447.

Moczek, A.P. (2009) Developmental plasticity and the origins of diversity: a case study on horned beetles. In: Ananthakrishnan, T.N.& Whitman, D. (eds.) *Phenotypic plasticity in insects: mechanisms and consequences*. pp. 81–134. Science Publishers, Inc., Plymouth, UK.

Moczek, A.P. (2010) Phenotypic plasticity and diversity in insects. *Philosophical Transactions of the Royal Society of London B* **365**, 593–603.
Moczek, A.P. & Cochrane, J. (2006) Intraspecific female brood parasitism in the dung beetle *Onthophagus taurus*. *Ecological Entomology* **31**, 1–6.
Moczek, A.P. & Emlen, D.J. (1999) Proximate determination of male horn dimorphism in the beetle *Onthophagus taurus* (Coleoptera: Scarabaeidae). *Journal of Evolutionary Biology* **12**, 27–37.
Moczek, A.P. & Emlen, D.J. (2000) Male horn dimorphisms in the scarab beetle, *Onthophagus taurus*: do alternative reproductive tactics favor alternative phenotypes. *Animal Behaviour* **59**, 459–466.
Moczek, A.P. & Nagy, L.M. (2005) Diverse developmental mechanisms contribute to different levels of diversity in horned beetles. *Evolution and Development* **7**, 175–185.
Moczek, A.P. & Nijhout, H.F. (2002) Developmental mechanisms of threshold evolution in a polyphenic beetle. *Evolution and Development* **4**, 252–264.
Moczek, A.P. & Nijhout, H.F. (2003) Rapid evolution of a polyphenic threshold. *Evolution and Development* **5**, 259–268.
Moczek, A.P. & Nijhout, H.F. (2004) Trade-offs during the development of primary and secondary sexual traits in a horned beetle. *American Naturalist* **163**, 184–191.
Moczek, A.P. & Rose, D.J. (2009) Differential recruitment of limb patterning genes during development and diversification of beetle horns. *Proceedings of the National Academy of Sciences USA* **106**, 8992–8997.
Moczek, A.P., Hunt, J., Emlen, D.J. & Simmons, L.W. (2002) Evolution of a developmental threshold in exotic populations of a polyphenic beetle. *Evolutionary Ecology Research* **4**, 587–601.
Moczek, A.P., Rose, D.J., Sewell, W. & Kesselring, B.R. (2006a) Conservation, innovation, and the evolution of horned beetle diversity. *Development Genes and Evolution* **216**, 655–665.
Moczek, A.P., Cruickshank, T.E. & Shelby, J.A. (2006b) When ontogeny reveals what phylogeny hides: gain and loss of horns during development and evolution of horned beetles. *Evolution* **60**, 2329–2341.
Moczek, A.P., Andrews, J., Kijimoto, T., Yerushalmi, Y. & Rose, D.J. (2007) Emerging model systems in evo-devo: horned beetles and the origins of diversity. *Evolution and Development* **9**, 323–328.
Moilanen, A. & Hanski, I. (2001) On the use of connectivity measures in spatial ecology. *Oikos* **95**, 147–152.
Moilanen, A. & Nieminen, M. (2002) Simple connectivity measures in spatial ecology. *Ecology* **84**, 1131–1145.
Møller, A.P. (1989) Ejaculate quality, testes size and sperm production in mammals. *Functional Ecology* **3**, 91–96.
Monaghan, M.T., Inward, D.J.G., Hunt, T. & Vogler, A.P. (2007) A molecular phylogenetic analysis of the Scarabaeinae (dung beetles). *Molecular Phylogenetics and Evolution* **45**, 674–692.
Monteith, G.B. & Storey, R.I. (1981) The biology of *Cephalodesmius,* a genus of dung beetles which synthesises 'dung' from plant material (Coleoptera: Scarabaeidae: Scarabaeinae). *Memoirs of the Queensland Museum* **20**, 253–277.
Montes de Oca, E. (2001) Escarabajos coprófagos de un scenario ganadero típico de la región de Los Tuxtlas, Veracruz, México: importancia del paisaje en la composicón de un gremio funcional. *Acta Zoológica Mexicana* **82**, 111–132.
Montes De Oca, E. & Halffter, G. (1998) Invasion of Mexico by two dung beetles previously introduced into the United States. *Studies on Neotropical Fauna and Environment* **33**, 37–45.
Montreuil, O. (1998) Phylogenetic analysis and paraphyly of Coprini and Dichotomiini (Coleoptera, Scarabaeidae): biogeographic scenario. *Annales de la Société Entomologique de France* **34**, 135–148.
Montreuil, O. (2003a) Contribution à l'étude des Canthonini Malgashes: Description de deux *Pseudoarachnodes* Lebis, 1953 (Coleoptera, Scarabaeidae). *Revue Française d'Entomologie* **25**, 113–116.
Montreuil, O. (2003b). Contribution à l'étude des Canthonini Malgashes Deuxième note: Description de deux Noveaux *Aleiantus* Olsoutieff, 1947 (Coleoptera, Scarabaeidae). *Revue Française d'Entomologie* **25**, 143–146.
Montreuil, O. (2004). Contribution à l'étude des Canthonini de Madagascar (3e note): Description de deux *Apotolamprus* Olsoutieff, et mises au point taxonomiques et nomenclatures (Coleoptera, Scarabaeidae). *Revue Française d'Entomologie* **26**, 67–72.
Montreuil, O. (2005a) Contribution à l'étude des Canthonini de Madagascar (5e note): Deux nouveaux *Aleiantus* Olsoutieff 1947 (Coleoptera, Scarabaeidae). *Revue Française d'Entomologie* **27**, 153–160.
Montreuil, O. (2005b) Nouveaux *Helictopleurus* d'Orbigny, 1915 de Madagascar et révision du ⟨⟨ groupe semivirens⟩⟩ sensu Lebis, 1960 (Insecta, Coleoptera, Scarabaeidae, Oniticellini). *Zoosystema* **27**, 123–135.

Montreuil, O. (2005c) Contribution à l'étude du genre *Helictopleurus* d'Orbigny, 1915 (Coleoptera, Scarabaeidae). *Bulletin de la Société entomologique de France* **110**, 373–376.

Montreuil, O. (2006) Contribution à l'étude des Canthonini de Madagascar (6e note): Mises au point taxonomiques et nomenclaturales et description de nouvelles espèces du genre *Arachnodes* Westwood 1947 (Coleoptera, Scarabaeidae). *Revue Française d'Entomologie* **28**, 97–110.

Montreuil, O. (2007) Nouveaux *Helictopleurus* d'Orbigny, 1915 (Coleoptera, Scarabaeidae). *Bulletin de la Société entomologique de France* **112**, 79–84.

Montreuil, O. (2008a) Contribution à l'étude des Canthonini de Madagascar (8e note): Nouveaux *Apotolamprus* Olsoufieff, 1947 (Coleoptera, Sacarabaeidae). *Revue Française d'Entomologie* **30**, 27–37.

Montreuil, O. (2008b) Révision du genre *Cambefortatus* Paulian, 1986 (Insecta, Coleoptera, Scarabaeidae). *Zoosystema* **30**, 641–650.

Montreuil, O. (2010) Première espèce du genre *Haroldius* Boucomont, 1914, à Madagascar, et redéfinition des Epilissini (Coleoptera, Scarabaeidae, Ateuchini). *Bulletin de la Société entomologique de France* **115**, 73–76.

Montreuil, O. & Viljanen, H. (2007) Contribution à l'étude des Canthonini de Madagascar (7e note): Mises au point taxonomiques et nomenclaturales dans le genre *Nanos* Westwood, 1947 (Coleoptera, Scarabaeidae). *Revue Française d'Entomologie* **29**, 1–10.

Moreno, C.E., Guevara, R., Sánchez-Rojas, G., Téllez, D. & Verdú, J.R. (2007) Community level patterns in diverse systems: A case study of litter fauna in a Mexican pine-oak forest using higher taxa surrogates and re-sampling methods. *Acta Oecologica* **33**, 73–84.

Morgan, K.R. (1987) Temperature regulation, energy metabolism and mate-searching in rain beetles (*Pleocoma* spp.), winter-active, endothermic scarabs (Coleoptera). *Journal of Experimental Biology* **128**, 107–122.

Morgan, K.R. & Bartholomew, G.A. (1982) Homeothermic response to reduced ambient temperature in a scarab beetle. *Science* **216**, 1409–1410.

Mousseau, T.A. & Dingle, H. (1991) Maternal effects in insect life histories. *Annual Review of Entomolology* **36**, 511–534.

Mousseau, T.A. & Fox, C.W. (1998) *Maternal Effects as Adaptations*, Oxford University Press, New York.

Mueller, K.P. & Labhart, T. (2010) Polarizing optics in a spider eye. *Journal of Comparative Physiology A* **196**, 335–348.

Müller, G.B. & Wagner, G.P. (1991) Novelty in evolution: Restructuring the concept. *Annual Review of Ecology and Systematics* **22**, 229–256.

Munro, Z.M. (1988) *Semiochemiese kommunikasie. Ontwikkeling en toepassing van analitiese tegnieke vir die bepaling van die chemise samestelling van die abdominale afskeiding van Kheper species.* Ph.D. Thesis, University of Stellenbosch, Stellenbosch, South Africa.

Muona, J. & Viramo, J. (1995) Coleoptera associated with the dung of the brown bear. *Entomologische Blätte* **91**, 3–12.

Naeem, S. (1998) Species redundancy and ecosystem reliability. *Conservation Biology* **12**, 39–45.

Naeem, S. (2008) Advancing realism in biodiversity research. *Trends in Ecology and Evolution* **23**, 414–416.

National Board of Agriculture (1920–1985) *Official Statistics of Finland: III, Annual statistics of Agriculture.* Helsinki.

Navajas, E. (1950) Manifestações de predatismo em Scarabaeidae do Brasil e alguns dados bionomicos de Canthon virens (Mannh.) (Col., Scarabaeidae). *Ciencia e Cultura* **2**, 284–285.

Nealis, V.G. (1977) Habitat associations and community analysis of south Texas dung beetles (Coleoptera Scarabaeinae). *Canadian Journal of Zoology* **55**, 138–147.

Neff, B.D. (2003) Decisions about parental care in responses to perceived paternity. *Nature* **422**, 716–719.

Neff, B.D. & Gross, M.R. (2001) Dynamic adjustment of parental care in response to perceived paternity. *Proceedings of the Royal Society of London B* **268**, 1559–1565.

Neuhaus, W. (1983) Die Ausbreitung von Pheromonen im Wind. Zoologische Jahrbücher. *Abteilung für allgemeine Zoologie und Physiologie der Tiere* 87, 443–453.

New, T.R. (2002) Should we believe in waterbabies? *Journal of Insect Conservation* **6**, 71–72.

Nichols, E., Larson, T., Spector, S., Davis, A.L., Escobar, F., Favila, M., Vulinec, K. & the Scarabaeinae Research Network. (2007) Global dung beetle response to tropical forest modification and fragmentation: a quantitative review and meta-analysis. *Biological Conservation* **137**, 1–19.

Nichols, E., Spector, S., Louzada, J., Larsen, T., Amequita, S., Favila, M.E. & the Scarabaeinae Research Network. (2008) Ecological functions and ecosystem services provided by Scarabaeinae dung beetles. *Biological Conservation* **141**, 1461–1474.

Nichols, E., Gardner, T., Peres, C.A. & Spector, S. (2009) Co-declines in large mammals and dung beetles: an impending ecological cascade. *Oikos* **118**, 481–487.

Nicholson, E., Mace, G.M., Armsworth, P.R., Atkinson, G., Buckle, S., Clements, T., Ewers, R.M., Fa, J.E., Gardner, T.A., Gibbons, J., Grenyer, R., Metcalfe, R., Mourato, S., Muuls, M., Osborn, D., Reuman, D.C., Watson, C. & Milner-Gulland, E.J. (2009) Priority research areas for ecosystem services in a changing world. *Journal of Applied Ecology* **46**, 1139–1144.

Nicolson, S.W. (1987) Absence of endothermy in flightless dung beetles from southern Africa. *South African Journal of Zoology* **22**, 323–324.

Nicolson, S.W. & Louw, G.N. (1980) Preflight thermogenesis, conductance and thermoregulation in the Protea beetle, *Trichostetha fascicularis* (Scarabaeidae: Cetoniinae). *South African Journal of Science* **76**, 124–126.

Nijhout, H.F. (1991) *The Development and Evolution of Butterfly Wing Patterns*. Smithsonian Institution Press, Washington, DC.

Nijhout, H.F. (1999) Control mechanisms of polyphenic development in insects. *BioScience* **49**, 181–192.

Nijhout, H.F. (2003a) The control of body size in insects. *Developmental Biology* **261**, 1–9.

Nijhout, H.F. (2003b) The control of growth. *Development* **130**, 5863–5867.

Nijhout, H.F. (2003c) Development and evolution of adaptive polyphenisms. *Evolution and Development* **5**, 9–18.

Nijhout, H.F. & Emlen D.J. (1998) Competition among body parts in the development and evolution of insect morphology. *Proceedings of the National Academy of Sciences USA* **95**, 3685–3689.

Nijhout, H.F. & Grunert, L.W. (2002) Bombyxin is a growth factor for wing imaginal disks in Lepidoptera. *Proceedings of the National Academy of Science USA* **99**, 15446–15450.

Nijhout, H.F.and Wheeler, D.E. (1996) Growth models of complex allometries in holometabolous insects. *American Naturalist* **148**, 40–56.

Nilsson, D.-E. (1989) Optics and evolution of the compound eye. In: Stavenga D.G.& Hardie R.C. (eds.) *Facets of vision*. Springer, Berlin, Germany.

Nilsson, D.-E. & Warrant, E.J. (1999) Visual discrimination: seeing the third quality of light. *Current Biology* **9**, 535–537.

Niogret, J., Lumaret, J.-P. & Bertrand, M. (2006) Semiochemicals mediating host-finding behaviour in the phoretic association between *Macrocheles saceri* (Acari: Mesostigmata) and *Scarabaeus* species (Coleoptera: Scarabaeidae). *Chemoecology* **16**, 129–134.

Niven, J.E., Anderson, J.C. & Laughlin, S.B. (2007) Fly photoreceptors demonstrate energy-information trade-offs in neural coding. *PLoS Biology* **5**, 828–840.

Noss, R.F. (1990) Indicators for monitoring biodiversity: a hierarchical approach. *Conservation Biology* **4**, 355–364.

Noss, R.F. (2004) Some suggestions for keeping national wildlife refuges healthy and whole. *Natural Resources Journal* **44**, 1093–1111.

Novina, C.D. & Sharp, P.A. (2004) The RNAi revolution. *Nature* **430**, 161–164.

Novotny, V., Miller, S.E., Hulcr, J., Drew, R.A.I., Basset, Y., Janda, M., Setliff, G.P., Darrow, K., Stewart, A.J.A., Auga, J., Isua, B., Molem, K., Manumbor, M., Tamtiai, E., Mogia, M. & Weiblen, G.D. (2007). Low beta diversity of herbivorous insects in tropical forests. *Nature* **448**, 692–695.

O'Carroll, D.C., Bidwell, N.J., Laughlin, S.B. & Warrant, E.J. (1996). Insect motion detectors matched to visual ecology. *Nature* **382**, 63–66.

O'Carroll, D.C., Laughlin, S.B., Bidwell, N.J. & Harris, E.J. (1997). Spatio-temporal properties of motion detectors matched to low image velocities in hovering insects. *Vision Research* **37**, 3427–3439.

O'Donald, P. (1980) *Genetic Models of Sexual Selection*. Cambridge University Press, Cambridge, UK.

O'Hea, N.M., Kirwan, L. & Finn, J.A. (2010) Experimental mixtures of dung fauna affect dung decomposition through complex effects of species interactions. *Oikos* **119**, 1081–1088.

Ocampo, F.C. (2004) Food relocation behavior and synopsis of the southern South American genus *Glyphoderus* Westwood (Scarabaeidae: Scarabaeinae: Eucraniini). *Coleopterists Bulletin* **58**, 295–305.

Ocampo, F.C. (2005) Revision of the southern South American endemic genus *Anomiopsoides* Blackwelder 1944 (Coleoptera: Scarabaeidae: Scarabaeinae: Eucraniini) with description of its food relocation behavior. *Journal of Natural History* **39**, 2537–2557.

Ocampo, F.C. & Hawks, D.C. (2006) Molecular phylogenetics and evolution of the food relocation behavior of the dung beetle tribe Eucraniini (Coleoptera: Scarabaidae: Scarabaeinae). *Invertebrate Systematics* **20**, 557–570.

Ocampo, F.C. & Philips, T.K. (2005) Food relocation behavior of the Argentinian dung beetle genus *Eucranium* Brullé and comparison with the southwest African *Scarabaeus* (*Pachysoma*) MacLeay (Coleoptera: Scarabaeidae: Scarabaeiniae). *Revista de la Sociedad Entomológica Argentina* **64**, 53–59.

Oldham, S., Böhni, R., Stocker, H., Broglio, W. & Hafen, E. (2000) Genetic control of size in Drosophila. *Philosophical Transaction of the Royal Society of London B* **355**, 945–952.

Olendorf, R., Rodd, F.H., Punzalan, D., Houde, A.E., Hurt, C., Reznick, D.N. & Hughes, K.A. (2006) Frequency-dependent survival in natural guppy populations. *Nature* **441**, 633–636.

Oliveira, R.F., Taborsky, M. & Brockman, H.J. (2008) *Alternative Reproductive Tactics: An Integrative Approach*. Cambridge University Press, Cambridge, UK.

Oppenheimer, J.R. (1977) Ecology of some dung beetles (Scarabaeidae: Coprinae) in two villages in West Bengal. *Records of the Zoological Survey, India* **72**, 389–398.

Orme, C.D.L., Davies, R.G., Olson, V.A., Thomas, G.H., Ding, T.S., Rasmussen, P.C., Ridgely, R.S., Stattersfield, A.J., Bennett, P.M., Owens, I.P.F., Blackburn, T.M. & Gaston, K.J. (2006) Global patterns of geographic range size in birds. *PLoS Biology* **4**, 1276–1283.

Orsini, L., Koivulehto, H. & Hanski, I. (2007) Molecular evolution and radiation of dung beetles in Madagascar. *Cladistics* **23**, 145–168.

Otronen, M. (1988) Intra- and intersexual interactions at breeding burrows in the horned beetle, *Coprophanaeus ensifer*. *Animal Behaviour* **36**, 741–748.

Otte, D. & Stayman, K.M. (1979). Beetle horns: some patterns in functional morphology. In: Blum, M.S.& Blum, N.A. (eds.) *Sexual selection and reproductive competition in insects*. pp. 259–292. Academic Press, New York.

Ovaskainen, O. (2004) Habitat–specific movement parameters estimated using mark–recapture data and a diffusion model. *Ecology* **85**, 242–257.

Ovaskainen, O. & Hanski, I. (2003) The species–area relationship derived from species-specific incidence functions. *Ecology Letters* **6**, 903–909.

Ovaskainen O. & Hanski I. (2004) From individual behaviour to metapopulation dynamics: unifying the patchy population and classic metapopulation models. *American Naturalist* **164**, 364–377.

Ovaskainen O., Rekola, H., Meyke, E. & Arjas, E. (2008a) Bayesian methods for analyzing movements in heterogeneous landscapes from mark-recapture data. *Ecology* **89**, 542–554.

Ovaskainen O., Luoto M., Ikonen I., Rekola, H., Meyke, E. & Kuussaari, M. (2008b) An empirical test of a diffusion model: predicting clouded apollo movements in a novel environment. *American Naturalist* **171**, 610–619.

Owen, D.F. & Owen, J. (1974) Species diversity in temperate and tropical Ichneumonidae. *Nature* **249**, 583–584.

Owen-Smith, N. (1988) *Megaherbivores*. Cambridge University Press, Cambridge, UK.

Palestrini, C. & Rolando, A. (2001) Body size and paternal investment in the genus *Onthophagus* (Coleoptera, Scarabaeoidea). *Journal of Zoology* **255**, 405–412.

Palmer, R.A. (2004) Symmetry breaking and the evolution of development. *Science* **306**, 828–833.

Palmer, T.J. (1978) A horned beetle which fights. *Nature* **274**, 583–584.

Pandit, S.N., Kolasa, J. & Cottenie, K. (2009) Contrasts between habitat generalists and specialists: an empirical extension to the basic metacommunity framework. *Ecology* **90**, 2253–2262.

Panganiban, G., Irvine, S.M., Lowe, C., Roehl, H., Corley, L.S., Sherbon, B., Grenier, J.K., Fallon, J.F., Kimble, J., Walker, M., Wray, G.A., Swalla, B.J., Martindale, M.Q. & Carroll, S.B. (1997) The origin and evolution of animal appendages. *Proceedings of the National Academy of Science USA* **94**, 5162–5166.

Panhuis, T.M., Butlin, R., Zuk, M. & Tregenza, T. (2001) Sexual selection and speciation. *Trends in Ecology and Evolution* **16**, 364–371.

Parker, A. (2003). *In The Blink of an Eye*. Perseus Publishing, Cambridge, MA.

Parker, G.A. (1970) Sperm competition and its evolutionary consequences in the insects. *Biological Reviews* **45**, 525–567.

Parker, G.A. (1979) Sexual selection and sexual conflict. In: Blum, M.S.& Blum, N.A. (eds.) *Sexual Selection and Reproductive Competition in Insects*, pp. 123–166. Academic Press, London.
Parker, G.A. (1982) Why are there so many tiny sperm? Sperm competition and the maintenance of two sexes. *Journal of Theoretical Biology* **96**, 281–294.
Parker, G.A. (1990a) Sperm competition games: raffles and roles. *Proceedings of the Royal Society of London B* **242**, 120–126.
Parker, G.A. (1990b) Sperm competition games: sneaks and extra-pair copulations. *Proceedings of the Royal Society of London B* **242**, 127–133.
Parker, G.A. (1998) Sperm competition and the evolution of ejaculates: Towards a theory base. In: Birkhead, T.R.& Møller, A.P. (eds.) *Sperm Competition and Sexual Selection*. pp. 3–54. Academic Press, London.
Parker, G.A. (2006) Sexual conflict over mating and fertilization: an overview. *Philosophical Transactions of the Royal Society of London B* **361**, 235–259.
Parker, G.A. & Ball, M.A. (2005) Sperm competition, mating rate and the evolution of testes and ejaculate sizes: a population model. *Biology Letters* **1**, 235–238.
Parker, G.A. & Begon, M. (1986) Optimal egg size and clutch size: effects of environment and maternal phenotype. *American Naturalist* **128**, 573–592.
Parker, G.A. & Partridge, L. (1998) Sexual conflict and speciation. *Philosophical Transactions of the Royal Society of London B* **353**, 261–274.
Parker, G.A., Ball, M.A., Stockley, P. & Gage, M.J.G. (1997) Sperm competition games: a prospective analysis of risk assessment. *Proceedings of the Royal Society of London B* **264**, 1793–1802.
Parry, L., Peres, C.A., Day, B. & Amaral, S. (2010) Rural–urban migration brings conservation threats and opportunities to Amazonian watersheds. *Conservation Letters*, **3**, 251–259.
Pärtel, M., Zobel, M., Zobel, K. & Van Der Maarel, E. (1996) The species pool and its relation to species richness – evidence from Estonian plant communities. *Oikos* **75**, 111–117.
Partridge, L. & Harvey, P.H. (1988) The ecological context of life history evolution. *Science* **241**, 1449–1455.
Partridge, L., Green, A. & Fowler, K. (1987) Effects of egg production and of exposure to males on female survival in *Drosophila melanogaster*. *Journal of Insect Physiology* **33**, 745–740.
Parzer, H.F. & Moczek, A.P. (2008) Rapid antagonistic coevolution between primary and secondary sexual characters in horned beetles. *Evolution* **62**, 2423–2428.
Paschalidis, K.M. (1974) *The genus* Sisyphus *Latr. (Coleoptera, Scarabaeidae) in Southern Africa*. MSc Thesis, Rhodes University, Grahamstown, South Africa.
Paulian, R. (1933) Essai d'une phylogenie des lamellicornes coprophages. *Entomologisches Nachrichtenblatt* **VII**, 103–107.
Paulian, R. (1945) *Coléoptère Scarabéides de l'Indochine. Première partie. Faune de l'Empire Français III*. Paris.
Paulian, R. (1975) On some mountain Canthonina [Coleoptera Scarabaeidae] from Madagascar. *Annales de la Société Entomologique de France* **11**, 221–252.
Paulian, R. (1993) Deux nouveaux *Haroldius* Boucomont de Borneo (Coleoptera: Scarabaeidae). *Revue Suisse de Zoologie* **100**, 169–173.
Paulian, R. & Lebis, E. (1960) *Faune de Madagascar 11: Inséctes, Coléoptères, Scarabaeidae, Scarabaeina et Onthophagini par R. Paulian, Helictopleurini par E. Lebis*. l'Institut de Reserche Scientifique, Tsimbazaza, Tananarive.
Pawar, S. (2003) Taxonomic chauvinism and the methodologically challenged. *Bioscience* **53**, 861–864.
Pearson, D.L. (1994) Selecting indicator taxa for the quantitative assessment of biodiversity. *Philosophical Transactions of the Royal Society of London B* **345**, 75–80.
Pearson, D.L. & Cassola, F. (1992) World-wide species richness patterns of tiger beetles (Coleoptera: Cicindelidae): indicator taxon for biodiversity and conservation studies. *Conservation Biology* **6**, 376–391.
Peck, S.B. & Forsyth, A. (1982) Composition, structure, and competitive behaviour in a guild of Ecuadorian rain forest dung beetles (Coleoptera: Scarabaeidae). *Canadian Journal of Zoology* **60**, 1624–1634.
Peck, S.B. & Howden, H.F. (1984) Response of a dung beetle guild to different sizes of dung bait in a Panamanian rainforest. *Biotropica* **16**, 235–238.
Percy, J.E. & Weatherston, J. (1974) Gland structure and pheromone production in insects. In: Birch, M. (ed.) *Pheromones*. pp. 11–34. North-Holland, Amsterdam.
Pereira, F.S. & Martínez, A. (1956) Os generous de Canthonini Americanos. *Revista Brasileira de Entomologia* **6**, 91–192.

Peres, C.A. & Dolman, P.M. (2000) Density compensation in neotropical primate communities: evidence from 56 hunted and non-hunted Amazonian forests of varying productivity. *Oecologia* **122**, 175–189.

Peres, C.A. & Lake, I.R. (2003) Extent of nontimber resource extraction in tropical forests: Accessibility to game vertebrates by hunters in the Amazon basin. *Conservation Biology* **17**, 521–535.

Peres, C.A. & Palacios, E. (2007) Basin-wide effects of game harvest on vertebrate population densities in Amazonian forests: implications for animal-mediated seed dispersal. *Biotropica* **39**, 304–315.

Pérez-Ramos, I.M., Marañón, T., Lobo, J.M. & Verdú, J.R. (2007) Acorn removal and dispersal by the dung beetle *Thorectes lusitanicus*: ecological implications. *Ecological Entomology* **32**, 349–356.

Pfennig, D.W., Rice, A.M. & Martin, R.A. (2007) Field and experimental evidence for competition's role in phenotypic divergence. *Evolution* **61**, 257–271.

Pfennig, D.W., Wund, M.A., Snell-Rood, E.C., Cruickshank, T., Schlichting, C.D. & Moczek, A.P. (2010). Phenotypic plasticity's impacts on diversification and speciation. *Trends in Ecology and Evolution* **25**, 459–467.

Philips, T.K. (2005) Phylogeny of the oniticelline and onthophagine scarabaeine dung beetles. *2005 ESA Annual Meeting, December 15–18, 2005, Ft. Lauderdale, FL*; abstract and poster handout at http://esa.confex.com/esa/2005/techprogram/paper_20479.htm.

Philips, T.K. & Bell, K.L. (2008) *Attavicinus*, a new generic name for the myrmecophilous dung beetle *Liatongus monstrosus* (Scarabaeidae: Scarabaeinae), *Coleopterists Bulletin* **62**, 67–81.

Philips, T.K., Scholtz, C.H. & Ocampo, F.C. (2002) A phylogenetic analysis of the Eucraniini (Scarabaeidae: Scarabaeinae). *Insect Systematics and Evolution* **33**, 241–252.

Philips, T.K., Pretorius, E. & Scholtz, C.H. (2004a) A phylogenetic analysis of dung beetles (Scarabaeidae: Scarabaeinae): unrolling an evolutionary history. *Invertebrate Systematics* **18**, 53–88.

Philips, T.K., Edmonds, W.D. & Scholtz, C.H. (2004b) A phylogenetic analysis of the New World tribe Phanaeini (Coleoptera: Scarabaeidae: Scarabaeinae): Hypotheses on relationships and origins. *Insect Systematics and Evolution* **35**, 43–63.

Pianka, E.R. (1966) Latitudinal gradient in species diversity: a review of concepts. *American Naturalist* **100**, 33–46.

Pilastro, A., Simonata, M., Bisazza, A. & Evans, J.P. (2004) Cryptic female preference for colorful males in guppies. *Evolution* **58**, 665–669.

Pineda, E., Moreno, C., Escobar, F. & Halffter, G. (2005) Frog, bat, and dung beetle diversity in the cloud forest and coffee agroecosystems of Veracruz, Mexico. *Conservation Biology* **19**, 400–410.

Pither, J. (2003) Climate tolerance and interspecific variation in geographic range size. *Proceedings of the Royal Society B* **270**, 475–481.

Pitnick, S. & Brown, W.D. (2000) Criteria for demonstrating female sperm choice. *Evolution* **54**, 1052–1056.

Pitnick, S., Brown, W.D. & Miller, G.T. (2000) Evolution of female remating behaviour following experimental removal of sexual selection. *Proceedings of the Royal Society of London B* **268**, 557–563.

Pitnick, S., Miller, G.T., Reagan, J. & Holland, B. (2001) Males' evolutionary responses to experimental removal of sexual selection. *Proceedings of the Royal Society of London B* **268**, 1071–1080.

Pizzari, T. & Birkhead, T.R. (2002) The sexually-selected sperm hypothesis: sex-biased inheritance and sexual antagonism. *Biological Reviews* **77**, 183–209.

Pliske, T.E. & Eisner, T. (1969) Sex pheromone of the Queen butterfly; Biology. *Science* **164**, 1170–1172.

Pluot-Sigwalt, D. (1991) Le système glandulaire abdominal des coléoptères coprophages Scarabaeidae: ses tendances évolutives et ses relations avec la nidification. *Annales de la Société Entomologique de France* **27**, 205–229.

Polak, M. (1996) Ectoparasitic effects on host survival and reproduction: the *Drosophila-Macrocheles* association. *Ecology* **77**, 1379–1389.

Polak, M. (1998) Effects of ectoparasitism on host condition in the *Drosophila-Macrocheles* system. *Ecology* **79**, 1807–1817.

Polak, M. & Simmons, L.W. (2009) Secondary sexual trait size reveals competitive fertilization success in *Drosophila bipectinata*. *Behavioral Ecology* **20**, 753–760.

Polak, M., Starmer, W.T. & Wolf, L.L. (2004) Sexual selection for size and symmetry in a diversifying secondary sexual character in *Drosophila bipectinata* Duda (Diptera: Drosophilidae). *Evolution* **58**, 597–607.

Pomfret, J.C. (2004) *The ecology and evolution of horns in South African dung beetles*. PhD thesis, Queen Mary, University of London, London.

Pomfret, J.C. & Knell, R.J. (2006a) Immunity and the expression of a secondary sexual trait in a horned beetle. *Behavioral Ecology* **17**, 466–472.

Pomfret, J.C. & Knell, R.J. (2006b) Sexual selection and horn allometry in the dung beetle *Euoniticellus intermedius*. *Animal Behaviour* **71**, 567–576.

Pomfret, J.C. & Knell, R.J. (2008) Crowding, sex ratio and horn evolution in a South African beetle community. *Proceedings of the Royal Society of London B* **275**, 315–321.

Pomiankowski, A., Iwasa, Y. & Nee, S. (1991) The evolution of costly mate preferences I. Fisher and biased mutation. *Evolution* **45**, 1422–1430.

Potten, C. & Wilson, J. (2004) *Apoptosis: the life and death of cells*. Cambridge University Press, Cambridge, New York.

Powell, J.R. (1997) *Progress and Prospects in Evolutionary Biology. The Drosophila model*. Oxford University Press, Oxford.

Preston, B.T., Stevenson, I.R., Pemberton, J.M. & Wilson, K. (2001) Dominant rams lose out by sperm depletion. *Nature* **409**, 681–682.

Price, D.L. & May, M.L. (2009) Behavioral ecology of *Phanaeus* dung beetles (Coleoptera: Scarabaeidae): review and new observations. *Acta Zoologica Mexicana* **25**, 211–238.

Proulx, S.R. (1999) Mating systems and the evolution of niche breadth. *American Naturalist* **154**, 89–98.

Proulx, S.R. (2002) Niche shifts and expansion due to sexual selection. *Evolutionary Ecology Research* **4**, 351–369.

Prpic N.M., Wigand, B., Damen, W.G. & Klingler, M. (2001) Expression of *dachshund* in wild-type and *Distal-less* mutant *Tribolium* corroborates serial homologies in insect appendages. *Development Genes and Evolution* **211**, 467–477.

Queller, D.C. (1997) Why do females care more than males? *Proceedings of the Royal Society of London B* **264**, 1555–1557.

Quinlan, M.C. & Gibbs, A.G. (2006) Discontinuous gas exchange in insects. *Respiratory Physiology and Neurobiology* **154**, 18–29.

Quintero, I. & Roslin, T. (2005) Rapid recovery of dung beetle communities following habitat fragmentation in Central Amazonia. *Ecology* **86**, 3303–3311.

Qvarnström, A. & Price, T.D. (2001) Maternal effects, paternal effects and sexual selection. *Trends in Ecology and Evololution* **16**, 95–100.

Rachinsky, A. & Hartfelder, K. (1990) Corpora allata activity, a prime regulating element for caste-specific juvenile hormone titre in honey bee larvae (*Apis mellifera carnica*). *Journal of Insect Physiology* **36**, 189–194.

Radford, A.N. & Ridley, A.R. (2006) Recruitment calling: a novel form of extended parental care in an altricial species. *Current Biology* **16**, 1700–1704.

Radwan, J. (2004) Effectiveness of sexual selection in removing mutations induced with ionization. *Ecology Letters* **7**, 1149–1154.

Raff, R. (1996) *The shape of life: genes development, and the evolution of animal form*. University of Chicago Press, Chicago, USA.

Rahagalala, P., Viljanen, H., Hottola, J. & Hanski, I. (2009) Assemblages of dung beetles using cattle dung in Madagascar. *African Entomology* **17**, 71–89.

Rainio, M. (1966) Abundance and phenology of some coprophagous beetles in different kinds of dung. *Annales Zoologici Fennici* **3**, 88–98.

Rasmussen, J.L. (1994) The influence of horn and body size on the reproductive behavior of the horned rainbow scarab beetle *Phanaeus difformis* (Coleoptera: Scarabaeidae). *Journal of Insect Behavior* **7**, 67–82.

Rassi, P., Alanen, A., Kanerva, T. & Mannerkoski, I. (2001) *Suomen lajien uhanalaisuus 2000*. [The 2000 Red List of Finnish Species. In Finnish with English summary]. Ministry of the Environment, Finnish Environment Institute, Helsinki.

Ratcliffe, B. (1980) Sloth associates. *Coleopterists Bulletin* **34**, 337–350.

Rauter, C.M. & Moore, A.J. (2002) Evolutionary importance of parental care performance, food resources, and direct and indirect genetic effects in a burying beetle. *Journal of Evolutionary Biology* **15**, 407–417.

Reinhold, K. (1999) Energetically costly behaviour and the evolution of resting metabolic rate in insects. *Functional Ecology* **13**, 217–224.

Reynolds, J.D., Goodwin, N.B. & Freckleton, R.P. (2002) Evolutionary transitions in parental care and live bearing in vertebrates. *Proceedings of the National Academy of Sciences USA* **357**, 269–281.

Rice, W.R. (1996) Sexually antagonistic male adaptation triggered by experimental arrest of female evolution. *Nature* **381**, 232–234.

Ricklefs, R.E. & Schluter, D. (1993) *Species diversity in ecological communities: historical and geographical perspectives*. The University of Chicago Press, Chicago.

Ricklefs, R.E. & Wikelski, M. (2002) The physiology/life-history nexus. *Trends in Ecology and Evolution* **17**, 462–468.

Ridley, M. (1978) Parental care. *Animal Behaviour* **26**, 904–932.

Ridsdill-Smith, T.J. (1986) The effect of seasonal changes in cattle dung on egg production by two species of dung beetles (Coleoptera: Scarabaeidae) in south-western Australia. *Bulletin Entomological Research* **76**, 63–68.

Ridsdill-Smith, T.J. (1988) Survival and reproduction of *Musca vetustissima* Walker (Diptera: Muscidae) and scarabaeine dung beetle in dung of cattle treated with avermectin B1. *Journal of the Australian Entomological Society* **27**, 175–178.

Ridsdill-Smith, T. (1991) Competition in dung insects. In: Bailey, W.J.& Ridsdill-Smith, J. (eds.) *Reproductive Behaviour in Insects*, pp. 264–292. Chapman and Hall, London.

Ridsdill-Smith, T.J. (1993a) Effects of avermectin residues in cattle dung on dung beetle (Coleoptera: Scarabaeidae) reproduction and survival. *Veterinary Parasitology* **48**, 127–137.

Ridsdill-Smith, T.J. (1993b) Asymmetric competition in cattle dung between two species of *Onthophagus* dung beetle and the bush fly, *Musca vetustissima*. *Ecological Entomology* **18**, 241–246.

Ridsdill-Smith, T. & Hall, G.P. (1984a) Beetles and mites attracted to fresh cattle dung in southwestern Australian pastures. *CSIRO Division of Entomology Report* **34**, 1–24. Canberra.

Ridsdill-Smith, T.J. & Hall, G.P. (1984b) Seasonal patterns of adult dung beetle activity in south-western Australia. *Proceedings of the 4th International Conference on Mediterranean Ecosystems*, pp. 139–140. The University of Western Australia, Perth.

Ridsdill-Smith, T.J.R. & Matthiessen, J.N. (1984) Field assessments of the impact of night-flying dung beetles (Coleoptera: Scarabaeidae) on the bush fly, *Musca vetustissima* Walker (Diptera: Muscidae), in south-western Australia. *Bulletin of Entomological Research* **74**, 191–195.

Ridsdill-Smith, T.J. & Matthiessen, J.N. (1988) Bush fly, *Musca vetustissima* walker (Diptera, Muscidae), control in relation to seasonal abundance of scarabaeine dung beetles (Coleoptera: Scarabaeidae) in southwestern Australia. *Bulletin of Entomological Research* **78**, 633–639.

Ridsdill-Smith, T.J. & Simmons, L.W. (2009) Dung beetles. In: Resh, V.H.& Carde, R.T. (eds.) *Encyclopedia of Insects* 2nd Ed. pp. 2003–2005. Elsevier, Burlington, MA.

Ridsdill-Smith, T.J., Hall, G.P. & Craig, G.F. (1982) Effect of population density on reproduction and dung dispersal by the dung beetle *Onthophagus binodis* in the laboratory. *Entomologia Experimentalis et Applicata* **32**, 80–85.

Ridsdill-Smith, T.J.R., Weir, T.A. & Peck, S.B. (1983) Dung beetles (Scarabaeidae, Scarabaeinae and Aphodiinae) active in forest habitats in southwestern Australia during winter. *Journal of the Australian Entomological Society* **22**, 307–309.

Ridsdill-Smith, T.J., Hayles, L. & Palmer, M.J. (1986) Competition between the bush fly and a dung beetle in dung of differing characteristics. *Entomologia Experimentalis et Applicata* **41**, 83–90.

Rodrigues, A.S.L., Andelman, S.J., Bakarr, M.I., Boitani, L., Brooks, T.M., Cowling, R.M., Fishpool, L.D. C., Da Fonseca, G.A.B., Gaston, K.J., Hoffmann, M., Long, J.S., Marquet, P.A., Pilgrim, J.D., Pressey, R.L., Schipper, J., Sechrest, W., Stuart, S.N., Underhill, L.G., Waller, R.W., Watts, M.E.J. & Yan, X. (2004) Effectiveness of the global protected area network in representing species diversity. *Nature* **428**, 640–643.

Roff, D.A. (1986) The evolution of wing dimorphism in insects. *Evolution* **40**, 1009–1020.

Roff, D.A. (1990) The evolution of flightlessness in insects. *Ecological Monographs* **60**, 389–421.

Roff, D.A. (1996) The evolution of threshold traits in animals. *The Quarterly Review of Biology* **71**, 3–35.

Roff, D.A. & Fairbairn, D.J. (2007) The evolution of trade-offs: where are we? *Journal of Evolutionary Biology* **20**, 433–447.

Rosenlew. H. & Roslin, T. (2008) Habitat fragmentation and the functional efficiency of temperate dung beetles. *Oikos* **117**, 1659–1666.

Rosenzweig, M.L. (1995) *Species Diversity in Space and Time*. Cambridge University Press, Cambridge.

Roslin, T. (1999) *Spatial ecology of dung beetles*. PhD thesis, University of Helsinki. http://ethesis.helsinki.fi/julkaisut/mat/ekolo/vk/roslin/spatiale.pdf

Roslin, T. (2000) Dung beetle movements at two spatial scales. *Oikos* **91**, 323–335.

Roslin, T. (2001a) Spatial population structure in a patchily distributed beetle. *Molecular Ecology* **10**, 823–837.

Roslin, T. (2001b) Large-scale spatial ecology of dung beetles. *Ecography* **24**, 511–524.

Roslin, T. & Heliövaara, K. (2007) *Suomen lantakuoriaiset. Opas santiaisista lantiaisiin.* [Dung beetles of Finland, in Finnish with English summary]. Helsinki University Press, Helsinki.

Roslin, T. & Koivunen, A. (2001). Distribution and abundance of dung beetles in fragmented landscapes. *Oecologia* **127**, 69–77.

Roslin, T., Avomaa, T., Leonard, M., Luoto, M. & Ovaskainen, O. (2009) Some like it hot: Microclimatic variation affects the abundance and movements of a critically endangered dung beetle. *Insect Conservation and Diversity* **2**, 232–241.

Rossel, S. (1993) Navigation by bees using polarized skylight *Comparative Biochemistry and Physiology* **104A**, 695–708.

Roth, J.P., Macqueen, A. & Bay, D.E. (1988) Scarab activity and predation as mortality factors of the buffalo fly, *Haematobia irritans exigua*, (Diptera: Muscidae) in central Queensland. *Southwestern Entomologist* **13**, 119–124.

Rowe, L. & Houle, D. (1996) The lek paradox and the capture of genetic variance by condition dependent traits. *Proceedings of the Royal Society of London B* **263**, 1415–1421.

Rowland, J.M. & Emlen, D.J. (2009) Two thresholds, three male forms result in facultative male trimorphism in beetles. *Science* **323**, 773–776.

Rundle, H.D. & Nosil, P. (2005) Ecological speciation. *Ecology Letters* **8**, 336–352.

Rundle, H.D., Chenoweth, S.F. & Blows, M.W. (2006) The roles of natural and sexual selection during adaptation to a novel environment. *Evolution* **60**, 2218–2225.

Saeki, Y., Kruse, K.C. & Switzer, P.V. (2005) Physiological costs of mate guarding in the Japanese beetle (*Popillia japonica* Newman). *Ethology* **111**, 863–877.

Samsom, R. & Brandon, R.N. (2007) *Integrating evolution and development: from theory to practice.* MIT Press, Cambridge, MA.

Samways, M.J. (1994) *Insect conservation biology.* Chapman and Hall, London.

Samways, M.J. (2002) Caring for the multitude: current challenges. *Biodiversity and Conservation* **11**, 341–343.

Samways, M.J. (2006) Insect extinctions and insect survival. *Conservation Biology* **20**, 245–246.

Samways, M.J. & Grant, P.B. (2007) Honing Red List assessments of lesser-known taxa in biodiversity hotspots. *Biodiversity Conservation* **16**, 2575–2586.

Santschi, F. (1913) Comments' orientent les fourmis. *Revue Suisse de Zoologie* **21**, 347–425.

Sasvàri, L. (1990) Feeding responses of mated and widowed bird parents to fledglings: an experimental study. *Ornis Scandinavica* **21**, 287–292.

Sato, H. (1988) Further observations on the nesting behaviour of a subsocial ball-rolling scarab, *Kheper aegyptiorum*. *Kontyû, Tokyo* **56**, 873–878.

Sato, H. (1997) Two nesting behaviours and life history of a subsocial African dung-rolling beetle, *Scarabaeus catenatus* (Coleoptera: Scarabaeidae). *Journal of Natural History* **31**, 457–469.

Sato, H. (1998a) Male participation in nest building in the dung beetle *Scarabaeus catenatus* (Coleoptera: Scarabaeidae): mating effort versus paternal effort. *Journal of Insect Behavior* **11**, 833–843.

Sato, H. (1998b) Payoffs of the two alternative nesting tactics in the African dung beetle, *Scarabaeus catenatus*. *Ecological Entomology* **23**, 62–67.

Sato, H. & Hiramatsu, K. (1993) Mating behaviour and sexual selection in the African ball rolling scarab *Khepher platynotus* Bates (Coleoptera: Scarabaeidae). *Journal of Natural History* **27**, 657–668.

Sato, H. & Imamori, M. (1986) Production of two brood pears from one dung ball in an African ball-roller, *Scarabaeus aegyptiorum* (Coleoptera, Scarabaeidae). *Kontyû, Tokyo* **54**, 381–385.

Sato, H. & Imamori, M. (1987) Nesting behaviour of a subsocial African ball roller *Kheper platynotus* (Coleoptera: Scarabaeidae). *Ecological Entomology* **12**, 415–425.

Sato, H. & Imamori, M. (1988) Further Observations on the Nesting Behavior of a Subsocial Ball-rolling Scarab, *Kheper aegyptiorum*. *Japanese Journal of Entomology* **56**, 873–878.

Scarabnet (2007) Standardized method for ecological collections of Scarabaeine dung beetles. Portal, Arizona, The Scarabaeinae Research Network.

Scarabnet (2009) ScarabNet Global Taxon Database Version 1.5 [November 2, 2009]. scarabnet.org.

Scheffler, P. (2005) Dung beetle (Coleoptera: Scarabaeidae) diversity and community structure across three disturbance regimes in eastern Amazonia. *Journal of Tropical Ecology* **21**, 9–19.

Schilman, P.E., Kaiser, A. & Lighton, J.R.B. (2008) Breath softly, beetle: continuous gas exchange, water loss and the role of the subelytral space in the tenebrionid beetle, *Eleodes subobscura*. *Journal of Insect Physiology* **54**, 192–303.

Schimpf, N.G., Matthews, P.G.D., Wilson, R.S. & White, C.R. (2009) Cockroaches breathe discontinuously to reduce respiratory water loss. *Journal of Experimental Biology* **212**, 2773–2780.

Schlichting, C.D. & Pigliucci, M. (1998) *Phenotypic evolution. A reaction norm perspective*. Sinauer, Sunderland, MA.

Schluter, D. (1988) Estimating the form of natural selection on a quantitative trait. *Evolution* **42**, 846–861.

Schluter, D. (1994) Experimental evidence that competition promotes divergence in adaptive radiation. *Science* **266**, 798–801.

Schmid, A. (1998) Different functions of different eye types in the spider *Cupiennius salei*. *Journal of Experimental Biology* **201**, 221–225.

Schmitt, C.B., Belokurov, A., Besançon, C., Boisrobert, L., Burgess, N.D., Campbell, A., Coad, L., Fish, L., Gliddon, D., Humphries, K., Kapos, V., Loucks, C., Lysenko, I., Miles, L., Mills, C., Minnemeyer, S., Pistorius, T., Ravilious, C., Steininger, M. & Winkel, G. (2008) *Global Ecological Forest Classification and Forest Protected Area Gap Analysis. Analyses and recommendations in view of the 10% target for forest protection under the Convention on Biological Diversity (CBD)* University of Freiburg, Freiburg, Germany.

Scholtz, C.H. (1989) Unique foraging behaviour in *Pachysoma* (=*Scarabaeus*) *striatum* Castelnau (Coleoptera: Scarabaeidae): an adaptation to arid conditions? *Journal of Arid Environments* **16**, 305–313.

Scholtz, C.H. (2000) Evolution of flightlessness in Scarabaeoidea (Insecta, Coleoptera). *Deutsche Entomologische Zeitschrift* **47**, 5–28.

Scholtz, C.H. & Chown, S.L. (1995) The evolution of habitat use and diet in the Scarabaeoidea: a phylogenetic approach. In: Pakaluk, J.& Slipiñski, A. (eds.) *Biology, Phylogeny and Classification of Coleoptera. Papers Celebrating the 80th Birthday of Roy A. Crowson*. pp. 355–374. Museum i Instytut Zoologii PAN, Warsaw.

Scholtz, C.H. & Howden, H.F. (1987) A revision of the southern African Canthonina (Coleoptera: Scarabaeidae: Scarabaeinae). *Journal of the Entomological Society of Southern Africa* **50**, 75–119.

Scholtz, C.H., Harrison. J. du G. & Grebennikov, V.V. (2004) Dung beetle (*Scarabaeus* (*Pachysoma*)) biology and immature stages: reversal to ancestral states under desert conditions (Coleoptera: Scarabaeidae)? *Biological Journal of the Linnean Society* **83**, 453–460.

Scholtz, C.H., Davis, A.L.V. & Kryger, U. (2009) *Evolutionary Biology and Conservation of Dung Beetles*. Pensoft Publishers, Sofia.

Scholtz, G. (2008) Scarab beetles at the interface of wheel invention in nature and culture? *Contributions to Zoology* **77**, 139–148.

Schwagmeyer, P.L. & Mock, D.W. (1993) Shaken confidence of paternity. *Animal Behaviour* **46**, 1020–1022.

Seddon, N., Merrill, R.M. & Tobias, J.A. (2008) Sexually selected traits predict patterns of species richness in a diverse clade of Suboscine birds. *American Naturalist* **171**, 620–631.

Seely, M., Henschel, J. & HamiltonIII, W. (2005) Long-term data show behavioural fog collection adaptations determine Namib Desert beetle abundance. *South African Journal of Science* **101**, 570–572.

Seymour, R.S., White, C.R. & Gibernau, M. (2009) Endothermy of dynastine scarab beetles (*Cyclocephala colasi*) associated with pollination biology of a thermogenic lily (*Philodendron solimoesense*). *Journal of Experimental Biology* **212**, 2960–2968.

Shahabuddin, Schulze, C.H. & Tscharntke, T. (2005) Changes of dung beetle communities from rainforests towards agroforestry systems and annual cultures in Sulawesi (Indonesia). *Biodiversity and Conservation* **14**, 863–877.

Shahabuddin, Hidayat, P., Manuwoto, S., Noerdjito, W.A., Tscharntke, T. & Schulze, C.H. (2010) Diversity and body size of dung beetles attracted to different dung types along a tropical land-use gradient in Sulawesi, Indonesia. *Journal Of Tropical Ecology* **26**, 53–65.

Shapiro, A.M. (1976) Seasonal polyphenism. *Evolutionary Biology* **9**, 259–333.

Shelby, J.A., Madewell R. & Moczek A.P. (2007) Juvenile hormone mediates sexual dimorphism in horned beetles. *Journal of Experimental Zoology B, Molecular and Developmental Evolution* **308B**, 417–427.

Sheldon, B.C. (2000) Differential allocation: tests, mechanisms and implications. *Trends in Ecology and Evolution* **15**, 397–402.

Sheldon, B.C. & Ellegren, H. (1998) Paternal effort related to experimentally manipulated paternity of male collared flycatchers. *Proceedings of the Royal Society of London B* **265**, 1737–1742.

Shepherd, B.L., Prange, H.D. & Moczek, A.P. (2008) Some like it hot: body and weapon size affect thermoregulation in horned beetles. *Journal of Insect Physiology* **54**, 604–611.
Sheppard, P.M. (1975) *Natural Selection and Heredity*. Hutchinson, London.
Sherratt, T.N. & Roberts, G. (1998) The evolution of generosity and choosiness in cooperative exchanges. *Journal of Theoretical Biology* **193**, 167–177.
Shingleton, A.W., Frankino, W.A., Flatt, T., Nijhout, H.F. & Emlen, D.J. (2007) Size and shape: the developmental regulation of static allometry in insects. *BioEssays* **29**, 536–548.
Shingleton, A.W., Mirth, C.K. & Bates, P.W. (2008) Developmental model of static allometry in holometabolous insects. *Proceedings of the Royal Society of London B* **275**, 1875–1885.
Shingleton, A.W., Estep, C.M., Driscoll, M.V. & Dworkin, I. (2009) Many ways to be small: Different environmental regulators of size generate different scaling relationships in *Drosophila melanogaster*. *Proceedings of the Royal Society of London B* **276**, 2625–2633.
Shubin, N., Tabin, C. & Carroll, S. (2009) Deep homology and the origins of evolutionary novelty. *Nature* **457**, 818–823.
Shuster, S.M. & Wade, M.J. (2003) *Mating Systems and Strategies*. Princeton University Press, USA.
Silfverberg, H. (2004) Enumeratio nova Coleopterorum Fennoscandiae, Daniae et Baltiae. *Sahlbergia* **9**, 1–111.
Simmons, L.W. (1992) Quantification of role reversal in relative parental investment in a bush cricket. *Nature* **358**, 61–62.
Simmons, L.W. (2001) *Sperm Competition and its Evolutionary Consequences in the Insects*, Princeton University Press, Princeton, MA.
Simmons, L.W. & Emlen, D.J. (2006) Evolutionary trade-off between weapons and testes. *Proceedings of the National Acadamy of Sciences USA* **103**, 16349–16351.
Simmons, L.W. & Emlen, D.J. (2008) No fecundity cost of female secondary sexual trait expression in the horned beetle *Onthophagus sagittarius*. *Journal of Evolutionary Biology* **21**, 1227–1235.
Simmons, L.W. & García-González, F. (2008) Evolutionary reduction in testes size and competitive fertilization success in response to the experimental removal of sexual selection in dung beetles. *Evolution* **62**, 2580–2591.
Simmons, L.W. & Kotiaho, J.S. (2002) Evolution of ejaculates: patterns of phenotypic and genotypic variation and condition dependence in sperm competition traits. *Evolution* **56**, 1622–1631.
Simmons, L.W. & Kotiaho, J.S. (2007a) The effects of reproduction on courtship, fertility and longevity within and between alternative male mating tactics of the horned beetle *Onthophagus binodis*. *Journal of Evolutionary Biology* **20**, 488–495.
Simmons, L.W. & Kotiaho, J.S. (2007b) Quantitative genetic correlation between trait and preference supports a sexually selected sperm process. *Proceedings of the National Academy of Science USA* **104**, 16604–16608.
Simmons, L.W. & Moore, A.J. (2009) Evolutionary quantitative genetics of sperm. In: Birkhead, T.R., Hosken, D.J.& Pitnick, S. (eds.) *Sperm Biology: An Evolutionary Perspective*, pp. 405–434. Academic Press, London.
Simmons, L.W., Tomkins, J.L. & Hunt, J. (1999) Sperm competition games played by dimorphic male beetles. *Proceedings of the Royal Society of London B* **266**, 145–150.
Simmons, L.W., Beveridge, M. & Krauss, S. (2004) Genetic analysis of parentage within experimental populations of a male dimorphic beetle, *Onthophagus taurus*, using amplified fragment length polymorphism. *Behavioural Ecology and Sociobiology* **57**, 164–173.
Simmons, L.W., Emlen, D.J. & Tomkins, J.L. (2007) Sperm competition games between sneaks and guards: a comparative analysis using dimorphic male beetles. *Evolution* **61**, 2684–2692.
Simmons, L.W., House, C.M., Hunt, J. & García-González, F. (2009) Evolutionary response to sexual selection in male genital morphology. *Current Biology* **19**, 1442–1446.
Siva-Jothy, M.T. (1987) Mate securing tactics and the cost of fighting in the Japanese horned beetle, *Allomyrina dichotoma* L. (Scarabaeidae). *Journal of Ethology* **5**, 165–172.
Siva-Jothy, M.T. & Tsubaki, Y. (1989) Variation in copulation duration in *Mnais pruinosa pruinosa* Selys (Odonata: Calopterygidae) 1. Alternative mate-securing tactics and sperm precedence. *Behavioural Ecology and Sociobiology* **24**, 39–45.
Sivinski, J. (1984) Sperm in competition. In: Smith, R.L. (ed.) *Sperm Competition and the Evolution of Animal Mating Systems*. pp. 86–115. Academic Press, London.
Šizling, A., Storch, D. & Keil, P. (2009) Rapoport's rule, species tolerances, and the latitudinal diversity gradient: geometric considerations. *Ecology* **90**, 3575–3586.

Skelley, P.E. & Gordon, R.D. (2002) Aphodiinae. In: Arnett, R.H., Jr., Thomas, M.C., Skelley, P.E. & Frank, J.H. (eds.) *American Beetles. Polyphaga: Scarabaeoidea through Curculionoidea*, pp. 42–48. CRC Press, Boca Raton, FL.

Slade, E.M., Mann, D.J., Villanueva, J.F. & Lewis, O.T. (2007) Experimental evidence for the effects of dung beetle functional group richness and composition on ecosystem function in a tropical forest. *Journal of Animal Ecology* **76**, 1094–1104.

Slatkin, M. (1980) Ecological character displacement. *Ecology* **61**, 163–177.

Slatkin, M. (1985) Gene flow in natural populations. *Annual Review of Ecology and Systematics* **16**, 393–430.

Slatkin, M. (1987) Gene flow and the geographical structure of natural populations. *Science* **236**, 787–792.

Slatkin, M. (1993) Isolation by distance in equilibrium and non-equilibrium populations. *Evolution* **47**, 264–279.

Slobodchikoff, C.N. & Wismann, K. (1981) A function of the subelytral chamber of tenebrionid beetles. *Journal of Experimental Biology* **90**, 109–114.

Smith, A.B.T., Hawks, D.C. & Heraty, J.M. (2006) An Overview of the Classification and Evolution of the Major Scarab Beetle Clades (Coleoptera: Scarabaeoidea) Based on Preliminary Molecular Analyses. In: Jameson, M.L., Ratcliffe, B.C. (eds.). *Scarabaeoidea in the 21st Century: A Festchrift Honoring Henry F. Howden. Coleopterists Society Monograph* **5**, 35–46.

Smith, C.C. & Fretwell, S.D. (1974) The optimal balance between size and number of offspring. *American Naturalist* **108**, 499–506.

Smith, H.G., Källander, H., Fontell, K. & Ljungström, M. (1988) Feeding frequency and parental division of labour in the double-brooded great tit *Parus major*. *Behavioral Ecology and Sociobiology* **22**, 447–453.

Smith, R.L. (1984) *Sperm Competition and the Evolution of Animal Mating Systems*. Academic Press, London.

Snell-Rood, E.C. & Moczek, A.P. (in press) Horns, hormones, and hox genes: the role of development in the evolution of beetle contests. In: Hardy, I.C.W. & Briffa, M. (eds.) *Animal Contests*. Cambridge: Cambridge University Press.

Snell-Rood, E.C. & Moczek, A.P. (in review) Insulin signaling as a mechanism undeerlying developmental plasticity and trait integration: the role of FOXO in a nutritional polyphenism. Heredity.

Snell-Rood, E.C., VanDyken, J.D., Cruickshank, T.E., Wade, M.J. & Moczek, A.P. (2010) Toward a population genetic framework of developmental evolution: costs, limits, and consequences of phenotypic plasticity. *BioEssays* **32**, 71–81.

Snell-Rood, E.C., Cash, A., Kijimoto, T., Andrews, J. & Moczek, A.P. (2011) Developmental reprogramming and alternative phenotypes: insights from the transcriptomes of horn-polyphenic beetles. *Evolution* **65**, 231–245.

Snijders, S.E.M., Dillon, P.G., O'Farrell, K.J.O. Diskin, M., Wylie, A.R.G., O'Callaghan D., Rath, M. & Boland, M.P. (2001) Genetic merit for milk production and reproductive rate in diary cows. *Animal Reproduction Science* **65**, 17–31.

Snyder, A.W. & Laughlin, S.B. (1975) Dichroism and absorption by photoreceptors. *Journal of Comparative Physiology* **100**, 101–116.

Sole, C.L. & Scholtz, C.H. (2010) Did dung beetles arise in Africa? A phylogenetic hypothesis based on five gene regions. *Molecular Phylogenetics and Evolution* **56**, 631–641.

Sole, C.L., Scholtz, C.H. & Bastos, A.D.S. (2005) Phylogeography of the Namib Desert dung beetles *Scarabaeus* (*Pachysoma*) MacLeay (Coleoptera: Scarabaeidae). *Journal of Biogeography* **32**, 75–84.

Sole, C.L., Bastos, A.D.S. & Scholtz, C.H. (2007) Do individual and combined data analyses of molecules and morphology reveal the generic status of '*Pachysoma*' MacLeay (Coleoptera: Scarabaeidae)? *Insect Systematics and Evolution* **38**, 311–330.

Sowig, P. (1995) Habitat selection and offspring survival rate in three paracoprid dung beetles: the influence of soil type and soil moisture. *Ecography* **18**, 147–154.

Sowig, P. (1996a) Duration and benefits of biparental brood care in the dung beetle *Onthophagus vacca* (Coleoptera: Scarabaeidae). *Ecological Entomology* **21**, 81–86.

Sowig, P. (1996b) Brood care in the dung beetle *Onthophagus vacca* (Coleoptera: Scarabaeidae): the effect of soil moisture on time budget, nest structure and reproductive success. *Ecography* **19**, 254–258.

Spaethe, J. & Chittka, L. (2003) Interindividual variation of eye optics and single object resolution in bumblebees. *The Journal of Experimental Biology* **206**, 3447–3453.

Spector, S. (2006a) Scarabaeine dung beetles (Coleoptera: Scarabaeidae: Scarabaeinae): An invertebrate focal taxon for biodiversity research and conservation. In: Jameson, M.L. & Ratcliffe, B.C. (eds.). *Scarabaeoidea in the 21st Century: A Festchrift Honoring Henry F. Howden, Coleopterists Society Monograph* **5**, 71–83.

Spector, S. (2006b) Scarabaeine dung beetles (Coleoptera: Scarabaeidae: Scarabaeinae): an invertebrate focal taxon for biodiversity research and conservation. *The Coleopterists Bulletin* **60**, 71–83.

Spector, S. & Ayzama, S. (2003) Rapid turnover and edge effects in dung beetle assemblages (Scarabaeidae) at a Bolivian Neotropical forest-savanna ecotone. *Biotropica* **35**, 394–404.

Spector, S. & Forsyth, A.B. (1998) Indicator taxa in the vanishing tropics. In: Mace, G.M., Balmford, A.& Ginsberg, J.R. (eds.) *Conservation in a Changing World*. Cambridge University Press, London.

Srinivasan, M.V. & Bernard, G.D. (1975) The effect of motion on visual acuity of the compound eye: a theoretical analysis. *Vision Research* **15**, 515–525.

Srinivasan, M.V., Zhang, S., Lehrer, M. & Collett, T. (1996) Honeybee navigation en route to the goal: visual flight control and odometry. *Journal of Experimental Biology* **199**, 237–44.

Srinivasan, M.V., Zhang, S.W., Altwein, M. & Tautz, J. (2000) Honeybee navigation: nature and calibration of the 'odometer'. *Science* **287**, 851–853.

Stavenga, D.G. (1992) Eye regionalization and spectral tuning of retinal pigments in insects. *Trends in Neuroscience* **15**, 213–218.

Stebnicka, Z.T. (1985) A new genus and species of Aulonocnemidae from India with notes on comparative morphology. *Revue Suisse Zoologique* **92**, 649–658.

Stebnicka, Z.T. & Skelley, P.E. (2009) A revision of the genus *Haroldiataenius* Chalumeau (Scarabaeidae: Aphodiinae: Eupariini). *Insecta Mundi* **62**, 1–16.

Steenkamp, H.E. & Chown, S.L. (1996) Influence of dense stands of an exotic tree, *Prosopis glandulosa* Benson, on a savanna dung beetle (Coleoptera: Scarabaeinae) assemblage in southern Africa. *Biological Conservation* **78**, 305–311.

Steiger, S., Peschke, K., Francke, W. & Müller, J.K. (2007) The smell of parents: breeding status influences cuticular hydrocarbon pattern in the burying beetle *Nicrophorus vesoilloides*. *Proceedings of the Royal Society of London B* **274**, 2211–2220.

Steiger, S., Franz, R., Eggert, A.-K. & Müller, J.K. (2008) The coolidge effect, individual recognition and selection for distinctive cuticular signatures in a burying beetle. *Proceedings of the Royal Society of London B* **275**, 1831–1838.

Stern, D.L. & Emlen, D.J. (1999) The developmental basis for allometry in insects. *Development* **126**, 1091–1101.

Stern, D.L. & Foster, W.A. (1996) The evolution of soldiers in aphids. *Biological Reviews* **71**, 27–79.

Stevens, G.C. (1989) The latitudinal gradient in geographical range: How so many species coexist in the tropics. *American Naturalist* **133**, 24–256.

Stickler, P.D. (1979) *The Ecology of the Scarabaeinae (Coleoptera) in the High Veld of South Africa*. Ph.D thesis. Rhodes University, Grahamstown, South Africa.

Stillwell, R.C., Blanckenhorn, W., Teder, T., Davidowitz, G. & Fox, C.W. (2010) Sex differences in phenotypic plasticity affect variation in sexual size dimorphism in insects: from physiology to evolution. *Annual Review of Entomology* **55**, 227–245.

Stone, G.N. & Willmer, P.G. (1989) Endothermy and temperature regulation in bees: a critique of 'grab and stab' measurement of body temperature. *Journal of Experimental Biology* **143**, 211–223.

Strausfeld, N.J. (1976) *Atlas of an insect brain*. Springer-Verlag, Berlin.

Stutt, A.D. & Siva-Jothy, M.T. (2001) Traumatic insemination and sexual conflict in the bed bug *Cimex lectularius*. *Proceedings of the National Academy of Sciences USA* **98**, 5683–5687.

Suding, K.N., Lavorel, S., Chapin, F.S., Cornelissen, J.H.C., Diaz, S., Garnier, E., Goldberg, D., Hooper, D.U., Jackson, S.T. & Navas, M.L. (2008) Scaling environmental change through the community-level: a trait-based response-and-effect framework for plants. *Global Change Biology* **14**, 1125–1140.

Suzuki, S. & Nagano, M. (2009) To compensate or not? Caring parents respond differentially to mate removal and mate handicapping in the burying beetle, *Nicrophorus quadripunctatus*. *Ethology* **115**, 1–6.

Svácha, P. (1992). What are and what are not imaginal discs: reevaluation of some basic concepts (Insecta, Holometabola). *Developmental Biology* **154**, 101–117.

Tack, A., Ovaskainen, O., Harrison, P. & Roslin, T. (2009) Competition as a structuring force in leaf miner communities. *Oikos* **118**, 809–818.

Tallamy, D.W. (1984) Insect parental care. *Bioscience* **34**, 20–24.

Tallamy, D.W. & Wood, T.K. (1986) Convergence patterns in subsocial insects. *Annual Review of Entomology* **31**, 369–390.

Terblanche, J.S., Clusella-Trullas, S. & Chown, S.L. (2010) Phenotypic plasticity of gas exchange pattern and water loss in *Scarabaeus spretus* (Coleoptera: Scarabaeidae): deconstructing the basis for metabolic rate variation. *Journal of Experimental Biology* **213**, 2940–2949.

Templeton, A.R. (1998) Nested clade analyses of phylogeographic data: testing hypotheses about gene flow and population history. *Molecular Ecology* 7, 381–397.

Thomas, C.D. & Kunin, W.E. (1999) The spatial structure of populations. *Journal of Animal Ecology* 68, 647–657.

Thomas, M.L. (2010) Detection of female mating status using chemical signals and cues. *Biological Reviews* in press, doi: 10.1111/j.1469–185X.2010.00130.x

Thomas, M.L. & Simmons, L.W. (2008) Sexual dimorphism in cuticular hydrocarbons of the Australian field cricket *Teleogryllus oceanicus* (Orthoptera: Gryllidae). *Journal of Insect Physiology* **54**, 1081–1089.

Thomas, M.L. & Simmons, L.W. (2009) Male dominance influences pheromone expression, ejaculate quality, and fertilization success in the Australian field cricket, *Teleogryllus oceanicus*. *Behavioral Ecology* **20**, 1118–1124.

Thompson, D.W. (1942) *On Growth and Form*. Cambridge University Press. Cambridge.

Thompson, J.N. (1984) Variation among individual seed masses in *Lomatium grayi* (Umbelliferae) under controlled conditions: magnitude and partitioning of the variance. *Ecology* **65**, 626–631.

Thornhill, R. (1983) Cryptic female choice and its implications in the scorpionfly *Harpobittacus nigriceps*. *American Naturalist* **122**, 765–788.

Tomkins, J.L. (1999) Environmental and genetic determinants of the male forceps length dimorphism in the European earwig *Forficula auricularia*. L. *Behavioral Ecology and Sociobiology* **47**, 1–8.

Tomkins, J.L. & Brown, G.S. (2004) Population density drives the local evolution of a threshold dimorphism. *Nature* **431**, 1099–1103.

Tomkins, J.L. & Hazel, W. (2007) The status of the conditional evolutionarily stable strategy. *Trends in Ecology and Evolution* **22**, 522–528.

Tomkins, J.L. & Moczek, A.P. (2009) Patterns of threshold evolution in polyphenic insects under different developmental models. *Evolution* **63**, 459–468.

Tomkins, J.L. & Simmons, L.W. (1996) Dimorphisms and fluctuating asymmetry in the forceps of male earwigs. *Journal of Evolutionary Biology* **9**, 753–770.

Tomkins, J.L. & Simmons, L.W. (2000) Sperm competition games played by dimorphic male beetles: fertilisation gains with equal mating access. *Proceedings of the Royal Society of London B* **267**, 1547–1553.

Tomkins, J.L., LeBas, N.R., Unrug, J. & Radwan, J. (2004a) Testing the status-dependent ESS: population variation in fighter expression in the mite *Sancassania berlesei*. *Journal of Evolutionary Biology* **17**, 1377–1388.

Tomkins, J.L., Radwan, J., Kotiaho, J.S. & Tregenza, T. (2004b) Genic capture and resolving the lek paradox. *Trends in Ecology and Evolution* **19**, 323–328.

Tomkins, J.L., Kotiaho, J.S. & LeBas, N.R. (2005) Matters of scale: positive allometry and the evolution of male dimorphisms. *American Naturalist* **165**, 389–402.

Tomkins, J.L., Kotiaho, J.S. & LeBas, N.R. (2006) Major differences in minor allometries: a reply to Moczek. *American Naturalist* **167**, 612–618.

Tomkins, J.L., LeBas, N.R., Witton, M.P., Martill, D.M. & Humphries, S. (2010) Positive allometry and the prehistory of sexual selection. *American Naturalist* **176**, DOI: 10.1086/653001.

Tomkins, J.L., Hazel, W.N., Penrose, M.A., Radwan, J. & LeBas, N.R. (in review) Habitat complexity drives the experimental evolution of a conditionally expressed secondary sexual trait.

Tregenza, T., Wedell, N. & Chapman, T. (2006) Introduction. Sexual conflict: a new paradigm? *Philosophical Transactions of the Royal Society of London B* **361**, 229–234.

Tribe, G.D. (1975) Pheromone release by dung beetles (Coleoptera: Scarabaeidae). *South African Journal of Science* **71**, 277–278.

Tribe, G.D. (1976) *The ecology and ethology of ball-rolling dung beetles (Coleoptera: Scarabaeidae)*. MSc thesis, University of Natal, Pietermaritzburg, South Africa.

Trivers, R.L. (1971) The evolution of reciprocal altruism. *Quarterly Review of Biology* **46**, 35–57.

Trivers, R.L. (1972) Parental investment and sexual selection. In: Campbell, B (ed.) *Sexual Selection and the Descent of Man*. pp. 136–179. Aldine Publishing, Chicago.

True, J.R. & Haag, E.S. (2001) Developmental system drift and flexibility in evolutionary trajectories. *Evolution and Development* **3**, 109–119.

Trumbo, S.T. (1997) Juvenile hormone-mediated reproduction in burying beetles: from behaviour to physiology. *Archives of Insect Biochemistry and Physiology* **35**, 479–490.

Tu, M.P., Yin, C.M. & Tatar, M. (2005) Mutations in insulin signaling pathway alter juvenile hormone synthesis in *Drosophila melanogaster. General and Comparative Endocrinology* **142**, 347–356.

Turchin, P. (1998) *Quantitative Analysis of Movement: Measuring and Modeling Population Redistribution in Animals and Plants*. Sinauer Associates, Sunderland.

Tylianakis, J.M., Rand, T.A., Kahmen, A., Klein, A.M., Buchmann, N., Perner, J. & Tscharntke, T. (2008) Resource heterogeneity moderates the biodiversity-function relationship in real world ecosystems. *Plos Biology* **6**, 947–956.

Tyndale-Biscoe, M. (1984) Adaptive significance of brood care of *Copris diversus* Waterhouse (Coleptera: Scarabaeidae). *Bulletin of Entomological Research* **74**, 453–461.

Tyndale-Biscoe, M. (1994) Dung burial by native and introduced dung beetles (Scarabaeidae). *Australian Journal of Agricultural Research* **45**, 1799–1808.

Tyndale-Biscoe, M. & Vogt, W.G. (1991) Effects of adding exotic dung beetles to native fauna on bush fly breeding in the field. *Entomophaga* **36**, 395–401.

Tyndale-Biscoe, M. & Vogt, W.G. (1996) Population status of the bush fly, *Musca vetustissima* (Diptera: Muscidae), and native dung beetles (Coleoptera: Scarabaeinae) in south-eastern Australia in relation to establishment of exotic dung beetles. *Bulletin of Entomological Research* **86**, 183–192.

Tyre, A.J., Tenhumberg, B., Field, S.A., Niejalke, D., Parris, K. & Possingham, H.P. (2003) Improving precision and reducing bias in biological surveys: estimating false-negative error rates. *Ecological Applications* **13**, 1790–1801.

Ukkonen, P. (1993) The post-glacial history of the Finnish mammalian fauna. *Annales Zoologici Fennici* **30**, 249–264.

Ukkonen, P. (1996) Osteological analysis of the refuse fauna in the Lake Saimaa area. *Helsinki Papers in Archaeology* **8**, 63–91.

Unlu, M., Morgan, M.E. & Minden, J.S. (1997) Difference gel electrophoresis: a single gel method for detecting changes in protein extracts. *Electrophoresis* **18**, 2071–2077.

Unrug, J., Tomkins, J.L. & Radwan, J. (2004) Alternative phenotypes and sexual selection: can dichotomous handicaps honestly signal quality? *Proceedings of the Royal Society of London B* **271**, 1401–1406.

Urban, M.C. (2004) Disturbance heterogeneity determines freshwater metacommunity structure. *Ecology* **85**, 2971–2978.

van der Pers, J.N.C. (1980) *Olfactory receptors in small ermine moths (Lepidoptera: Yponomeutidae): Electrophysiology and morphology*. Ph.D. Thesis, University of Groningen, Netherlands.

Van Dyken, J.D. & Wade, M.J. (2010) On the evolutionary genetics of conditionally expressed genes. *Genetics* **184**, 557–570.

van Nouhuys, S. (2005) Effects of habitat fragmentation at different trophic levels in insect communities. *Annales Zoologici Fennnici* **42**, 433–447.

van Rensburg, B.J., McGeoch, M.A., Matthews, W., Chown, S.L. & van Jaarsveld, A.S. (2000) Testing generalities in the shape of patch occupancy frequency distributions. *Ecology* **81**, 3163–3177.

Vaz-de-Mello, F.Z. (2007a) *Revisión taxonómica y análisis filogenético de la tribu Ateuchini (Coleoptera: Scarabaeidae: Scarabaeinae)*. D.Sc. Thesis, Instituto de Ecología, A.C., Xalapa, México.

Vaz-de-Mello, F.Z. (2007b) Revision and phylogeny of the dung beetle genus *Zonocopris* Arrow 1932 (Coleoptera: Scarabaeidae: Scarabaeinae), a phoretic of land snails. *Annales de la Société Entomologique de France* **43**, 231–239.

Vaz-de-Mello, F.Z. (2008) Synopsis of the new subtribe Scatimina (Coleoptera: Scarabaeidae: Scarabaeinae: Ateuchini), with descriptions of twelve new genera and review of *Genieridium*, new genus. *Zootaxa* **1955**, 1–75.

Vaz-de Mello, F.Z. & Edmonds, W.D. (2007) *American Genera and Subgenera of the Subfamily Scarabaeinae (Coleoptera: Scarabaeidae) (version 2.0 English)*. The Scarabaeinae Research Network.

Vences, M., Wollenberg, K.C., Vieites, D.R. & Lees, D.C. (2009) Madagascar as a model region of species diversification. *Trends in Ecology and Evolution* **24**, 456–465.

Verdú, J.R. & Galante, E. (1999) Larvae of *Ataenius* (Coleoptera: Scarabaeidae: Aphodiinae): generic characteristics and species descriptions. *European Journal of Entomology* **96**, 57–68.

Verdú, J.R. & Galante, E. (2001) Larval morphology and breeding behavior of the genus *Pedaridium harold* (Coleptera: Scarabaeidae). *Annals of the Entomological Society of America* **94**, 596–604.

Verdú, J.R. & Galante, E. (2002) Climatic stress, food availability and human activity as determinants of endemism patterns in the Mediterranean region: the case of dung beetles (Coleoptera, Scarabaeoidea) in the Iberian Peninsula. *Diversity and Distributions* **8**, 259–274.

Verdú, J.R. & Galante, E. (2004) Behavioral and morphological adaptations for a low-quality resource in semi-arid environments: dung beetles (Coleoptera, Scarabaeoidea) associated with the European rabbit (*Oryctolagu cuniculus* L.). *Journal of Natural History* **38**, 705–715.

Verdú, J.R., Diaz, A. & Galante, E. (2004) Thermoregulatory strategies in two closely related sympatric *Scarabaeus* species (Coleoptera: Scarabaeinae). *Physiological Entomology* **29**, 32–38.

Verdú, J.R., Arellano, L. & Numa, C. (2006) Thermoregulation in endothermic dung beetles (Coleoptera: Scarabaeidae): effect of body size and ecophysiological constraints in flight. *Journal of Insect Physiology* **52**, 854–860.

Verdú, J.R., Arellano, L., Numa, C. & Micó, E. (2007a) Roles of endothermy in niche differentiation for ball-rolling dung beetles (Coleoptera: Scarabaeidae) along an altitudinal gradient. *Ecological Entomology* **32**, 544–551.

Verdú, J.R., Lobo, J.M., Numa, C., Pérez-Ramos, I.M., Galante, E. & Marañón, T. (2007b) Acorn preference by the dung beetle, *Thorectes lusitanicus*, under laboratory and field conditions. *Animal Behaviour* **74**, 1607–1704.

Verdú, J.R., Casas, J.L., Lobo, J.M. & Numa, C. (2010) Dung beetles eat acorns to increase their ovarian development and thermal tolerance. *PLoS One* **5**, e10114.

Vessby, K. & Wiktelius, S. (2003) The influence of slope aspect and soil type on immigration and emergence of some northern temperate dung beetles. *Pedobiologia* **47**, 39–51.

Via, S., Gomulkiewicz, R., de Jong, G., Scheiner, S.M., Schlichting, C.D. & Van Tienderen, P. (1995) Adaptive phenotypic plasticity: consensus and controversy. *Trends in Ecology and Evolution* **10**, 212–217.

Viljanen, H. (2004) *Diet specialization among endemic forest dung beetles in Madagascar.* MSc Thesis, University of Helsinki, Helsinki.

Viljanen, H. (2009a) *Dung beetle communities in Madagascar.* PhD thesis, University of Helsinki.

Viljanen, H. (2009b) Life history of *Nanos viettei* (Paulian) (Coleoptera: Scarabaeidae: Canthonini), a representative of an endemic clade of dung beetles in Madagascar. *The Coleopterists Bulletin* **63**, 265–288.

Viljanen, H., Escobar, F. & Hanski, I. (2010a) Low local but high beta diversity of tropical forest dung beetles in Madagascar. *Global Ecology and Biogeography* **19**, 886–894.

Viljanen, H., Wirta, H., Montreuil, O., Rahagalala, P., Johnson, S. & Hanksi, I. (2010b) Structure of local communities of endemic dung beetles in Madagascar. *Journal of Tropical Ecology* **26**, 481–496.

Villalba, S., Lobo, J.M., Martín-Piera, F. & Zardoya, R. (2002) Phylogenetic relationships of Iberian dung beetles (Coleoptera: Scarabaeinae): insights on the evolution of nesting behavior. *Journal of Molecular Evolution* **55**, 116–126.

Vinson, J.R. (1951) Le cas des Sisyphes Mauriciens (Insectes, Coléoptères). *Proceedings of the Royal Society of Arts and Sciences of Mauritius* **1**, 105–121.

von Frisch, K. (1949) Die Polarisation des Himmels licht als orienterender Faktor bei den Tanzen der Bienen. *Experentia* **5**, 142–148.

Vulinec, K. (1997) Iridescent dung beetles: a different perspective. *Florida Entomologist* **80**, 132–141.

Vulinec, K., Edmonds, W.D. & Mellow, D.J. (2003) Biological and taxonomic notes on a rare phanaeine dung beetle, *Phanaeus alvarengai* Arnaud (Coleoptera: Scarabaeidae). *Coleopterists Bulletin* **57**, 353–357.

Waage, J.K. (1979) Dual function of the damselfly penis: sperm removal and transfer. *Science* **203**, 916–918.

Wachmann, E. & Schrörer, W.D. (1975) Zur Morphologie des Dorsal- und Ventralauges des Taumelkäfers *Gyrinud substriatus* (Steph.) (Coleoptera, Gyrinidae). *Zoomorphologie* **82**, 43–61.

Wade, M.J. (1998) The evolutionary genetics of maternal effects. In: Mousseau T.A.& Fox, C.W. (eds.) *Maternal Effects as Adaptations*. pp. 5–21. Oxford University Press, Oxford.

Wade, M.J. & Shuster, S.M. (2002) The evolution of parental care in the context of sexual selection: A critical reassessment of parental investment theory. *American Naturalist* **160**, 285–292.

Wake, D.B. (1999) Homoplasy, homology and the problem of 'sameness' in biology. In: Bock, G.R.& Cardew, G. (ed.) *Homology*. pp. 24–46. John Wiley and Sons, London.

Wake, D.B. (2003) Homology and homoplasy. In: Hall, B.K.& Olsen, W.M. (ed.) *Keywords and concepts in evolutionary developmental biology*, pp. 191–200. Harvard University Press, Cambridge, MA.

Walker, B.H. (1992) Biodiversity and ecological redundancy. *Conservation Biology* **6**, 18–23.

Wallace, A.R. (1891) *Natural Selection and Tropical Nature*. McMillan, London.

Wallace, M.M.H. & Tyndale-Biscoe, M. (1983) Attempts to measure the influence of dung beetles (Coleoptera, Scarabaeidae) on the field mortality of the bush fly *Musca vetustissima* Walker (Diptera: Muscidae) in southeastern Australia. *Bulletin of Entomological Research* 73, 33–44.

Walsh, G.C., Cordo, H.A., Briano, J.A., Gandolfo, D.E. & Logarzo, G.A. (1997) Laboratory culture of beneficial dung scarabs. *Journal of Economic Entomology* **90**, 124–129.

Wardhaugh, K.G., Holter, P. & Longstaff, B. (2001) The development and survival of three species of coprophagous insect after feeding on faeces of sheep treated with controlled-release formulations of ivermectin or albendazole. *Australian Veterinary Journal* 79, 125–132.

Warrant, E.J. (1999a) Seeing better at night: life style, eye design and the optimum strategy of spatial and temporal summation. *Vision Research* **39**, 1611–1630.

Warrant, E.J. (1999b) Vision in Dim Light. In: Lehrer, M. (ed.) *Orientation and communication in arthropods*. Birkhäuser Verlag, Basel, Switzerland.

Warrant, E.J. & McIntyre, P.D. (1990a) Limitations to resolution in superposition eyes. *Journal of Comparative Physiology A* **167**, 785–803.

Warrant, E.J. & McIntyre, P.D. (1990b). Screening pigment, aperture and sensitivity in the dung beetle superposition eye. *Journal of Comparative Physiology A* **167**, 805–815.

Warrant, E.J. & McIntyre, P.D. (1991) Strategies for retinal design in arthropod eyes of low F-number. *Journal of Comparative Physiology A* **168**, 499–512.

Warrant, E.J. & McIntyre, P.D. (1993) Arthropod eye design and the physical limits to spatial resolving power. *Progress in Neurobiology* **40**, 413–461.

Warrant, E.J. & McIntyre, P.D. (1996) The visual ecology of pupillary action in superposition eyes. *Journal of Comparative Physiology A* **178**, 75–90.

Wasik, B.R., Rose, D.J. & Moczek, A.P. (2010) Beetle horns are regulated by the Hox gene, *Sex combs reduced*, in a species- and sex-specific manner. *Evolution & Development* **12**353–362.

Waterhouse, D.F. & Sands, D.P.A. (2001) Classical biological control of arthropods in Australia. *Australian Centre for International Agricultural Research Monograph* 77, 560 pp.

Watson, N.L. & Simmons, L.W. (2010a) Mate choice in the dung beetle *Onthophagus sagittarius*: are female horns ornaments? *Behavioral Ecology* **21**, 424–430.

Watson, N.L. & Simmons, L.W. (2010b) Reproductive competition promotes the evolution of female weaponary. *Proceedings of the Royal Society of London B* 277, 2035–2040.

Weatherhead, P.J., Montgomerie, R., Gibbs, H.L. & Boag, P.T. (1994) The cost of extra-pair fertilizations to female red-winged blackbirds. *Proceedings of the Royal Society of London B* **258**, 315–320.

Weckström, M. & Laughlin, S.B. (1995) Visual ecology and voltage-gated ion channels in insect photoreceptors. *Trends in Neuroscience* **18**, 17–21.

Wehner, R. (1984) Astronavigation in insects. *Annual Review of Entomology* **29**, 277–298.

Wehner, R. (1997) The ant's celestial compass system: In *Orientation and Communication in Arthropods* (ed. M. Lehrer), pp. 145–185. Birkhäuser, Basel, Switzerland.

Wehner, R. (2001) Polarization vision-a uniform sensory capacity? *Journal of Experimental Biology* **204**, 2589–2596.

Wehner, R. & Labhart, T. (2006) Polarisation vision. In: Warrant, E.& Nilsson, D.-E. (eds.) *Invertebrate vision*. pp. 291–248. Oxford University Press, Oxford.

Weiss, K.M. & Fullerton, S.M. (2000) Phenogenetic drift and the evolution of genotype-phenotype relationships. *Theoretical Population Biology* **57**, 187–195.

Werren, J.H., Gross, M.R. & Shine, R. (1980) Paternity and the evolution of male parental care. *Journal of Theoretical Biology* **82**, 619–631.

West-Eberhard, M.J. (1983) Sexual selection, social competition, and speciation. *Quarterly Review of Biology* **58**, 155–183.

West-Eberhard, M.J. (1989) Phenotypic plasticity and the origins of diversity. *Annual Review of Ecology and Systematics* **20**, 249–278.

West-Eberhard, M.J. (2003) *Developmental Plasticity and Evolution*. Oxford University Press, New York.

Westneat, D.F. & Sherman, P. (1993) Parentage and the evolution of parental behaviour. *Behavioral Ecology* **4**, 66–77.

Wheeler, D.E. (1986) Developmental and physiological determinants of caste in social hymenoptera: evolutionary implications. *American Naturalist* **128**, 13–34.

Wheeler, D.E. (1991) Developmental basis of worker caste polymorphism in ants. *American Naturalist* **138**, 1218–1238.

Wheeler, D.E. & Nijhout, H.F. (1983) Soldier determination in *Pheidole bicarinata*: effect of methophrene on caste and size within castes. *Journal of Insect Physiology* **29**, 847–854.

White, C.R., Blackburn, T.M., Terblanche, J.S., Marais, E., Gibernau, M. & Chown, S.L. (2007a) Evolutionary responses of discontinuous gas exchange in insects. *Proceedings of the National Academy of Sciences USA* **104**, 8357–8361.

White, C.R., Cassey, P. & Blackburn, T.M. (2007b) Allometric exponents do not support a universal metabolic allometry. *Ecology* **88**, 315–323.

White, E. (1960) The natural history of some species of Aphodius (Col., Scarabaeidae) in the northern Pennines. *Entomologist's Monthly Magazine* **96**, 25–30.

Whitlock, M.C. & Agrawal, A.F. (2009) Purging the genome with sexual selection: reducing mutation load through selection on males. *Evolution* **63**, 569–582.

Whittingham, L.A. & Lifjeld, J.T. (1995) High paternal investment in unrelated young: extra-pair paternity and male parental care in house martins. *Behavioral Ecology and Sociobiology* **37**, 103–108.

Whittingham, L.A., Taylor, P.D. & Robertson, R.J. (1992) Confidence of paternity and male parental care. *American Naturalist* **139**, 1115–1125.

Wiens, J.A. (1989) Spatial scaling in ecology. *Functional Ecology* **3**, 385–397.

Wiens, J.A. (1990) Ecology 2000: an essay on future direction in ecology. *Revista Chilena de Historia Natural* **63**, 309–315.

Wilcox, B.A. (1984) *In situ* conservation of genetic resources: determinants of minimum area requirements. In: McNeely, J.A.& Miller, K.R. (eds.) *National Parks, Conservation and Development. Proceedings of the World Congress on National Parks*. Smithsonian Institution Press, Washington, DC.

Wilkins, A.S. (2002) *The Evolution of Developmental Pathways*. Sinauer Associatec Inc, Sunderland, MA.

Wilkinson, D.G. (1998) *In situ hybridization. A practical approach*. Oxford University Press. Oxford.

Wilkinson, G.S. & Reillo, P.R. (1994) Female choice response to artificial selection on an exaggerated male trait in a stalk-eyed fly. *Proceedings of the Royal Society of London B* **255**, 1–6.

Wilkinson, G.S. & Taper, M. (1999) Evolution of genetic variation for condition-dependent traits in stalk-eyed flies. *Proceedings of the Royal Society of London B* **266**, 1685–1690.

Williams, G.C. (1966) *Adaptation and Natural Selection*. Princeton University Press, Princeton, NJ.

Willig, M.R., Kaufman, D.M. & Stevens, R.D. (2003) Latitudinal gradients of biodiversity: Pattern, process, scale, and synthesis. *Annual Review of Ecology, Evolution and Systematics* **34**, 272–309.

Wilmé, L., Goodman, S.M. & Ganzhorn, J.U. (2006) Biogeographic evolution of Madagascar's microendemic biota. *Science* **312**, 1063–1065.

Wilson, S.R. & Prodan, K.A. (1976) The synthesis and stereochemistry of cascarillic acid. *Tetrahedron Letters* **1976**, 4231–4234.

Winkler, D.W. (1987) A general model for parental care. *American Naturalist* **130**, 526–543.

Winkler, D.W. & Wallin, K. (1987) Offspring size and number: a life history model linking effort per offspring and total effort. *American Naturalist* **129**, 708–720.

Wirta, H. (2009) Complex phylogeographic patterns, introgression and cryptic species in a lineage of Malagasy dung beetles (Coleoptera: Scarabaeidae). *Biological Journal of Linnean Society* **96**, 942–955.

Wirta, H. & Montreuil, O. (2008). Evolution of the Canthonini Longitarsi (Scarabaeidae) in Madagascar. *Zoologica Scripta* **37**, 651–663.

Wirta, H., Orsini, L. & Hanski, I. (2008) An old adaptive radiation of forest dung beetles in Madagascar. *Molecular Phylogenetics and Evolution* **47**, 1076–1089.

Wittlinger, M., Wehner, R. & Wolf, H. (2006) The ant odometer: stepping on stilts and stumps. *Science* **30**, 1965–1967.

Wolf, J.B., Brodie, E.D. III, Cheverud, J.M., Moore, A.J. & Wade, M.J. (1998) Evolutionary consequences of indirect genetic effects. *Trends in Ecology and Evolution* **13**, 64–69.

Wolf, J.B., Brodie, E.D. III.,& Moore, A.J. (1999) Interacting phenotypes and the evolutionary process. II. Selection resulting from social interactions. *American Naturalist* **53**, 254–266.

Wolf, L., Ketterson, E.D. & NolanJr, V. (1990) Behavioural response of female dark-eyed juncos to the experimental removal of their mates: implications for the evolution of male parental care. *Animal Behaviour* **39**, 125–134.

Wolfner, M.F. (1997) Tokens of love: Functions and regulation of *Drosophila* male accessory gland products. *Insect Biochemistry and Molecular Biology* **27**, 179–192.

Woods, H.A. & Smith, J. (2010) Universal model for water costs of gas exchange by animals and plants. *Proceedings of the National Academy of Sciences USA* **107**, 8469–8474.

Wright, J. (1998) Paternity and paternal care. In: Birkhead, T.R.& Møller, A.P. (eds.) *Sperm Competition and Sexual Selection*. pp. 189–205. Academic Press, London.

Wright, J. & Cuthill, I. (1990) Manipulation of sex differences in parental care: the effect of brood size. *Animal Behaviour* **40**, 462–471.

Wright, J. & Dinglemanse, N.J. (1999) Parents and helpers compensate for experimental changes in the provisioning effort of others in the Arabian babbler. *Animal Behaviour* **58**, 345–350.

Wright, J. & Muller-Landau, H.C. (2006) The uncertain future of tropical forest species. *Biotropica* **38**, 443–445.

Wright, J.P., Naeem, S., Hector, A., Lehman, C., Reich, P.B., Schmid, B. & Tilman, D. (2006) Conventional functional classification schemes underestimate the relationship with ecosystem functioning. *Ecology Letters* **9**, 111–120.

Wright, S. (1920) The relative importance of heredity and environment in determining the piebald pattern of guinea pigs. *Proceedings of the National Academy of Sciences USA* **6**, 320–332.

Wyatt, T.D. (2003) *Pheromones and Animal Behaviour.* Cambridge University Press, Cambridge, UK.

Xia, X. (1992) Uncertainty of paternity can select against paternal care. *American Naturalist* **139**, 1126–1129.

Yasui, Y. (1997) A 'good-sperm' model can explain the evolution of costly multiple mating by females. *American Naturalist* **149**, 573–584.

Ybarrondo, B.A. & Heinrich, B. (1996) Thermoregulation and response to competition in the African Dung Beetle *Kheper nigroaeneus* (Coleoptera: Scarabaeidae). *Physiological Zoology* **69**, 35–48.

Yokoyama, K., Kai, H., Koga, T. & Aibe, T. (1991) Nitrogen mineralization and microbial-populations in cow dung, dung balls and underlying soil affected by paracoprid dung beetles. *Soil Biology and Biochemistry* **23**, 649–653.

Young, O.P. (1981) The attraction of neotropical Scarabaeinae (Coleoptera: Scarabaeidae) to reptile and amphibian fecal material. *Coleopterists Bulletin* **35**, 345–348.

Zeh, D.W. & Smith, R.L. (1985) Parental investment by terrestrial arthropods. *American Zoologist* **25**, 785–805.

Zeil, J. (1983) Sexual dimorphism in the visual system of flies: The compound eyes and neural superposition in Bibionidae (Diptera). *Journal of Comparative Physiology* **150**, 379–393.

Ziani, S. (1994) Un interessante caso di teraologia simmetrica in *Onthophagus (Paleonthophagus) fracticornis* (Coleoptera, Scarabaeidae). *Bollettino dell'Associazione Romana di Entomologia* **49**, 165–167.

Ziani, S. & Gharakhloo, M.M. (2010) Studies on palearctic *Onthophagus* associated with burrows of small mammals. IV. A new Iranian species belonging to the *furciceps* group (Coleoptera, Scarabaeidae, Onthophagini). *ZooKeys* **34**, 33–40.

Zobel, M. (1992) Plant species coexistence – the role of historical, evolutionary and ecological factors. *Oikos* **65**, 314–320.

Zobel, M. (1997) The relative role of species pools in determining plant species richness: an alternative explanation of species coexistence? *Trends in Ecology and Evolution* **12**, 266–269.

Zunino, M. (1983) Essai preliminaire sur l'evolution des armures genitales des Scarabaeinae, par rapport a la taxonomie du groupe et a l'evolution du comportement di nidification. *Bulletin de la Societé Entomologique de France* **88**, 531–542.

Zunino, M. (1985) Las relaciones taxonomicas de los Phanaeina (Coleoptera, Scarabaeinae) y sus implicaciones biogeograficas. *Folia Entomológica Mexicana* **64**, 101–115.

Subject Index

Ecology and Evolution of Dung Beetles, First Edition. Edited by Leigh W. Simmons and T. James Ridsdill-Smith.

Taxonomic Index

Ecology and Evolution of Dung Beetles, First Edition. Edited by Leigh W. Simmons and T. James Ridsdill-Smith. Published 2011 by Blackwell Publishing Ltd.